Ius Gentium: Comparative Perspectives on Law and Justice

Volume 95

Ius Gentium is a book series which discusses the central questions of law and justice from a comparative perspective. The books in this series collect the contrasting and overlapping perspectives of lawyers, judges, philosophers and scholars of law from the world's many different jurisdictions for the purposes of comparison, harmonisation, and the progressive development of law and legal institutions. Each volume makes a new comparative study of an important area of law. This book series continues the work of the well-known journal of the same name and provides the basis for a better understanding of all areas of legal science.

The *Ius Gentium* series provides a valuable resource for lawyers, judges, legislators, scholars, and both graduate students and researchers in globalisation, comparative law, legal theory and legal practice. The series has a special focus on the development of international legal standards and transnational legal cooperation.

More information about this series at http://www.springer.com/series/7888

Evanson Chege Kamau
Editor

Global Transformations in the Use of Biodiversity for Research and Development

Post Nagoya Protocol Implementation Amid Unresolved and Arising Issues

Springer

Editor
Evanson Chege Kamau
Faculty of Law
University of Bremen
Bremen, Germany

ISSN 1534-6781 ISSN 2214-9902 (electronic)
Ius Gentium: Comparative Perspectives on Law and Justice
ISBN 978-3-030-88710-0 ISBN 978-3-030-88711-7 (eBook)
https://doi.org/10.1007/978-3-030-88711-7

This Springer imprint is published by the registered company Springer Nature Switzerland AG.
The registered company address is: Gewerbestrasse 11, 6330 Cham, Switzerland

Foreword

It has been 10 years already since the international community agreed in Nagoya, Japan on a groundbreaking agreement, which aims to ensure the fair and equitable sharing of benefits arising from genetic resources. This has been ten years during which the underlying principles of equity, justice and fairness found in the Convention on Biological Diversity (CBD) have been put at the forefront of international discussions. The Nagoya Protocol, the result of years of negotiations amongst Parties of the CBD, presents means of implementation for Article 15 of the Convention, the main Article addressing the third objective of the Convention. The adoption of the Nagoya Protocol demonstrates how countries can come together in an innovative spirit to uphold the principles of solidarity, equality and fairness.

The Nagoya Protocol is a legal instrument through which Parties commit to having clear rules at the national level on access to genetic resources and associated traditional knowledge, and on ensuring the sharing of benefits, including through putting in place compliance measures. Parties are also required to share information through the Access and Benefit-Sharing Clearing-House (ABS-CH), which is accessible to all.

The 10th anniversary of the adoption of the Protocol is an excellent opportunity for the international community to look at what has been accomplished, what challenges have been encountered on the way and what are the next steps for this important agenda. There are numerous successes in its implementation to celebrate. At the time of the writing of this foreword, 130 countries are parties to the Protocol. Almost 70 of those Parties had shared on the ABS Clearing-House their legal framework on access and benefit-sharing, and close to 2000 international recognized certificates of compliance have been published on the ABS Clearing-House.

However, implementation of the Nagoya Protocol also faces several challenges. The Protocol provides flexibility for countries to implement it in light of their national circumstances. Therefore, implementation at the national level has taken many paths in terms of scope, type of measures taken and approaches. But implementation is also dynamic, not stagnant, and there is a need for a constant dialogue with all stakeholders involved to ensure that implementation of the Protocol achieves its objectives and contributes to conservation and sustainable use of biodiversity. This book will present a few examples.

As a community, we need to exchange ideas and views on what we can do to advance on pending and emerging issues so that together we can move in the right direction. This book aims to contribute to those discussions by presenting various case studies on how countries are implementing the different obligations of the Protocol, and in what way some unresolved and emerging issues are being addressed. I hope that perusing through those pages will help the reader gain insight on potential ways to address those issues to let us advance together on achieving our common objective.

CBD, Montreal, QC, Canada Elizabeth Maruma Mrema

Preface

Almost twenty years have passed since the Nagoya Protocol was adopted on 29 October 2010 and six years since it entered into force on 12 October 2014. The instrument has been termed a game changer in the quid pro quo relationship between providers of genetic resources and associated traditional knowledge and users of such resources and knowledge in what is termed as access and benefit-sharing. It was agreed to bindingly enforce the obligations of parties in order to operationalize a system that had failed to achieve its goals since the adoption of the Convention on Biological Diversity in 1992. The Convention entered into force in 1993. With the implementation of the Nagoya Protocol being several years old, it was time to examine how countries were coping with the new rules of the Protocol in terms of complying with it, how research and development were reacting to the new national laws and practice, whether new challenges had cropped up during the implementation and which solutions had been found. The research project titled "New ABS legislation and practice and their compliance with the Nagoya Protocol", from which the results of this book emanate, was conceived mainly with this in mind. The aim was to offer an opportunity for cross-cutting learning from the implemented measures as well as solutions to unresolved issues. In order to give a picture that represents the current implementation situation, country case studies were selected from five continents of the world: Africa, the Americas, Asia, Australia and Europe. The research project started in February 2017 and was concluded in May 2021. It is acknowledged that the project was funded by the German Research Foundation (DFG), which I hereby greatly appreciate!

I am very proud of my team for the hard work accomplished to ensure that these results are achieved and for the remarkable cooperation. It cannot be forgotten that the pandemic (COVID-19) came at a very sensitive time of drafting, carrying out revisions and completing the final bits of the research. Thankfully the entire team remained resilient and showed great commitment and determination to deliver. I appreciate all of you! My very special appreciation goes to Prof. Gerd Winter, Dr. Luciana Silvestri, Dr. Chris Lyal, Thomas Greiber, Prof. Christine Godt and

Dr. Marcelin Mahop who formed a small team to review the draft papers. Without you, the burden upon me would have been extremely big. Thank you very much!

This project was steered through three events which were the main milestones—a kick-off workshop (19 April 2017), an international conference (19–21 September 2018) and a review workshop (27–28 June 2019). It is acknowledged that the review workshop was funded by the German Federal Agency for Nature Conservation (Bundesamt fuer Naturschutz, BfN), which I hereby appreciate! Likewise, I thank all participants of the international conference of 2018 for all the useful comments and feedback and our special guests, Dr. Joachim Lohse, the Senator for Environment, Housing and Transport of the Hanseatic State of Bremen and the Dean of the Law Faculty of the University of Bremen, Prof. Lorenz Kähler, for their great support.

Finally, my thanks go to the Secretary of the Research Center for European Environmental Law (FEU), Anna Himmelskamp, for extraordinary logistical support throughout the research project, and to my student assistants Christin Reinke, Talline Koerner and Michel Barongo Chege for their great support in research and organizational tasks.

This book does not only give insight into the current status of national implementation of the Nagoya Protocol but offers solutions for horizontal cross-fertilization of national ABS laws and practice as well as for unresolved issues. It is noteworthy that it also attempts to propose solutions to questions on ABS based on real-life and hypothetical cases. Whilst it contributes to the existing knowledge of the former issues, it is hoped that it will likewise provoke practice and litigation in the area of ABS, an area I strongly believe has reached a stage of solving relevant cases.

Bremen, Germany
September 2021

Evanson Chege Kamau

Contents

Part IV Unresolved Issues and Solutions to Implementation Challenges

Editor and Contributors

About the Editor

Evanson Chege Kamau Dr. iur., is Senior Research Fellow at the Research Centre for European Environmental Law (FEU), University of Bremen. He is also an Adjunct Lecturer at the CvO University of Oldenburg. His area of expertise is public international law, European law and intellectual property law. His research focuses on genetic resources, environment, development and traditional rights on biological resources and traditional knowledge. Kamau has over 17 years of working experience in the area of ABS in which he has published widely. For many years he has also worked as a freelance legal consultant for governments, African Union (hired by IUCN and Capacity Building Initiative resp.), private industries and research organizations. echege@uni-bremen.de.

List of Contributors

Erwin Beck University of Bayreuth, Bayreuth, Germany

Markus Burchardi Carl von Ossietzky University of Oldenburg, Faculty II, Institute for Legal Studies, Oldenburg, Germany

Geoffrey C. Burton Jean Shannon & Associates, Sydney, NSW, Australia

Jorge Cabrera Medaglia University of Costa Rica, Law Faculty and Lead Counsel on Biodiversity Law, CISDL, San Pedro Montes de Oca, Costa Rica

Ah Young Cho Seoul National University School of Law, Seoul, South Korea

Cuc Dang Thu Nguyen Ministry of Natural Resources and Environment, Ha Noi, Viet Nam

Ellen Frederichs Division Div I 1.5 "Competent National Authority for the Nagoya Protocol", German Federal Agency for Nature Conservation, Bonn, Germany

Christine Godt Carl von Ossietzky University of Oldenburg, Faculty II, Institute for Legal Studies, Oldenburg, Germany

Thomas Greiber Division Div I 1.5 "Competent National Authority for the Nagoya Protocol", German Federal Agency for Nature Conservation, Bonn, Germany

Ashenafi Ayenew Hailu ABS National Focal Point and Researcher at Genetic Resources Access and Benefit Sharing Directorate of the Ethiopian Biodiversity Institute, Addis Ababa, Ethiopia

Evanson Chege Kamau Faculty of Law, University of Bremen, Bremen, Germany

Jae-Hyup Lee Seoul National University School of Law, Seoul, South Korea

Christopher H. C. Lyal Department of Life Sciences, Natural History Museum, London, UK

Marcelin Tonye Mahop Sustainability Research Institute (SRI), School of Earth and Environment (SEE), Faculty of Environment, University of Leeds, Leeds, UK

Lilian Massini Mozini São Bernardo do Campo, Brazil

Maria Victoria Cabrera Ormaza Universidad Espíritu Santo-Ecuador, International Labour Organization, Samborondón, Ecuador

Luciana C. Silvestri Institute of Human, Social and Environmental Sciences, National Scientific and Technical Research Council (CONICET), Mendoza, Argentina

Trang Thi Huong Tran Guihou Ministry of Natural Resources and Environment, Ha Noi, Viet Nam

Tu Ba Nguyen Ministry of Natural Resources and Environment, Ha Noi, Viet Nam

Morten Walløe Tvedt Molde University College (HiMolde), Molde, Norway

China Williams Royal Botanic Gardens, Kew, Kew, UK

Gerd Winter Research Center for European Environmental Law, University of Bremen, Bremen, Germany

Abbreviations and Acronyms

3TG	Tin, Tantalum, Tungsten and Gold
ABNJ	Areas Beyond National Jurisdiction
ABRS	Australian Biological Resources Study
ABS CH/ABSCH	Access and Benefit Sharing Clearing-House
ABS TC	ABS Technical Committee
ABS	Access and Benefit-Sharing
AC	Advisory Committee
ACT	Australian Capital Territory
AHTEG	Ad Hoc Technical Expert Group
ARIPO	Intellectual Property Organization
Art.	Article
ASEAN	Association of Southeast Asian Nations
aTK	Associated traditional knowledge
AVH	Australasian Virtual Herbarium, and formerly the Australian Virtual Herbarium
BABS Regulations	Bioprospecting, Access and Benefit-Sharing Regulations
BBNJ	Biodiversity Beyond National Jurisdiction
BCA	Nature and Biodiversity Conservation Agency
BCP	Biocultural Protocol
BfN	Bundesamt für Naturschutz
BL	Biodiversity Law
BNDES	National Bank for Economic and Social Development
BR	Biological resource(s)
BS	Benefit-sharing
BSA	Benefit-Sharing Agreement
BTF	Bioprospecting Trust Fund
c.	clause
CA	Competent Authority
CAN	Andean Community of Nations
CBD	Convention on Biological Diversity
CETAF	Consortium of European Taxonomic Facilities

CGEN	Genetic Heritage Governing Council
CGIAR	Consultative Group of International Agricultural Research
Ch./ch.	Chapter
CHAR	Council of Heads of Australasian Herbaria
CHM	Clearing House Mechanism
CISG	UN Convention on Contracts for the International Sale of Goods
CITES	Convention on International Trade in Endangered Species of Wild Fauna and Flora
CJEU	Court of Justice of the European Union
CK	Community Knowledge
cl.	clause
CNA(s)	Competent National Authority(ies)
COA	Nuevo Código Orgánico del Ambiente en Recursos Naturales, Energía e Infraestructura
COFEMA	Federal Council of the Environment
COGTA	Cooperative Governance and Traditional Affairs
COMIFA	Commission des Forêts d'Afrique Centrale (Central African Forest Commission)
CONAGEBIO	National Commission for the Biodiversity Management
CONCYTEC	National Council for Science, Technology and Technological Innovation
CONICET	National Scientific and Technical Research Council
COP	Conference of the Parties (to the CBD)
COP-MOP	Meeting of the Parties to the Nagoya Protocol
CPs	Checkpoints
CS	Cabinet Secretary
CSIRO	Commonwealth Scientific Industrial and Research Organisation
CSR	Corporate social responsibility
CV	Curriculum vitae
DAFF	Department of Agriculture, Forestry and Fisheries
DD	Due diligence
DDD	Declaration of Due Diligence
DEA	Department for Environmental Affairs
DEFF	Department of Environment, Forestry and Fisheries
DFG	German Research Foundation
DG	Director-General
DNA	Deoxyribonucleic Acid
DoI	Department of Immigration
DSI	Digital Sequence Information
DSMZ	Deutsche Sammlung von Mikroorganismen und Zellkulturen (German Collection of Microorganisms and Cell Cultures)

DST	Department of Science and Technology
DVS	Department of Veterinary Services
DWS	Department of Water and Sanitation
E.g.	Exempli gratia (for example)
EAC	East Africa Community
EBI	Ethiopian Biodiversity Institute
ECPTA	Eastern Cape Parks and Tourism Agency
ECSF	Estación Científica san Francisco (Ecuador)
EEA	European Economic Area
EEZ	Exclusive Economic Zone
EFCCC	Environment, Forest and Climate Change Commission
EFTA	European Free Trade Area
EMCA	Environment Management and Coordination Act
EPA	Environmental Protection Authority
EPBC Act regulations	Environment Protection Biodiversity Conservation Act regulations
EPBC Act	Environment Protection Biodiversity Conservation Act
ERUDEF	Environment and Rural Development Foundation
EU	European Union
FAO	Food and Agriculture Organization
FCMA	The Forest Conservation and Management Act
FLEGT	Forest Law, Enforcement, Governance and Trade Action Plan of the European Union
Fn/fn	Footnote
FNRB	National Fund for the Benefit-Sharing
GDPR	General Data Protection Regulation of the European Union
GEF	Global Environment Facility
GIZ	German Technical Cooperation Agency
GMAC	Genetic Modification Advisory Committee
GMO(s)	Genetically modified organism(s)
GoC	Government of Cameroon
GPA	General Access Procedures
GPS	Global Positioning System
GR	Genetic resource(s)
GRAIN	Genetic Resources Action International
GRSD	Genetic Resource Sequence Data
GSD	Genetic Sequence Data
GSPC	Global Strategy for Plant Conservation
GTI	Global Taxonomy Initiative
iA	Issuing Authority
IARC	International Agricultural Research Centre(s)
ICJ	International Court of Justice
ICMBio	Chico Mendes Institute for Biodiversity Conservation

ICSWGSB	International Civil Society Working Group on Synthetic Biology
ID (No)	Identification (number)
IGBRs	Indigenous genetic and biological resources
IKS	Indigenous Knowledge Systems
ILA	International Law Association
ILC	International Law Commission
ILCs	Indigenous and local communities
ILO	International Labour Organization
IMCE	Interministerial Committee on Environment
INABIO	Instituto Nacional de Biodiversidad (National Biodiversity Institute)
INBio	National Biodiversity Institute
INDECOPI	National Institute for the Defense of Competence and Protection of Intellectual Property
INIA	National Institute of Agrarian Innovation
INSDC	International Nucelotide Sequence Database Consortium
INTA	National Institute of Agricultural Technology
IP	Intellectual Property
IPR	Intellectual property rights
IPRs	Intellectual property rights
IRCC	Internationally Recognized Certificate of Compliance
ITPGRFA/IT	International Treaty on Plant Genetic Resources for Food and Agriculture
IUCN	International Union for Conservation of Nature
KALRO	Kenya Agricultural and Livestock Research Organization
KECOBO	Kenya Copyright Board
KEPHIS	Kenya Plant Health Inspectorate Service
KFS	Kenya Forest Service
KIPI	Kenya Industrial Property Institute
KWS	Kenya Wildlife Service
LA(s)	Lead agency(ies)
MAA	Material Acquisition Agreement
MAAE	Ministerio del Ambiente y Aqua del Ecuador, formerly MAE (Ministerio del Ambiente del Ecuador)
MARD	Ministry of Agriculture and Rural Development
MAT	Mutually Agreed Terms (in Brazil known as Benefit-sharing Agreement)
MEAs	Multilateral Environmental Agreements
MEC	Member of the Executive Council
MINAE	Ministry of Environment and Energy
MINAGRI	Ministry of Agriculture
MinMEC	Ministers and Members of Executive Councils
MITECO	Ministry for Ecological Transition

MLS	Multilateral System
MO	Monitoring Organization
MoC	Memorandum of Collaboration
MONRE	Ministry of Nature Resources and Environment
MOST	Ministry of Science and Technology
MOSTE	Ministry of Science, Technology and the Environment
MoU	Memorandum of Understanding
MRV	Monitoring, Reporting and Verification of CO_2 Emissions from Large Vessels
MSA	Material Supply Agreement
MSBP	Millennium Seed Bank Partnership
MTA	Material transfer agreement
NACOSTI	National Commission for Science, Technology and Innovation
NBS	National Biodiversity Strategy
NBSAP	National Biodiversity Strategy and Action Plan
NCA	National Competent Authority
NCBD	National Committee on Biological Diversity
NCBI	National Center for Biotechnology Information
NCI	Nature and Culture International/Natura y Cultura Internacional
NEMA	National Environment Management Authority
NEMBA	National Environmental Management: Biodiversity Act
NEMLA	National Environmental Management Laws Amendment Act
NES	National Environment Secretariat
NFP	National Focal Point
NGO(s)	Non-governmental Organization(s)
NMK	National Museums of Kenya
No.	Number
NP	Nagoya Protocol
NSBA	National Spatial Biodiversity Assessment
NSD	Nucleotide Sequence Data
NT	Northern Territory
NTFPs	Non-Timber Forest Products
OECD	Organization for Economic Co-operation and Development
OED	Oxford English Dictionary
OFC	Overseas Fieldwork Committee
ORF	Open Reading Frame (portion of a DNA molecule that, when translated into amino acids, contains no stop codons)
Para./para.	Paragraph(s)
PGRFA	Plant Genetic Resources for Food and Agriculture
PI(s)	Principal Investigator(s)

PIC	Prior informed consent
PIN	Personal identification number
PIP	The WHO Pandemic Influenza Preparedness Framework
PM	Provisional Measure 2.186-16/2001 (Brazil)
PMSEIC	Prime Minister's Science and Engineering Council
PP	Passport
PRODUCE	Ministry of Production
PTKCEA	Protection of Traditional Knowledge and Cultural Expressions Act
R&D	Research and development
r./rr.	Regulation(s)
RBS	Regional Biodiversity Strategy for the Tropical Andean Countries
REDD+	Reducing Emissions from Deforestation and Forest Degradation
Reg.	Regulation
RNA	Ribonucleic Acid
s./ss.	Section(s)
SA	South Africa
SANBI	South African National Biodiversity Institute
SANParks	South African National Parks
SBSTTA	Subsidiary Body on Scientific, Technical and Technological Advice
SCBD	Secretariat of the Convention on Biological Diversity
SE&SD	Secretariat of Environment and Sustainable Development
SENADI	Servicio Nacional de Derechos Intelectuales (National Service of Intellectual Property Rights)
SENASA	Argentine Agri-Food Health and Quality Service
SENESCYT	Secretaría Nacional de Educación superior, Ciencia, Tecnología e Innovación/National Secretariat for High Level Education
SERFOR	National Forestry Service
SINAC	National System of Conservation Areas
SisGen	National Genetic Heritage and Associated Traditional Knowledge Management System
SMTA	Standard Material Transfer Agreement
SUIA	Unique System of Environmental Information
SUMA	Unique System of Environmental Management
TK	Traditional Knowledge
TO	Technical Office of the CONAGEBIO
TOPS	Threatened or Protected Species
TWN	Third World Network
UN	United Nations
UNCLOS	UN Convention on the Law of the Sea

UNCTAD	United Nations Conference on Trade and Development
UNDP	United Nations Development Programme
UNEP	United Nations Environment Programme
UNESCO	United Nations Educational, Scientific and Cultural Organization
UNFCCC	United Nations Framework Convention on Climate Change
UNGA	General Assembly of the United Nations
UPOV	International Union for the Protection of New Varieties of Plants Convention
US	United States
USDA	United States Department of Agriculture
USFDA	United States Food and Drug Administration
UTPL	Universidad Tecnica Particular de Loja
VACNE	Viet Nam Association for Conservation of Nature and Environment
WCMA	Wildlife Conservation and Management Act 2013
WEF	Wildlife Endowment Fund
WHO	World Health Organization
WIPO	World Intellectual Property Organization
WRTI	Wildlife Research and Training Institute
WWF	World Wide Fund for Nature

List of Figures

Chapter 20

List of Tables

List of Boxes

List of Addendums

Part I
Introduction

Transformations in International Law on Access to Genetic Resources and Benefit-Sharing and Domestic Implementation. Introduction, Synthesis, Observations, Recommendations and Conclusions

Evanson Chege Kamau

Abstract The chapter describes the new rules on access to genetic resources and associated traditional knowledge (aTK). The Nagoya Protocol, an instrument of the Convention on Biological Diversity (CBD), created globally binding rules in order to operationalize its third objective, i.e. the fair and equitable sharing of the benefits arising from the utilization of genetic resources (GR) and aTK. The chapter starts by describing the new transformations that have changed the landscape for research and development (R&D) based on genetic resources and aTK. Further, it identifies implementation issues that were unresolved in the negotiations leading to the adoption of the Protocol and those that have emerged during the implementation phase, and shows which challenges they present for the implementation process. An example of such issues is the current disagreement between the providers of GR and their users as to whether Digital Sequence Information falls under the definition of 'genetic resources' and consequently the scope of the access and benefit-sharing (ABS) legislation. Besides, it summarizes each chapter. For chapters with case study examination the focus is laid on how the laws are coping with the ABS obligations of the Protocol, how the salient (unresolved and emerging) issues identified are addressed and whether conformity with the Nagoya Protocol is paid attention to. For general themes, challenges, opportunities and lessons from practical experiences are identified and solutions for enriching the implementation process proposed. Finally, the chapter produces some synthesis, observations and recommendations and reaches some conclusions.

Keywords Transformations in international law on access to genetic resources · Access and benefit-sharing · Nagoya Protocol · Domestic implementation · New legislation and practice

---

E. C. Kamau (✉)
Faculty of Law, University of Bremen, Bremen, Germany
e-mail: echege@uni-bremen.de

E. Chege Kamau (ed.), *Global Transformations in the Use of Biodiversity for Research and Development*, Ius Gentium: Comparative Perspectives on Law and Justice 95, https://doi.org/10.1007/978-3-030-88711-7_1

1 About the Subject

Prior to the CBD, access to genetic resources was unrestricted, uncontrolled and free as they were considered a common heritage of mankind.[1] The users of such resources who were mainly located in developed countries did not have any obligation to share the gains of research and development (R&D) with countries from where they were taken, which are mainly developing countries—so-called provider States.[2] That led to a number of repercussions including the overuse of biodiversity[3] and an imbalance in the conservation burden which had to be borne by provider States alone.[4] The General Assembly resolution 1803 (XVII) of 14 December 1962 declared that States have permanent sovereignty over (their) natural resources, i.e. the right of States to decide freely and independently on the use and exploitation of their natural resources.[5] This became the springboard from which debates on among others the deterrence of the misappropriation of genetic resources (GR) and associated traditional knowledge (aTK) of developing countries[6] and the fair distribution of wealth were launched, eventually leading to negotiations for a treaty.[7] The Convention on Biological Diversity (CBD) was adopted in 1992 in Rio de Janeiro, Brazil, and entered into force on 29 December 1993. The Convention overturned the 'common heritage of mankind' principle, reaffirmed the (sovereign) rights of States over their natural resources and declared that "... the authority to determine access to genetic resources rests with the national governments and is subject to national legislation" (Art. 15.1). The Convention's three main objectives are the conservation of biological diversity, the sustainable use of its components, and the fair and equitable sharing of the benefits arising from the utilization of genetic resources (Art. 1 CBD). Access to GR and aTK and benefit-sharing are built on the third objective which comprises the focus of this book. According to the formula foreseen under article 15 CBD, access to GR is subject to the prior informed consent (PIC) of the contracting party to the CBD that is providing such GR, unless that party determines otherwise (Art. 15.5). Where granted, access shall be on mutually agreed terms (Art. 15.4). Access to aTK is subject to the PIC or approval and involvement of indigenous and local communities that hold such knowledge (Art. 8 (j)). Users of GR must share in a fair and equitable way benefits arising from the utilization with their providers upon mutually agreed terms (Art. 15.7). Due to *inter alia* poor implementation of access and benefit-sharing (ABS) as a result of e.g. too restrictive provider

[1] Kamau (2014), p. 143; Reichman et al. (2016), p. 39ff.

[2] Kamau, ibid.

[3] The international debates since the early 1980s implicated intellectual property rights especially patents for the erosion of genetic resources, biodiversity loss and biopiracy of traditional knowledge. For more see Dutfield and Suthersanen (2020), p. 463ff. Also Reichman et al. (2016), p. 52ff.

[4] Kamau (2014).

[5] Scholtz (2008), p. 288; Cabrera Medaglia and Welch (2018), p. 182ff.

[6] See also Reichman et al. (2016), p. 43.

[7] Cabrera Medaglia and Welch (2018), p. 181ff.

measures and too loose or no compliance measures on the side of users, this formula became somehow a failure, and led to a standoff between the two sides.[8] To alleviate the situation, the international community under the intergovernmental umbrella institution known as the Conference of the Parties (COP) initiated a process to push parties to respect and obey their ABS obligations under the CBD.[9] This led to the adoption of a binding instrument, the Nagoya Protocol on Access to Genetic Resources and the Fair and Equitable Sharing of Benefits Arising from their Utilization to the Convention on Biological Diversity (NP).[10]

The adoption of the Nagoya Protocol in October 2010 and its entry into force in October 2014 became the game changer in the law on access to GR and aTK and benefit-sharing. With it the obligations of the CBD attained an obligatory and binding nature. On the provider side States were taken to task to establish concrete measures in order to facilitate access to GR and aTK as well as create legal certainty, clarity and transparency for users. In the same vein, on the user side States were obliged to put compliance measures in place in order to ascertain that domestic ABS requirements of the providing party are complied with in their jurisdictions and benefits arising from the utilization of accessed GR and aTK are shared (with providers) in a fair and equitable manner. The contributions in this volume focus on the two dynamics. The said measures are considered as key in achieving what is referred to as the ABC (Access, Benefit-sharing and Compliance) of ABS. To that end, countries will need to have ABS frameworks and the required institutional structures (national focal points, competent national authorities, checkpoints) in place and concurrently organize or reorganize their administrative procedures. This has transformed the ABS landscape. But, the Nagoya Protocol (NP) does not prescribe a strict formula of implementation; it gives States leeway as far as the implementation approach they should employ is concerned. For example, they are to decide whether to take legislative, administrative or policy measures, whether to revise or amend old measures, or develop new ones, whether to develop stand-alone laws or integrate ABS provisions in other existing laws, for instance in environmental laws. Likewise, it gives them a big discretion in deciding what and what not to regulate, in taking a narrow or broad interpretation of terms, establishing far-reaching or modest obligations, in determining how to organize administrative procedures, etc. They are also free not to establish any ABS measures altogether for access to their GR and/or aTK. What seems important to the NP, however, is that, if provider/access measures are established, they must meet, *inter alia*, the standards of article 6.3 of the Nagoya Protocol—legal certainty, clarity and transparency. User measures must ensure compliance with domestic ABS laws of provider States and

[8]Kamau (2014), pp. 146–158.

[9]Ibid. For details see Kamau et al. (2010).

[10]For details of the process leading to the Nagoya Protocol see Kamau et al. (2010), pp. 248–250. For an in-depth explanation of and commentary on the Nagoya Protocol see Greiber et al. (2012) and Morgera et al. (2014).

mutually agreed terms—in order to guarantee benefit-sharing as required by articles 5 and 6 NP. In implication, the new post Nagoya Protocol measures are diverse.

Apart from the NP obligations other issues also play out in the implementation process and have the potential of either slowing or crippling, or facilitating it as well as determining the level at which it can be operationalized effectively and efficiently. Such include arising and unresolved issues e.g. the question on (Digital Sequence) Information (DSI) based on GR; possible limits (cut-off points) to provider rights; rights over GR initially accessed as bulk commodities and later utilized in research and development (R&D); availability of sectorial practices; practical tools e.g. well drafted contracts; and the possibility to resolve disputes in court.

2 About the Book

The volume therefore delves on a post Nagoya Protocol study to interrogate some of these issues which are dealt with under three parts of the book.

1. Under **Part II** post NP measures on access and benefit-sharing and domestic implementation are presented. This is the largest part of the book and occupies chapters 2–16. It examines case studies of legislation and practice featuring five continents: South America, Africa, Asia, Australia and Europe. The focus is on establishing how domestic laws have implemented the obligations of the Protocol and how they are coping, their *de jure* and *de facto* compliance with the NP and how they are dealing with arising and unresolved issues, which is critical for learning across the different regimes. In total 15 national legislations and the EU Regulation and their practices have been examined. Only Costa Rica and Australia have not ratified the NP. From the 16 case studies 12 concern post Nagoya Protocol measures with 10 adopted (Argentina, Brazil, Ecuador, Viet Nam, Malaysia, South Korea, Spain, France, Cameroon and the EU) and 2 drafts (Peru and Ethiopia). Of the remaining 4 case studies (Costa Rica, Kenya, South Africa and Australia) old measures are still in operation although some have undertaken a few revisions after the NP (Costa Rica and South Africa) and in Australia one of its States, Queensland, has enacted a post NP law.
2. Following an investigation of the implementation of the NP by national legislations it is important to examine how they are affecting the practical ABS landscape. Therefore under **Part III** covering chapters 17-20 we look at the experiences and lessons from the application of post NP measures by pure biodiversity researchers, an enforcement agency and a botanical garden on the user side. The pre and post NP activities of the researchers (as users) and a botanical garden (as both a user and intermediary) show the changes the new access requirements and procedures have brought in practice and how they have eased or made restrictive the work of research. On the whole the situation seems to have become more difficult. However, their effort to create trust is identified as a critical tool to ameliorate the situation. Besides, the practice of the botanical

garden has shown that having a good policy to guide the behavior of its researchers and their activities as well as having in place exemplary agreements can by far alleviate difficulties with stakeholders and also add to confident dealings. On the other hand activities of an enforcing agency are presented to show how checks are conducted in order to establish violations by users and ensure benefit-sharing. The example of a competent national authority of a European Union Member State, which must take action according to the due diligence compliance concept of the Regulation (EU) 511/2014, is taken as an example. Another question examined under Part III is how successful this concept is in effecting compliance. For that, its origin and nature are also investigated.

3. Affecting the implementation of ABS and its effectivity are not only how the obligations of the NP are transposed nationally and what stakeholders are doing to make it work but likewise critical issues that have arisen and that are unresolved. Some of the most important currently are the issues on DSI, lack of well-thought contractual solutions and lack of litigation. The relevant questions are discussed under **Part IV** in chapters 21–23 and solutions to identified challenges suggested.

3 Further Research Questions

There are many questions than this book can handle requiring research and for which solutions are needed. Four that are considered critical in making ABS dynamic, workable and practical shall be named.

1. Alternative models of ABS.

The current system of exchange of GR and aTK and sharing of benefits is still performing dismally[11] 29 and 10 years since the adoption of the CBD and the NP respectively. Even with the new laws attempting to reach conformity with the NP, access and benefit-sharing (ABS) retain a myriad of challenges. The monetary benefit-sharing project in particular, which is consequential for conservation and sustainable use of biodiversity, has been poorly realized and is considered by some authors as having failed.[12] Our hypothesis is that the core of these challenges lies with the bilateral *quid pro quo* arrangement. Besides, bilateralism is considered a poor fit to modalities of access to, use of and benefit-sharing from DSI[13] which is a basis for the real source of (monetary) benefits, and the ownership of which is currently a major issue of contention between providers and users. A solution to these challenges could be alternative models with the substantial question of linking

[11] Prip and Rosendal (2015).

[12] Muller et al. December (2019); Winter (2021).

[13] Laird et al. (2020); Scholz et al. (2020).

or delinking sovereign rights and benefit-sharing. The theme is very topical and timely given the current discussions about multilateral benefit-sharing in several fora, and the decisions that will be taken over the next few years. Indeed, deliberations in this regard have started at the CBD level.

2. Cut-off points

Ad infinitum provider rights are said to have negative effects on R&D.[14] Notably in the agricultural breeding sector the value chain can be protracted and involve genetic resources from different sources.[15] This can have a number of consequences identified here being two: (1) Due to multiple stages of reproduction within a single (long) value chain sets of different ABS contractual obligations accumulate continuously all of which will apply to the breeding pool and the products developed thereof.[16] This does not only enlarge the bundle of contracts to be monitored, but maintains contracts for which *de facto* obligations have been blurred. Concerning blurred obligations this could be the case e.g. regarding obligations to share benefits arising from utilization if the final commercial variety contains no components of the original sequence of a GR exchanged under the relevant contract. (2) The original contribution of the GR may reduce progressively diminishing its influence.[17] At the end it might be doubtful that there will still be left any benefits to be shared.[18] In such cases it could be alluded that the benefit-sharing rights of the provider no longer existed. For this Winter (2019)[19] suggests the application of the general legal principle "de minimis non curat practor" and in reference to the stipulation of article 5 NP that benefit-sharing shall be fair and balanced. However, the question should be treated differently if the original trait and its functions are still noticeable even after a long value chain.[20] Concerning the first challenge, a study is required to search for ways of establishing cut-off points of ABS rights in the value chain and regarding the second, to determine which level of contribution could still be considered for benefit-sharing.

3. Practical approaches for improving ABS

At the present stage of the implementation of the NP, the EU Regulation and the national implementation law, it is apparent that a number of legal and practical questions remain unanswered. The answers to these questions are indispensable for legally secure implementation of the instruments in question, such as the availability of necessary information on the actors concerned, protection of secrets, data protection and access to information, burden of proof, content and limits of due diligence

[14] Schloen (2019).

[15] Schloen (2019, p. 128).

[16] Ibid.

[17] Winter (2019, p. 109).

[18] Schloen (2019).

[19] Winter (2019).

[20] Ibid.

obligations, treatment of DSI, advice to users, enforcement of access agreements (mutually agreed terms, MAT), control of benefit-sharing, planning of administrative controls, powers for administrative orders, use of sanctions, cooperation with foreign enforcement authorities and the ABS Clearing House, legal protection of users and providing institutions. On a level of doctrinal systematization it can be asked, how the individual problems that arise can be condensed to more general overarching problems, such as due diligence of users, institutional networking in the multi-level system, the relationship between administrative and contractual instruments, the relationship between the rights of disposal over GR as material and as information, etc. It can also be explored how the practical implementation of the ABS regime is to be assessed with regard to pertinent principles of constitutional and international law, in particular freedom of research and of enterprise, the sovereign rights of the provider states, and fair compensation of non-monetary and monetary benefits.

With a view to improving implementation practices the role of standardization and harmonization of prior informed consent (PIC) and MAT procedures can be further investigated. Currently parties to the NP have different legal, policy and administrative frameworks implementing the Protocol at the national level. As a consequence, the international ABS landscape is marked by highly divergent national ABS procedures and obligations leading to a continuous implementation problem for users. Standardization and harmonization can be done at the global level but more realistically would be at the regional level. This is critical in addressing the prevailing lack of legal certainty, clarity and transparency and would be very supportive to users in implementation of ABS and compliance in practice.

Related to standardization and harmonization is the possibility of developing specialized ABS instruments under article 4.4 NP. Such specialized instruments have the potential to create uniform ABS conditions for a specific set of genetic resources, specific types of utilization or in other contexts. Thus, if standardization and harmonization are not possible under the Nagoya Protocol, specialized instruments provide an option to push for standardization in certain sectors and/or amongst e.g. a coalition of the willing who understand the problem of individualized national ABS approaches.

4. The role of informal suppliers of GR as important ABS players

Parataxonomists, private land owners, traders and research partners act as intermediaries of genetic resources. With the coming into force of the NP, users of genetic resources from such intermediaries are required to prove that the material utilized in R&D was legally accessed from the country of origin of such resources and that MAT were established. That means while accessing GR from an informal intermediary, a user has to ask for corresponding certification. Unlike most mainstream intermediaries, in particular *ex-situ* collections which possess structures for certifying access of material for further research and monitoring downstream use, or can easily adjust to such requirements, informal intermediaries lack such structures. The following questions therefore arise: If materials are accessed from such intermediaries, how will a user be able to prove that the ABS law of the provider was complied with? How will such intermediaries cope with the need to provide users with evidence that such material was legally accessed? What format of certification would be considered acceptable to (user) checkpoints and enforcement agencies? How can access be organized in the provider country to hinder trade with genetic resources belonging to the provider state and/or indigenous and local communities and also to prevent overharvesting? These questions need to be confronted. The Malaysian Act, for instance, creates a formal opening for suppliers of GR to get involved in the ABS process as long as they are named by the permit applicant and appointed to take the biological resource and/or aTK on their behalf and thus acknowledged as "authorized intermediaries".[21] It could be examined using such examples how that would operate in practice. Other questions could probably be answered by looking at existing certification systems e.g. of the Forest Stewardship Council (FSC) and Marine Stewardship Council (MSC) and drawing lessons from them.

These questions are suggested as important themes for further research and will not be discussed in more detail in this book.

Presented below are the summaries of individual contributions covered in the book and synthesis, observations, recommendations and general conclusions made at the end of this introductory chapter.

4 Summaries of Contributions

PART II: Post Nagoya Protocol Measures on Access and Benefit-Sharing and Domestic Implementation

A. *South American Continent*

In Chapter "Access and Benefit-Sharing Regime in Argentina: Experiences and Perspectives", Luciana C. Silvestri probes the framework for the management of biological resources in Argentina. Argentina has rich and high endemic biodiversity making it an attractive destination for research and development (R&D) based on genetic resources. Besides, in recent years it proactively utilizes its own biodiversity making it a *de facto* user. Hence, it is for the interest of the country to have a comprehensive and effective access and benefit-sharing (ABS) legal framework. The author, however, indicates the contrary. Although a national post Nagoya Protocol (NP) legislation was adopted in October 2019 (Administrative Decision No. 410 of 2019) to rectify the ambiguity created by previous laws and establish uniform minimum standards in all 23 provinces, the law did not create the desired environment for a flawless exchange of genetic resources (GR) and sharing of benefits. This is attributed to the fact that, although the Constitution also foresees that the federal government (national congress) shall enact a formal law that sets minimum

[21] Malaysian Act of 2017, s. 4.

environmental legal standards in order to guarantee a uniform level of environmental protection across the country, it also gives ownership of natural resources to the provinces and hence the competency to define the terms and conditions for access to and utilization of their resources. This has resulted to lack of measures in more than half of the provinces as well as divergent measures and varying requirements and procedures in the 10 provinces which have so far enacted ABS rules. In addition, Administrative Decision No. 410 does not regulate associated traditional knowledge (aTK) and also excludes domesticated or cultivated species from its scope thus neglecting the constitutional rights of the indigenous peoples. It is also void of compliance measures for utilization of GR and aTK from other countries while these are being utilized in Argentina. Likewise, it does not address the issue of digital sequence information (DSI) in spite of Argentina being of the opinion that DSI falls under the scope of GR, and thus makes no contribution to the current debate in this issue. However, it has also brought about some innovations e.g. by including important definitions spelling out the requirements, procedures and authorities responsible, including for which GR. Another vital innovation is the introduction of differentiated conditions and procedure for non-commercial research the effectiveness of which could be minimized by the fact that the Decision introduces checking for access authorizations including for taxonomic purposes. The current situation in Argentina is caused by a lack of national formal law that according to the Constitution must be issued by the national congress. The current administrative Decision No. 410 of 2019 establishes minimum common standards for ABS which betters the situation at the national level but is not binding to the provinces. Its effect and success depend on their willingness to adhere to it a situation which can lead to legal uncertainty and unclarity. This and other factors including the neglect of aTK and lack of user compliance measures are indications that the existing legal situation in Argentina is still below NP standards.

In chapter "Brazilian Biodiversity Law: Challenges and Opportunities for Industries and Research Institutions", Lilian Massini Mozini looks at the Brazilian Biodiversity Act, Federal Law No. 13,123/2015, which regulates access to Brazilian genetic heritage, associated traditional knowledge (aTK) and their utilization and benefit-sharing therefrom and is elaborated by the implementing Decree No. 8,772/2016. Mozini states that it was necessary to adopt a new law which was able to counteract the failures of the previous law, the Provisional Measure No. 2.186/2001, the complex and bureaucratic environment it created leading to the circumvention of the established ABS measures by users. Some of the regime's innovations include the elaboration of key definitions. One of the terms defined that lacks uniform international understanding though being part and parcel of the ABS system is 'access'. From this definition two main factors are revealed: (1) 'Access' according to the Brazilian regime implies not physical taking of a specimen, but rather an intellectual operation involving research and development (R&D) activities. This means that any such activity on Brazilian genetic heritage wherever it may be located falls under the scope of the regime. That includes genetic heritage found in *ex-situ* conditions as long as they originated in Brazilian *in-situ* conditions. (2) Specimens acquired for bulk uses but later used in R&D are also covered by the legislation. In its elaboration of terms, 'research' and 'technological development' also differentiate work on genetic heritage meant for or not meant for economic exploitation. Consequently the new law also establishes different procedures. Interestingly foreigners cannot undertake access and hence these procedures must be executed by a local entity. Therefore, a foreign entity must enter into partnership with a national institution, which will also be responsible for other formalities including notification of finished products developed from the access whether they are to be marketed abroad or in Brazil, application for intellectual property right, commercialization of the intermediate product, or publication of results. The new Brazilian law differentiates itself from most provider States' measures in a number of ways: (1) It does not only address access to genetic resources (genetic heritage) and aTK and benefit-sharing from their uses but also delves into issues regarding access to and transfer of technology and economic exploitation of finished product or reproductive material and the fair and equitable sharing of the benefits from their exploitation. (2) It allows foreign recipients of samples of genetic heritage to make transfers to third persons provided that these agree to comply with the requirements of the law before the transfer is made and thus introduces into the ABS legislation a viral instead of a come-back clause. (3) It differentiates between prior informed consent (PIC) for aTK of identifiable and unidentifiable origin thus subjecting only the former to the PIC of the indigenous and traditional peoples. Apropos aTK, such communities and peoples must be acknowledged in publications for their contribution to conservation of genetic heritage as well as the origin of aTK indicated. (4) It only requires benefit-sharing if there is economic exploitation of reproductive material. (5) It allows the choice of benefits to be shared whether monetary or non-monetary benefits including from economic exploitation. (6) It is articulate that benefits deposited in the National Fund for Benefit-Sharing shall be used for the purposes of conservation and sustainable use of genetic heritage and aTK. As the author conclusively remarks, although the new biodiversity law presents some difficulties e.g. for the intermediate products' industry, which finds it hard to understand the nature and responsibilities regarding the development of such products, and to users in general in the operations of the electronic registration system, SisGen (the National System of Genetic Resource Management and Associated Traditional Knowledge), it has introduced a new way of management of ABS that gives users more flexibility. The suggestions made in this book will help improve the ABS situation further.

In Chapter "Towards Mutual Supportiveness Between the Nagoya Protocol and the Andean Sub-regional ABS Regime: The Cases of Ecuador and Peru", Maria Victoria Cabrera Ormaza analyzes the access and benefit-sharing (ABS) regimes of Ecuador and Peru. Both countries belong to the so-called megadiverse countries. They are also members of a pre-NP sub-regional common regime on ABS established under Decision 391 (1996) of the Commission of the Cartagena Agreement of the Andean Community of Nations (CAN) upon which their ABS rules and principles are built. As a result their ABS legal frameworks reflect its complexity and discretionary power accorded to the State, which was seen as a way of controlling the presumed high economic value of their genetic resources (GR) and expected economic benefits. This can be seen e.g. in its definition of the term 'access' which is open-ended and encompasses any form of use leading to multiple negotiations and conclusion of contracts during the entire duration of use. Besides, it does not allow access to foreigners unless they involve an officially recognized national support institution or individual who must participate in the project from its planning phase to conclusion. However unlike Brazil, the applicant is permitted to negotiate the terms of access and benefit-sharing. To ameliorate the situation for non-commercial research though the regime introduces the so-called contrato marco, a 'framework agreement', as a special access procedure for such research. This has been further regulated by both Ecuador and Peru. As the author notes, Decision 391 was a milestone in the development of a multilateral ABS regime, but it is ambiguous and lacks guidance on post-access monitoring of the value chain. During the debates preceding the implementation of the Nagoya Protocol (NP), Decision 391 was still considered in Ecuador as the basic ABS framework to be complemented by the NP whilst in Peru it did not feature in the debates. Consequently, Ecuador in its Executive Decree (No. 905) of 2011 basically just adopted the same stringent requirements and procedures of Decision 391 and elaborated them. An implementing regulation of 2017 established a one-stop counter with the aim of streamlining procedures in order to improve the situation, but no procedure exists as yet on access to GR for non-commercial research or associated traditional knowledge (aTK). On the other hand although Peru's law Supreme Decree No. 003-2009-MINAM that transposed Decision 391 reflected similar stringiness, its draft of 2019 for a new regulation to implement both Decision 391 and the NP shows some difference. Besides introducing new definitions including from the NP as well as reformulating those adopted from Decision 391 it only requires authorization for non-commercial research to access GR from the competent authority and eliminates the requirement to conclude a framework agreement. The applicant must only identify at least three non-monetary benefits for the country of origin. It also designates checkpoints with functions as required by the NP which is recognition of its obligations to control value chain for use of other countries' GR and aTK in Peru. ABS regulation in both Ecuador and Peru still presents many challenges due to their strict attachment to Decision 391, which could probably explain why no contract for commercial purposes has been concluded up-to-date. A fair improvement seems evident once this bond is avoided. This suggests that to have an effective and efficient ABS regulation and to fully comply with the NP parties to Decision 391 must either develop new domestic regimes independent from it or initiate a radical process to completely transform it in order to reconcile the two systems. Until now the process to revise it has failed.

In chapter "New ABS Legislation and Practice in Compliance with the Nagoya Protocol: Current Situation and Perspectives in Costa Rica", Jorge Cabrera Medaglia examines the access and benefit-sharing framework of Costa Rica. The Biodiversity Law No. 7788 (BL) and the ABS regulations of 2003 and 2007 precede the Nagoya Protocol (NP), but the Costa Rican regime has been applauded as having one of the most complete and functioning frameworks through which a huge amount of experience necessary for providing relevant lessons on how to implement ABS in practice has been produced thus justifying its choice. This is testified by the over

650 permits issued between 2004 and 2020 and numerous negotiated and concluded agreements. It is also quite innovative. For example, a label was created recently to certify that a product has been produced in consistency with the permit and national regulatory framework. Likewise, an applicant can request that a certificate of legal provenance for export purposes be granted, which, although is not an internationally recognized certificate of compliance, can provide legal certainty about the rightful acquisition and transfer of the materials. Nonetheless, some post NP work has been done to improve and clarify the ABS situation, *viz.* the regulations of 2003 and 2007 were revised in 2019 by Decree No. 41591-MINAE, a Memorandum of Understanding was signed in 2014 to clarify and address grey areas as well as achieve a common understanding on key areas regarding access to plant genetic resources for food and agriculture (PGRFA) falling under the multilateral system (MLS) of the International Treaty (IT) and the Decree No. 39341-MINAE was adopted in 2016 to establish procedures for imposition of sanctions for illegal access. In addition, a bill has been submitted to parliament to amend several provisions of the BL, which includes exclusion of most basic research from an access permit requirement and instead creating a notification procedure. Based on the public domain approach established by the law the State controls the entire ABS process by means of the procedure established in Chapter V of the BL, *albeit* the customary rules of the communities and any existing *sui generis* intellectual rights must be taken into account when access is taking place in indigenous territories. The rules established by the regime apply equally to all types of research despite the fact that 'basic research', 'bioprospecting' and 'commercial use' are considered as different activities, but the requirements for commercial purposes are very detailed and royalties mandatory. Only a lean exemption from the obligation to obtain an access permit is provided to basic research activities by Decree No. 31514 of 2003 (regulations) in its amendments of 2019. The regime applies equally to foreign and local entities and does not oblige the former to cooperate with a local institution, but a foreign entity if domiciled abroad must have a national representative for the sake of receiving notifications from the Technical Office (TO). It has a full online platform/system for access for which the National Commission for Biodiversity Management (CONAGEBIO), which is the Competent National Authority, provides a guiding manual for users. In spite of its forerunner position, however, the regime still needs improvements. The validity of 3 years for permits despite the conditions of periodic reporting, monitoring and control, for instance, is considered very short bearing in mind that commercial investigations can take many years before a final product is produced. Although the permits can be renewed, the uncertainty related to the danger of failure of renewal and the burden of the work connected to this still persist. It is also not very bold in committing shared benefits to conservation of biodiversity. Although it mentions the percentage to be shared it does not make it an obligation to invest such shares into conservation. Besides, it establishes registration of *ex-situ* collections as providers of GR but still requires users to negotiate PIC with the collections. This installs a kind-of double PIC negotiation procedure: between the collection and the initial provider of GR and between the collection and the user. In addition, it has no measures for compliance with ABS legislations of other countries

when GR and/or aTK are being utilized in Costa Rica. To bring the regime into full compliance with the NP and incorporate some of its innovative provisions more revision work is necessary. For that Costa Rica maybe also needs to ratify the Protocol in order to get the intuition for the urge.

B. *African Continent*

In Chapter "The South African ABS Regime: New Wine in Old Wine Skins?", Evanson Chege Kamau discusses the regulation of access to genetic resources, traditional knowledge associated to such resources and benefit-sharing in South Africa. In putting its regime in place South Africa had a high demand on its biodiversity for research and development (R&D) by both local and foreign industries, demand for local non-industrial uses, unconstrained access under the notion of common heritage for humankind and the encroachment on associated traditional knowledge (aTK) to contend with. It was therefore necessary for the country to develop a comprehensive biodiversity policy to regulate its use and aTK in line with the international legal framework. Apart from the Act of 2004 regulating access and benefit-sharing (ABS), regulations thereto have been developed (2015) and guidelines for both providers and users (2012). A number of revisions have been done on some of the instruments including after the adoption as well as entry into force of the Nagoya Protocol (NP), *albeit* not in response to the need for compliance with it. But work is underway to bring them into full compliance with the NP. The regime has a number of interesting elements, for example: (1) It has a distinct way of defining terms *viz.* 'indigenous genetic and biological resources', 'bioprospecting', 'any other kind of research' and 'traditional use or knowledge', which makes it clear to what exactly and to which activities it applies. From the definition of 'indigenous genetic and biological resources' in the amendment Act of 2013, for example, it is clear that 'genetic information' and any export of biological resources including by traders for bioprospecting purposes also fall under the scope of the regime; (2) It separates phases of bioprospecting giving ease to early stages of R&D—where the nature and extent of any actual or potential commercial or industrial exploitation in relation to the project is not sufficiently clear or known to begin the process of commercialization—in terms of permits and benefit-sharing; (3) It establishes a clear institutional framework; and (4) It provides a fixed timeframe for permit issuing authorities to decide on applications with consequences for failure to meet the set deadlines. But the regime also has some flaws. For instance there are many scattered bits of statutes which make it burdensome to understand the regime with certainty yet the guidelines of 2012 are outdated; the rules meant to differentiate requirements and procedures for the sake of facilitation of non-commercial research are a bit obscure and confusing; model MTA and BSA are provided without an option to expunge, replace or rectify any of the clauses or use another agreement e.g. one proposed by the applicant; and it does not have compliance measures for R&D undertaken in South Africa based on other countries' genetic resources and/or aTK. Among others it is recommended that a kit is developed which brings together the rules, requirements and procedures into one document with references to the different relevant laws and that the guidelines are updated. Currently it is left to wonder whether the new changes which have been continually made fit/sit well with the old law or are like new wine in old wine skins.

In Chapter "The Post Nagoya Protocol ABS Regime in Cameroon: Exploring the Extent to Which Ongoing Policy, Regulatory Developments and ABS Practices Uphold the Obligations of the Protocol", Marcelin Tonye Mahop presents the current situation in the regulation of access to genetic resources, associated traditional knowledge and sharing of benefits arising from their utilization in Cameroon. Cameroon's rich biological diversity has been subjected to different threats hence the importance for an effective legislation which is able to promote its conservation and sustainable use and benefit-sharing from its utilization. Until recently Cameroon was regulating access and benefit-sharing (ABS) through sectoral laws but now a new law has been adopted and it is hoped that the draft implementing instruments which were developed along the law will be adopted soon. The author discusses this new law and the draft instruments. One of the greatest strengths of the framework is that it closely involves the indigenous and local communities in the ABS process including through the use of their bio-cultural protocols and making them the managers of the benefits shared for, among others, the conservation of biodiversity. Of course it is doubtable that such communities have the ability to manage as well as apply scientific non-monetary benefits. Besides, the law makes it a condition to publish the permit issued together with PIC and MAT on the ABS Clearing-House which should help to create certainty of legal access for users. Also according to the author, the elements of the PIC developed within the new framework align themselves with the minimum requirements of an internationally recognized certificate of compliance in accordance with article 17.4 of the Nagoya Protocol (NP). Likewise, it differentiates between permits for commercial and non-commercial purposes, *albeit* the difference made for access is only the exemption from PIC and MAT for fundamental research and research and development undertaken locally by national research institutions without transfer of materials abroad. In addition, it provides clauses on resolution of conflicts. However, it also has several weaknesses. For instance it says that "previously accessed resources" fall under its scope but does not define a temporal scope. This makes the scope extremely broad but it is assumed that could have been intentional. Also it complicates the permit issuing process by creating a double authorization procedure which requires the applicant to seek authorization or consent to negotiate prior informed consent (PIC), which is a pre-condition for the obtainment of a permit. Likewise, it fails to address the issue of mutual recognition and enforcement of foreign judgements and arbitral awards. In addition, established checkpoints do not embrace the functions of checkpoints as foreseen by article 17 NP, but are focused on monitoring and surveillance of the utilization of Cameroonian genetic resources. This is evident too in the way the draft framework empowers the ABS committee created as an advisory body to the CNA to monitor the value chain of genetic resources utilized abroad including those held in international *ex-situ* collections without an afterthought on how possible that is. That kind of too much focus on control comprises one of the major weaknesses of the regime.

In Chapter "The Ethiopian Access and Benefit-Sharing Regime: Stringent with a Purpose", Ashenafi Ayenew Hailu and Evanson Chege Kamau discuss the regime regulating access to genetic resources and associated traditional knowledge and benefit-sharing in Ethiopia. Being one of the high-ranking biodiversity rich countries of the world Ethiopia in passing its access and benefit-sharing legislation aimed *inter alia* to ensure that the country and its people benefit from its biodiversity, terminate uncontrolled access to it and the knowledge of the local communities, and encourage conservation and sustainable use. Ethiopia regulates access and benefit-sharing (ABS) through a pre Nagoya Protocol (NP) framework which though developed to implement the main pre NP international legal framework was greatly influenced by a Teff biopiracy case. Its stringent measures and a more prudently regulated legal regime to control ABS was therefore intentional. This is reflected, for example, in the condition that foreign institutions and competent national authorities guarantee compliance of users or accept liability for enforcement of their obligations. Nonetheless, the current regime has been revised in order to bring it into compliance with the NP, in particular in regards to user and fully compliant monitoring measures, as well as to expunge measures that are hard to implement. As this book was written a draft law had been sent to the Council of Ministers for scrutiny before submission to parliament for adoption. Therefore, the authors examine the current regime in parallel with the revised draft law. This is useful in particular in establishing the level of compliance of the to-be regime with the NP. For instance, the old law did not have user (compliance) measures but these have been included in the revised draft law; the old law included the wording 'genetic information' in its definition of 'genetic resources' and believed that this included 'digital sequence information' (DSI) but the draft law uses the latter term in the revised definition of 'genetic resources', which makes it clear that DSI falls under its scope. The authors also look at the strengths and weaknesses of the revised draft law. Among strengths other than those mentioned above the draft law is clear concerning bulk resources used eventually in research and development (R&D), which will then fall under its scope, and establishes a temporal scope from 27 February 2006; it maintains a kind-of one-stop shop with Ethiopian Biodiversity Institute (EBI) being the national focal point and the competent national authority; although the old law also differentiates between access for commercial and non-commercial purposes and their requirements and procedures the terms 'commercial' and 'non-commercial' were not defined, but have been defined now in the draft law; it exempts plant genetic resources for food and agriculture under Annex I of the International Treaty from its regulation unlike the old law; and it is specific concerning allocation of benefits for conservation and sustainable use purposes. The main weakness of the Ethiopian regime including the revised draft law is its restrictive and protective nature which tries to regulate every aspect touching on genetic resources in an attempt to close every window of misappropriation and misuse and guarantee the country's sovereign rights over its resources. The revised draft law's architecture, however, is out to *inter alia* promote research and the fair and equitable sharing of benefits. This it is hoped will contribute to the conservation and sustainable use of biological diversity in line with the Nagoya Protocol.

In Chapter "Abracadabra! Or When and How Will the Kenyan ABS Law Be Born?", Evanson Chege Kamau presents the Kenyan access and benefit-sharing regime. As one of the megadiverse countries Kenya in developing an access and benefit-sharing (ABS) regime aspired to advance conservation and sustainable utilization of its biodiversity, contribute to fair and equitable sharing of benefits arising therefrom and curb its overexploitation and unregulated bioprospecting. In the process a number of challenges have been encountered the main one having a negative effect on the legislation being lack of a concerted approach by the different lead agencies leading to fragmentation and sectoralism. ABS is still regulated through a pre Nagoya Protocol (NP) biodiversity framework legislation of 2006 and sectoral laws. Without need to say the biodiversity legislation is one of the post CBD laws passed with the aim of instilling control rather than enabling research. Though it is an old law the choice of the case study is justified by the fact that sectoral laws have always played a major role in regulation of ABS in Kenya of which one has been revised and one legislated post NP. The author examines the interplay between the relevant laws and how ABS functions and is affected as a result. It is established that, despite some of the sectoral laws being post NP legislations, the complexity and unclarity of the regime persist or is occasionally hiked. Apart from the procedural complexes and challenges the regime has other weaknesses, *inter alia*: The definitions of 'biological/genetic resources' as well as activities contemplated under the definition of 'access' are very broad; the biodiversity legislation does not provide complaint channels against applications granted by the competent national authority although opposed by the public, *albeit* the Constitution of 2010 as well as Environmental Management and Coordination Act grant standing on public matters; it is not clear to which fund the benefits shared shall be put or how they shall be shared; there are no user compliance measures; designated checkpoints have no functions in line with the NP; county governments have been created in the new Constitution but their roles in ABS are not concretized; and terms are used inconsistently across the different laws. Maybe the strength of the Kenyan regime lies in the approach of the new practice of finding ways to ameliorate the challenges created by the relevant laws, although the author warns that such an approach can also cause legal uncertainty e.g. if applied measures are later challenged. Existing challenges are a cause for legal uncertainty, unclarity and intransparency which at the end are a stumbling block to Kenya's initial aspirations. There is need for the law to catch up with practice although the many years of attempts to revise the biodiversity law of 2006 as well as develop a stand-alone law without tangible results makes questionable the ability of these efforts to bring forth a "healthy product".

C. *Asian Continent*

In Chapter "The New Law and Practice on ABS in Viet Nam: Innovations and Compliance with the Nagoya Protocol", Tran Thi Huong Trang Guihou, Nguyen Ba Tu and Nguyen Dang Thu Cuc look at the regime regulating access to genetic resources, associated traditional knowledge and benefit-sharing (ABS) arising from their utilization in Viet Nam. In response to conserve and sustainably use its rich biodiversity, being one of the world's biodiversity hotspots and based on the key role biodiversity plays to its development Viet Nam has promulgated relevant policies in line with international treaties. After becoming party to the Nagoya Protocol (NP) it adopted a new law, Decree No. 59 of 2017, replacing Decree No. 65 of 2010 which had been issued in order to elaborate and guide the implementation of the Biodiversity Law (BL) of 2008, but its provisions were still limited and inconcrete to be applied directly into practice. During its seven years lifetime (2010–2017) there was no applicant for a permit of access. Decree No. 59 was adopted in order to deal with these shortcomings, implement the NP and clarify the conformity situation/question with the NP and the ABS provisions of the BL. The Decree was developed in collaboration with a number of international partners and therefore the Viet Nam case study like the Argentinian and Malaysian also foreshows the role of foreign international partners in the development of domestic laws. This is not the focus of the case study but just to mention, this could maybe explain whether the interests of foreign entities are better represented following such collaborations (according to the authors one of the aims of the regime is to "create a system of ABS for foreign entities"), or whether the rights of the relevant country are muffled, or if the resulting laws are well balanced. One of the notable characteristics of the regime is that its model contract also serves as the prior informed consent and mutually agreed terms. Its weaknesses include the very broad definition of the term 'genetic resources' (in the BL) that puts biological resources in general under scope, but lacks a list of activities it considers as 'utilization'; there are no measures to regulate associated traditional knowledge but a guidance document on the same is being developed; the practice allows supply of genetic resources (GR) by "informal" intermediaries (parataxonomists) but users may not interact directly with them which is a cause for legal uncertainty; it does not exempt plant genetic resources for food and agriculture under Annex I of the International Treaty, *albeit* Viet Nam is not party to the treaty; it does not establish user compliance measures; and it lacks a financial mechanism to receive and distribute shared benefits as well as guide on the use of benefits for biodiversity conservation and sustainable use purposes. It also has strengths, e.g. it establishes a clear temporal scope of 1 July 2009 when the BL came into effect; it gives the possibility of regularization for users who accessed GR from this date up to the date Decree No. 59 entered into force and would wish to continue utilizing the GR but had not obtained a permit/license; it defines clearly which authority is responsible for which GR with clear procedures and has a relatively short timeline for application decisions to be taken both for non-commercial and commercial purposes; it establishes an online application procedure; it offers an option to share benefits either in monetary form or non-monetary also for commercial uses. In general the law looks relatively comprehensive and many of its attributes comply with the NP, but the regime is prone to implementation challenges due to potential conflicts within the regime as some provisions of the BL became outdated with the issuing of Decree No. 59. Whilst the latter is more specific on the subject area, it must be consistent with the former being a by-law. It is recommended that the BL be revised in order to eradicate these conflicts and simultaneously the opportunity be whisked to fix other shortcomings of the regime.

In Chapter "The Fastest Animals Are Not the Fastest Over Time: Malaysia Adopts a Comprehensive ABS Legislation After a Long Steady Effort", Evanson Chege Kamau discusses the Malaysian legislation on access to genetic resources, associated traditional knowledge (aTK) and benefit-sharing. Despite being a megadiverse country and ratifying the Convention on Biological Diversity (CBD) early followed by an immediate and a focused initiative to implement it, Malaysia did not have an access and benefit-sharing (ABS) legislation at a national/Federal level until the adoption of the Act of 2017. Efforts at the national level had produced the first national/Federal draft framework as early as 1999 but this study could not establish with certainty why such a long delay followed afterwards. Collaboration with international partners that began in 2010 seems to have played a decisive role in the final success. The Malaysian Act applies to genetic and biological resources and information thereof, associated traditional knowledge (aTK) and, contrary to the CBD, human genetic resources. Besides its application to human genetic resources other peculiar features are its retroactive application to biological resources (BR) and aTK if no benefit-sharing agreement (BSA) had been entered into prior to its entry into force and the possibility for the competent authority (CA) to deny an access permit if the jurisdiction of the user does not have effective user compliance measures. Although BR for bulk uses are exempted from its application the provisions of the Act will be triggered if they are later used in research and development (R&D) obliging the person involved in the activity to enter into a BSA with the resource provider and receive a permit from the competent authority. Among its strengths, unlike most traditional provider countries' legislation, the Malaysian Act establishes user compliance and monitoring measures; defines critical terms e.g. non-commercial research even though the procedure is not differentiated; provides a provision to direct benefits deposited in the fund to conservation and sustainable use of biodiversity; establishes a clear temporal scope; exempts specialized instruments for ABS from its scope; recognizes the rights of the indigenous and local communities (ILCs) as providers of BR on land to which they have a right as established by law and to their aTK and establishes that the process of obtaining prior informed consent and concluding benefit-sharing agreements must occur according to the customary laws and practices, protocols and procedures of such communities; and foresees as well as provides solutions for situations where the representation of ILCs does not exist or where the aTK is shared by more than one community. However, it misses a mark concerning the requirements of article 18 of the Nagoya Protocol as it has not addressed the questions of access to justice in case of disputes arising from mutually agreed terms or mutual recognition and enforcement of foreign judgements. Also, it does not say to which jurisdiction disputes shall be subjected to or give options for dispute resolution, or which will be the applicable law. Without denying that the new law has slightly more flaws, observation is made that the law is clear and to a great extent gives a sense of legal certainty, which the author attributes to the long duration and steady effort taken by Malaysia to legislate.

In Chapter "Access and Benefit-Sharing Law and Policy in South Korea", Jae-Hyup Lee and Ah Young Cho examine the new law regulating the access and use of genetic resources in South Korea. Until the adoption of the Act on access to and utilization of genetic resources and benefit-sharing of 2017 (Act), Korea did not have a comprehensive law to address ABS. The laws and regulations for the protection of the environment and species, some predating the CBD, did not protect biological diversity directly, were fragmented and were not sufficient to implement the Nagoya Protocol (NP). Unlike most case studies examined in the book, Korea is more a user of genetic resources (GR) than a provider with the biotechnology industry having grown into a major industry. Besides, a large amount of GR is imported from overseas due to a lack of domestic alternative resources, insufficient supply and a high cost of domestic GR. Therefore, the delay in ratifying the NP, which only took place in 2017, and putting access and benefit-sharing (ABS) measures in place is criticized by the authors as having disadvantaged local industries because provider countries would be reluctant to export GR to Korea. However, this could have been a calculated move to avoid or delay compliance with measures of provider countries as it was also evidenced in other industrialized countries. The current Act likewise exhibits some signs of alignment with user perspectives, e.g. it does not mention derivatives and digital sequence information (DSI), establishes a temporal scope of 17 August 2018 which is post its entry into force and a list of the Regulation (EU) No 511/2014 is reproduced in a guideline published in 2019 as examples of activities considered as utilization. Some of the strengths of the Act are that it exempts GR subject to international treaties on ABS from its scope which is in line with article 4 NP; by defining the term 'benefit' to mean monetary or non-monetary it opens the opportunity for parties to agree on any kind of benefits including from commercial uses; it establishes user compliance measures including for tracking and enforcing in line with articles 15, 16 and 5 NP; it designates checkpoints in line with article 17 NP; it eliminates the requirement for an application, prior informed consent or conclusion of a benefit-sharing agreement for access to GR and requires only a report to the head of the competent national authority (CNA); it creates a short duration of 30 days from the date of receipt of the report for the CNA to inform the person seeking access of the outcome; it has given power for waiver of the requirement of a report or simplification of procedures; has provided for expeditious access for utilization related to human, plant or animal health in line with article 8 (a) and (b) of the NP; and has installed a simple and easy integrated online reporting system. However, it also has weaknesses, e.g. no distinction is made between research of a commercial and non-commercial nature; two national focal points have been established contrary to article 13 NP which states "Each Party shall designate a national focal point on access and benefit-sharing" and unlike any other party to the NP; no sanctions are offered against violations relating to foreign GR utilized in South Korea thus weakening enforcement; and except mentioning traditional knowledge in the declaration of the purpose (Art. 1) of the Act and in definitions (Art. 2) there are no measures to regulate associated traditional knowledge but its argued that the mention might be just precautionary as such knowledge is non-existent in Korea; it exempts privately held GR from its scope and thus leaves a gap concerning regulation of such resources. But the greatest weaknesses of the regime maybe is the fragmentation of relevant laws and national authorities with jurisdiction over ABS issues thus making it hard for users to understand the procedures, increasing administrative tasks and raising costs and management burdens. It is advised that the government issues guidelines for users, researchers and companies with a clear interpretation of representative cases which they can use as a basis for consultation and negotiation.

D. *Australian Continent*

In Chapter "ABS in Australia: A Story of Early Success and Faltering Progress", Geoffrey Burton describes the situation with the access and benefit-sharing (ABS) legislation in Australia. Being a megadiverse country with an extensive body of traditional knowledge associated with genetic resources and also a growing biotechnology and life sciences industry, Australia is a developed-economy user of genetic resources (GR) and associated traditional knowledge (aTK) and a rich provider of the same. An interesting observation would be to see how it reconciles the interests of both sides domestically as well as address them internationally making it a good choice for our study in spite of not having a post Nagoya Protocol (NP) federal legislation. It is apparent that Australia became party to the Convention on Biological Diversity (CBD) immediately when it entered into force on 29 December 1993. However, the implementation domestically of its ABS measures by including a general provision in the biodiversity law of 1999 did not detail how ABS should be undertaken, an issue which was left to yet to be developed subordinate legislation (regulations). The top-down process which was initiated to do this, the so-called Voumard Inquiry of 2000, had to ensure the enactment of a robust legislation that addressed the different stakeholder interests and likewise that social and economic benefits of the use of genetic material and products derived from Australia's biological diversity accrue to Australia. Another thing that makes complex the implementation of ABS measures is the form of government being federal with 6 sovereign States and 2 States that are accorded self-governance, and strong indigenous peoples' rights which the Government at the same time must protect. The inquiry made recommendations to the Government of the legal framework and legal scheme advocated based on principles it developed in response to stakeholder concerns. Parallel to the inquiry and the development of a draft law, Australia was actively involved in the development and adoption of the Bonn Guidelines of 2002. Being consistent with the Government's ABS policy, the emerging draft Environment Protection Biodiversity Conservation (EPBC) Act regulations and the national Strategy of 1996, the national Government in an agreement with all States and territories agreed on a uniform implementation in the same year. The so-called Nationally Consistent Approach comprised 14 principles and 11 common elements which were to be reflected in each jurisdiction's subsequent ABS law and administration. The draft law was finalized in 2005 and included in the regulations of the Biodiversity Act as "Access to biological resources in Commonwealth areas" and had the aim of giving effect to the Bonn Guidelines and the Nationally Consistent Approach. It serves as a model ABS law with the purpose of among others conservation of biological resources and their sustainable use, equitable benefit-sharing, respecting indigenous peoples' traditional knowledge, and providing legal certainty. For non-commercial research the permit is free and for commercial purposes only a nominal fee of AU$50 is charged. For the former there are arrangements for facilitated access and exemption from benefit-sharing agreement (BSA). For the latter a permit and BSA are necessary but the approval of applications is provided rapidly within 2–10 days. A permit of access from the territories of the indigenous peoples is granted on condition of evidence of a BSA between them and the prospective user based on prior informed consent (PIC) and mutually agreed terms. The current system has been functioning successfully for 14 years with the granting of permits taking place at the rate of 2–3 permits per month. Australia also went on to solve the challenges of operationalization of the regime e.g. the uncertainty created by international and domestic debates on the scope of GR, the difficulty in verifying the legal provenance of GR, disclosure of use of GR to the indigenous peoples and proper grant of their PIC, and facilitation of access. Further improvements to the regime have been introduced as recently as 2020 through an amendment to the biodiversity law in order to devolve Commonwealth environmental approvals down to states and territories. Nonetheless, Australia has not yet ratified the Nagoya Protocol (NP) although it signed it in 2012. Options for complying with the NP were published and the initial progress towards ratification made in 2014 but no further development of a model system for implementing the NP has occurred since then. It is noted that Australia's ratification of the NP is dependent on continued pressure from its biotechnology research communities and from its state and territory governments, which is indicated by some of them undertaking their own initiatives. The State of Queensland, for example, passed the Biodiscovery and Other Legislation Amendment Act on 20 August 2020, with its amendments coming into force on 30 September 2020.

E. *European Continent*

The book enters a new section of studies examining the regulation of ABS in the European continent. The focus of the studies is limited to the European Union (EU). The EU transposes international treaties as a block. Following this pattern the EU implemented the Nagoya Protocol through Regulation (EU) 511/2014.[22] The EU adopted the Regulation on 16 April 2014 and it applies as of 12 October 2014, which is the date of entry into force of the Nagoya Protocol (NP). It aims to implement the compliance pillar of the NP in the EU which means that it does not regulate access to genetic resources (GR), associated traditional knowledge (aTK) and related benefit-sharing in the Member States (MS) of the EU. However, it gives them leeway to implement provider/access measures. The Regulation has direct effect in MS and therefore each MS must implement it nationally. Up to date only 5 out of the 27 MS have put in place (in addition to the obligatory user measures) provider measures for

[22] For a thorough scrutiny of the implementation ('due diligence') model of the Regulation and suggestion for an alternative ('integrative') model see Godt et al. (2020).

access to GR and aTK, i.e. Spain, France, Croatia, Malta, Bulgaria. Two more, Finland and Sweden, have provider measures only for associated traditional knowledge. This section examines Regulation (EU) 511/2014 which consists of the compliance regime of the EU and then presents two case studies of Spain and France as examples of MS with a dual provider and user State role.

In Chapter "The ABS Compliance Regime of the European Union", Gerd Winter scrutinizes the regime of compliance with the obligations of the Nagoya Protocol in the European Union. The genetic potential of organisms living *in-situ* within the borders of most legal systems of the EU Member States (MS) are considered in principle as *res nullius*, i.e. free goods. This implies that both foreign and domestic researchers are free to access and utilize them in research and development without any restrictions as well as obligations on benefit-sharing. It is not surprising that Regulation (EU) 511/2014 that transposes the Nagoya Protocol (NP) in the EU only considered user measures for complying with provider measures. Apart from excluding provider measures, the Regulation does not include transfer of relevant technologies, capacity building and appropriate funding which are also objectives of the CBD and its NP, a failure Winter criticizes. The Regulation uses basic duties as instruments to ensure that access to genetic resources (GR) and associated traditional knowledge (aTK) and their utilization take place in a legal manner/in compliance with provider access and benefit-sharing (ABS) measures. These duties are placed on users of GR and administrative oversight. Besides, there are additional provisions on registered collections and recognized best practices with the aim of simplifying the obligations of users and administrative bodies. How this system is meant to function from access to benefit-sharing is summarized in a table by the author. The Regulation enforces its obligations on users through its concept of compliance called due diligence. Users must ascertain that first, they exercise due diligence and second, they declare due diligence at specified stages. This burden is alleviated if the GR is accessed from a registered collection as due diligence is considered to have been exercised. Administrative/enforcement authorities on the other hand must conduct checks to ensure that users are complying or have complied of which the author suggests must go beyond the normal practice and extend to the bringing on the market of products developed on the basis of accessed genetic resources, as well as to the sharing of benefits therefrom. Administrative/enforcement authorities are also empowered and mandated to receive due diligence declarations from users. In addition, other important issues examined concern the scope of application of the Regulation—material, personal, geographical, temporal. Regulation (EU) 511/2014 is an exemplary step forward of user States which tries to ensure compliance with the ABS regimes of provider countries but the compliance scheme it establishes suffers a number of shortcomings in practice e.g. in regard to declarations, basic duties of users and checking for which the author suggests ways of solving.

In Chapter "Access and Benefit-Sharing Regime of Spain: Striking the Right Balance Between Its Interests as a Provider and a User of Genetic Resources", Luciana C. Silvestri presents the regime of access and benefit-sharing of Spain. Spain is the most biodiverse country in Europe with one of the greatest cultural diversity endowing it with a wealth of traditional knowledge associated with genetic resources (aTK). At the same time, Spain has a fast-growing biotechnology industry and is a Member State of the European Union making it an important user country. Thus, Spain's interests cut across both provider and user sides. It supports effective access and benefit-sharing (ABS) regimes that enable providers to benefit from the utilization of their genetic resources (GR) and aTK while facilitating research and development by users and simultaneously strong (user) compliance regimes which are able to curb violations against provider interests/measures. This bestows on Spain a double role as a provider and a user and has put in place provider measures and, in implementing its obligations based on Regulation (EU) 511/2014 and Commission Implementing Regulation (EU) 2015/1866, user measures. The features of both provider and user measures are evident in its ABS regime. In regard to provider measures, for example, prior informed consent (PIC) must be obtained, mutually agreed terms (MAT) established and an access permit sought for access to *in-situ* or *ex-situ* GR from the date of entry into force of Royal Decree No. 124/2017, i.e. 15 March 2017, as a general rule. Spain consists of 17 autonomous communities each of which is responsible for granting PIC and establishing MAT for access to GR from their jurisdictions and, in addition, authorizing access if the GR is endemic to their community/territory. There are variations of the rule e.g. if the GR is located in the territory of more than one community, or falls under the jurisdiction of the State, or is intended for non-commercial purposes or for commercial purposes. It is also notable that the duration for review of an application for commercial purposes is relatively long taking up to six months. Concerning user measures the temporal scope, for instance, follows the usual trend of user countries' approach. Besides, it does not apply retroactivity in regards to entities that at the date of entry into force of Royal Decree No. 124/2017 had GR already under their control even if their activities concern new utilization, it exempts pure scientific biodiversity research from its regulation, and it embraces user compliance measures built on the concept of due diligence as foreseen by the EU instruments on ABS. It can be observed that Spain has managed to comply with the obligations of the NP and the EU Regulation and has even been more ambitious than the latter, a sign that ABS has a chance of being efficiently and successfully implemented in Spain. Although it will take some years to assess its effects the author concludes rightly that the current regime of Spain may serve as a model for countries with a dual provider and user role.

In Chapter "The Post Nagoya Protocol ABS Regime in France: Exploring the Extent to Which It Upholds the Obligations of the Protocol", Marcelin Tonye Mahop explores the regime that domesticates the Nagoya Protocol on access and benefit-sharing (ABS) and also, since France is a Member State of the European Union (EU), implements European Union law on ABS according to Regulation (EU) 511/2014 and Commission Implementing Regulation (EU) 2015/1866 in France. France considers itself a megadiverse country based on its unique position in Europe and the rich biodiversity found in its overseas territories. France is equally a major user of genetic resources (GR) being a technologically developed country. Therefore like Spain, it is committed to implementing these agreements and laws in order to consolidate its position as a provider of GR and a user of GR and associated traditional knowledge (aTK). France formulated its domestic ABS regulatory framework to comply with the NP following its ratification. France comprises the mainland and the overseas departments and regions, but according to the Constitution the ABS law Code de l'Environnement (Loi No 2016-1087 of 2016), its implementing Decree No 2017-848 of 2017 and the model contract for utilization of aTK annexed thereto would equally apply to mainland France as well as the overseas regions and territories. There is a second model contract for access and utilization of GR which was promulgated as a separate implementing regulation (Arrêté (Order) of 2017). With two different approaches for access it installs provider measures for access to GR. Through a "declarative approach" access for non-commercial uses is allowed without need for an access permit. In the contrary, an "authorization approach" is used for access for commercial uses with the applicant required to obtain an access permit after the relevant authority is satisfied with the application and the applicant having first successfully negotiated a benefit-sharing agreement (BSA)/contract which is based on the PIC of the communities. A BSA is meant to materialize the prior informed consent of the relevant authority, which will either be the national competent authority or the relevant authority of an overseas territory. A special application process is foreseen for utilization of aTK of the communities of inhabitants of Guyana and Wallis and Futuna Islands. The access permit contains the usual provider conditions on third party transfers, change of intent etc. In fulfilling its obligations under Regulation (EU) 511/2014 France also puts in place user measures following the concept of due diligence. User characteristics include the application of the ABS regime from the date of its entry into force on 9 August 2016 and non-applicability to resources accessed as bulk commodities and used later for research and development. It is strange that it subjects subsequent access to resources held in collections that were constituted before that date and any new uses of those resources to the law as a highly industrialized EU State, but again it shows its provider inclination. Although the regime is clear, contains transparent procedures and is generally *de jure* compliant with the NP, it lacks a mechanism for bringing corrective and punitive actions against users for non-compliance. This has the potential of lowering the degree of its compliance and hence should be corrected.

Part III: Implementation Experiences and Lessons

In Chapter "Post Nagoya Protocol Experiences of Basic Research in Ecuador", Erwin Beck looks at the circumstances and experiences of doing research in Ecuador for basic biodiversity research before and after the ratification of the Nagoya Protocol. This is reported from first-hand information of the author's work within German research groups which have been engaged in ecosystem studies in South Ecuador since 1997. The groups cooperate with local (Ecuadorian) partner institutions under Memorandums of Cooperation since the last 22 years and are mainly funded by the German Research Foundation (DFG). The implementation of the regulations of the Nagoya Protocol (NP) turned out to be a difficult and still ongoing process in Ecuador. This, according to the group's experience, is based on a number of reasons, *inter alia*, that its 24 provincial governments practice a relatively high level of autonomy in particular with respect to biodiversity issues. As a result, the research group's projects located in different provinces are accorded different treatment in administrative supervision, including in regard to details of permits. With the implementation of the NP, 6 focal points have been recorded. Competences are not centralized and there is a shortage of experts in the Federal ministries. Biological research as a field reaching into several administrative responsibilities is also not easy to handle by the authorities. Three governmental authorities are relevant for granting access to genetic resources (GR) and permission for basic/academic/non-commercial research with biological materials. The Secretaría Nacional de Educación superior, Ciencia, Tecnología e Innovación (SENESCYT) is the authority competent for issuing research permits. Concerning biological materials the Ministry of Water and Environment (Ministerio de Agua y Ambiente del Ecuador, MAAE) (and its provincial branches) has to agree and decide whether the project has a commercial purpose. If it does have a commercial purpose it will need a so-called framework contract (contrato marco) with prior informed consent (PIC), mutually agreed terms (MAT) and an Internationally Recognized Certificate of Compliance (IRCC). If it is only for basic research then it could benefit from facilitated procedure of granting research permission in line with article 8 (a) NP. This differentiation is also valid for Ecuadorian research institutions. Considered as an interim regulation, the authority of granting access to GR and issuing research permits is currently with MAAE with counselling by the National Biodiversity Institute (Instituto Nacional de Biodiversidad, INABIO). For the research group's projects in South Ecuador, ministerial supervision has been transferred to the provincial branches of MAAE, however, permission to transport biological materials between provinces ('movilización') requires also approval by INABIO. MAAE is competent for export permits of biological materials and also for activities in the scope of RED+. Prior to the adoption of regulations to ratify the NP the groups worked on group research permits. Thereafter each subproject had to obtain a separate research permission with the responsibility for compliance given to the speaker of the entire group in cooperation with an official of INABIO. Besides, research permits were initially issued according to funding duration of the projects but after 2006 they were valid for one year only. In addition, annual reports and annual applications were obligatory. Other obligations have been maintained in the new post NP contract following the ratification of the NP in 2017. In the post NP era the responsibility of ensuring conformity with the permission has been taken from the local partner to INABIO under its framework contract on access to GR with MAAE which was established in 2016. The agreement allows *inter alia* access to GR exclusively for academic research purposes to increase basic knowledge, excludes other types and aims of research, possession and utilization for other purposes, and excludes access to traditional knowledge. The group received its post NP research permit from INABIO (with agreement by MAAE) and was registered under the title "RESPECT: Environmental Changes in Biodiversity Hotspot Ecosystems of South Ecuador: Response and Feedback Effects" in June 2018. The rights and obligations of the group are defined in individual permits. Accordingly, research is only granted on those biological resources which have been specified in advance in the group's projects. Export and mobilization permits can be applied for under these permits. To show the obligations of former research permits and changes following the incorporation of the research group in the framework contract the author makes reference to his own projects carried out in the period of 2013–2018 and 2018–2020 mentioning concrete conditions and obligations. Since his project was a subproject of the program of a research group some points of general importance were also mentioned in the research permit. For example, the mobilization document could be used in the export process if the German University was declared as the final destination of the samples. For the research unit RESPECT (2018–2020) its sub-projects are legally viewed as projects of INABIO ("Genetic Biodiversity of Ecuador") under the framework contract which, however, expired in July 2019. The application form contains information on details to be provided while the framework contract contains more general provisions. Most importantly, it authorizes the named coordinators of the projects (in joint responsibility with the INABIO) for the execution of the collection, manipulation and access to biological resources. They are responsible to ensure that all persons working in the projects perform in compliance with the contract and that they are familiar with its provisions. Any change of the coordinators must be immediately reported to MAAE. The contract allows fresh material to be analyzed only in Ecuador but for specified analysis (e.g. stable isotope signals) dry samples may be exported subject to a mobilization permit from the relevant provincial branch of MAAE and an export permit by the Federal MAAE in Quito, applied for the group by INABIO. Whilst GR rights are under the sovereign rights of Ecuador, property rights of the authors over their data is recognized by the current legislation. The framework contract does not mention digital sequence information. Except capacity building for the staff of MAAE and the sharing of results it does not impose other forms of benefit-sharing. It is presupposed that this is because the contract was negotiated on the basis of an Ecuadorian institution with 7 Ecuadorian and only 1 external group. After the termination of the framework contract in 2019 the group works on the basis of a "simple" research permit which does not involve access to GR. In spite of constraints arising from the implementation of the NP a number of things have been critical in the groups success in doing research in Ecuador, among them the group's declaration to abstain from any bioprospecting from the onset and in the unlikely case its commitment to leave any promising organism with the Ecuadorian partner, collaboration with several partners which has developed into a sustainable fundament of mutual support, and the trust built over the years based on transparency.

In Chapter "Rights Over Genetic Resources and Ways of Monitoring the Value Chain: A Case Study from the Royal Botanic Gardens, Kew", China Williams describes Kew's experiences monitoring the utilization of genetic resources down the value chain to ensure that the access requirements and national measures of provider countries are adhered to and benefits are shared. Ensuring that the measures of the party providing genetic resources (GR) are observed and benefits are shared has been a major challenge for users and providers alike because "use and actors change along the value chain as the resources are used in different ways". In spite of the Nagoya Protocol (NP) obliging all parties to put in place compliance and monitoring/checking measures as well as establishing the ABS Clearing-House, there are still challenges and variations in implementation. Such challenges vary according to the kind of intermediaries involved and their roles. This raises practical questions, for instance, how should restrictions and conditions imposed by the provider of the material stay linked with the material and how can the value chain be tracked? How should future users be bound by terms of what was initially a bilateral agreement? These questions are addressed through the practices and experiences of Kew. Research undertaken at Kew does not generally have a commercial application; it is largely fundamental, exploratory, research focused on enhancing understanding of plant and fungal diversity. A small area of research focusses on the exploration of phytochemicals and their biological activity. With new ABS legislations following the adoption of the Convention on Biological Diversity (CBD) and resulting challenges, Kew developed an Access and Benefit Sharing (ABS) Policy in 1998. The policy was designed to ensure that all material brought into Kew (whether collected on fieldwork or from other institutions or individuals) has been legally acquired on mutually agreed terms (MAT), that it is used and supplied on terms that are consistent with those under which it was acquired, and that benefits arising from use are fairly and equitably shared. To this end, Kew staff at the newly created ‘CBD Unit’ started working with science staff to create templates for more formal collaborative agreements with partners, with specific terms on the acquisition, use and supply of genetic resources. The author reports that in October 2019 Kew had over 140 active research agreements with a wide range of international partners in over a hundred countries. These agreements ensure that Kew is working legally, within the national and international laws and regulations, and have become a vital tool in ensuring agreed best practice is followed. These agreements include Memoranda of Collaboration (MoC) and model Access and Benefit Sharing Agreement (ABSA). MoC are considered low risk and are the primary agreements used by Kew and its partners whilst the model ABSA represents a more formal style of agreement and is used where the material is considered more ‘high risk’. The latter confirms that prior informed consent (PIC) was obtained and MAT have been negotiated. Kew’s ABSAs are usually signed, or counter signed by the government authority in the partner country responsible for granting access. Apart from the MoCs and ABSAs, Kew has developed a suite of other model agreements to be used to ensure that terms and conditions relating to the use of the plant material it has accessed, used or supplied, are monitored and passed on to third parties. The main terms of model agreements used by Kew for monitoring the value chain include conditions on use of material, purpose of acquisition (non-commercial), transfer to third parties, benefit-sharing and reporting. Any material brought back from overseas fieldwork trips by Kew’s scientific staff is incorporated into Kew’s diverse collections according to terms and conditions of permits and agreements with provider countries. Likewise, fieldwork is planned well in advance in accordance with the Kew’s ABS policy to ensure that all aspects of legal collection are adequately prepared. In addition to having its own ABS policy, Kew ensures that its internal policies and processes are in line with sectoral best practice in the area. All these efforts attempt to fill the existing policy gap on both user and provider sides. Their aim is to set out, what is fair and equitable, standard setting and working with others in the sector to ensure there is an agreed understanding of what is ‘best practice’. Kew has taken commendable steps to ensure compliance with provider measures and benefit-sharing but work is still needed to improve long-term use tracking to which a number of recommendations have been made.

In Chapter "First Experiences in the Implementation of the EU ABS Regulation in Germany", Thomas Greiber and Ellen Frederichs look at the implementation of Regulation (EU) No 511/2014 in Germany, which transposes the Nagoya Protocol (NP) compliance provisions, or so-called user measures, in the European Union (EU). The NP compliance provisions are articles 15 and 16 on prior informed consent (PIC) and mutually agreed terms (MAT) requirements of the country providing GR and aTK, article 17 on monitoring the value chain meant to support articles 15 and 16, article 18 on compliance with MAT and articles 19 and 20 to promote the development of compliance support tools. Except article 18 which is left to the individual MS to implement, the EU Regulation implements these provisions through the so-called due diligence (DD) system. This is comprised of the obligation of users to exercise DD and file different DD declarations, the obligation of competent authorities of EU MS to check user compliance with the DD obligations, and a voluntary EU register of collections and recognition of best practices by the European Commission. Germany is a NP party since 2016. It does not regulate access to GR and arising benefits and therefore does not require PIC and MAT for GR collected *in-situ* but there are the general public and private law restrictions to be observed. Concerning access to *ex-situ* GR it would depend on their *in-situ* origin and whether there are any conditions attached thereto. As such Germany focuses on the compliance obligations on the basis of the EU Regulation (EU) No 511/2014. Germany has implemented the NP and Regulation (EU) No 511/2014 through the German Act Implementing the Obligations under the Nagoya Protocol and Transposing Regulation (EU) No 511/2014 of 2016. The Act designates the Federal Agency for Nature Conservation (Bundesamt für Naturschutz—BfN) as the competent national authority (CNA). The BfN is empowered to conduct compliance checks, receive DD declarations, handle applications for registration of collections, provide relevant advice to users in Germany, and cooperate with CNAs of provider states to ensure user compliance. In addition, the Act regulates other possible measures that the BfN can undertake in order to address situations of non-compliance. The national focal point (NFP) is the Federal Ministry for the Environment, Nature Conservation and Nuclear Safety (Bundesministerium für Umwelt, Naturschutz und nukleare Sicherheit—BMU). The authors give an overview of the activities of the BfN to implement the Act and the EU Regulation in the period of 2016 to 2019. In Germany, declarations are mandatory at both stages of research funding and product development. Once the BfN has received a DD declaration it checks its completeness, timeliness and plausibility. A number of difficulties have been encountered with checking the plausibility due to unstandardized forms of agreements and permits and also because these are issued in different languages. It is therefore vital to think about harmonization to which Internationally Recognized Certificates of Compliance (IRCC) are critical. After checking, the BfN in collaboration with the NFP submits information provided/declared by users as required by Annexes II and III of the Implementing Regulation

(EU) 2015/1866 to the ABS Clearing-House after which it is published as a checkpoint communiqué. A copy of this is forwarded to the provider country which may open a bilateral communication between the BfN and the (competent) authorities of the provider country. If the provider country reports irregularities in the communiqué the BfN may become active again to try and investigate the allegations. This kind of communication was helpful is bringing further clarification in the first DD declaration/checkpoint communiqué. Germany was the first country to receive DD declarations in the EU in mid-2018. These became the first checkpoint communiqués in the world. By December 2019 BfN had received 11 DD declarations and from the 17 checkpoint communiqués 10 were from Germany. All 11 fall in the first category of DD declarations, i.e. at the stage of research funding. The lack of DD declarations at the stage of commercialization is perceived to be due to first, the temporal scope of the EU Regulation which means it only applies to GR accessed after 12 October 2014 and yet the final development of a product can take long and second, because products can be developed on the basis of old materials which fall out of its temporal scope. But how effective are the DD declarations as a tool to monitor compliance with the NP in practice and do they not only lead to bureaucracy? The authors observe that users making DD declarations have nothing to hide while those who violate the DD obligation will not file the declarations. That makes checkpoints only a means to confirm compliance or to detect negligent non-compliance. Thus, first experiences with DD declarations indicate that their ability to detect severe and urgent cases of ABS non-compliance is limited. Apart from receiving DD declarations, checkpoints have a duty to undertake compliance checks in order to verify whether DD obligations are being/have been fulfilled according to the two triggers foreseen by the EU Regulation: periodically reviewed risk-based plans and substantiated concerns. The German Act specifies the rights of BfN concerning compliance checks and stipulates sanctions. BfN began its first risk-based checks in 2018. The experiences by the BfN point to a number of challenges in enforcing the DD obligations. There is a lack of relevant ABS/user data which could facilitate *a priori* selection and reliable assessment of institutions falling within the scope of the Regulation in connection to their activities. The scope requirements of the Regulation that must be fulfilled for it to apply give many options for users to find their way out of the scope. In this case putting the burden of proof on the competent authorities the EU Regulation thus makes the determination of its applicability a real difficulty. Besides, the highly differentiated, small-sized interpretation of what is to be considered as utilization or not adds complexity to compliance checks and limits the scope of the Regulation. In addition, the use of bio-innovation as a marketing strategy gives misleading information of institutions/ companies involved in *de facto* bio-based R&D. Thus, in conclusion, compliance checks show that clarifying the applicability of the EU Regulation is not only challenging but an integral part of the control process which cannot be anticipated. In spite of the challenges mentioned above compliance checks by the BfN have led to the detection of due diligence violations in both cases of risk-based compliance checks and substantiated concerns. These have led to administrative processes requiring retroactive remedy as a first step failure to which the institutions will be sanctioned. Additionally to monitoring compliance the EU Regulation aims to support users with the implementation of their DD obligations. Hence, it envisages a register of collections and best practices. Application for registration as a registered collection is submitted to the BfN. By the end of 2019 (and in the entire EU) only one collection had been approved as a registered collection, i.e. the Leibniz Institute DSMZ—German Collection of Microorganisms and Cell Cultures GmbH. However, despite this first success, no other collection has applied for registration in Germany possibly due to a lack of awareness about the instrument of registration including its legal requirements, lack of financial resources, fear of increased bureaucracy etc. Hence, BfN has undertaken a series of awareness-raising and capacity-building activities. In the same vein associations of users or other interested parties may have their combination of procedures, tools or mechanisms, developed and overseen by them recognized as best practices. The MS provide comments to the application and the Commission makes the decision. From 3 applications submitted by the end of 2019 only 1 succeeded while the other 2 applicants did not follow up their applications after comments by MS and the Commission's request for improvements. Besides awareness-raising for the collections, BfN has undertaken general awareness-raising campaigns to promote and encourage information, awareness-raising and training to help stakeholders to understand the Protocol, ABS and the EU DD system as foreseen in article 13 etc. As a general conclusion the authors state, *inter alia*, that it is far from clear whether the implementation of the NP and the EU Regulation has facilitated R&D activities in the EU or at least created legal clarity and hence certainty, whereby industry and public research indicate the opposite. However, in spite of the criticism on both ABS sides of the provider and the EU and noting a number of positive developments that have already been provoked by the initiation of the implementation of the EU Regulation, it is acknowledged that the NP and EU Regulation are young and require more time and implementation experiences to reach their full operationalization.

In Chapter "Due Diligence and the Regulation of Transnational Economic Activity: Regulation (EU) No 511/2014 Compared to Other EU Due Diligence Schemes", Christine Godt and Markus Burchardi examine the genealogy of due diligence (DD) and compare its use in the European Union's (EU) ABS Regulation No. 511/2014 to other EU Regulations that likewise built on DD as a concept. With this concept Regulation (EU) 511/2014 stipulates a self-standing duty of care concerning lawful access unlike earlier thoughts which had envisioned a direct enforcement of provider States' rules by the user State. This self-standing duty functions as a 'regulatory hinge-joint', mitigating what would otherwise have been seen as an encroachment on the territoriality principle, restricting user States' sovereignty. DD had already been embraced in other policy areas before the adoption of the EU Regulation, e.g. in the regulation of trade in tropical timber (EU Timber Regulation 995/2010). Today various regulations subscribe to it, covering a wide range of subjects. Following an inquiry into the genealogy of DD the authors examine the insertion of the remaining legacy of the 'forerunner models of DD' as a 'legal transplant' into a new EU legal environment with questions whether the transplant changes the environment or *vice versa* and how those changes look like having in mind that the EU Regulation contains further elements of industry self- and co-regulation. A comparative study of 'parallel' Regulations (tropical timber, maritime transport emissions, data protection, conflict minerals/metals) is carried out asking: What is the amount of self- and co-regulation, i.e. what role is foreseen for the regulator and the user (industry/importers)? How is risk distributed? What needs to be done in order to produce compliance? When are efforts exhausted? When do national competent authorities intervene? These questions are asked with a view to the 'crystallization point' of converging interests in Regulation (EU) 511/2014, namely the formulation "or discontinue [use]" in its article 4.5. It is concluded that Regulation (EU) 511/2014 first, leans more towards a command and control style of regulation; common features of self- and co-regulation are underdeveloped. In essence, the Regulation embraces a strong role for competent authorities and a higher risk of compliance failure for industry (=higher duties). Second, "best practices" come with extremely limited monitoring and enforcement duties. Third, Regulation (EU) 511/2014 does not separate procedural obligations and substantial obligations, which is particularly troubling for crystallization. Article 4.5 links procedure and substance in quite a novel fashion. Its formulation is understood by the EU Commission as an "obligation of result" which according to the authors means procedural efforts to produce compliance may not last indefinitely. Hence, they ask themselves: When does the procedural element come to an end and the competent national authority (CNA) steps in? Since the "duty to discontinue" is integrated into the program of duties in article 4, one may also say that it has the effect of 'proceduralizing' what would otherwise have amounted to a straightforward prohibition—meaning there is leeway for CNA discretion and room to produce compliance for users that are found to be incompliant. This leeway is a legacy of the "New Approach". Consecutively they produce a number of insights. For the actual content of EU DD, it can be said that it neither derives from the model forerunners nor from any particular legal family. It is also not simply a management tool. It must be read autonomously, with the intent of the EU lawmaker in mind. Apart from that, some legacy remains most notably with regard to the standard of care, which can be split up into two dimensions: the 'objective standard of care' (or what a user 'ought to do' to produce compliance) and the 'subjective standard of care' (what a user 'should have known' at a certain point in time, i.e. when the user could have known that the use is non-compliant). While irrelevant for the procedural duties, it makes sense to take certain user attributes (role as global player, prior dealings with provider State authorities) into account with regard to the substantive duty ("to discontinue"). The authors conclude that ultimately, it will be up to the Court of Justice of the European Union (CJEU) to determine the exact extent of the standard of care under article 4 Regulation (EU) 511/2014.

Part IV: Unresolved Issues and Solutions to Implementation Challenges

In Chapter "Digital Sequence Information on Genetic Resources and the Convention on Biological Diversity", Christopher H C Lyal discusses the issue of digital sequence information (DSI), based on existing arguments for its inclusion or exclusion from the scope of the Convention on Biological Diversity (CBD) and its Nagoya Protocol (NP). Discussions on this matter are full of misunderstandings, the terminology is confused and legal views are conflicting. The following questions thus arise: Do sovereign rights of the provider country extend to DSI? If yes, is DSI within the scope of the CBD and NP? Would the NP apply to DSI held in foreign databases? What exactly is DSI? According to the jurisdictional scope of contracting parties under article 4 of the CBD based on their sovereign rights as foreseen in article 3, it is the components of biodiversity envisaged under article 4 that come under the sovereign rights of parties. This is reinforced by article 15 CBD which reaffirms sovereign rights of states over their natural resources and their authority to determine access to genetic resources (GR) subject to national legislation. Further, the GR provided by parties should be those for which the providing country is either a country of origin or has acquired in accordance with the Convention. Therefore this relationship between the scope of the CBD and sovereign rights are critical to interpreting DSI and resolving the questions on scope. The conceptual framework of the CBD and its NP anticipate access to physical GR followed by research and development (R&D) resulting to benefits to be shared fairly and equitably. This model, which the author dubs "simple linear model" has little relationship "to the complex interactions that often characterize supply and value chains of physical genetic resources", not to mention modalities of sharing 'DSI'. The term is not used anywhere in the Convention or under definitions in article 2, which also seem to restrict the article's scope to physical material. However, there are views that a "teleological interpretation of the Protocol (and the CBD) suggests that benefits generated through the use of DSI must be shared 'by the terms of the Nagoya Protocol'". One argument for such inclusion would be that the definition of GR under article 2 CBD needs to be dynamic in order to meet the overall object of access and benefit-sharing (ABS). But an agreement on whether DSI falls under the scope of CBD partly depends on understanding what it comprises. This term was developed in the CBD process and is not used anywhere else and the ultimate decision on its scope can only come from negotiations under the CBD. Work within the CBD began after the Conference of Parties (COP) 13 in 2016 and is ongoing with technical groups and studies trying to deal with different issues relating to DSI. Examining what DSI actually is the author, considering the present challenges and the questions initially raised above, looks at the terminology and scope of the concept according to processes under the CBD and scholarly work. Under the question of DSI as a component of ABS the following are interrogated: sovereign rights as reflected in the CBD and NP; whether DSI falls under the sovereign rights of the country providing GR; and whether DSI can be within the scope of the CBD. Finally, the questions concerning implementation of national legislation covering DSI looking at its inclusion in prior informed consent and mutually agreed terms when GR are accessed, assertion of sovereign rights over DSI held in databases, and rights of benefit-sharing based on further/persistent processes in line with article 5 (1) NP are examined. As the study shows, the disagreements concerning the issue are currently so profound and the outcomes of the parallel processes going on under different treaties and organizations are varying. In exploring the way forward a number of suggestions have been made which include the critical importance of developing a benefit-sharing system to avert the pending threat of provider countries restricting generation of DSI. For this a bilateral benefit-sharing mechanism is unlikely to be practicable. Options can be the Global Multilateral Benefit-Sharing Mechanism envisaged under article 10 NP or a separate mechanism which supports other fora where the question of DSI is presently being discussed. However, this also invokes other questions that will need answers.

In Chapter "Lessons from Writing Binding and Enforceable ABS Contracts. A Contract Solution to Digital Sequence Data in ABS", Morten Walløe Tvedt focuses on the question, how ABS can be implemented in practice to become a functional legal tool for meeting the obligations of the fair and equitable benefit-sharing objective of the Convention on Biological Diversity (CBD) and its Nagoya Protocol (NP). The author is of the opinion that until there is a statutory system in place to promote the third objective of the CBD on benefit-sharing, contracts remain the functional legal tool for securing public benefits. Implementation of international law obligations occurs by tools of domestic legal instruments and a contract is a practical tool that can make access and benefit-sharing (ABS) work in practice thus providing legal certainty and sufficient flexibility to cater for each individual situation. A contract, for instance, is able to secure the rights of provider countries over findings of non-commercial research which are eventually used for commercial purposes. It will need in principle to have clear terms and conditions to become binding and enforceable on the user and have a format which is recognized in other countries to become binding in other jurisdictions. Besides, an access and benefit-sharing (ABS) contract must comprehensively regulate the relationship between parties and all aspects. Also, it will need to resolve fundamental challenges of ABS e.g. in regard to long value chains. In addition, it needs to anticipate the possible paths of value chain and establish clear legal consequences on non-monetary benefit-sharing. A contract likewise has the ability of dealing with the two scoping challenges (potential loopholes) of ABS based on biotrade (biological bulk) and digitalization of genetic information. As Lyal says in this volume, discussions on whether 'digital sequence data' fall under the scope of the CBD are still ongoing in different fora, however, the author believes that "there are practical solutions to regulate these subject matters under the current regime by a contractual approach". A contract does not rely on the outcome of the discussions on scope but depends on mutual acceptance from its parties and is able to set a global standard for this kind of data. Further, a contract is able to contemplate the fast growing and future technologies as well as ways of sharing data. For a contract to be effective what is referred to as "Golden rules for contract drafting" must be taken into consideration while drafting. Contract obligations must be clear, dynamic and not narrow. A contract should opt for positive regulation other than prohibition of some actions e.g. not to commercialize, seek intellectual property rights or access traditional knowledge, obligations which are hard to enforce in contracts, but rather clearly regulate actions by formulating enforceable consequences should these activities take place. These should be regulated in the original contract as come-back clauses are not effective or enforceable. Therefore, a core issue is how contract clauses can be formulated in order to embrace relevant scenarios and translate them into binding language. A change in the manner of drafting contracts is proposed with need for improvement of the level of details and bridging of the language in both law and in research. This is deemed a critical step in making ABS work in practice even in the era of data technology in applied research and development.

In Chapter "Cases and Questions in Application of ABS Regimes", Gerd Winter and Evanson Chege Kamau endeavor to offer solutions for questions on ABS based on real life cases and hypothetical ones following existing legal documents. Stakeholders in the access and benefit-sharing (ABS) process have been confronted with many questions concerning the regulation of access to genetic resources and associated traditional knowledge (aTK), their utilization and sharing of benefits arising therefrom. Besides guidance provided by the EU Commission, ABS literature is mute on issues concerning practice. Until now, stakeholders faced with such inquiries have been forced to look for answers from national competent authorities and/or experts in the area. Looking at queries that have been asked by researchers, research institutions, commercial organizations and intermediary enterprises as well as inferred scenarios, the authors construct cases and suggest legal solutions. The answers proposed are not based on any judgement of courts but on the authors' opinion. Cases touch on varying matters on ABS and are clustered under two groups: issues concerning provider States and issues concerning user States. Most of the issues examined concern user States. We start with the former. The issues interrogated concern: (1) Cooperation between local and foreign research the question asked in essence being: In conditions requiring cooperation for access to be permitted, what kind of collaboration is acceptable? Should the local scientist accept an offer only to perform facilitating duties or should he/she insist on substantial participation in research? The authors conclude that the latter is what should be understood with such a condition and is what the local scientist (or, what the authors should have added, the relevant national competent authority) should insist on; (2) Multiple permits: If provider State's measures require an applicant for a permit for access to genetic resources (GR) to obtain several other permits and to establish mutually agreed terms (MAT) as a condition for grant of such a permit, can an authority facilitate the procedure for non-commercial research in accordance with article 8 (a) of the Nagoya Protocol (NP) by integrating all procedures of the relevant organs into one? It is concluded that the authority can, if it is mandated through executive powers to organize administrative procedures. If not, this can only be done if streamlined procedures are established by explicit legislative order; (3) Publication of research results. The main issue concerns the dilemma of allowing sequenced data by non-commercial research to be placed in public domain e.g. through publication in databases and their fate in regard to the possibility of their eventual use for commercial purposes. Should the competent authority hinder access by denying a permit? No, as this will have a negative impact on the provider State in regard to possible future benefit-sharing opportunities. However, the competent authority should insist that results be published with a conditional proviso that benefits from commercialization of products resulting from the utilization of the data are to be shared with the provider based on prior *bona fide* negotiation; and 4) Consent of indigenous communities. The authors want to know, what advice can a lawyer E give to a company C that plans to collect plant material for utilization from a territory occupied by indigenous communities? Forgetting other conditions of the State, if C would like to use traditional knowledge of the communities, their PIC must be obtained. Also, their PIC must be obtained for access to GR if the communities have the established right to grant access to such resources. Besides, E must advice on the number of communities to be approached if that is necessary, their customary laws, community protocols and procedures and may offer his services concerning the issue. The case may change if no information is required from the communities or if the national law either does not regulate aTK, or if the provider country considers traditional knowledge a national heritage, or if communities have no established right to grant access to GR. In these cases the PIC of the communities will not be required, *albeit* this might be a violation by the host country of their CBD/NP access and benefit-sharing rights ensuing from the Indigenous and Tribal Peoples Convention, 1989 (No. 169). The issues concerning user States are discussed mainly based on the perspective of the ABS regime of the European Union (EU) which is built on the ABS Regulation (EU) 511/2014. Generally, issues dealt with involve the scope of application of the EU Regulation (concerning both GR/aTK and R&D); when use of GR can be considered as 'utilization' or not; when PIC should be obtained, MAT established or benefit-sharing agreement concluded etc. They are so broad and will hence be only mentioned without describing the cases or briefly explaining the relevant issue. These include the question on: (1) "Temporal scope" concerning GR and aTK accessed before the entering into force of the NP, from a country that is not party to the NP, or which has no ABS regime: the EU Regulation does not apply. Thus, a user who accessed GR/aTK before will not be under any obligation under EU law even for uses continued on the old GR. However, if the country from where the GR/aTK was accessed establishes retrospective measures, the user is advised to respect the law of the provider e.g. by entering into a benefit-sharing agreement, *albeit* the EU competent authority in the relevant user State cannot enforce this; (2) "Geographical scope" concerning GR collected from a CBD/NP party with an ABS regime without permission and utilized in a non-CBD/NP party by a researcher R from the latter in cooperation with a researcher A from an EU Member State. R is not under any obligation of his/her country, but nevertheless should comply with the measures of the provider country. A is expected to exercise due diligence according to EU law; (3) "Material scope" related to screening of a number of plants/herbs with the question: When does this qualify as utilization? If EU law is to be considered, screening is divided into three categories with two being considered as utilization. Concerning source countries, the rule can only be established following consultations with relevant authorities. In regard to data from screening which is published in databases the authors advice that it should be considered as utilization because it might be a subject of future research and development (R&D); (4) "Information stored at and taken from public databases or print media" looking at different scenarios relating to MAT conditions concerning published data and publication of R&D results; (5) "Contributions or not to product development" delving into questions if research is being carried on or with the GR and thus deciding when the obligation to share benefits is triggered, and "research without modification" and qualifications for PIC and MAT requirements; (6) Whether the uses made of the GR are bulk or they comprise utilization examining the question of "change of intent from consumption to R&D", "known properties and new R&D" and "propagation and cultivation of GR"; (7) Technological application interrogating issues on "studying mechanisms of a lawfully acquired GR" and "purchase of derivatives" from an intermediary outside the EU and the EU Commission's criterion of continuum in R&D; (8) Regularization in regard to "*ex post* PIC and MAT" (i.e. rectification of unlawful situation), and "transfer to third parties of unlawfully obtained GR"; (9) Due diligence in ascertaining provider State requirements in case of a "unresponsive provider State" (i.e. when should the user be considered as having exercised due diligence in regard to a provider State that does not respond to inquiries concerning its ABS requirements?) and "ABS clearance declaration" (i.e. concerning a country that does not regulate ABS); (10) Benefit-sharing in relation to "ABS and patenting", "ABS and plant variety protection" and "sharing of benefits accruing to the buyer of a patent on accessed GR information"; and (11) Cut-off points in regard to "processing of raw materials for subsequent incorporation into a product". The questions dealt with and the solutions offered by the authors show that a stage of maturity has now been reached in ABS whereby practice can begin solving relevant issues.

5 New Legislation and Practice and Status of Implementation of the Nagoya Protocol. Synthesis, Observations, Recommendations and Conclusion

The book grapples mainly with the new environment of accessing and utilizing genetic resources and traditional knowledge in R&D as well as sharing benefits therefrom. The post Nagoya Protocol laws regulating this scenario can now be referred to as the second-generation ABS legislation since the adoption of the CBD in 1992. The architecture of the first-generation legislation was influenced by reactionary approach for what was perceived as past violations of the rights of (mainly developing) provider countries, great expectation of 'green gold' money, lack of binding obligations for compliance in (mainly developed) user countries, and a lack of detailed, concrete and obligatory measures on ABS in the CBD. The ensuing atmosphere was not conducive for R&D based on GR and aTK and the expected benefits remained a dream yet to come true.[23] The Nagoya Protocol was meant to heal this situation. Whilst implementation is being undertaken in order to comply with the obligations of the NP the underlying thought and intention of the negotiators of the instrument (NP) should not be lost/forgotten. That is, to deal with (or eliminate) the challenges faced in access and benefit-sharing caused by *inter alia* legal uncertainty, unclarity and intransparency of provider measures and lack of

[23] Ruiz Muller (2015) attributes the failure of achieving benefit-sharing to the bilateral system and advocates for "bounded openness".

compliance measures in user countries, and thus ensure the fair and equitable benefit-sharing. We therefore ask ourselves: Have the new laws and practice managed to be *de jure* and *de facto* compliant with the Nagoya Protocol? Besides complying with the Protocol, which other challenges are being faced in the implementation and how are they being surmounted, or which solutions are contemplated? How are arising/unresolved issues affecting the ABS landscape?

Our book has shown that compliance with the NP should not be one of "form" but "fact". Many provider legislations, for instance, have implemented the word of the instrument in order to conform, without having in mind how the measures taken will facilitate access, which is its spirit. Conformity only to the law does not necessarily ease access. A Nagoya Protocol compliance check depicts most measures as having conformed to the requirements of its provisions, but do not solve the question as to whether those who intend to access GR/aTK are able to navigate through the process with ease, without delay, with certainty etc. A juxtaposition of the Ethiopian and Kenyan regimes may serve as a good example. The former's regime is intentionally strict, but is clear and streamlined. The latter's regime does not intend to be restrictive but is fragmented and oblique. Comparing how many permits have been issued by the two it is evident that Ethiopia experiences and executes many times more requests for access than Kenya.[24] The standard of *de facto* compliance should hence be the effectiveness of the regime. In relation to this the role of procedures, predictability, timeliness etc. is critical. Below a number of observations and recommendations have been made.

Like the old situation, fragmentation of measures is still a problem with some new laws. The resulting complexity is still widely observed. This is in particular true in Federations (Argentina), and where ABS powers are delegated to provinces or counties (South Africa, Argentina, Kenya). Centralized regulation and/or standalone laws seem to put this situation under control and are recommended.

Although centralization eases access occasionally the consultation right of the indigenous and local communities is thereby sometimes ignored and thus violated. In trying to resolve access challenges it is advised not to lose sight of this right.

Where law complicates or aggravates the process, practice can still help. This is exhibited by the case of Kenya. This practice, however, is not without flaw. As noted, without a legal backing the uncertainty the practice aims to rout out returns. Besides, most persons who might be interested in accessing GR/aTK depend on published laws, regulations and guidelines and would not know about such temporal, not legally prescribed measures. Therefore we recommend that any well-functioning requirements and procedures which are not yet foreseen in law should be established by explicit legislative order.

---

[24] Since the adoption of the NP in 2010 Ethiopia has granted 983 permits for research purposes and 13 for commercial purposes (with 13 access agreements signed), *pers. comms* (15 November 2020) with Ashenafi Ayenew Hailu, NFP, EBI. Kenya on the other hand has granted 148 permits in the same period, all for non-commercial purposes, *pers. comms* (7 January 2021) with Joyce Imende, ABS Desk Officer, NEMA.

Following on the same example, it has been demonstrated that the timing of the measures applied in practice matters. The said measures of the Kenyan practice come at the end of the process, i.e. during the review of the access application. That means, it is important to locate points in the procedure where most challenges are encountered and offer the measures/solutions there. In this particular case, however, it could be an indication that the relevant agencies are still not ready to give up their mandates for the sake of explicit streamlining. This is also the case in South Korea. In such a case procedural streamlining could help but the most effective option which we recommend is one of the variants of material integration.[25] For the latter, some stakeholders will have to give up their ABS powers.

Lack of uniformity creates a big burden for researchers. A person interested in undertaking research activities in several countries will have a massive task of understanding the varying laws, requirements and procedures. Alternatively he/she would be subjected to consultancies by lawyers and experts, which can raise the research budget exponentially, at times to levels which e.g. basic biodiversity research cannot afford. Whereas the NP does not prescribe a stiff approach to implementation—leaving room for parties to implement it in light of their national circumstances—a certain level of harmonization e.g. of PICs and MATs would be a big asset for the success of the ABS regime. In the same vein, harmonization of terms is important. The use of terms across regimes is presently a calamity. At times the same term is given varying meanings in different ABS laws within the same domestic regime (e.g. Kenya).

Some providers extend the scope of application of their legislation to activities conducted abroad involving their GR/aTK (Ethiopia, Costa Rica, Cameroon, South Africa). This approach was initially influenced by lack of compliance measures in user countries. However, this is ineffective and also against the international law principle of territoriality. The model of user measures examined in this book and representing user States, i.e. the EU ABS Regulation, nonetheless depict weaknesses which do not encourage providers to give up their approach. The Regulation, for instance, establishes a temporal scope as the date of entry into force of the NP. This might be interpreted as undercutting provider States in terms of opportunities for benefit-sharing. But since the NP does not exclude GR and aTK accessed prior to its entry into force, the provider countries can still impose benefit-sharing obligations by establishing a temporal scope which covers the duration starting from the entry into force of the CBD and requiring users of such GR and/or aTK to conclude benefit-sharing agreements with the source country. At least the EU Regulation does good to caution Member States' users to respect MATs established with providers of GR and aTK. The same can be done concerning benefit-sharing from Digital Sequence Information (DSI) relating to previously accessed GR. For DSI relating to post NP genetic resources, conditions on use and benefit-sharing can be included in MAT. Many provider countries are asserting their rights over DSI and although

[25] See Kamau and Winter (2009), p. 371ff. For a summary of details see Kamau (Ch. on Kenya) in this volume.

discussions are still ongoing to establish whether DSI falls under the scope of the CBD/NP, it has been demonstrated that the will of parties can still be effectively expressed and protected under the law. Therefore, MAT can be concretized in contracts. For that well formulated contracts have been recommended.

But, of course, contracts cannot solve problems of formerly accessed GR/aTK and related DSI. Besides, retrospective approach without the intervention by user States pushes the provider back to the weak position as a result of the territoriality principle. Without enforcement by the user States, such rules are likely to remain largely ineffective, especially where users do not need to conduct supplementary access to the GR. Furthermore compliance checks, e.g. in Germany, have shown that it is difficult to establish when the GR were accessed if the user claims to have accessed them prior to the NP. In addition, the EU due diligence concept of compliance is also having its share of challenges probably due to its unique nature which is different from other concepts of due diligence in other sectors; or due to its novel approach of trying to enforce foreign law. It implies that even where GR and aTK are considered to fall under the scope of the user regime, compliance still relies much on good faith by users. Thus, putting more effort in developing cooperation between providers, users and user countries, intensifying communication between competent authorities of both sides under article 17 NP might be more productive.

Likewise, trust plays a big role in operationalizing ABS. This has been demonstrated by the case of a German researcher working in Ecuador and of RBG Kew. The possibility to have registered collections under the EU ABS system is also supposed to foster the same.

Most provider countries have implemented only access measures. This approach probably stems from the old attitude that traditional provider States are not users. The NP's approach is that every party is both a provider and a user. For example, whereas Argentina has in the past being referred to as a provider country, its exponential growth in biotechnology industries show that it is equally a user. In any case each party must establish measures to counteract violations of other parties' requirements in its jurisdiction. Partial implementation needs to be addressed.

At times provisions of the NP are implemented improperly, for example, in regard to functions of checkpoints under article 17.1 (a). Most provider States' laws have established checkpoints to monitor compliance with their own legislation/regulations. Of course each party has the prerogative to do that, but the functions on monitoring compliance with measures of other parties have to be included.

Capacity-building is required in provider countries in regard to the implementation of user measures and generally the proper implementation of the instrument. The role GEF and UNDP projects have played to develop legislation e.g. in Viet Nam, Argentina, Malaysia is evident and speaks for more international cooperation, resources and experts to build local skills.

New issues are challenging the ABS system e.g. on: (1) DSI (issue mentioned above); (2) Limit of the duration of provider rights as most provider States do not foresee any end to their rights. Users on the other hand feel that if provider rights are *ad infinitum* the burden on R&D will be so heavy and therefore they solicit for cut-off points.[26] Among the case studies presented in this volume only the case study of Viet Nam tries to establish an interpretative possible cut-off point based on the legislation. Kamau and Winter in the chapter on cases suggest number and kinds of utilization as a possible criterion; and (3) GR initially accessed as bulk commodities. The measures of provider countries can be circumvented by users accessing such resources on their territories. Providers have been of the opinion that such access and utilization should be subject to benefit-sharing obligations and a few countries have regulated rights thereon in their new legislation (Malaysia, Brazil, Ethiopia). The EU law has also addressed this question in the guidance document of the Commission on the scope of application and core obligations of Regulation (EU) No 511/2014[27] and is exemplary. It states that "... if and when research and development is carried out on genetic resources which originally entered the EU as commodities, the intended use has changed and such new use falls within the scope of the EU ABS Regulation (provided the other conditions for application of the Regulation are also met)".[28] Consequently "... the user is expected to contact the provider country and clarify whether requirements to obtain prior informed consent and establish mutually agreed terms apply to this utilization of such genetic resources (and if yes, obtain the necessary permits and establish mutually agreed terms)".[29] (4) The use of public domain data from non-commercial research for commercial purposes with provider States requiring (with some already imposing obligations on) the depositor of data to publish it with conditions of further use and databases to flag those conditions for subsequent access. But databases might be reluctant to establish the necessary infrastructure and to take up the ensuing checking duties, whereas the researchers are afraid that databases might be unwilling to accept any data attached to such conditions.[30]

These observations seem to depict only challenges with the ABS system. However as the study has shown, there is also progress. For example, with the new legislation facilitation of access is not only being extended to considerations foreseen under article 8 (a) NP but also to those under article 8 (b) (Cameroon, Viet Nam). Besides, understanding seems to be growing that any research which aims at putting results in the public domain (i.e. which does not aim at making profit) should be considered as non-commercial research—even when the results have a commercial potential (e.g. Viet Nam). Also, more legislation is acknowledging the need to link-up the third objective of the CBD with its other two objectives (Ethiopia, Brazil, Malaysia). Furthermore, new legislation is also laying ground in areas that ABS legislation never ventured into before e.g. regarding the legal involvement of

[26] E.g. International Chamber of Commerce (17 December 2012), p. 6. For examples of challenges such rights cause on the agricultural breeding sector and suggested cut-off points see Kamau et al. (2018), pp. 30–34.

[27] Commission Guidance document on the scope of application and core obligations of Regulation (EU) No 511/2014, OJ 2021/C 13/01.

[28] Ibid, p. 9.

[29] Ibid, p. 10.

[30] Implied in Beck in this volume. Also *pers. comms* (anonymous, during our international conference on ABS in Bremen), 20 September 2018.

informal suppliers/intermediaries of biological/genetic resources in the ABS process (e.g. in Viet Nam and Malaysia)—and not just through interpretation of laws (e.g. Costa Rica, Cameroon)—which is a positive step in raising certainty for users. Very important is also the observation that ABS has now reached a stage of litigation which is a good opportunity to put it into practice and start developing case law in the area in order to shape the ABS landscape.

There are more positive observations which will not be reiterated in detail because they have recently been discussed in Kamau (2019).[31] They include the many new post Nagoya Protocol legislations; a big number of States that have become parties to the NP totaling 130 countries to date; establishment of (clear) scopes; attempts to define terms; attempts to comply with the NP access and benefit-sharing rules; existence of already functioning compliance measures; exemption of specialized ABS instruments; establishment of national authorities and definition of their functions; and a demonstrated high number of executed access requests, issued permits and their notification to the ABS Clearing-House.[32] This shows that in spite of challenges, ABS is growing and advancing and hence the challenging areas should be researched in order to have solutions which can transform it.

There is no better way to conclude this book than with the words of the Executive Secretary of the CBD, Ms. Elizabeth Maruma Mrema, cited from her foreword in this volume: "... implementation is also dynamic, not stagnant, and there is a need for a constant dialogue with all stakeholders involved to ensure that implementation of the Protocol achieves its objectives and contributes to conservation and sustainable use of biodiversity".

References

Books and Book Chapters

Dutfield G, Suthersanen U (2020) Global intellectual property law, 2nd edn. Edward Elgar, UK and USA

Godt C, Šušnjar D, Wolff F (2020) Umsetzung des Nagoya Protokolls in EU- und nationales Rechts. Ein Alternativvorschlag zur Umsetzung der EU VO 511/2014, 1. Auflage. Nomos, Baden-Baden

Greiber T et al (2012) An explanatory guide to the Nagoya Protocol on access and benefit-sharing. IUCN, Gland, Switzerland

Kamau EC (2014) Valorisation of genetic resources, benefit sharing and conservation of biological diversity. In: Dilling O, Markus T (eds) Ex Rerum Natura Ius? – Sachzwang und Problemwahrnehmung im Umweltrecht, Nomos, pp 143–173

[31] See Kamau (2019), pp. 7–10. The book is open access and can be seen at https://www.bfn.de/fileadmin/BfN/service/Dokumente/skripten/Skript564.pdf.

[32] See foreword by the Executive Secretary of the CBD, Ms. Elizabeth Maruma Mrema, in this volume.

Kamau EC (2019) Implementation of the Nagoya Protocol. Fulfilling new obligations among emerging issues. Bundesamt für Naturschutz (BfN), Bonn. https://www.bfn.de/fileadmin/BfN/service/Dokumente/skripten/Skript564.pdf. Accessed 16 Nov 2020

Kamau EC, Winter G (2009) Genetic resources, traditional knowledge & the law. Solutions for access and benefit sharing. Earthscan, London

Morgera E, Tsioumani E, Buck M (2014) Unraveling the Nagoya Protocol. A commentary on the Nagoya Protocol on access and benefit-sharing to the Convention on Biological Diversity. Brill Nijhoff

Muller MR (2015) Genetic information as natural resources. Implications for the Convention on Biological Diversity. Routledge, London and New York

Reichman JH, Uhlir PF, Dedeurwaerdere T (2016) Governing intergrated genetic resources, data, and literature. Cambridge University Press

Schloen M (2019) The persistence of ABS contractual obligations in the context of agricultural breeding. In: Kamau EC (ed) Implementation of the Nagoya Protocol. Fulfilling new obligations among emerging issues. BfN-Skripten 564, pp 125–129. https://www.bfn.de/fileadmin/BfN/service/Dokumente/skripten/Skript564.pdf

Winter G (2019) ABS regulation in the European Union. In: Kamau EC (ed) Implementation of the Nagoya Protocol. Fulfilling new obligations among emerging issues. BfN-Skripten 564, pp 105–110. https://www.bfn.de/fileadmin/BfN/service/Dokumente/skripten/Skript564.pdf

Articles

Cabrera Medaglia J, Welch F-P (2018) Current status and future research agenda on benefit-sharing in international sustainable development law. J Korean Law 17:179–216

Kamau EC, Fedder B, Winter G (2010) The Nagoya Protocol on access to genetic resources and benefit sharing: what is new and what are the implications for provider and user countries and the scientific community? LEAD J 6(3):246–262. http://www.lead-journal.org/2010.htm. Accessed 11 Dec 2020

Laird S, Wynberg R, Rourke M, Humphries F, Ruiz Muller M, Lawson C (2020) Rethink the expansion of access and benefit sharing. Science 367(6483):1200–1202

Prip C, Rosendal GK (2015) Access to genetic resources and benefit-sharing from their use (ABS) – state of implementation and research gaps. Report number: FNI Report 5/2015. https://doi.org/10.13140/RG.2.2.22287.18088

Scholtz W (2008) Common heritage: saving the environment for humankind or exploiting resources in the name of eco-imperialism? Comp Int Law J South Afr 41(2):273–293

Scholz AH, Hillebrand U, Freitag J, Cancio I, Dos S, Ribeiro C, Haringhuizen G, Oldham P, Saxena D, Seitz C, Thiele T, van Zimmeren E (2020) Finding compromise on ABS & DSI in the CBD: requirements & policy ideas from a scientific perspective. https://doi.org/10.13140/RG.2.2.35180.80001

Winter G (2021) Problems and solutions of access to genetic resources and benefit sharing: a theoretical perspective - Part II. LEAD J 17(0):1–15. http://www.lead-journal.org/content/a1706.pdf

Legislations

Convention on Biological Diversity, 1992. https://www.cbd.int/doc/legal/cbd-en.pdf. Accessed 14 Dec 2020

Commission Guidance document on the scope of application and core obligations of Regulation (EU) No 511/2014 of the European Parliament and of the Council on the compliance measures for users from the Nagoya Protocol on Access to Genetic Resources and the Fair and Equitable Sharing of Benefits Arising from their Utilisation in the Union, OJ 2021/C 13/01

Nagoya Protocol on Access to Genetic Resources and the Fair and Equitable Sharing of Benefits Arising from their Utilization to the Convention on Biological Diversity, 2010. https://www.cbd.int/abs/doc/protocol/nagoya-protocol-en.pdf. Accessed 14 Dec 2020

General Assembly resolution 1803 (XVII) of 14 December 1962. https://legal.un.org/avl/ha/ga_1803/ga_1803.html. Accessed 14 Dec 2020

Online Material

International Chamber of Commerce (17 December 2012) Comments on draft EU Regulation on access to genetic resources and the fair and equitable sharing of benefits arising from their utilization in the Union, Document No. 450/1078. https://iccwbo.org/content/uploads/sites/3/2012/12/Comments-on-Draft-EU-Regulation.pdf. Accessed 5 Jan 2021

Kamau EC et al (March 2018) New and arising issues in implementing the Nagoya Protocol. Workshop presentations and discussions, Forschungsstelle fuer Europaesches Umweltrecht (FEU), University of Bremen. https://www.uni-bremen.de/fileadmin/user_upload/fachbereiche/fb6/feu/FEU/ABS/ABS_PDFs/New_and_arising_issues_in_implementing_the_Nagoya_Protocol_Workshop_presentations_and_discussions_Kamau_et_al..pdf, 59 pp. Accessed 7 Jan 2021

Muller MR, Vogel JH, Angerer K, Pauchard N (December 2019) Access to genetic resources and benefit-sharing in the post-2020 global biodiversity framework. https://trade4devnews.enhancedif.org/en/op-ed/access-genetic-resources-and-benefit-sharing-post-2020-global-biodiversity-framework

Evanson Chege Kamau Dr. iur., is Senior Research Fellow at the Research Centre for European Environmental Law (FEU), University of Bremen. He is also an Adjunct Lecturer at the CvO University of Oldenburg. His area of expertise is public international law, European law and intellectual property law. His research focuses on genetic resources, environment, development and traditional rights on biological resources and traditional knowledge. Kamau has over 17 years of working experience in the area of ABS in which he has published widely. For many years he has also worked as a freelance legal consultant for governments, African Union (hired by IUCN and Capacity Building Initiative resp.), private industries and research organizations. echege@uni-bremen.de.

Part II
Post Nagoya Protocol Measures on Access and Benefit-Sharing and Domestic Implementation

Access and Benefit-Sharing Regime in Argentina: Experiences and Perspectives

Luciana C. Silvestri

Abstract Argentina, a rich biodiversity country, has shown interest in regulating access to its genetic resources (GR) and in the distribution of benefits arising from their utilization, a regime internationally known as ABS (Access and Benefit-Sharing). Nevertheless, until recently, the country had not advanced steadily or strategically in the matter. On the other hand, provinces, which are competent to grant access to the GR found within their jurisdictions, have enacted rules which greatly differ from one another in every possible aspect. As a result, access to GR in Argentina has become a challenging effort. Newly adopted Decision No. 410 of 2019 of the Secretariat of Environment and Sustainable Development, which sets minimum common standards on ABS at the national level, has the capacity to change history. However, its success and effects in reality are yet to be seen. This chapter describes how ABS has been regulated in Argentina so far and which changes have occurred with the new regulation. It focuses on the strengths and challenges the system presents. Lastly, the chapter speculates on the possible effects&spi2;the recently adopted regulation could have in the access to GR.

Keywords Access and benefit-sharing legislation · Genetic resources · Argentina · Decision No. 410 of 2019 of the Secretariat of Environment and Sustainable Development

L. C. Silvestri (✉)
Institute of Human, Social and Environmental Sciences, National Scientific and Technical Research Council (CONICET), Mendoza, Argentina
e-mail: lsilvestri@mendoza-conicet.gob.ar

E. Chege Kamau (ed.), *Global Transformations in the Use of Biodiversity for Research and Development*, Ius Gentium: Comparative Perspectives on Law and Justice 95, https://doi.org/10.1007/978-3-030-88711-7_2

1 Background

Argentina presents marked latitudinal and altitudinal diversity, ranking high among countries with a high number and variety of existing ecoregions: 18 in all.[1] Argentina's great environmental and geographic complexity translates into a wide variety of ecosystems, species and GR. For instance, regarding species, more than 10,000 vascular plants—20% of which are endemic[2]—, 385 mammals[3] and over 1000 birds[4] have been recorded. Argentina is also known for its genetic richness, making it an attractive destination for the development of natural resources and research.

Considering the above, Argentina has shown interest in regulating access to its GR and in the distribution of benefits arising from their utilization, a regime internationally known as ABS (Access and Benefit-Sharing). The country considers itself a biodiversity provider. For this reason, it showed high interest in and played an active role during the negotiations which led to the adoption of the Nagoya Protocol (NP). During those negotiations, Argentina advocated a regime which would include effective control of and compliance with the ABS rules of resource-provider countries in resource-user countries.

Furthermore, in recent years, Argentina has also become an important user of GR, mostly of its own resources. The highest demand comes from, undoubtedly, the national scientific community focused on the study of biodiversity and of traditional knowledge associated to its utilization (aTK). Most of these researchers work in public universities and state organizations devoted to technology and research such as the National Scientific and Technical Research Council (CONICET) and the National Institute of Agricultural Technology (INTA). In addition, a large group of biotechnological companies have for over four decades been actively working in human health care, animal health care and reproduction, inoculants, seeds, plant micropropagation and industrial supplies. By 2015, 201 companies were registered as biotechnological companies, raising the country's ranking to sixteenth position in the world.[5]

This would explain why Argentina recognizes itself, though timidly, as an incipient user of GR. It also sheds light on why Argentina has pointed out, on several occasions, the difficulties that may arise from obscure and complex access regimes.[6]

From this dual reality, yet placing a greater emphasis on its role as a provider of GR, and from a top-down perspective, Argentina has regulated access to its GR and participation in the benefits derived from their utilization.

---

[1] Secretaría de Ambiente y Desarrollo Sustentable (2018).

[2] Zuloaga and Belgrano (2015), p. 992.

[3] Ojeda et al. (2012), p. 43.

[4] López Lanús et al. (2008), p. 10.

[5] Ministerio de Ciencia, Tecnología e Innovación Productiva (2016), p. 8.

[6] Secretariat of the Convention on Biological Diversity (2007), p. 4.

2 ABS Regulatory Framework

2.1 Argentine National Regulatory Framework

2.1.1 Regulations

Argentina ratified the Convention on Biological Diversity (CBD) and the NP through Laws No. 24375 of 1994 and No. 27246 of 2015, respectively.

As far as the environment is concerned, the Argentine Constitution establishes that the provinces have original ownership of the natural resources existing in their territories.[7] There is a unanimous interpretation acknowledging that, although the Magna Carta does not explicitly refer to 'GR', these are subsumed under the more comprehensive term 'natural resources'. Therefore, the provinces have competency to define the terms and conditions for access to and utilization of their resources.

In regard to the competence to regulate ABS matters, the Constitution establishes that it is the responsibility of the federal government to adopt minimum legal environmental standards.[8] These common criteria are believed to guarantee a uniform level of environmental protection across the country and set the conditions necessary for such purpose.[9] They have to be adopted through a formal law dictated by the National Congress. All provinces and jurisdictions are bound by them and are entitled to implement them with more but not less stringent measures.[10]

ABS issues in Argentina are considered to be a matter related to the environment; as such, the federal government, through the national congress, has to adopt an ABS legal framework that includes minimum standards on the issues. However, no such law has been adopted.

Instead, different administrative measures have been taken at the national level. The first one is Administrative Decision No. 226 of 2010, adopted by the Argentine Secretariat of Environment and Sustainable Development (SE&SD). The Decision, a pre-Nagoya legally-binding measure that only covered access to GR when they were to be later exported or imported,[11] resembled a customs norm. This was the most important piece of regulation on ABS until October 2019 when it was derogated.[12]

Even though at its time it was central to the regulatory system, the Decision did not establish common minimum standards on ABS nor did it regulate critical elements of an ABS regime. For example, although it mentioned concepts such as prior informed consent (PIC) and mutually agreed terms[13] (MAT), it did not regulate or develop them. On the contrary, it simply required that the said elements existed, in

---

[7] Argentine Constitution article 12.

[8] Argentine Constitution article 41.

[9] Law No. 25675/2002 article 6.

[10] Argentine Constitution article 41.

[11] Administrative Decision No. 226/2010 article 1.

[12] Administrative Decision No. 410/2019 article 16.

[13] Administrative Decision No. 226/2010 article 1 and appendix 1.

whatever form and content; and that this was proved before the Secretariat for export purposes. Furthermore, the Decision did not establish a minimum common understanding on important concepts either. For example, it did not provide a definition of GR, users, access, utilization or commercial and/or non-commercial research, which hindered the interpretation and application of the regulation. As a result of its incomplete and somehow rudimentary nature, there was no guidance to the provinces for the adoption of their own rules on ABS, a situation that has led to high disparity in the content of provincial rules on the issue and to ABS having been regulated in only half of Argentine provinces as it will be explained below.

The second piece of regulation on ABS at the national level is Administrative Decision No. 208 of 2011, adopted by the Federal Council of the Environment (COFEMA). Even though it is a post-Nagoya measure, it does not include any of the NP's developments. It merely reasserts the provinces' ownership rights on the GR found in their jurisdictions and urges scientific institutions to conduct scientific research on genetic diversity only when a permit has been granted by the competent authority.[14]

A third piece of regulation brought in at the national level is the Scientific Research Regulations, which are part of Decision No. 81 of 2016 on national parks, adopted after the NP. The Regulations are applicable to all kinds of research conducted in national parks, including research on natural and cultural heritage.[15] Consequently, they do not constitute ABS-specific rules. Even so, their provisions might prove relevant for the collection, transfer and utilization of the genetic material obtained from national parks. Lastly, in October 2019, the SE&SD adopted Administrative Decision No. 410 of 2019 that regulates basic common standards for the access and utilization of GR in Argentina.

The new Decision No. 410 of 2019 of the Secretariat of Environment and Sustainable Development is a last-minute attempt to solve the precarious legal situation created by Decision No. 226 of 2010, honor the compromises assumed by Argentina under a Global Environment Facility (GEF)/United Nations Development Program (UNDP) funding project,[16] and avoid what seems to be a very long debate at the Congress to sanction a biodiversity law that could set minimum legal standards on ABS as mandated by the National Constitution. With the derogation of Decision No. 226 of 2010, Decision No. 410 of 2019 has now become the most important piece of regulation on ABS at the national level.

---

[14] Administrative Decision No. 208/2011 article 1. The Constitution sets competences on environmental issues. The jurisdiction over natural resources, including GR, belongs to the provinces which decide in their provincial rules who the competent authority to grant PIC and permit is.

[15] Decision No. 81/2016 article 1.

[16] Due to the absence of a national law setting minimum common standards for ABS issues, the Argentine government applied in 2014 for international funds provided in the framework of a GEF/UNDP project. ARG/14/G54 Project "Promoting the application of the Nagoya Protocol on access to genetic resources and equitable sharing of the benefits deriving from their utilization in Argentina". Decision No. 410/2019 was a result of this project.

2.1.2 Scope of the Regulation

Administrative Decision No. 410 of 2019 covers GR.[17] Derivatives are also included in the regulation,[18] but it does not specify whether these are included in the scope when they are derived directly from the GR and included in the MAT or when accessed alone. Both concepts (GR and derivatives) conform to the definitions of the CBD and the NP.

The following resources are excluded from the Decision:[19] (a) resources of human origin; (b) resources derived from domesticated or cultivated species; (c) the use of biological resources not involving the utilization of their genetic and/or biochemical components; and (d) plant GR used for food and agriculture that are included in Annex I of the International Treaty of Plant Genetic Resources for Food and Agriculture.

The reasons for excluding from the present regime GR derived from domesticated or cultivated species are unknown and appear questionable. Such broad exemption entails guaranteeing free access to them without the need to share the benefits derived from their utilization with indigenous and local communities whose efforts and contribution could be the reason behind the existence and conservation of some of these species. On the contrary, even though former Decision No. 226 of 2010 also exempted cultivars from the application of the regime,[20] in practice these were not exempted when wild species or native, local or landrace varieties were involved in their development.

Decision No. 410 of 2019 presents a novel list of definitions which is likely to be very useful for the implementation of the regime. For example, 'commercial purposes' is defined as those purposes destined to obtain a development or product for commercialization or sale, or a patent or product to which access restrictions will be applied by means of intellectual or industrial property rights. On the other hand, research is considered 'non-commercial' when the activities do not pursue the commercialization of a product or process, or the protection thereof by means of intellectual property rights.[21]

Lastly, it should be pointed out that no rule at the national level refers to the nature or the ownership of the genetic information contained in GR. There exists a loophole in the legislation. Nevertheless, it is worth highlighting that, over the last months, Argentina has developed a position on the issue, particularly on the so-called digital sequence information (DSI) on genetic resources.

In regard to the above, experts and national authorities consider that DSI should fall under the concept of 'GR' for the purposes of the application of the international

[17] Administrative Decision No. 410/2019 articles 1 and 3.

[18] Administrative Decision No. 410/2019 article 3.

[19] Administrative Decision No. 410/2019 article 6.

[20] Administrative Decision No. 226/2010 article 4.

[21] Administrative Decision No. 410/2019 article 3.

regime on ABS.[22] They also understand that the correct term would be 'genetic information', with no reference to the 'digital' aspect, which entails a single form of information transmission, and this may not necessarily apply to all cases. The word 'sequence' should also be dispensed with, as it unreasonably restricts the information on which the international regime would be applicable to the information strictly contained in the sequences of nucleotides in DNA or RNA or those of amino acids in a protein. Thus, it would not contemplate other types of information which may be contained in biomolecules, such as the tridimensional structure, and which would also be susceptible of the application of the international regime.

2.1.3 Institutional Framework and Access Procedures

Decision No. 410 of 2019 provides for the obligation to require PIC and to establish MAT for access to GR.[23] Such PIC, named 'access authorization', is granted by the competent authorities; that is to say, by the provinces, the Autonomous City of Buenos Aires and the federal government; each within the scope of their competences.[24]

The provinces and the City of Buenos Aires authorize access to the GR found within their territorial jurisdictions while the federal government will do so with respect to the resources found in maritime areas subject to Argentine jurisdiction, in the waters of Argentina's Exclusive Economic Zone and in the Argentine continental shelf beyond twelve (12) nautical miles measured from the base lines defined by the pertinent national legislation.[25]

Once the competent authority has issued the access authorization, the enforcement authority, which is the Secretariat of Environmental Policy on Natural Resources of the SE&SD,[26] grants the compliance certificate and notifies its content to the Access and Benefit-Sharing Clearing-House.[27] If resources are to be exported

[22] Second meeting of the Experts Inter-ministerial Group on the use of Digital Sequence Information (DSI) of genetic resources held on 5 February, 2018 in Buenos Aires to discuss Argentina's stance for the Meeting of the Ad Hoc Technical Expert Group (AHTEG) on Digital Sequence Information (DSI) on Genetic Resources which was held from 13 to 16 February 2018 in Montreal, Canada. In one of the meetings during COP 14 at Sharm El-Sheikh, Egypt, in 2018 the Argentinian delegation argued that DSI was in the scope of GR. They based their arguments on the interpretation of the Vienna Convention, recognition of genetic information, works prior to the convention which referred to GR not just as material and reference to DSI in early discussions and negotiations (notes of EC Kamau, *pers. comms*, August 2020).

[23] Administrative Decision No. 410/2019 article 1.

[24] Administrative Decision No. 410/2019 article 4.

[25] Administrative Decision No. 410/2019 article 5.

[26] Administrative Decision No. 410/2019 article 2.

[27] Administrative Decision No. 410/2019 article 4.

from Argentina, an export permit shall be requested from the enforcement authority.[28]

Administrative procedures are executed by means of different important appendices included in the Decision. Annex II is related to the access form for the use of GR for commercial purposes. Annex IV has to do with requesting authorization for access to GR for non-commercial purposes. Annex V includes the necessary form for requesting the compliance certificate, and Annex VI relates to the authorization for the export of resources.

The Decision includes a facilitated procedure to obtain access to GR when purely scientific purposes are pursued.[29] This procedure involves the non-requirement of establishing MAT (Annex III). On the contrary, it is compulsory to establish MAT when access to GR is requested for commercially focused research, and they must be established with the competent authority (Annex I).

The adoption of differentiated and facilitated procedures for the access to GR for non-commercial purposes has been welcome by the national scientific community, who complained that former Decision No. 226 of 2010 unnecessarily overcomplicated this type of research as all applicants wishing to access GR, no matter what their intention was, had to comply with a unique list of requisites and go along the same lengthy administrative procedures as research intended for commercial purposes.

Finally, those who want to export GR may also need to comply with a different set of obligations. These target phytosanitary and conservation concerns, amongst other issues. In the former case, for example, an applicant wanting to access GR for export purposes might also need to obtain a phytosanitary certificate if the resources fell under products and byproducts of vegetal origin.[30] The certificate is issued by the Argentine Agri-Food Health and Quality Service (SENASA).

Then, if the case of export is GR consisting of live or dead specimens, products or byproducts of wild flora species included in the appendices of the Convention on International Trade in Endangered Species of Wild Fauna and Flora (CITES), a flora export certificate issued by the SE&SD is needed.[31] On the other hand, if export involves GR consisting of live animals or products or byproducts derived from wild fauna, it is compulsory to obtain a fauna export certificate[32] from the same Secretariat. In this case, national rules foster the conservation of species.

[28] Administrative Decision No. 410/2019 article 14.

[29] Administrative Decision No. 410/2019 article 8.

[30] Decision No. 409/1996 of the former Argentine Institute of Vegetable Hygiene and Quality.

[31] Administrative Decision No. 477/2018 of the Ministry of the Environment and Sustainable Development article 1.

[32] Decision No. 437/2006 of the SE&SD and Regulatory Decree No. 666/1997 of Fauna Law No. 22421.

2.1.4 Access Permits

Decision No. 410 of 2019 sets common standards, applicable across all provinces and the City of Buenos Aires, regulating the requirements necessary to request access to GR. It also prescribes the minimum content to be included by competent authorities in access permits and in ABS contracts.

In regard to the information/elements that must be provided to the competent authorities in order to require access to GR—either for commercial or non-commercial purposes—, Administrative Decision No. 410 of 2019 sets the following:[33] (a) data of the applicant; (b) a description of the type of genetic material to be accessed; (c) information on the geographical location where the material/resources will be collected/accessed; (d) activities to be undertaken; and (e) specifications on possible subsequent utilization of GR by a third party. In addition, applicants who pursue non-commercial use of GR must accompany a sworn statement indicating that:[34] (a) GR will only be used for such purposes; (b) if, during the course of the research, a possibility arises for use of the resources for commercial purposes, a new application will be made setting forth such purposes; and (c) the GR will not be facilitated to any non-authorized person.

The minimal contents that access permits for commercial and non-commercial purposes must include are set forth in Appendices I and III of Administrative Decision No. 410 of 2019, respectively. Both permits must include information on the following extremes:[35] (a) date of granting; (b) data of the individual or institution in favor of which the permit is granted; (c) a description of the GR covered under the permit; (d) the purposes for which the permit is granted (non-commercial/commercial); (e) a statement that the permit is issued as proof of the fact that MAT have been established for the utilization of the specified GR (only in permits for commercial purposes); and (f) a statement indicating that if, during the course of the research, a possibility arises for use of the resources for commercial purposes, a new application must be made setting forth such purposes (only in permits for non-commercial purposes).

Finally, ABS contracts—which are only mandatory in case access to GR pursues commercial purposes—must include the following minimal content:[36] (a) objective of the contract; (b) rights and obligations of the parties; (c) benefit-sharing terms; (d) specifications on the possibility to transfer the material to a third party; (e) confidentiality terms; and (f) dispute settlement arrangements.

[33] Administrative Decision No. 410/2019, annex I (commercial purposes) and III (non-commercial purposes).

[34] Administrative Decision No. 410/2019, annex III.

[35] Administrative Decision No. 410/2019. appendices I and III.

[36] Administrative Decision No. 410/2019, annex I.

2.1.5 Traditional Knowledge

The Argentine legal framework does not provide for the protection of aTK on the utilization of the biodiversity found in the hands of indigenous and local communities as required by article 7 of the NP. There exists no rule at the national level requiring the PIC of these communities for access to such knowledge. Neither does the system provide for a legal obligation to fairly and equitably share with them the benefits derived from the use of aTK. Previously, Decision No. 226 of 2010, in its limited scope—only covered access to GR when they were to be later exported or imported—mandated the submission of elements ensuring respect for, and preservation and maintenance of the knowledge, innovations and practices of indigenous and local communities when authorization was requested to export GR.[37] No definitions were included in the Decision as to which such elements were, or what type of evidence had to be submitted, what procedures had to be followed, what inquiries had to be made, which communities had to be addressed, or any other kind of considerations that are critical to obtain PIC and to guarantee a legitimate access to aTK. However, despite of its rudimentary nature, at least Decision No. 226 of 2010 included a provision that contemplated the protection of aTK. Current Decision No. 410 of 2019 does not. At present, the absence of national rules protecting aTK is absolute. This gap not only does not comply with the NP, but is also unconstitutional as the Constitution acknowledges the ethnic and cultural pre-existence of Argentine indigenous peoples and guarantees respect to their identity as well as their participation in the management of their natural resources and other interests affecting them.[38] Thus, the regulation does not also comply with articles 7 and 12 NP.

2.1.6 Compliance Measures

Decision No. 410 of 2019 does not include compliance provisions of the nature required by the article 15 of the NP. It does, however, appoint the Secretariat of Environmental Policy on Natural Resources of the SE&SD, as a checkpoint. Its function is to receive information related to the access authorizations, the PIC, the source of the GR, the establishment of MAT and/or the utilization of GR, as applicable.[39]

This control, as it can be interpreted, will be carried out as follows: when foreign GR are imported into Argentina, an authorization shall be requested before the Secretariat of Environmental Policy on Natural Resources of the SE&SD, pursuant

[37] Administrative Decision No. 226/2010 appendix 1, item 6. This is a requirement of article 8 j CBD.

[38] Argentine Constitution article 75.1.

[39] Decision No. 410/2019 article 12.

to the form in Appendix VII (valid for the import of GR), which contains a requirement to submit all of the aforementioned documentation.[40]

The selected checkpoint seems to create a new complexity with respect to the range of activities to be controlled, as absolutely all the cases of imported GR into the country will be checked, no matter their intended use. According to this, even the import of GR for taxonomic research purposes will have to be authorized. The checking process seems, therefore, somewhat extreme and it might hinder or halt, in the best-case scenario, research on biodiversity for non-commercial purposes, which, on the other hand, the CBD promotes.

In addition to this, the kind of mechanism selected to control the legality of the access and utilization of GR from other jurisdictions is questionable. Other vigilance mechanisms in which the actual economic potential of the resource is activated might be more effective for such purpose. For instance, if they were implemented during research financing, commercialization, patent application, or product development. Finally, this checkpoint already proved ineffective under Decision No. 226 of 2010. The Decision mandated a similar obligation:[41] to request an authorization for the import of GR and to present along with it evidence of the existence of the PIC and of a benefit-sharing agreement. The mentioned control was never implemented due to the lack of an institutional structure that would allow for that type of monitoring activity. As the institutional capacities of the SE&SD remain weak, there are good reasons to believe that this type of control will fail again.

Finally, it should be pointed out that the Decision does not contain any clause providing for the protection of the traditional knowledge associated with the utilization of biodiversity when such knowledge is used in Argentina and pertains to indigenous and/or local communities settled in other countries. Neither is there any project being conducted to that end. Therefore, the country is in breach of article 16 of the NP.

2.2 *Provincial Legal and Regulatory Framework*

2.2.1 Legislation and Regulations

At the provincial level, 10 out of 23 provinces have enacted ABS rules. All of them were adopted before Administrative Decision No. 410 of 2019 was passed. Consequently, none of them includes any of the new developments this Decision has brought.

Provincial rules are applicable to GR found in their respective jurisdictions. Each of them greatly differs from the other in every possible aspect. The reason behind

[40]Decision No. 410/2019 article 15.

[41]Administrative Decision No. 226/2010 appendix 1.

this dissimilarity was the lack, until recently, of a national regulation setting minimum regulatory standards on ABS.

The provinces that have adopted ABS rules are: (a) Misiones (Executive Decree No. 474 of 2002 regulating Provincial Law No. 3337 on conservation and sustainable use of biological diversity and its components); (b) Neuquén (Provincial Law No. 2503 of 2005 on the access to genetic and biochemical resources); (c) Santa Cruz (Provincial Law No. 2993 of 2007 on access to genetic and biochemical resources); (d) Tierra del Fuego (Administrative Decision No. 570 of 2012 of the Secretariat of Environment and Sustainable Development establishing the legal regime for conducting research that entails access to genetic material); (e) Catamarca (Administrative Decision No. 90 of 2012 of the Secretariat of Environment and Sustainable Development establishing the legal regime for conducting research that directly or indirectly involves natural and/or genetic resources); (f) Jujuy (Administrative Decision No. 15 of 2013 of the Environment Secretariat establishing the regime for the access to provincial biodiversity); (g) San Luis (Provincial Law No. 9-0851 of 2013 on access and registry of genetic and biochemical resources pertaining to provincial biodiversity and Executive Decree No. 8804 of 2015 that further regulates the provincial law mentioned before); (h) Entre Ríos (Administrative Decision No. 1721 of 2014 of the General Directorate of Natural Resources establishing the regulation on access to genetic resources and its derivatives); (i) Formosa (Administrative Decision No. 40 of 2015 of the Ministry of Production and Environment regulating the requirements for access to biological resources and its derivatives); and (j) Buenos Aires (Administrative Decision No. 19 of 2019 of the Ministry of Agroindustry regulating access to genetic resources).

A second category of rules comprises those enacted by provinces that only enable future regulation of ABS issues. No further legal developments have occurred so far. These laws pertain to the provinces of: (a) La Rioja (Provincial Environmental Law No. 7801 of 2005); (b) La Pampa (Provincial Environmental Law No. 1914 of 2000); and (c) Rio Negro (Provincial Law No. 2600 of 1993: Law of preservation of the natural heritage and genetic resources).

A third group of provinces has opted to regulate scientific research within their jurisdictions but not ABS as such. These regulations could be applicable to activities related to access to GR. This is the case of the provinces of: (a) Salta (Administrative Decision No. 91 of 2005 of the Secretariat of Environment and Sustainable Development regulating scientific research); and (b) Mendoza (Administrative Decision No. 1243 of 2012 of the Natural Renewable Resources Agency regulating scientific research within natural protected areas).

Lastly, there is a group of provinces that has regulated the acquisition of bulk biological resources for scientific research purposes. Even though these rules do not focus on GR as such, they still may prove applicable to some activities related to the access and utilization thereof. These provinces are: (a) Corrientes (Administrative Decision No. 845 of 2013 of the Natural Resources Management Agency, regulating the collection of biodiversity for scientific purposes); (b) Córdoba (Law No. 7343 of 1985 establishing the guiding principles for the preservation, conservation, defense and improvement of the environment) and (c) Tucumán (Administrative Decision No. 87 of 2017 of the Agency of Wild Flora, Fauna and Soils, regulating scientific activities on terrestrial and aquatic fauna).

Apart from the obligations emerging from the provincial rules specifically regulating ABS or research on biodiversity, or the very collection of biological material, there exists a host of obligations arising from provincial rules which may also have an impact on the activities of access and utilization of GR. They regulate, for example, each in a different way, the obligation of obtaining a transit guide for the transportation of the biological material within the province and between provinces,[42] or the obligation of getting a permit for the collection or exploitation of the biological material, which are additional to the specific permit for access to GR. This is the case of the San Luis legislation,[43] for example.

2.2.2 Scope of the Rules

A focus on the first group of provinces—the ones that actually do have an ABS regime—reveals that the scope of the regulatory framework of Neuquén, Jujuy, San Luis and Santa Cruz covers both genetic and biochemical resources. Formosa, instead, includes all biological resources and their derivatives. Provincial rules are applicable to all kinds of access to genetic and biochemical resources, no matter whether they will be used to conduct commercial or non-commercial research, or utilized inside or outside the country, or accessed by a national or a foreign user.

In regard to the geographical scope, it is worth mentioning that Jujuy's legislation covers not only genetic and biochemical resources found within its jurisdiction, but also those that have been collected in Jujuy and are now conserved in *ex-situ* collections outside its boundaries. This provision might be contested under the territoriality principle of the law.

Exemptions to the applicability of the ABS regime are only found in the legal frameworks of the provinces of Neuquén, Jujuy and Santa Cruz; which expressly exclude human GR from their respective laws. In addition to that, Jujuy's law excludes non-native and exotic GR.

Provincial rules do not include definitions for the terms user, utilization, commercial and non-commercial research. Only Jujuy, Neuquén and San Luis have defined GR. In Jujuy, these are characterized in the same way the CBD defines biological resources. On the other hand, Neuquén and San Luis provide definitions for GR which are very much in line with that of the CBD. Regarding the notion of access, the rules of Neuquén, San Luis and Formosa understand that it refers to the acquisition of GR with the intention of doing any type of research and/or

[42] Administrative Decision No. 40/2015 appendix 6 (Formosa) and Executive Decree No. 474/2002 article 46 (Misiones), for example. The provincial transit guides guarantee the legal origin and tenure of the biological material.

[43] Law No. 9-0851/2013 article 11.

development. Possible activities range from bioprospection to education, conservation, and monitoring, amongst others.

2.2.3 Institutional Framework and Access Procedures

In those provinces that have actually adopted ABS specific rules, the competence for providing PIC for the access to GR corresponds to the pertinent environmental provincial authority. In Entre Ríos, for example, is the General Directorate of Natural Resources;[44] in Formosa, the Directorate of Natural Resources and Management;[45] and, in Jujuy, the Directorate of Genetic Resources and Protection of Biodiversity.[46]

Provincial rules on ABS greatly differ from one another on the degree to which access procedures have been regulated. Some of them, for example Neuquén and Santa Cruz, have no regulation on procedures for access to GR despite the fact that their laws date back to 2005 and 2007, respectively.[47] In these provinces, users who want to access GR are faced with an extra burden of difficulties as procedures have not been established.

On the contrary, other provinces, such as Jujuy, Formosa, Entre Ríos, Tierra del Fuego and San Luis, have regulated in a detailed manner the procedure and the requirements needed to access GR found in their jurisdictions. It can be observed that the more recent the rule is, the greater its degree of development and of consistency with the NP.

Some provincial rules, especially those that have been recently developed, such as those of Tierra del Fuego, Entre Ríos and Formosa, require foreign applicants wanting to access GR to jointly apply with an Argentine scientific institution.[48] Compliance with this requirement has to be proved by submitting evidence of an agreement signed between the local and the foreign research institutions.

None of these provinces has a facilitated procedure to access GR for purely scientific research purposes. Consequently, provincial rules on ABS hinder efficient access to GR even for basic research, such as studies on molecular systems pursued for taxonomic, systematic, phytogeographic, biogeographic, evolutionary and/or genetic conservation reasons.

[44] Administrative Decision No. 1721/2014 article 4.

[45] Administrative Decision No. 40/2015 article 4.

[46] Administrative Decision No. 15/2013 article 5.

[47] Law No. 2503/2005 (Neuquén) and Law No. 2993/2007 (Santa Cruz).

[48] Administrative Decision No. 570/2012 article 2 (Tierra del Fuego); Administrative Decision No. 1721/2014 article 3 (Entre Ríos) and Administrative Decision No. 40/2015 appendix 2, article 3 (Formosa).

2.2.4 Access Permits

Access permits are to be obtained from the authority of the province where the GR are accessed. Only in the case where resources are collected from a national park must the access permit be obtained from the national park authority.[49]

Conditions applicable to access permits in provincial rules include the following: (a) permits are personal and cannot be transferred. This means the permit is issued for the benefit of a specific person and it cannot be alienated or disposed of, in any way.[50] Still, if GR are to be transferred to a third party, a case-by-case authorization can be obtained; (b) obligation to deposit a sample of the collected biological material in a collection belonging to the province where the sample was obtained, or to the country;[51] (c) obligation to hand in a report on activities undertaken, including research and developments;[52] (d) obligation to acknowledge, in any scientific publication, the geographical origin of the GR;[53] and (e) if access to the GR involves the collection of any biological resources, a collection permit is required by certain provincial rules.[54]

2.2.5 Benefit-Sharing

Provincial rules set that MAT are to be established between the user of GR and the province that grants access to them. In practice, however, when GR are utilized for purely scientific research purposes, MAT are established with the providing scientific institution. If this is the case, the province still has to grant its PIC.

If the GR are obtained from an *ex-situ* collection, the agreement is concluded with it. When this is the case, there is disagreement on whether the provinces should still grant their PIC for access or not, as it could be considered that the province still holds original ownership of its resources.

Neuquén, Jujuy, San Luis, Formosa, Santa Cruz, Entre Ríos, Misiones and Tierra del Fuego thoroughly regulate the fair and equitable distribution of benefits obtained from the utilization of GR. Amongst the benefits to be distributed are upfront payments. San Luis, for example, requires an upfront payment to give consent for access to GR. Educational institutions and those conducting scientific research

[49] Decision No. 81/2016 article 2.

[50] Administrative Decision No. 1721 of 2014 appendix 3 (Entre Ríos) and Executive Decree No. 474/2002 article 53 (Misiones).

[51] Administrative Decision No. 1721/2014 appendix 3 (Entre Ríos).

[52] Administrative Decision No. 1721 of 2014 appendix 1 (Entre Ríos); Law No. 9-0851/2013 article 18 (San Luis); Law No. 2503 of 2005 article 14 (Neuquén) and Administrative Decision No. 570/2012 appendix 1, item 4 (Tierra del Fuego).

[53] Administrative Decision No. 1721/2014 appendix 1 (Entre Ríos).

[54] Law No. 2503/2005 article 8 (Neuquén) and Law No. 9-0851/ 2013 article 11 (San Luis).

pursuing purely scientific intentions are exempted from this charge.[55] The amount of payment is determined by law on a year-by-year basis. Neuquén also requires an upfront payment that even institutions that conduct scientific research with non-commercial intentions have to satisfy.[56] Jujuy also requires an upfront payment in order to grant access to GR regardless of the intended purpose. The value of the charge is a fixed amount equivalent to 75 liters of petrol (for Argentineans) and 150 liters (for foreigners).[57]

In addition to upfront payments, several provinces establish unrealistic percentages of royalties that have to be shared with them when commercial and/or industrial research and development are conducted on GR. Neuquén, for instance, requires that a minimum of 20% of the tax base of the gross income provincial tax is shared with the province.[58] This provision does not acknowledge the fact that willingness to pay for raw GR is low.[59] It also wrongly assumes that the user pays taxes in Neuquén, or even in Argentina. Entre Ríos goes in the same direction and requires that 50% of all utilities be shared with the province.[60] San Luis also requires a distribution of royalties.[61] The percentage of utilities to receive is determined by law on a yearly basis. Finally, Tierra del Fuego also requires a symmetrical distribution of economic benefits.[62]

Other provincial rules have not established a percentage of monetary benefits to be distributed, but, on the contrary, simply require that when economic benefits are likely to be generated, an agreement is negotiated between the province and the user of the GR. This is the case of Jujuy.[63]

On the contrary, some provincial legal frameworks also focus on non-monetary benefits. These are more likely to occur and usually take place in the short and medium term. San Luis, for instance, promotes scientific research collaboration and joint technological development as benefits to be negotiated whenever possible.[64] In the same line of thought, some provincial rules establish that foreigners have to apply for access in association with a national or provincial scientific research

[55] Law No. 9-0851/ 2013 article 13.

[56] Law No. 2503/2005 article 11.

[57] Administrative Decision No. 15/2013 article 7. At present, that would be equivalent to some 70 and 140 euros respectively.

[58] Law No. 2503/2005 article 11.

[59] Correa (2005), p. 458.

[60] Administrative Decision No. 1721/2014 article 6.

[61] Law No. 9-0851/2013 article 14.

[62] Administrative Decision No. 570/2012 article 2.

[63] Administrative Decision No. 15/2013 article 22.

[64] Law No. 9-0851/2013 article 16.

institution. This is the case of Tierra del Fuego,[65] Entre Ríos,[66] Jujuy[67] and Formosa.[68]

3 Assessment

The analysis of the Argentine ABS regime shows that no law setting minimum common environmental standards on ABS has been dictated by the National Congress. This gap constitutes a breach to the National Constitution that mandates such obligation. In turn, ABS issues have been regulated at the administrative level through different administrative decisions.

Until October 2019, the existing pieces of regulation at the national level, to wit: Administrative Decision No. 226 of 2010 of the SE&SD, now derogated, Administrative Decision No. 208 of 2011 of the COFEMA and the National Park Scientific Research Regulations, included in Decision No. 81 of 2016 on National Parks, did not set minimum environmental standards on ABS. On the contrary, they just reassembled a customs norm, reasserted the provinces' ownership rights over their respective GR, and regulated all kinds of scientific activity under national parks, respectively.

As a result of the rudimentary state of development of the ABS national framework, there was no guidance until October 2019, which could have served as a basis for the provinces to adopt rules on the issue in a strategic and concerted manner. Consequently, at the provincial level, the overwhelming proliferation and disparity of laws and regulations seems to be the rule and the most urgent issue to be addressed. Their content, when the rules exist, varies greatly.

Finally, recently adopted Administrative Decision No. 410 of 2019 of the SE&SD, which sets minimum common standards on ABS, positively advances the regime at the national level. It lacks, however, the required legal hierarchy. It is just an administrative decision and not a formal law enacted by the National Congress. Therefore, a law on the issue is still missing. Moreover, even though in practical terms, the new administrative decision advances towards a solution to the ABS conundrum, provinces will not be bound by it unless they expressly adhere to the Decision. If they do not, the administrative Decision will just be a worthless scrap of paper.

65 Administrative Decision No. 570/2012 article 2.

66 Administrative Decision No. 1721/2014 article 3.

67 Administrative Decision No. 15/2013 article 2.

68 Administrative Decision No. 40/2015 appendix 2.3.

4 Conclusions

In spite of Argentina's rich biodiversity and of the existence of a highly qualified scientific community dedicated to research and development in the field, the country had not advanced steadily or strategically in the development of its own ABS national regulatory framework until recently. The main cause of this setback was the absence of a national regulation setting the minimum standards on the issue. As a result, provinces have not had any guidance on the adoption of their own rules on the matter.

Newly adopted Decision No. 410 of 2019 of the SE&SD, which sets minimum common standards on ABS at the national level, has the capacity to change history. Finally, the provinces have the needed guidance to develop their own regulatory frameworks in a concerted manner. Decision No. 410 of 2019 is indeed an important development. However, its success and effects in reality are yet to be seen.

References

Bibliographical References

Correa C (2005) Do national access regimes promote the use of genetic resources and benefit sharing? Int J Environ Sustain Dev 4(4):444–462

López Lanús B, Grilli P, Di Giacomo A, Coconier E, Banchs R (eds) (2008) Categorización de las aves de la Argentina según su estado de conservación. Aves Argentinas AOP, Buenos Aires

Ministerio de Ciencia, Tecnología e Innovación Productiva (MINCYT) (2016) Las Empresas de Biotecnología en Argentina, http://www.argentina.gob.ar/sites/default/files/est_bio_las-empresas-de-biotecnologia-en-argentina-2016.pdf. Accessed 30 Oct 2019

Ojeda R, Chillo V, Diaz Isenrath G (eds) (2012) Libro rojo de los mamíferos amenazados de la Argentina. Sarem, Mendoza

Secretaría de Ambiente y Desarrollo Sustentable (2018) Informe del estado del ambiente 2018, http://www.informe.ambiente.gob.ar. Accessed 30 Oct 2019

Secretariat of the Convention on Biological Diversity (2007) Document UNEP/CBD/WG-ABS/5/INF/2 Compilation of submissions by Parties on experiences in developing and implementing article 15 of the Convention at the national level and measures taken to support compliance with prior informed consent and mutually agreed terms, http://www.cbd.int/kb/record/meetingDocument/36704?Subject=ABS. Accessed 30 Oct 2019

Zuloaga F, Belgrano M (2015) The catalogue of vascular plants of the Southern Cone and the Flora of Argentina: their contribution to the World Flora. Rodriguésia 66(4):989–1024

Legal References

Argentina. Argentine National Constitution, 23 August 1994. Official Journal of Argentina No. 27759

Argentina. Law No. 24375 that approves the Convention on Biological Diversity, 6 October 1994. Official Journal of Argentina No. 27991

Argentina. Decision No. 409 of the former Argentine Institute of Vegetable Hygiene and Quality on procedures for inspection and verification of plant products in export, import and international transit, 30 September 1996. Official Journal of Argentina No. 28532

Argentina. Decree No. 666 setting rules for the protection and conservation of wild fauna, the rational use of wildlife and the import, export and interprovincial trade thereof, 18 July 1997. Official Journal of Argentina No. 28695

Argentina. Law No. 25675 that establishes the National Environmental Policy, 6 November 2002. Official Journal of Argentina No. 30036

Argentina. Administrative Decision No. 437 of the Argentine Secretariat of Environment and Sustainable Development that establishes rules for the import, export and re-export of specimens, products or byproducts of wild fauna species, 12 May 2006. Official Journal of Argentina No. 30914

Argentina. Administrative Decision No. 226 of the Argentine Secretariat of Environment and Sustainable Development on access to genetic resources, 28 April 2010. Official Journal of Argentina No. 31892

Argentina. Administrative Decision No. 208 of Federal Council of the Environment that reaffirms the rights of the Provinces over their natural and genetic resources, 16 March 2011., http://cofema.ambiente.gob.ar/?aplicacion=normativa&IdNorma=1305&IdSeccion=32 (accessed 16 March 2020)

Argentina. Law No. 27246 that approves the Nagoya Protocol on access to genetic resources and equitable sharing of the benefits deriving from their utilization, 24 November 2015. Official Journal of Argentina No. 33283

Argentina. Decision No. 81 of the National Parks Administration that sets Scientific Research Regulations, 18 May 2016. Official Journal of Argentina No. 33423

Argentina. Administrative Decision No. 477 of the Ministry of the Environment and Sustainable Development that establishes rules for the import, export and re-export of specimens, products or byproducts of wild flora species included in the appendices of the Convention on International Trade in Endangered Species of Wild Fauna and Flora (CITES), 23 August 2018. Official Journal of Argentina No. 33940

Argentina. Administrative Decision No. 410 of the Argentine Secretariat of Environment and Sustainable Development that regulates basic common standards for the access and utilization of genetic resources in Argentina. 22 October 2019. Official Journal of Argentina No. 34225

Buenos Aires. Administrative Decision No. 19 of the Ministry of Agroindustry regulating access to genetic resources, 13 February 2019. Official Journal of the Province of Buenos Aires No. 28466

Catamarca. Administrative Decision No. 90 of the Secretariat of Environment and Sustainable Development establishing the legal regime for conducting research that directly or indirectly involves natural and/or genetic resources, 16 August 2012. https://www.argentina.gob.ar/sites/default/files/recopilacion_normativa_rrgg_2019.pdf (accessed 16 March 2020)

Córdoba. Law No. 7343 establishing the guiding principles for the preservation, conservation, defense and improvement of the environment, 29 August 1985., https://www.argentina.gob.ar/sites/default/files/recopilacion_normativa_rrgg_2019.pdf (accessed 16 March 2020)

Corrientes. Administrative Decision No. 845 of the Natural Resources Management Agency establishing rules for the collection of biodiversity for scientific purposes, 21 of September 2013., https://www.argentina.gob.ar/sites/default/files/recopilacion_normativa_rrgg_2019.pdf (accessed 16 March 2020)

Entre Ríos. Administrative Decision No. 1721 of the General Directorate of Natural Resources establishing the regulation on access to genetic resources and its derivatives, 15 October 2014., https://www.argentina.gob.ar/sites/default/files/recopilacion_normativa_rrgg_2019.pdf (accessed 16 March 2020)

Formosa. Administrative Decision No. 40 of the Ministry of Production and Environment regulating the requirements for access to biological resources and its derivatives, 28 December 2015. Official Journal of the Province of Formosa No. 10456

Jujuy. Administrative Decision No. 15 of the Environment Secretariat establishing the regime for the access to provincial biodiversity, 25 March 2013., https://www.argentina.gob.ar/sites/default/files/recopilacion_normativa_rrgg_2019.pdf (accessed 16 March 2020)

La Pampa. Provincial environmental Law No. 1914, 2 February 2001. Official Journal of the Province of La Pampa No. 2408

La Rioja. Provincial environmental Law No. 7801, 29 March 2005. Official Journal of the Province of La Rioja No. 10259

Mendoza. Administrative Decision No. 1243 of the Natural Renewable Resources Agency regulating scientific research within natural protected areas, 7 December 2012. Official Journal of the Province of Mendoza No. 29306

Misiones. Executive Decree No. 474 regulating Provincial Law No. 3337 on conservation and sustainable use of biological diversity and its components. 22 May 2002., https://www.argentina.gob.ar/sites/default/files/recopilacion_normativa_rrgg_2019.pdf (accessed 16 March 2020)

Neuquén. Provincial Law No. 2503 on the access to genetic and biochemical resources, 25 November 2005. Official Journal of the Province of Neuquén Annex No. 2956

Río Negro. Provincial Law No. 2600 on the preservation of the natural heritage and genetic resources, 14 April 1993. Official Journal of the Province of Río Negro No. 3056

Salta. Administrative Decision No. 91 of the Secretariat of Environment and Sustainable Development regulating scientific research, 28 February 2005. Official Journal of the Province of Salta No. 17081

San Luis. Provincial Law No. 9-0851 on access and registry of genetic and biochemical resources pertaining to provincial biodiversity, 14 August 2013. Official Journal of the Province of San Luis No. 14096

San Luis. Executive Decree No. 8804 that further regulates Law No. 9-0851 of 2013 on access and registry of genetic and biochemical resources pertaining to provincial biodiversity, 18 November 2015. Official Journal of the Province of San Luis No. 14453

Santa Cruz. Provincial Law No. 2993 on access to genetic and biochemical resources, 18 September 2007. Official Journal of the Province of Santa Cruz No. 4122

Tierra del Fuego. Administrative Decision No. 570 of the Secretariat of Environment and Sustainable Development establishing the legal regime for conducting research that entails access to genetic material, 4 September 2012., https://www.argentina.gob.ar/sites/default/files/recopilacion_normativa_rrgg_2019.pdf (accessed 16 March 2020)

Tucumán. Administrative Decision No. 87 of the Agency of Wild Flora, Fauna and Soils that regulates scientific activities on terrestrial and aquatic fauna, 9 May 2017., https://www.argentina.gob.ar/sites/default/files/recopilacion_normativa_rrgg_2019.pdf (accessed 16 March 2020)

United Nations. Convention on Biological Diversity, 5 June 1992. UNTS, vol. 1760, I-30619

United Nations. Nagoya Protocol on Access to Genetic Resources and the Fair and Equitable Sharing of Benefits Arising from their Utilization to the Convention on Biological Diversity, 29 October 2010. UNTS, vol. N/D, I-30619

Luciana C. Silvestri is an environmental law researcher at the National Scientific and Technical Research Council (CONICET) in Argentina. She served as a legal and technical adviser to the Spanish Ministry of Environment for access and benefit-sharing (ABS) international negotiations. She was one of the experts contributing to the GEF UNEP IUCN Regional Project on ABS for Latin American and the Caribbean, where she focused on ensuring effective legal ABS frameworks and adequate institutional capacities in eight countries of the region. She also served as an expert to the Informal Advisory Committee to the ABS Clearing-House of the Nagoya Protocol. She has extensive expertise in international cooperation, international negotiations, biodiversity protection and ABS. lsilvestri@mendoza-conicet.gob.ar.

Brazilian Biodiversity Law: Challenges and Opportunities for Industries and Research Institutions

Lilian Massini Mozini

Abstract Federal Law No. 13.123/2015, known as the Brazilian Biodiversity Act, establishes rules for access to Brazilian genetic heritage, associated traditional knowledge and benefit-sharing. It is the second legislation adopted in Brazil since 2000 to regulate this subject. The Law brought about a new regime of access to and utilization of genetic heritage, highlighting simplified procedures for users who conduct research and technological development. This chapter presents the scope and instruments, including the declarations by users, administrative acts and agreements concerning access and benefit-sharing arising from the commercialization of products. It will also look at the actual registrations and notifications carried out by companies and research institutions in the National Genetic Heritage and Associated Traditional Knowledge Management System (SisGen) since it was released in 2017.

Keywords Brazil · Genetic resources · Benefits-sharing · Law 13.123/2015

1 Development of Legislation in Brazil

1.1 Status of Biodiversity

Brazil is one of the largest countries in the world, occupying almost half of South America. In 6 terrestrial biomes and 3 marine ecosystems, it has the greatest biodiversity of species in the world, hosting between 15 and 20% of all terrestrial and marine species.[1] To date, 49315 species are recognized for Brazil's flora (native, cultivated and naturalized).[2]

[1] Ministry of Environment (2020); CBD (2020).

[2] Flora do Brasil (2020).

L. M. Mozini (✉)
São Paulo, Brazil
e-mail: lmozini@ambiente-global.com

E. Chege Kamau (ed.), *Global Transformations in the Use of Biodiversity for Research and Development*, Ius Gentium: Comparative Perspectives on Law and Justice 95, https://doi.org/10.1007/978-3-030-88711-7_3

In order to catalogue the Brazilian biodiversity flora species, there is the REFLORA/CNPq Programme since 2010 with the objective to rescue and make available images and information concerning Brazilian plants deposited chiefly in overseas herbaria through an on-line facility, the Reflora Virtual Herbarium.[3] Also, the images and information concerning Brazilian plants are being published by the Brazilian Flora 2020,[4] which is a database available to the scientific community and the general public. This is one of the most important sources of research on the botanical origin of species found in Brazil today, including for the purposes of Law 13.123/2015.

The country has an important number of local and traditional communities that possesses knowledge on fauna and flora species—there are more than 200 indigenous peoples, what makes Brazil a megadiverse country from a cultural perspective.[5] The importance of the protection of such knowledge was recognized by the Government according to the Law 13.123/2015 and the traditional knowledge associated to genetic heritage is considered as part of Brazilian cultural heritage (Art. 8). Also, mechanisms are in place to allow for the participation of traditional knowledge holders in decision-making processes, for example, the Genetic Heritage Management Council (known as CGEN, in Portuguese), which is the competent national authority at the federal level, has 3 representatives of traditional peoples (Decree 8.772/2016, article 7, IV).

1.2 Historical Background

International conventions on natural resources and their conservation, to which Brazil is a signatory, constitute a decisive step in the national legal and political order because they define the scope of public policies and legal statutes. Brazil, a civil law country, has extensive environmental legislation to regulate the use and appropriation of natural resources. The country has several instruments related to the sustainable use of biodiversity, at states' and federal levels.

The Federal Constitution of 1988 (Brazil 1988) places the environment as one of its fundamental principles. Article 225 reads: "All have the right to an ecologically balanced environment, which is an asset of common use and essential to a healthy quality of life, and both the Government and the community shall have the duty to defend and preserve it for present and future generations".

The Convention on Biological Diversity (CBD) was ratified in 1994 by Legislative Decree No. 2/94, issued by the National Congress, and incorporated into the Brazilian legal system by Federal Decree No. 2.519/1998 (Decree No. 2.159/1998).

---

[3] Ibid.

[4] The list is available at: http://floradobrasil.jbrj.gov.br/reflora/listaBrasil/ConsultaPublicaUC/ConsultaPublicaUC.do#CondicaoTaxonCP.

[5] CBD (2020).

In doing so, Brazil undertook to adopt internally the principles proposed by the Convention. Various political and legal measures have been implemented in the country to promote the implementation of the CBD.

In addition, the country signed the Nagoya Protocol in 2011. It was ratified by the national congress in 2020, and entered in force in Brazil in June, 2021.

In June 2000, the executive branch of the federal government issued a Provisional Measure on access to both Brazilian genetic heritage and associated traditional knowledge (aTK) with biodiversity as well as benefit-sharing arising from the commercialization of products deriving from such access. In 2001, the Provisional Measure (PM) was renumbered as PM No. 2.186/2001 (Provisional Measure No. 2.186/2001)[6] and reissued 15 consecutive times.[7]

The issuance of the PM was followed by the bringing of several bills of law before the national congress with the purpose of replacing it. In June 2014, Bill of Law No. 7.735, which was drafted by the Brazilian presidency, was put forward. After passing the two legislative houses, it was approved in April 2015 and published after presidential sanction on 21 May 2015. Known as 'the new statute on biodiversity', Federal Law No. 13.123/2015 (Law No. 13.123/2015) became effective on 17 November 2015. It is elaborated on by Federal Decree No. 8.772/2016 (Decree No. 8.772/2016).

1.3 Provisional Measure No. 2.186-16/2001: Aims and Intentions

PM was fiercely criticized after its publication for several reasons. Firstly, it was not submitted to the National Congress for deliberation, being rather a unilateral act of the executive branch of the federal government. Secondly, it did not address many issues, such as access to traditional knowledge belonging to more than one community or appropriate mechanisms for a fair and equitable benefit-sharing. The adoption of the 2015 federal biodiversity statute aimed to replace the PM and improve the regulation of genetic heritage in Brazil.

---

[6] Regulates item II from paragraph 1 and paragraph 4 of article 225 of the Constitution, Articles 1, 8, item "j", 10, item "c", 15 and 16, items 3 and 4 of the Convention on Biological Diversity, provides for the access to genetic heritage, protection and access to associated traditional knowledge, benefit-sharing and access to and transfer of technology for its conservation and use, and makes other provisions.

[7] Provisional measures must be transformed into a law through a National Congress vote, but until 2001 they could be reissued unrestrictedly. With the constitutional amendment n. 32/2001, the extension was limited to a single time, and provisional measures should be voted on for conversion into law within 60 days, under penalty of loss of effectiveness. The PM issued prior to amendment 32/2001 were kept in force until the definitive deliberation of the National Congress and this is why Provisional Measure No. 2.186-16/2001 became a "permanent" measure.

The main objectives of PM were to regulate rights and obligations relating to:

- access to the genetic heritage of the national territory, the continental shelf and the exclusive economic zone for the purpose of scientific research, technological development or bioprospecting;
- access to traditional knowledge associated with genetic heritage, which are relevant to the conservation of biological diversity, to the integrity of Brazilian genetic heritage, and the use thereof;
- fair and equitable sharing of the benefits derived from the economic exploitation of genetic heritage and associated traditional knowledge; and
- access to technology and transfer of technology for the conservation and use of biological diversity.

Under the PM, access to Brazilian genetic heritage was only allowed if the federal government granted prior authorization, with the use and commercialization of such resources being subject to inspection, restrictions and a benefit-sharing obligation.

It was difficult for companies and research institutions to comply with the PM because access activities required prior authorization from the national authority, a prior consent agreement executed with the body or entity responsible for the area[8] where the genetic resource samples were collected, and a benefit-sharing agreement (the latter being required if there was an intention of economic exploitation).

According to the PM (Art. 27), the owner of the public or private area or the representative of the indigenous community and the official Indian Affairs body, or the representative of the local community is the party of the contract for use of genetic heritage and benefit-sharing and should receive the benefits arising from the genetic heritage access. On the other hand, the federal Constitution provides that the environment is a common use property of the people (Art. 225). For this reason, the law that replaced this PM recognized that genetic heritage is a common use property of the people (Law 13.123/2015, article 1, I), with the Union becoming the beneficiary of benefit-sharing, except in cases of access to traditional knowledge, where the beneficiary of the benefit-sharing is the holder of such knowledge (e.g. indigenous population, traditional community or traditional farmer).

In addition, prior authorization had to be granted by the national authority, a body that was composed of various government authorities. This made it difficult to get a project approved, since applications frequently led to further questions and the process must be reanalyzed by this body. The various government authorities discussed the terms of prior consent and whether the benefit-sharing was fair and equitable, so that the parties of the envisaged contracts frequently had to explain and present additional statements to support their proposals.

---

[8] According to article 16, para 9°, the prior consent shall be given by the indigenous community involved, after consulting the official Indian Affairs body, when the access occurs in indigenous lands or the competent body, when access occurs in protected area; or the owner of the private area, when the access occurs there or the National Defense Council, when the access takes place in an area essential for national security or the maritime authority, when the access takes place in Brazilian jurisdictional waters, on the continental shelf and in the exclusive economic zone.

Nevertheless, in the last years before the PM was repealed, especially from 2012 to 2015, there was an increase in the number of applications for authorization. Furthermore, applications were processed with greater speed.

A survey on the national authority's website[9] dated November 2015 shows the existence of 1106 applications for authorization of access to genetic heritage and/or aTK as well as for registration of depository institutions. About 76% of the total corresponded to applications for authorizations from research institutes, whereas businesses accounted for only 24% of the applications.

According to the national authority's annual report, released in 2015,[10] 388 authorizations were granted during that year. Also, during 2015, 159 contracts for use of genetic heritage and benefit-sharing were approved by the national authority. Interesting to note that from 2004 to 2014, another 138 contracts were approved, a smaller number than in 2015.

On the other hand, the annual report published in 2014[11] shows that the applications for authorizations had a very extensive processing time, on average 501 days between the application and approval. It means the research or the technological development could only occur after this time. It is a very long period, considering the interests of the business sector and the research institutions. It is important to note that the body made up a task force during 2015 with the aim to deliberate on all process before the new law entered into force. That is the reason for the impressive number of authorizations and contract approvals that year.

The data show that only a few companies applied for authorizations prior to access to genetic heritage, although it is evident that a lot of products containing genetic heritage from Brazilian biodiversity were developed and commercialized. Given the complex bureaucracy for obtaining prior authorization and the need for greater legal certainty for companies and research institutions which access genetic heritage and aTK, it became clear that there was a need to replace the PM with more effective legislation democratically legitimized by approval of the national congress.

For this reason, Federal Law No. 13.123/2015 aimed at improving and speeding up administrative procedures and more legal certainty associated with the use of genetic heritage and aTK.

2 The ABS Framework in Brazil

2.1 The Law 13.123/2015

Federal Law No. 13.123/2015, known as the Biodiversity Act, entered into force on November 17, 2015 and repealed PM No. 2.186-16/2001. It establishes rules for

---

[9]The Data presented in this study was updated until November 2015 in the CGEN's website.

[10]Ministry of Environment (2015).

[11]Ministry of Environment (2014).

access to Brazilian genetic heritage, access to aTK, and the sharing of benefits with a view to preserve and sustainably use Brazilian biodiversity.

Federal Law No. 13.123/2015 and implementing Decree No. 8.772/2016 provide for rights and obligations related to the following activities: (i) access to the country's genetic heritage and traditional knowledge associated to genetic heritage; (ii) access to technology and to the transfer of technology; (iii) economic exploitation of finished product or reproductive material; (iv) fair and equitable sharing of the benefits arising from economic exploitation of finished products or reproductive material; (v) shipping abroad intended for access to genetic heritage (Art. 1).

The Law 13.123/2015 provides some legal definitions (Art. 2) with the following key concepts, among others:

- 'genetic heritage': genetic information from plants, animals, and microbial species, or any other species, including substances originating from the metabolism of these living organisms;
- 'access to genetic heritage': research or technological development carried out on genetic heritage samples;
- 'access to associated traditional knowledge': research or technological development carried out on traditional knowledge associated to genetic heritage that makes possible or facilitates access to genetic heritage, even if obtained from secondary sources such as: street markets, publications, inventories, films, scientific articles, registries and other forms of systematization and record of associated traditional knowledge;
- 'research': experimental or theoretical activity carried out on genetic heritage or associated traditional knowledge with the objective of building new knowledge by means of a systematic process that creates and tests hypothesis, describes and interprets fundamentals of observed phenomena and facts;
- 'technological development': systematic work on genetic heritage or associated traditional knowledge based on existing procedures resulting from research or from practical experience carried out with the objectives of developing new materials, products or devices, or improving or developing new processes, for economic exploitation;
- 'shipment': transfer of a sample of genetic heritage, intended for access, to an institution located abroad, in which responsibility for the sample is transferred to the recipient institution;
- 'reproductive material': plant propagating material or animal reproductive material from any genus, species or crop, coming from sexual or asexual reproduction;

According to the Brazilian legislation, access occurs when activities of research or technological development are performed, even if the activities are carried out on a sample of Brazilian origin found in *ex-situ* collections, for example a biological collection.

2.2 *Scope of Law*

2.2.1 Geographical

Brazilian law covers the national territory, the continental shelf, territorial waters, or the exclusive economic zone (Art. 1, I) and must be complied with by users[12] accessing genetic heritage and/or associated traditional knowledge situated therein.

2.2.2 Material

The law applies if there is access to the country's genetic heritage, found in *in-situ* conditions, including domesticated species and spontaneous populations, or kept in *ex-situ* conditions, as long as found in *in-situ* conditions within the national territory, on the continental shelf, on territorial waters, or in the exclusive economic zone, as well if there is access to traditional knowledge associated with genetic heritage (Art. 1).

In the case of micro-organisms, they are considered as existing in the national territory if they were isolated from substrata of the national territory, the territorial sea, the exclusive economic zone or the continental shelf (Decree 8.772/2916, article 1, para. 1).

From a reading of the definition of 'access' together with 'genetic heritage' as cited above access is not conceived as the physical taking of a specimen but rather as an intellectual operation of conducting research and development activities. This implies that access also covers cases where a specimen was acquired for bulk uses and only later on became an object of research and development.

Once access is planned to take place, an obligation to register the access activity with the National Genetic Heritage and Associated Traditional Knowledge Management System—SisGen—arises. SisGen is an electronic system maintained and operated by Genetic Heritage Governing Council (known as CGEN in Portuguese) for the management of access and shipment, notification of finished products and benefit-sharing records (see further below).

2.2.3 Personal Scope

Both access to genetic heritage and/or aTK and the shipping of genetic heritage samples are subject to the requirements imposed by the law per its article 11. This being so, legislation allows access activities to be carried out by a foreign institution only if it is associated to a national institution.

---

[12]"a natural or legal person, that accesses genetic heritage or associated traditional knowledge, or economically explores a finished product or reproductive material originated from access to genetic heritage or associated traditional knowledge"—Article 2, XV of Law 13.123/2015.

Accordingly, SisGen has determined that only national institutions or Brazilian natural persons can register access and samples shipping (CGEN Ordinance n° 01/2017). So, foreign natural persons are not allowed to carry out access to genetic heritage or aTK (Law 13.123/2015, article 11, para. 1).

Similarly, if access is made by a legal person based abroad, the register must be made by the national scientific and technological research institution, public or private, indicating the partnership with the foreign institution (Law 13.123/2015, article 12, II and Decree 8.772/2016, article 22). If access results in the development of finished products to be manufactured and marketed in Brazil or abroad, the manufacturing company is responsible for the notification of products and the sharing of benefits under Chapter V of Law 13.123/2015. If there are other results, like as shipment, or the application of any intellectual property right, or the commercialization of the intermediate product, or publication of results, partial or final, in scientific or communication media, the registration also must be made by the national institution (Law 13.123/2915, article 12).

Although Brazilian authorities have no jurisdiction to supervise and punish institutions that carry out access activities and commercialize products with Brazilian biodiversity ingredients abroad, Law 13.123/2015 establishes a joint and several liability mechanisms seeking to ensure benefit-sharing. In this sense, article 17 para. 7 reads:

> In the case the finished product or reproductive material has not been produced in Brazil, the importer, subsidiary, associate, affiliate, partner, or commercial representative of a foreign producer in national territory or in the territory of a country that Brazil has an agreement with for this purpose, will be jointly liable with the manufacturer of the finished product or the reproductive material for benefit-sharing.

Therefore, even if the finished product is developed and manufactured outside Brazil, the importer, or a national company of the same economic group (subsidiary) can be jointly and severally liable for the sharing of benefits.

2.2.4 Accessing *ex-situ* or *in silico* Genetic Heritage

The Law defines *ex-situ* conditions as "conditions in which genetic heritage is kept outside its natural habitat" (Art. 2, XXVII) in collections and allows access to genetic heritage that occurs in *ex-situ* samples provided they are found in *in-situ* in the national territory, on the continental shelf, in the territorial sea and in the exclusive economic zone (Art, 1, I). It also provides that *ex-situ* conservation of samples of genetic heritage found in *in-situ* condition will be preferably carried out in national territory (Art. 14).

According to the definition of genetic heritage (Art. 2, I), *in silico* digital information may be considered to be in the scope of the law, although this concept was not expressly included because the Law defines genetic heritage as genetic information from species. So, access to *in silico* genetic heritage for the purposes of research or technological development will oblige the user to register access and, in case of commercialization of final product or reproductive material, to make notification and share benefits.

In this sense, the Decree regulating the Biodiversity Law provides for the procedures and information needed for the registration at SisGen by users, indicating the source of genetic heritage, even if the sample or information were accessed in *ex-situ* or *in silico* conditions.

With regard to access to the samples obtained in *ex-situ* conditions, the user is expected to also report if the access was made *in silico* (Decree 8.772/2016, article 22, II, f, 1). In this case, the user must indicate the database where the sample is filed, in addition to presenting the filing code and link to the internet database access site. If possible, the user shall also indicate the original location data *in-situ* of the sample, such as geographical coordinates, municipality, state and date of collection (Decree 8.772/2016, article 22, § 1°, II).

On the other hand, by virtue of article 107, single paragraph of Decree No. 8.772/16, the reading or consultation of information on the genetic origin available in national and international databases does not constitute access to genetic heritage. It is important to clarify that the law does not consider as access to genetic heritage the reading or consultation of information, so it is possible to consult *in silico* genetic heritage without performing access activities. However, if the user performs research or technological development activities aiming to build new knowledge or developing new materials, products or devices with *in silico* genetic heritage, the access will take place and the user (company or research institution) will have to comply with the law. In these cases, the register must be done before the application of any intellectual property right, or the commercialization of the intermediate product, or publication of results, partial or final, in scientific or communication media, or the notification of a finished product or reproductive material developed as result of the access (Art. 12). In cases where there is economic exploitation of reproductive material or finished products resulting from access to genetic heritage, *ex-situ* collections can be beneficiaries of benefit-sharing (Decree 8.772/2016, article 51, IX), that is, agreements of benefit sharing can be made between the Union and the institution using the genetic heritage in which the value of the sharing is destined to an *ex-situ* collection in order to promote the conservation and maintenance of this collection.

2.2.5 Exceptions to Legislation

The law does not apply to access to human genetic heritage (Art. 4). Access to genetic heritage and to associated traditional knowledge for practices that are harmful to the environment, to cultural reproduction and to human health, and for the development of biological and chemical weapons is also prohibited (Art. 5).

According to Decree 8.772/2016 and CGEN Technical Guideline,[13] certain activities, when not part of research or technological development, do not constitute access to genetic heritage, such as:

- Affiliation or paternity test, sexing technique and karyotype or DNA analysis and other molecular analyzes aimed at the identification of a species or specimen;
- Clinical diagnostic exams and tests for the direct or indirect identification of etiologic agents or hereditary pathologies in an individual;
- Extraction by means of grinding, pressing or bleeding resulting in fixed oils;
- Purification of fixed oils resulting in a product whose characteristics are identical to those of the original raw material;
- Tests that aim to measure rates of mortality, growth or multiplication of parasites, pathogens, pests and vectors of diseases;
- Comparison and extraction of genetic information available in national and international databases;
- Processing of extracts, physical separation, pasteurization, fermentation, pH evaluation, total acidity, soluble solids, bacterial and yeast counts, molds, fecal coliforms and totals of the genetic heritage samples; and
- Physical, chemical and physicochemical characterization for the determination of food nutritional information and also:
- Technical reports that include inventory, survey or monitoring of genetic heritage for the purpose of environmental licensing, potential assessment for exploration of natural resources or actions for recovery and environmental replanting of degraded areas;
- Identification or confirmation of the taxonomic identification of the genetic heritage to be embedded into the archive of an *ex-situ* collection;
- Physical, chemical, physicochemical or biochemical characterization of extracts, waxes, butters and oils;
- Quality control tests of products derived from access to genetic heritage or associated traditional knowledge, as well as proficiency assays performed in laboratories; and
- Performing tests that use genetic heritage exclusively in the condition of target organisms.

By determining that such activities are not considered as access to genetic heritage, these activities are exempt to be registered in SisGen, mainly activities carried out by the academic sector. In addition, for the business sector, the manufacturing of some raw materials is not considered as access, which means it is not necessary to register in SisGen (for example raw materials for use in the cosmetic and/or food industry, especially fixed oils, essential oils, butters and simple extracts).

[13] CGEN Technical Guideline No. 9/2018.

2.2.6 Shipment Abroad of Genetic Heritage Samples

The shipment abroad of part or all of living or dead organisms of animal, plant, microbial or other organisms that is intended for access to genetic heritage is also subject to the provisions of Law 13.123/15 (Art. 11, § 2°; 12, IV and § 2).

The national institution that intends to ship samples for access activities by an institution located abroad must register this in a transfer register, in which it must indicate the type, quantity and way of packaging of the samples to be sent, and must conclude a material transfer agreement (MTA) with the recipient institution.

The recipient institution, in turn, may access the genetic heritage of the samples and pass the samples to third parties, provided that it agrees to comply with all the requirements of the law prior to the transfer. It is important to note that the minute of the MTA is regulated by a Resolution.[14]

On the other hand, the law also allows the Brazilian institution to send samples of genetic heritage for services abroad as part of research or technological development, in which the Brazilian user bears responsibility for the sample. It is necessary to prove (by a contract, for example) that the foreign company will provide a service (tests or specialized technical activities) (Law 13.123/2015, article 2°, XXX and Decree 8.772/2016 article 24) and mention these services in SisGen access register.

3 Institutional Framework

3.1 The Genetic Heritage Management Council

As already mentioned, the Genetic Heritage Management Council (known as CGEN, in Portuguese) is the competent national authority at the federal level. It is a collegiate body linked to the Ministry of the Environment of deliberative, normative, advisory and appellate council, responsible for coordinating the development and implementation of policies for managing the access to genetic heritage and aTK and benefit-sharing. It is composed by representatives from different entities and bodies of the federal public administration with jurisdiction over the different actions, in a maximum of 60% and by representatives from members of society, in no less than 40%, ensuring parity of business and academic sectors, and indigenous peoples, traditional communities and traditional farmers (Decree 8.772/2016, article 6).

[14]Resolution No. 12/2018 of CGEN. The Ministry of Environment released an English MTA version: https://www.mma.gov.br/images/arquivo/80043/resolucoes/Resolution_12_TTM_english_version_nova.pdf (accessed 21 September 2020).

CGEN[15] is composed of the *Plenary*, which meets periodically, with 21 members; *Thematic and Sectorial Chambers*, which may be created aiming to discuss specific issues related to system users that are subsequently referred to the Plenary for deliberation; and the *Executive Secretariat*, which is responsible, among other topics, for the management of SisGen.

CGEN is responsible for: (i) establishing technical rules; (ii) creating guidelines for compliance and elaboration of the benefit-sharing agreement; (iii) monitoring access activities to genetic heritage and associated traditional knowledge; (iv) deliberating on accreditation of national institutions that keep an *ex-situ* collection; (v) function as the higher instance of appeal to decisions of infringement appeals, etc.

Until October, 2021, CGEN had edited its own internal law, 23 resolutions, 10 technical orientations and 57 deliberations.[16] It also had 26 meetings since 2016.[17]

3.2 SisGen

SisGen is an electronic system implemented, maintained and operated by the CGEN Executive Secretariat and was instituted by Decree 8.772/2016 (Chapter IV). It entered in force in November, 2017.

According to the law, registration must be carried out prior to shipment, or the application of any intellectual property right, or the commercialization of the intermediate product (example: a raw material for cosmetic or pharmaceutical industry, such as a fragrance, flavor), or publication of results, partial or final, in scientific or communication media, or the notification of a finished product or reproductive material developed as a result of the access (Law 13.123/2015, article 12, § 2).

The registry of access or shipment of genetic heritage or aTK is a mandatory declaration document of access or shipment of genetic heritage or aTK activities (Law 13.123/2015, article 2, XII), which must be made in SisGen by the user, in the cases indicated above. A product notification is a declaration document required prior to economic exploitation of a finished product or reproductive material originating from access to genetic heritage or to aTK in which the user declares compliance with the requirements of the law and indicates the modality of benefit-sharing, when applicable, to be established in the benefit-sharing agreement (Law 13.123/2015, article 2, XIX). This notification must also be made directly on SisGen.

---

[15]The jurisdiction and composition of CGEN are described in article 6, paragraph 1 of Law No. 13.123/15 and Chapter II of Decree No. 8.772/2016.

[16]All acts can be found in the Ministry of Environment website: https://antigo.mma.gov.br/patrimonio-genetico/conselho-de-gest%C3%A3o-do-patrim%C3%B4nio-gen%C3%A9tico/nrmas-do-cgen.html#delibera%C3%A7%C3%B5es (accessed 21 October 2021).

[17]The memories of meeting may be consulted in this website: https://www.mma.gov.br/patrimonio-genetico/conselho-de-gest%C3%A3o-do-patrim%C3%B4nio-gen%C3%A9tico/reunioes.html (accessed 21 October 2021).

On the other hand, the law also establishes an authorization for access or shipment (Art. 2, XIV), which is an administrative act which allows, under specific conditions, access to genetic heritage or associated traditional knowledge and shipment of genetic heritage. Some specific activities require prior authorization, such as: (i) access to genetic heritage or associated traditional knowledge in a crucial national security area may only occur after formal consent granted by the National Defense Council and (ii) access to genetic heritage or associated traditional knowledge in national waters, on the continental shelf, and in the exclusive economic zone may only occur after formal consent granted by the maritime authority (Law 13.1232015, article 13). The prior consent and authorization are issued directly by SisGen, after these bodies' approval (Decree 8.772/2016, Chapter IV, Section IV).

3.3 The Operation of SisGen Since Its Availability

Since SisGen was released, companies, researchers and research institutions are able to register their genetic heritage and/or aTK access, shipment or products. Considering data available up to 26 May 2020, SisGen has a large number of register of access to aTK and/or genetic heritage, register of shipment and product notifications, as we can verify on the website[18] provided by the Ministry of the Environment:

- 52,829 access activities to genetic heritage and/or associated traditional knowledge were registered, of which 1622 (3.1%) were from private companies; 35,316 (66.8%) from public and private research institutions and 15,891 (30.1%) from natural persons (researchers). Of these, 83.2% refer to access to genetic heritage, 12.2% of genetic heritage combined with associated traditional knowledge, and 4.4% access to associated traditional knowledge. This demonstrates a great tendency of the academic sector in the development of research on samples of genetic heritage, as the access to the associated traditional knowledge happens in less quantity of researches.
- We also identified the existence of 2700 notifications of products for agriculture activities, reproductive material and finished products for other industries (example: cosmetics, pharmaceutical, etc.) in SisGen. Of these, 4.8% correspond to the agricultural sector (agribusiness inputs and medicines), 87.6% correspond to the cosmetic sector, medicine and others subject to the law and 7.6% correspond to reproductive material. From these products, 15.4% will allocate benefits to monetary benefit-sharing modality (National Fund for Benefit Sharing) and 41.5% will allocate benefits to non-monetary benefit-sharing projects through MAT to be signed by the Union or—in cases of access to associated traditional knowledge—by the provider community itself.

[18]The Ministry of Environment provides the register data on its website: https://sisgen.gov.br/paginas/publicidade.aspx (accessed 21 September 2020).

4 Access Procedures and Agreements for Access to Genetic Heritage or Associated Traditional Knowledge

For access to genetic resources and aTK, it is necessary to fulfill the SisGen forms, following all rules previously mentioned.

Also, in line with the Nagoya Protocol, Law 13.123/15 requires, in specific cases, prior informed consent (PIC) and mutually agreed terms (MAT). In addition, benefit-sharing will be due when there is an economic exploitation of reproductive material—whether used for agriculture—or a finished product—when intended to other economic sectors. We now analyze the content of access and notification register and each instrument on the basis of the applicable legal provisions.

4.1 Content of the Access and Notification Registry

The user of the genetic heritage and/or associated traditional knowledge should make the registration stating (Decree 8.772/2016, article 22):

- activities of technological research or development, including: description of activity and objectives; application sector (if technological development); expected or obtained results; responsible persons in charge, including partner institutions, if any; period of activities; scientific name of the specimen accessed; origin of the samples (acquired under *in-situ*, *ex-situ*, *in silico*, trade or intermediate product—raw material) conditions; information about partner institutions in Brazil or abroad, if any.
- Previous access registration number, which allows a traceability system of activities;
- Prior informed consent, in cases of access to associated traditional knowledge;
- Request for confidentiality;

Statement of adequacy in the event of legal exemption or non-benefit sharing. After registration, the research institutions and companies are able to file their patent applications, publish and disclose results resulting from access, send samples abroad or market intermediate products.

If institutions intend to market reproductive material or finished products, such as medicines, cosmetics, sanitizing products etc., they must also notify their products prior to marketing. For this purpose, they must submit (Decree 8.772/2016, article 34):

- commercial identification of the product and the application sector;
- information on whether the genetic heritage or associated traditional knowledge is one of the key elements of value adding to the product;
- scope of marketing;
- product registration number with the competent bodies;

- expected date for the beginning of marketing;
- indication of the benefit sharing modality;
- access registration number which gave rise to the product;
- submission of the benefit-sharing agreement, when appropriate.

After the conclusion of the register of access and/or shipment and the notification of product or reproductive material, the user may request CGEN to issue an internationally recognized certificate of compliance, that will serve as proof that activities on genetic heritage or aTK have been carried out in accordance with Law and Decree (Decree 8.772/2016, article 6, single paragraph). It is an important mechanism to allow the Nagoya Protocol effectiveness in Brazil.

4.2 *Prior Informed Consent*

Prior informed consent is a legal instrument defined by the law as "formal consent, previously granted by indigenous population or traditional community according to their uses, customs and traditions or community protocols" (Art. 2, item VI). It is required when the user intends to access aTK of identifiable origin, that is, in those situations where it is possible to identify the origin of knowledge and link it to at least one indigenous population, traditional community or traditional farmer. In the contrary, access to genetic heritage does not require the presentation of the PIC.

Evidence of PIC can occur at the discretion of the indigenous population, traditional community or traditional farmer. It can be: a signed prior consent; a registered audiovisual consent; a statement from the official governing body; or adherence to the provisions set forth by community protocol (Law 13.123/2015, article 9).

It is important to highlight that access to aTK from unidentifiable origin does not require PIC, according to the law (Art. 9, para. 2). Also, access to genetic heritage of local traditional variety or landrace, or to locally adapted breed or creole breed for agricultural[19] activities, encompasses access to aTK of unidentifiable origin that originated the variety or breed and does not require prior consent from indigenous people, traditional community, or traditional farmer who creates, develops, holds or preserves the variety or breed (Law 13.123/2015, 9, para. 3).

The provider communities of aTK have the right guaranteed by law to have, among others, their contribution to the development and conservation of genetic heritage recognized in any form of publication or use; to have the origin of access to aTK indicated in publications, uses, holdings and disclosures; to receive benefits from economic exploitation by third parties of aTK (Law 13.123/2015, article 10). These rights should be included in the PIC. The Law and its Decree do not specify additional clauses.

[19] Agricultural activities include: activities of producing, processing and commercializing food, beverages, fibers, energy and planted forests (article 2° XXIV, Law 13.123/2015).

4.3 Mutually Agreed Terms

As mentioned, registration must be carried out prior the notification of a finished product or reproductive material developed as a result of access. The notification is one of the requirements for economic exploitation of a finished product or reproductive material resulting from access to genetic heritage or to aTK, The second one is the presentation of the benefit-sharing agreement in 365 days from the moment of notification, except in cases involving access to the aTK of identifiable origin (Law 13.123/2015 article 16).

The parties to the agreement shall be, according to the law (Art. 25):

- In case of economic exploitation of finished products or reproductive material originating from access to genetic heritage and associated traditional knowledge of unidentifiable origin: the federal government, represented by the Ministry of the Environment, and those who economically explore finished products or reproductive material originating from access.
- In case of economic exploitation of finished products or reproductive material arising from access to traditional knowledge of identifiable origin: the provider of the associated traditional knowledge and those who economically explore finished products or reproductive material originating from access.

It is important to note that the law focuses on benefits arising from economic exploitation of finished products or reproductive material arising from access. Other non-commercial utilizations of genetic heritage or aTK, for example for a scientific research, are not subject to the benefit-sharing modalities (as detailed in the next item), although there are other benefits, such as advancing science and increasing knowledge about national genetic heritage.

Essential clauses of the benefit-sharing agreement, without prejudice to others that may be established in a regulation, are those that provide for: products subject to economic exploitation; term of duration; benefit-sharing modality; rights and responsibilities of the parties; intellectual property right; termination; penalties; and forum in Brazil, according to article 26 of the law.

The Ministry of Environment edited the Ordinance 144/2021 on 23 April 2021, which provides for the form of benefit-sharing in the non-monetary modality in cases of access to genetic heritage. It also establishes the procedures to be adopted for proposing, analyzing and signing the non-monetary benefit-sharing agreement by the Union. This Ordinance will be applied in cases of exploitation of finished product or reproductive material arising from access to genetic heritage when the user chooses the non-monetary benefit-sharing modality.

The Ordinance also proposes a model for the benefit-sharing agreement, establishing the main clauses, such as: conditions for the benefit-sharing; list of finished products or reproductive materials that are related in the agreement; duration of benefit-sharing; responsibilities of parties; etc. It is necessary to fill a form with the main information of the project that the user intends to perform with the benefit-sharing amounts. This form must indicate: actions to be performed; deadlines, budget, place and community that will be benefit with the project; expected results, etc. The proposal must be evaluated and approved by the Ministry of Environment.

4.4 Benefit-Sharing of Products Developed Under the Law

4.4.1 Benefit-Sharing: Application and Exceptions

Pursuant to article 17 of Law 13.123/2015, the benefits resulting from economic exploitation of finished product or reproductive material arising from access to genetic heritage of species found in *in-situ* conditions or to aTK, even if produced outside the country, will be shared in a fair and equitable way by the manufacturer. In the case of a finished product, the genetic heritage or aTK component must be one of the key elements of value adding to the product.

To have the benefit-sharing obligation, after developing the finished product (e.g. personal hygiene or perfumery products, drugs) or a reproductive material and before commercializing it, the manufacturer must identify if the genetic heritage or the aTK is one of the key elements of value adding to the product, which means elements whose presence in the finished product is determinant to its functional characteristics or to its marketing appeal (Law 13.123/2015, articles 2 XVIII and 17).

To determine whether the inclusion of a genetic heritage element is a key element of the finished product its contribution to the "functional characteristics" and/or "marketing appeal" should be analyzed. The Decree (Art. 43, para. 3) establishes that the functional characteristics are the characteristics that determine the main purposes, improve the action of the product or expand its list of purposes. In the same sense, marketing appeal happens when there are references to genetic heritage or aTK, its origin or the differentials arising therefrom, related to a product, product line or brand, in any visual or auditory media, including marketing campaigns or prominence on the product label.

For example, if the genetic heritage is highlighted, written in full or if there are images referring it as a claim or part of the name of the product: 'chestnut shampoo' or 'activator of curls - contains *cupuaçu* butter', the marketing appeal will be presented.

On the other hand, the use of genetic heritage, exclusively as excipients, vehicles or other inert substances, which do not determine functionality, will not be considered determinant for the existence of functional characteristics (Decree 8.772/2016, article 43, para. 4). When it comes to personal hygiene, perfumery, and cosmetics product, the genetic heritage will not be considered as key elements of value adding to the product when used exclusively for the structuring of the formula, being responsible for stability, consistency or physical appearance.[20]

[20]CGEN Technical Guideline 2/2017.

There are other exceptions to the benefit-sharing obligations:

- the manufacturers of intermediate products and developers of processes originating from the access to genetic heritage or associated traditional knowledge along the production chain (Law 13.123/2015 article 17, para 2);
- micro-businesses, small businesses, micro individual entrepreneurs, and—traditional farmers and their cooperatives with annual gross revenue equal to or lower than the upper limit established in Complementary Law No 123, dated December 14, 2006 (Law 13.123/2015, article 17, § 5°);
- the substance originating from the metabolism of microorganism will not be considered determinant for the existence of the functional characteristics when it is identical to the substance of fossil origin already existing and used (Decree 8.772/2016, article 43, § 5°). So, the use of this substance is not subject to benefit-sharing obligations.

4.4.2 Benefit-Sharing Modalities

The 2015 biodiversity statute differs from PM insofar as it establishes clear and standardized rules for benefit-sharing. While the PM provided that benefit-sharing should occur by means of fair and equitable negotiation between the provider and the user, article 17 of the biodiversity statute prescribes that benefit-sharing may be monetary or non-monetary. Under article 20, monetary benefit-sharing should represent 1% of the annual net revenue obtained from economic exploitation of finished products or reproductive material arising from access to genetic heritage, except when reduced for up to 0.1% by sectoral agreement.[21]

The monetary benefits must be deposited in the National Fund for Benefit-Sharing (FNRB), which is a Fund under the Ministry of Environment with the purpose of strengthening the genetic heritage and the associated traditional knowledge and promoting their sustainable uses (Law 13.123/2015, Chapter VII). The FNRB shall be managed by a committee composed of representatives of both the federal government and traditional communities. The Fund entered in force in February 2020 and it is maintained by the National Bank for Economic and Social Development (BNDES, in Portuguese). Users can deposit the amount of monetary benefits to be shared in the account since then.

Regarding aTK, if its origin is identifiable, the provider is entitled to receive benefits according to MAT. Thus, the parties (user and provider, such as traditional

[21] Law 13.123/2015, article 21 establishes: "In order to foster competitiveness of the beneficiary sector, the Federal Government may, at the request of the interested party and according to regulations, celebrate sectoral agreements to reduce the amount of benefit-sharing for up to 0.1% of the annual net revenue obtained from economic exploitation of finished products or reproductive material arising from access to genetic heritage or associated traditional knowledge from an unidentifiable origin". The sectoral agreements rules are detailed in the Decree, Chapter V, Section VI.

community, traditional farmer, indigenous population) shall negotiate the benefit-sharing in a fair and equitable way, meeting clarity, loyalty and transparency parameters. It is also necessary to allocate half of the benefits determined for monetary modality to the National Fund in order to ensure the remuneration of the other communities having the same traditional knowledge (article 24). On the other hand, when the finished product or reproductive material results from access to aTK of unidentifiable origin the benefits should be deposited into the monetary benefit-sharing fund (Law 13.123/2015, article 23).

In case of access to genetic heritage, the user may choose, at his own discretion, one of the modalities of benefit-sharing. If the user chooses the non-monetary modality, the benefit should be equivalent to 75% of the monetary benefit-sharing amount (Art. 22), which means the user will pay 0.75% of the net revenue resulting from product commercialization. This modality includes, *inter alia*:

- projects for conservation or sustainable use of biodiversity, or for protection and maintenance of knowledge, innovations, or practices of indigenous peoples, traditional communities or traditional farmers, preferable at the site where the species occurs in *in-situ* conditions or where the sample was obtained, when the original location cannot be specified;
- capacity building of human resources in topics related to the conservation and sustainable use of genetic heritage or associated traditional knowledge; and
- distribution of products free of charge in social programs (Law 13.123/2015, article 19,II), among others.

In addition, under Federal Decree No. 8,772/2016 (Art. 51), the amount of benefit-sharing should be allocated to the following areas: conservation units; indigenous lands; remaining territories of quilombos; family farmers settlements; traditional territories under Federal Decree No. 6,040/2007; national public research and development institutions; priority areas for conservation, sustainable use and sharing of benefits of Brazilian biodiversity, as defined by the Ministry of Environment; activities related to the safeguarding of associated traditional knowledge; *ex-situ* collections maintained by accredited institutions; indigenous populations, traditional communities and traditional farmers.

Therefore, non-monetary benefit-sharing projects may be carried out preferably at the site where the species occurs in *in-situ* conditions or where the sample is obtained (Law 13.123/2015, article 19, II, 'a'). On the other hand, in cases of access to aTK, the project can be freely negotiated with the providing community. From the above follows that the biodiversity statute allows companies to invest benefit-sharing resources in their own supply chain. The investment made can be disclosed by the company in marketing actions. In addition, the benefit-sharing project avoids payments to the National Fund. Moreover, there are no provisions as to how to allocate such resources until now.

It is important to note that the monetary benefit-sharing referred may, at the user's discretion, be deposited directly in the FNRB without the need of a benefit-sharing agreement (Law 13.123/2015, article 25, para. 4).

5 Regime's Strengths and Weaknesses

The Biodiversity Law brought a new form of use and registration of access to genetic heritage, simplifying procedures for users who conduct research and technological developments. In this sense, by allowing registration to be made before the beginning of commercialization of products or shipping samples abroad or requests for protection of intellectual property or the publication of research results, the law gave users more flexibility.

On the other hand, the intermediate products' industry faces some difficulties to understand the nature and responsibilities regarding the development of such products, considering there are some exceptions to access to genetic heritage register obligations (as demonstrated in item 2.2.5). As some activities are exempt from access register, the criteria of contributions of genetic heritage access can vary according to the interpretation of the law and the company may face questioning by the national authority.

Considering that some *ex-situ* collections (mainly outside Brazil) were constituted before Law 13.123/2015 entered in force, access to genetic heritage maintained in these collections may cause questions concerning the compliance scope by the user countries. These could happen because the rules for traceability of the place of collection were not in force and there was no obligation to indicate the place of origin of samples. Some CGEN's resolutions were approved to facilitate the register of samples or allow the user to ship samples, even when the user does not have all information required by SisGen. For example, Resolution 13/2018 establishes an alternative form to register samples obtained *in silico* conditions, so the user will have to inform the URL where the sample is deposited, and the database must be open to access by the Brazilian State.[22]

In addition, users face some difficulties during operations in SisGen, as some situations cannot so far be performed. As an example, foreign institutions are unable to notify products arising from access to genetic heritage and/or aTK because it is necessary to present the registration number of the company in the Brazilian Federal Reserve, which not all foreign companies have. Moreover, the method of benefit-sharing calculation is based on the concept of net revenue in accordance with Brazilian tax legislation, which may not be applied in situations of international manufacturers. Despite that, the Ministry of Environment Ordinance 143/2020, published on 31 March 2020, establishes that to the declaration of net revenue that requires currency conversion, the official exchange rate will be used according to the closing of the last day of the year, applying the PTAX rate issued by Central Bank of Brazil.

Finally, despite the publication of Ordinance 144/2021 establishing the guidelines for approval of the benefit-sharing agreement, the Ministry of the Environment has so far not approved any request submitted by users.

[22]CGEN Resolution 13/2018.

6 Conclusion

Brazil is going to be a Party to the Nagoya Protocol after the Presidential Sanction, although it has adopted federal standards since 2000 in order to promote the protection of genetic heritage, associated traditional knowledge and benefit-sharing.

Also, the law determines that the activities related to the genetic heritage or aTK described in international agreements approved by the national congress and enacted by the federal senate, when performed under such international agreements, must comply with the conditions set therein (Art. 46).

In Brazil, only a small part of the users of Brazilian biodiversity complied with PM because administrative procedures were time-consuming and required several previous and costly steps, such as obtaining prior consent for all types of access, negotiation of a benefit-sharing contract with the provider (i.e. owner of the material or the area that would provide it), and access authorization granted by CGEN.

The law aimed to simplify administrative procedures by establishing an electronic registration system, with the duty of previous registration being imposed only in relation to shipping of genetic heritage samples, development of intermediate products (raw materials for industry), commercialization of finished products or reproductive material or application for intellectual property. This has been followed by a remarkable increase in registrations of independent researchers and research institutions, since these account for almost 97% of the registrations made until May 2020.

Businesses, which have the greatest potential to share the benefits resulting from access to Brazilian biodiversity, still accounts for the lowest percentage of registers of access to genetic heritage/aTK to the system. Nevertheless, about 57% of the products notified with SisGen up to May 2020 are subject to benefit-sharing payments, which shall allow for the maintenance of aTK and the conservation and use of the Brazilian biodiversity.

So far, the Ministry of the Environment has not made available data on the amounts already deposited in the FNRB account, which was created in February 2020. The fund's management committee is expected to create the rules for the use of that resource by the beneficiaries described by Decree 8.772/2016 (Art. 51).

Also, it is important to highlight that in cases of access to genetic heritage maintained in *ex-situ* conditions, even *in silico* conditions, it may not be possible to identify the presence of Brazilian genetic heritage, especially when the sample or information has been stored in a public database or when the place of origin is not indicated. These situations will challenge the register of access and, eventually, the benefit-sharing arising from commercialization of products resulting from these types of access.

Current legislation has included the requirement for foreign institutions to complete a benefit-sharing agreement for products even if produced abroad. In this sense, we can see an intention of the regulation to internationalize its rules. The Ministry of the Environment is also developing a new version of SisGen for access by foreign institutions, aiming at the notification of foreign products and enhancing benefit-sharing from institutions outside Brazil.

Efforts still need to be made to enhance and improve the Biodiversity Law, since several points are still unclear and need to be defined by the government, among which we can highlight: improvement of SisGen to receive notifications of products by foreign companies; improvement of SisGen so that it is possible to indicate samples coming from abroad in the access register; establishing rules for the use of amounts already deposited in the fund by the beneficiaries and providers of TK. With these advances, compliance with legislation by more users can be ensured, including foreigners, making it possible to increase benefit-sharing in areas of conservation and sustainable use of Brazilian biodiversity.

Acknowledgments Henry Philippe Ibañez de Novion—Environmental Analyst at the Ministry of Environment (DPG/SBio). He was Director of the Department of Genetic Resource, Secretariat of Biodiversity of the Ministry of Environment. Vice-Chairman of the Genetic Resource Management Council—CGEN and General Coordinator of Department of Genetic Resource's Regulatory Affairs and Benefit Sharing.

Maira Smith—General Coordinator of Normative Acts and Decision-Making Processes of the Department of Support to the Genetic Resource Management Council (DCGen) of the Ministry of Environment.

References

Legal Texts

Constitution of the Federative Republic of Brazil 1988

Decree 2.519/ 1998, promulgates de Convention on Biological Diversity

Federal Decree 8.772 2015

Federal Law 13.123 2015

Legislative Decree n° 02/1994., https://www2.camara.leg.br/legin/fed/decleg/1994/decretolegislativo-2-3-fevereiro-1994-358280-publicacaooriginal-1-pl.html (accessed 3 November 2020)

Ordinance n° 01/2017., http://www.in.gov.br/materia/-/asset_publisher/Kujrw0TZC2Mb/content/id/19351478/do1-2017-10-13-portaria-n-1-de-3-de-outubro-de-2017-19351409 (accessed 21 September 2020)

Ordinance n° 144/2021., https://www.in.gov.br/en/web/dou/-/portaria-mma-n-144-de-22-de-abril-de-2021-315696906 (accessed 21 October 2021)

Provisional Measure 2.186-16/2001

Resolution CGEN 12/2018, approves the Material Transfer Agreement Model, https://www.in.gov.br/materia/-/asset_publisher/Kujrw0TZC2Mb/content/id/45993008/do1-2018-10-19-resolucao-cgen-n-12-de-18-de-setembro-de-2018-45992843#:~:text=1%C2%BA%20Aprovar%20o%20modelo%20de,do%20Anexo%20I%20desta%20Resolu%C3%A7%C3%A3o.&text=Cl%C3%A1usulas%20adicionais%2C%20de%20interesse%20espec%C3%ADfico,Resolu%C3%A7%C3%A3o%20ou%20na%20legisla%C3%A7%C3%A3o%20pertinente (accessed 21 October 2021)
Resolution CGEN 13/2018, establishes an alternative way of registering in the SisGen the samples that have been obtained in silico, https://www.mma.gov.br/images/arquivo/80043/resolucoes/res13-cgen.pdfhttps://www.in.gov.br/materia/-/asset_publisher/Kujrw0TZC2Mb/content/id/45993017/do1-2018-10-19-resolucao-cgen-n-13-de-18-de-setembro-de-2018-45992861 (accessed 21 October 2021)
Technical Guideline n° 02/2017., https://www.gov.br/mma/pt-br/assuntos/biodiversidade/patrimonio-genetico/conselho-de-gestao-do-patrimonio-genetico-cgen-1/agenda/reunioes-anteriores (accessed 21 Ocotber 2021)
Technical Guideline No. 09/2018., https://mma.gov.br/images/arquivo/80043/orientacao-tecnica/ot9-cgen.pdfhttps://www.gov.br/mma/pt-br/assuntos/biodiversidade/patrimonio-genetico/conselho-de-gestao-do-patrimonio-genetico-cgen-1/normas-do-cgen/pdf/OrientaoTcnican9.pdf (accessed 21 Ocotber 2021)

Links

Convention on Biological Diversity (2020) Brazil Main Details, https://www.cbd.int/countries/profile/?country=br#facts (accessed 3 November 2020)
Flora do Brasil 2020 under construction (2020) Jardim Botânico do Rio de Janeiro, https://www.in.gov.br/materia/-/asset_publisher/Kujrw0TZC2Mb/content/id/45993008/do1-2018-10-19-resolucao-cgen-n-12-de-18-de-setembro-de-2018-45992843#:~:text=1%C2%BA%20Aprovar%20o%20modelo%20de,do%20Anexo%20I%20desta%20Resolu%C3%A7%C3%A3o.&text=Cl%C3%A1usulas%20adicionais%2C%20de%20interesse%20espec%C3%ADfico,Resolu%C3%A7%C3%A3o%20ou%20na%20legisla%C3%A7%C3%A3o%20pertinente (accessed 3 November 2020)
Ministry of Environment (2014) CGEN Activities Report 2014, https://www.mma.gov.br/images/arquivo/80043/relatorio-cgen-2014.pdf (accessed 21 September 2020)
Ministry of Environment (2015) CGEN Activities Report 2015, https://www.mma.gov.br/images/arquivo/80043/documentos/relatorio%202015%20cgen.pdf (accessed 21 September 2020)
Ministry of Environment – Biodiversity (2020) https://www.gov.br/mma/pt-br/assuntos/biodiversidade/patrimonio-genetico/conselho-de-gestao-do-patrimonio-genetico-cgen-1/agenda/reunioes-anteriores (accessed 03 November 2020)
Ministry of Environment (2021) CGEN legislation, https://www.gov.br/mma/pt-br/assuntos/biodiversidade/patrimonio-genetico/conselho-de-gestao-do-patrimonio-genetico-cgen-1/normas-do-cgen (accessed 21 October 2021)
SisGen (2020), https://sisgen.gov.br/paginas/pubpesqatividade.aspx (accessed 21 September 2020)
http://floradobrasil.jbrj.gov.br/reflora/listaBrasil/ConsultaPublicaUC/ConsultaPublicaUC.do#CondicaoTaxonCP (accessed 3 November 2020)

Lilian Massini Mozini has a Bachelor's degree in Law and a Master's in Public Administration and Management from the University of Aveiro (Portugal), specializing in sustainable development economy. She has been acting as a consultant for national and foreign companies in environmental law, development of strategies for biodiversity access, innovation, products development, Prior Informed Consent and Mutually Agreed Terms (MAT). She negotiated and approved more than 30 prior consent and MAT for access of genetic resources and associated traditional knowledge with traditional communities in Brazil since 2014. She is currently focused on enabling the approval of benefit-sharing projects by the Ministry of the Environment, developing strategies on genetic heritage access and ensuring compliance with the national law for companies. lmozini@ambiente-global.com.

Towards Mutual Supportiveness Between the Nagoya Protocol and the Andean ABS Regime: The Cases of Ecuador and Peru

Maria Victoria Cabrera Ormaza

Abstract Ecuador and Peru are parties to the Nagoya Protocol to the Convention on Biological Diversity. Before, in 1996, they became part of a sub-regional regime on access to genetic resources within the Community of Andean Nations. Article 4 of the Nagoya Protocol prevents State parties to pursue specialized access and benefit-sharing (ABS) instruments, which run counter to the objectives of the Protocol. Conceived as a binding instrument to prevent misappropriation of genetic resources, the Andean regime has proven complex and over-regulatory, failing to promote research and development on genetic resources in the region. At the domestic level, Ecuador and Peru have been confronted by a number of pre-established rules, which appear to give the State a disproportionate bargaining power *vis-à-vis* the user, rendering the implementation of the Protocol's balanced approach a difficult task. This chapter identifies areas where Decision 391 and the Nagoya Protocol can be mutually supportive and provides examples of how the existing ABS legal framework in Ecuador and Peru can help reconcile possible areas of conflict.

Keywords Community of Andean Nations · Decision 391 · Nagoya Protocol · Genetic resources · Traditional knowledge

---

Views expressed by the author do not reflect the views of the International Labour Organization.

M. V. C. Ormaza (✉)
Universidad Espíritu Santo-Ecuador, International Labour Organization, Samborondón, Ecuador
e-mail: mvcabrer@uees.edu.ec

E. Chege Kamau (ed.), *Global Transformations in the Use of Biodiversity for Research and Development*, Ius Gentium: Comparative Perspectives on Law and Justice 95, https://doi.org/10.1007/978-3-030-88711-7_4

1 Introduction

Ecuador and Peru are two megadiverse countries located in a strategic geographical area comprising the Pacific coast, the Andean mountain range and the Amazonian rainforest.[1] Both countries are bound by the Common Sub-regional Regime on Access to Genetic Resources that operates within the framework of the Community of Andean Nations. This regime is contained in the Decision 391 adopted in 1996 by the Commission of the Cartagena Agreement,[2] an organ of the Community of Andean Nations (CAN). Implementation of this regime, on which the domestic access and benefit-sharing (ABS) legislation of Ecuador and Peru is based, appears to be problematic. Over the years, Decision 391 has been regarded by some as a complex regulatory framework that hinders promotion of research on genetic resources (GR) in the region.[3] Scientists, for example, have expressed their concern about cumbersome procedures for access to GR for non-commercial research, arbitrary fees for research permits, and high expectations of providers for benefits.[4]

Both, Ecuador and Peru, have ratified the Nagoya Protocol on Access and Genetic Resources and the Fair and Equitable Sharing of Benefits Arising from their Utilization to the Convention on Biological Diversity (NP).[5] The Protocol constitutes a multilateral framework aimed at achieving fair and equitable use of the genetic components of the biodiversity and the associated traditional knowledge (aTK), in line with article 15 of the Convention on Biological Diversity. It sets out three requirements, which together make the basis for a multilateral ABS regime:

1. access to GR for their utilization shall be subject to the prior informed consent (PIC) of the State party providing such resources (Art. 6 para. 1);
2. benefits arising from the utilization of GR shall be shared in a fair and equitable manner with the provider country according to mutually agreed terms (Art. 5 para. 1); and
3. access to the traditional knowledge associated with GR is subject to the PIC of the indigenous and local communities that are holders of such knowledge, who shall also participate in a fair and equitable manner in the benefits arising from the utilisation of their knowledge, based on mutually agreed terms (Art. 5 para. 2 and 7).

[1] See CBD Country profiles.

[2] The CAN is a sub-regional intergovernmental organization created to promote economic development and cooperation among Andean countries. See the Agreement for the Integration of the Andean Sub-region (Cartagena Agreement) of 1969, Art. 1.

[3] Ruiz (2008a, b), p. 140; Ribadeneira (2017), p. 195; Silvestri (2016), p. 77; Gómez-lee (2012), p. 59.

[4] Beck and Rodríguez (2017).

[5] The Protocol was adopted on 29 October 2010 after 6 years of negotiation. It entered into force on 12 October 2014, after achieving 50 ratifications. Peru ratified the Protocol in 2014 whereas Ecuador did so in 2017.

Additionally, article 8 (a) of the Protocol calls on State parties to create conditions to promote and encourage research by introducing "simplified measures on access for non-commercial research purposes, taking into account the need to address a change of intent for such research".

Decision 391 was a pioneer in ABS regulation.[6] In fact, it addresses many of the issues which are now reflected in the NP. For instance, it recognizes the role of local communities as holders of traditional knowledge associated with GR, provides for a special procedure for non-commercial research, sets forth an access procedure and defines the core elements of an access agreement. Pursuant to article 4 (2) of the NP State parties can continue implementing other relevant ABS agreements to the extent they are supportive and do not run counter to the objectives of the Protocol. Legally speaking, mutual supportiveness requires that international rules are understood and applied as reinforcing each other with a view to fostering harmonization and complementarity.[7]

The purpose of this chapter is to explore the content of Decision 391 and the domestic ABS legislation in Ecuador and Peru in light of the requirements of the NP. In doing so, it tries to identify areas where Decision 391 and the Protocol could be mutually supportive. It also examines how the ABS legislation of such countries have dealt with potential areas of conflict.

2 'Mutual Supportiveness'

In 2018, the Conference of the Parties to the CBD serving as the Meeting to the Parties to the NP took note of a study into criteria to identify a specialized international access and benefit sharing instrument, and a possible process for its recognition. The study was prepared jointly by Elisa Morgera, Dr. Stephanie Switzer and Dr Elsa Tsioumani at the request of the Executive Secretariat of the CBD. According to that study, 'mutual supportiveness' has both an interpretation and law-making dimension[8] and requires specialized ABS instrument to meet the following criteria: "(1) consistency with biodiversity conservation and sustainable use objectives; (2) fairness and equity in the sharing of benefits; (3) legal certainty with respect to access to genetic resources or traditional knowledge and to benefit-sharing; (4) contribution to sustainable development, as reflected in internationally agreed goals; and (5) other general principles of law including good faith, effectiveness and legitimate expectations".[9]

Based on the idea of 'mutual supportiveness', article 4 (2) of the Protocol could be read as a call for State parties to engage in a harmonized understanding of the

[6]Ribadeneira (2017), p. 195.
[7]Pavoni (2010), p. 650.
[8]UN Doc. CBD/SBI/2/INF/17 (2018), para. 18.
[9]Ibid, para. 5.

Protocol and specialized ABS instruments, in order to ensure 'fruitful interactions'.[10] Arguably, it is for domestic law and, in particular, national legal operators to ensure such harmonious relationship.[11] In certain cases, States have taken 'innovative approaches' to try to reconcile various international legal obligations.[12]

3 The Andean ABS Regime: Decision 391

3.1 Objectives and Principles

The Andean Common Regime for Access to Genetic Resources was established in 1996 by the Decision 391 of the Commission of the Cartagena Agreement. It operates within the Andean Community of Nations (CAN) which is currently composed of four countries: Ecuador, Colombia, Bolivia and Peru.[13] Decisions of the Commission of the Cartagena Agreement are part of the so-called "Andean law", which is directly applicable on CAN member States.[14]

In line with the NP, Decision 391 aims to contribute to the conservation and sustainable use of biodiversity in line with the CBD. It recognizes the strategic importance of GR and attributes them a great economic value.[15] Indeed, an important element that marked the negotiations of the Decision was the expectation of Andean countries to obtain speedy economic retributions for the utilization of GR. This contrasted sharply with their intention to regulate to the extent possible access and eventual exploitation of GR.[16] In addition, Decision 391 aims at strengthening integration and scientific cooperation among CAN member States.[17] However, the fact that all Andean countries are providers of similar GR could put them in a situation of competition for research initiatives from third countries.

The principle of State's sovereignty over GR and their derivatives is anchored in the preamble as well as in article 5 of Decision 391.[18] Unlike the NP, which remains silent as regards ownership over GR, Decision 391 consecrates the State's ownership over the GR originating in its territory, including derivatives of such GR, along with the right to regulate their utilization.[19] Arguably, the inclusion of derivatives under the coverage of the sovereignty principle, may exclude the possibility for a user to

---

[10] Ibid, para. 39.

[11] A/CN.4/L.682 13 (April 2006), para. 282.

[12] Megiddo (2018).

[13] Venezuela was bound by the regime until 2006.

[14] Treaty creating the Tribunal of the Andean Community, Art. 3.

[15] Decision 391, preamble.

[16] Caillaux et al. (1999), pp. 10, 11.

[17] Decision 391, preamble.

[18] Ibid, Art. 5.

[19] Ibid, Art. 6.

claim ownership over a product that has been developed through the utilization of GR originating from an Andean country.

3.2 Scope

Decision 391 applies to GR with respect to which a CAN member State is the country of origin; as well as to GR of migratory species, which, due to natural reasons can be found in the territory of such State.[20] It also covers the derivatives of such GR and the associated intangible component.[21] Human genetic resources and their derivatives are explicitly excluded from the coverage of the Decision.[22] The same holds true for the exchange of GR, their derivatives of the associated intangible component when it takes place within indigenous, afro-descendant and local communities within the CAN for their own consumption, based on customary practices.[23]

Given its inter-governmental character and the subject matter of regulation, the Decision could qualify as a specialized ABS instrument within the meaning of article 4 of the Protocol.

3.3 Relevant Definitions

Decision 391 defines 'access' as "the obtaining and utilization of genetic resources, whether located in *ex-situ* or *in-situ* conditions, its derivatives, or its intangible component" for purposes which include, but are not limited to, "research, biological prospection, conservation, industrial application and commercial exploitation".[24] As noted by Manuel Ruiz, this is an open-ended definition that could include any form of utilization of GR.[25]

Although the NP does not define 'access', it provides a definition of 'utilization of genetic resources' as "research and development on the genetic and/or biochemical composition of genetic resources". Should 'utilization' be understood as part of 'access' under the Protocol, Decision 391's understanding of 'access' would be broader than that of the Protocol. This could result into different legal consequences, particularly in relation to the obligation to share benefits.

---

[20]Ibid, Art. 3.

[21]Ibid.

[22]Ibid, Art. 4.

[23]Ibid.

[24]Ibid, Art. 1, para. 1.

[25]Ruiz (2008a, b), p. 27.

Decision 391 defines GR as “every material of biological nature with genetic information of real or potential value”.[26] The term ‘derivatives of GR’ is defined as “a molecule, combination or mixture of natural molecules, including crude extracts of living or dead organisms of biological origin, which stem from the metabolism of living beings”.[27]

An ‘intangible component’ is, according to Decision 391, “all know-how, innovation or individual or collective practice, with a real or potential value, that is associated with the genetic resource, its by-products or the biological resource that contains them, whether or not protected by intellectual property regimes”.[28] It has been argued that genetic data—also referred to as Digital Sequence Information—may be covered by the definition of intangible component, thus falling within the scope of the Andean ABS regime.[29] However, a contextual reading of Decision 391 and its drafting history[30] suggests that access and utilization of genetic data was not conceived by Andean countries as a process that could be separated from the access to the genetic material.

3.4 Access Procedure

In line with the NP, Decision 391 establishes an access procedure that requires PIC of the country of origin in conjunction with negotiation of mutually agreed terms (MAT) between the provider country and the user. Like article 6 of the NP, Decision 391 calls on CAN member States to ensure that procedures and acts of government authorities relating to access be clear, efficient and in accordance with the rule of law.

The access procedure set forth in Decision 391 is clear, though complex. It starts with the submission of a request for access to the competent national authority (CNA) of the country of origin.[31] Every potential user of GR is required to involve a national support institution in his or her research activity on GR, and to provide the name of such institution in the access request.[32] The request is subject to technical and legal evaluations as defined by the respective domestic legislation of the country of origin.[33]

[26] Decision 391, Art. 1.

[27] Ibid.

[28] Ibid.

[29] Deplazes-Zemp (2018), p. 89; CBD/DSI/AHTEG/2020/1/5 29 (29 January 2020), p. 16.

[30] Caillaux et al. (1999).

[31] Decision 391, Art. 26.

[32] Ibid, Arts 26 and 41.

[33] Ibid, Art. 29.

Once the access request is approved by the CNA, the applicant shall negotiate the terms of an access agreement with the provider country.[34] This contract shall define the terms of access and the rules of distribution of benefits.[35] During the whole access procedure, the applicant is required to engage into negotiation and conclusion of a number of related contracts, including with the national support institution, the owner of the biological resources that contains the genetic material, the *ex-situ* conservation center and the local community holder of the associated intangible component.[36] This requirement could be seen as problematic from the point of view of efficiency.

3.5 Access to GR for Non-commercial Research

In line with article 8 of the NP, article 36 of Decision 391 provides for a facilitated access procedure for non-commercial research. Under this provision, the CNA of the country of origin can conclude 'framework agreements' (*contratos marco*) with universities, research centres or individual researchers, subject to national legislation. On this basis, Ecuador and Peru have been granting permissions for foreign researchers to carry out research activities on GR with non-commercial intention, in collaboration with local universities.

3.6 Access to Traditional Knowledge

Decision 391 recognizes the right of indigenous, Afro-descendant and local communities to decide on the use of their traditional knowledge and practices associated with GR,[37] which, as explained before, fall within the concept of 'intangible component'. This recognition was critical during the negotiations of Decision 391[38] and represents one of the areas where the Protocol and the Andean regime are clearly mutually reinforcing. Decision 391 does not establish a procedure to obtain the communities' consent. Arguably, this opens the door for the application of article 12 of the Protocol, which requires State parties to take into consideration indigenous and local communities' customary laws, community protocols and procedures, as applicable, with respect to traditional knowledge associated with genetic resources.

---

[34] Ibid, Art. 32 to 47.

[35] Ibid, Art. 33 and 35.

[36] Ibid, Art. 41.

[37] Ibid, Art. 7.

[38] Caillaux (1999), p. 7.

3.7 Monitoring

Decision 391 fosters sub-regional cooperation in monitoring the use of GR.[39] This was supposed to be operationalized through the Andean Committee on Genetic Resources, whose functioning has been declining over the years.[40] The Committee was abolished in 2014 and its functions were absorbed by the Ad-Hoc Andean Committee on Intellectual Property.[41]

In relation to non-compliance with regulatory requirements, Decision 391 states that, subject to domestic law, CNA can impose sanctions on users such as fines, confiscation, closure of premises and prohibition to subsequently access GR.[42]

3.8 Situation Post-Nagoya

Following the adoption of the NP, negotiations with a view to revise Decision 391 have taken place since 2012.[43] This process has represented an opportunity for CAN member States to address their concerns about the applicability of the Andean ABS regime and the necessity to adapt their provisions to the current realities of scientific work.[44] Bolivia has suggested that the emphasis of the Decision should shift from regulation to facilitating access.[45] Together with Peru, it has proposed to redefine access as "obtaining and/or utilization" instead of 'obtaining and utilization', but such proposal has not been supported by other member States.[46] Negotiations have also addressed the issues of 'synthetized products' and 'genetic information'. In this respect, all CAN countries have supported the inclusion of such elements within the scope of Decision 391.[47] Likewise, there is consensus on the need to establish a regional inventory of GR.[48] Despite this, governments have disagreed on the way forward. Whereas Bolivia and Ecuador have proposed to involve experts of international organizations in the process of revision, Colombia

[39]Decision 391, Art. 10 (2).

[40]Before its abolition, in 2012, CAN had announced the reactivation of the Committee.

[41]Andean Council of Ministers, Decision 797 of 2014.

[42]Decision 391, Art. 47.

[43]Acts of the Third Meeting of the Ad-Hoc Andean Committee on Intellectual Property of the Andean Community of Nations, 9 September 2016, p. 2. At the time of writing this chapter, Colombia is the only CAN member State that has not ratified the Protocol.

[44]Acts of the Third Meeting of the Ad-Hoc Andean Committee on Intellectual Property of the Andean Community of Nations, 9 September 2016, p. 3.

[45]Ibid, p. 7.

[46]Ibid, p. 6.

[47]Ibid, p. 12.

[48]Ibid, p. 16.

and Peru have considered that this should be dealt by CAN State members alone.[49] Since 2016, negotiations have not been resumed.

In 2017, the Andean Parliament, a directly-elected advisory body of the CAN Council of Ministers approved a ‘normative framework’ to safeguard GR and aTK, and improve application of Decision 391. Decisions of the Andean Parliament are, however, not binding on CAN member States. The 2017 framework reflects, to a larger extent, the spirit of Decision 391, but it has some relevant innovations. It excludes from the scope of Decision 391 GR that are located in ‘strategic areas’, a wording that was left undefined.[50] It also establishes that any consultation process with the indigenous communities’ holders of traditional knowledge should be in line with the principles contained in ILO Convention on Indigenous and Tribal Peoples (No. 169). More interestingly, it proclaims the right of the scientific community “to have clear, easy and non-discriminatory rules for accessing genetic resources and associated traditional knowledge”.[51] A reading and application of Decision 391 in light of the Andean Parliament’s framework could reinforce mutually supportiveness between the Andean regime and the Protocol.

4 The Ecuadorian ABS Framework

4.1 Legal Context

Ecuador has a civil law system, which means that laws are codified in statutes or in administrative measures, including executive decrees. Ratified international treaties are part of the national legal system but subordinated to the Constitution.[52] Interestingly, the Ecuadorian Constitution consecrates regional integration as a national strategic objective. Moreover, it states that promoting the political, cultural and economic integration within the Andean region is a principle governing the State’s international relations. Decision 391 could be seen as instrumental in the achievement of this goal.

Ecuador’s first national ABS regulation is contained in the 2011 Executive Decree (No. 905). The Decree’s main purpose is to help implementation of Decision 391. The Decree has to be read in conjunction with the Organic Environmental Code and the Organic Code on the Social Economy of Knowledge. The latter introduces some modifications as regards the CNA.

---

[49] Ibid, p. 7.

[50] Andean Normative framework of Safeguards for Genetic Resources and Traditional Knowledge, adopted by the Andean Parliament in 2017, Art. 6 (5).

[51] Ibid, Art. 12 (2).

[52] Constitution of Ecuador, Art. 425.

The NP was ratified by Ecuador in December 2017, following its approval by the Parliament and a favorable opinion of the Constitutional Court.[53] The issue of misappropriation of GR was repeatedly raised during debates in the Parliament around ratification. It was pointed out that the NP could provide solutions for such problem.[54] Notably, instead of being neglected, Decision 391 continued to be referred to as the main ABS legal framework, whereas the NP was seen as complementary.[55]

4.2 *Recognition of the Value of GR*

Specific provisions on GR can be found in the Constitution as well as in the environmental and intellectual property legislation. The Constitution recognizes and values the country's 'genetic heritage' as an inalienable property of the State.[56] It includes it into the list of national strategic sector subject to the State's "exclusive decision and control".[57] It also prohibits any form of appropriation of traditional collective knowledge as well of GR containing the biodiversity.[58]

The Organic Environmental Code also reaffirms the strategic value of GR.[59] It specifies that the State shall administer and control access to GR, as well as to their derivatives and related synthetic products.[60] In addition, the Organic Code on the Social Economy of Knowledge, Creativity and Innovation states that the generic heritage cannot be privatized.[61]

4.3 *Relevant Definitions*

Article 6 of Executive Decree No. 905 reflects some of the definitions contained in Decision 391, including the definition of access and GR. In addition, it defines genetic material as "every material of plant, animal, microbial or other origin that contains functional unities of heredity or molecules of deoxyribonucleic acid (DNA) or ribonucleic acid (RNA) with decisive information of the hereditary character of

[53] Constitutional Court of Ecuador, Decision No. 011-13-DTI-CC, Case No. 0023-11-T1 (25 April 2013).

[54] Act 328-A, 18 July 2017, National Assembly Ecuador, p. 25.

[55] Act 328-A, 18 July 2017, National Assembly Ecuador, pp. 11 and 28.

[56] Constitution of Ecuador, Art. 408.

[57] Ibid, Art. 313.

[58] Ibid, Art 322.

[59] Organic Environmental Code, Art. 73.

[60] Ibid.

[61] Organic Code on the Social Economy of Knowledge, Art. 4 (16).

the offspring". It incorporates a definition of bioprospection as the "systematic search, classification and research for commercial purposes of the sources of new chemical components, genes, proteins, microorganisms and other products with actual or potential economic value present in the biodiversity". Following this definition, bioprospection cannot be treated as non-commercial research.

4.4 Rules on Benefit-Sharing

According to Executive Decree 905, benefits encompass "monetary or non-monetary benefits, including transfer of technology or royalties, among others, obtained from the utilization of GR, their derivatives, their applications and subsequent commercialization", but the Decree doesn't clarify when benefit-sharing ends.

With regard to benefit-sharing, the Ecuadorian ABS regulation seems to provide little room for negotiation. The Constitution provides that the State is entitled to at least the same amount of benefits to be perceived by an enterprise exploiting the State's natural resources.[62] This rule has been reaffirmed by the Organic Code on the Social Economy of Knowledge, Creativity and Innovation as applicable to GR.[63] It is arguable that this rule may put the country at a competitive disadvantage in the region.[64]

4.5 Competent National Authorities

Executive Decree No. 905 reflects the Protocol's rule according to which access to GR is subject to the prior and informed consent of the country of origin of the GR, which shall be issued through the CNA.[65] Under Executive Decree No. 905, the Ministry of the Environment was the CNA for the purpose of ABS. The situation changed in 2016 with the entry into force of the Organic Code for the Social Economy of Knowledge, Creativity which designates the National Secretariat for High Level Education (SENESCYT) as the entity in charge of formulating rules and principles for the conduction of scientific research in Ecuador.[66] This Code also confers on the SENESCYT the mandate to authorize access to GR for

[62] Constitution of Ecuador, Art. 408.

[63] Organic Code of the social economy for knowledge, creativity and innovation, adopted in 2016, Art. 73.

[64] These views were expressed by Diego Inclán and Lenin Nunez, Director and Legal Advisor, respectively, of the National Biodiversity Institute in Ecuador in interviews conducted in March 2018.

[65] Executive Decree No. 905 of 2011, Art. 6.

[66] Organic Code of the Social Economy for Knowledge, Creativity and Innovation of 2016, Art. 47.

non-commercial research based on the criteria given by the Ministry of the Environment.[67] The competence for granting access to GR and their derivatives for commercial research as well as for negotiating the corresponding benefits rests with the National Biodiversity Institute (INABIO),[68] an institution established in 2014 under the direction of the Ministry of the Environment.[69]

4.6 Access Procedure

The Organic Environmental Code clarifies that rights over the biological resources do not imply rights over their genetic component. Hence, any kind of permission granted by the Ministry of the Environment for research on natural species does not authorize research on GR or their derivatives.[70]

Building on Decision 391, Executive Decree No. 905 sets forth the requirements and steps for accessing GR, with the respective timelines. In line with Decision 391, the user is required to involve a national support institution defined by the Decree as "the national legal person dedicated to scientific or technical biological research that accompanies the user and participates with him or her in the access activities".[71] From the submission of the access application to the issuance of the resolution granting access the procedure should last approximately 100 days.[72] It should be noted that, in case of unjustified delays, the user can resort to a legal action known as 'administrative silence' (*silencio administrativo*). According to this action, any petition to the public administration that is not decided within the timelines indicated in the law can be considered as decided in favor of the applicant.[73]

Executive Decree No. 905 also indicates the different State institutions involved in the evaluation of the access application, which include, among others, the national intellectual property office and the entity in charge of indigenous peoples' issues.[74] An important innovation to the access procedure was brought by the implementing regulation of the Organic Code for the Social Economy of Knowledge, Creativity. This regulation introduces the virtual 'one-stop counter model' (*ventanilla única virtual*) for processing access application with commercial or non-commercial purposes.[75] Under this model, which has the aim of reducing transaction costs and

[67] Ibid, Art. 68.

[68] Ibid, Art. 69.

[69] Executive Decree No. 245 of 2014.

[70] Organic Environmental Code of 2017, Art. 72.

[71] Executive Decree No. 905, Art. 6.

[72] Executive Decree No. 905 of 2011, Arts. 16 to 24.

[73] Organic Administrative Code, Art. 209.

[74] Executive Decree No. 905, Art. 11.

[75] Implementing regulation of the Organic Code of the Social Economy for Knowledge, Creativity and Innovation of 2017, Art. 25.

streamlining the access procedure, once the request for access is submitted, the CNA shall internally request for opinion of the different relevant entities, without requiring further intervention of the user.

Once the resolution granting access is issued, the user and the CNA shall agree on the terms of the access contract. According to Executive Decree No. 905, such agreement shall include clauses regarding distribution of monetary and non-monetary benefits, conditions for the collection of the genetic material, access to information, intellectual property rights, third parties' cooperation, confidentiality, guarantees and means of compliance, dispute settlement, among many others.[76] By the time of this writing, no indication has been found regarding the conclusion of a single access agreement for commercial research.

4.7 *Access to Traditional Knowledge*

With respect to traditional knowledge, article 57 of the Ecuadorian Constitution recognizes the collective right of indigenous peoples to maintain, protect and develop their collective ancestral knowledge and science, including the GR containing biodiversity and agro-biodiversity. Though it does not explicitly recognize indigenous peoples' ownership over their traditional knowledge, it prohibits its appropriation by third parties. It is important to note that, by virtue of the International Labour Organization (ILO) Convention on Indigenous and Tribal Peoples (No.169), which has been ratified by Ecuador, the State has the obligation to consult indigenous peoples whenever consideration is given to legislative and administrative measures that may affect them directly. The obligation to consult may cover draft legislation on aTK.

In line with the NP and Decision 391, Executive Decree No. 905 requires the user seeking to access the intangible component associated with GR to obtain the free and informed consent of the community holder of such component.[77] In connection with this, article 94 of the Organic Code for the Social Economy of Knowledge states that the 'legitimate holders' of traditional knowledge associated with the biodiversity shall participate in the benefits arising out of a product or procedure in proportion to the contribution that such knowledge made to the development of that product or procedure. Furthermore, article 526 of the same Code provides that the distribution of benefits between the holder of the aTK and the user shall be established through MAT. The procedure for accessing aTK has not yet been defined by the law. Notably, by 2019, the National Service of Intellectual Property Rights (SENADI)

[76] Executive Decree 905, Art. 30.

[77] Executive Decree No. 905, Art. 29.

has registered 204 types traditional knowledge from indigenous communities, including those associated to biological resources.[78]

4.8 Procedure for Non-commercial Research

In 2015, the Ministry of the Environment adopted a legal framework to regulate access to GR for non-commercial purposes, with a view to give effect to article 36 of Decision 391. This regulation is contained in the Ministerial Agreement No. 034, which is in force since March 2015. It covers GR and their derivatives contained in the flora, wildlife, microorganisms, fungus and any other organism, with the exceptions provided in Decision 391.[79] It provides guidance for the negotiation and conclusion of a 'framework agreement' between the State and an individual or institution accredited to do scientific research in Ecuador[80] by the SENESCYT.

In practice, the applicant can conclude with the CNA a single framework agreement for a research program, which automatically authorizes the user to access GR for a specific research project that is related to that program.[81] A framework agreement could be agreed for a period of 3 years and could be extended to a total of 5 years.[82]

The rights granted under a framework agreement are not transferable.[83] Users' obligations include the duty to submit to the competent authority annual reports as well as other information related to the research as required by that authority; the obligation to store, for a minimum period of 5 years, following the expiration of the frame agreement, all information concerning access to GR undertook under the program; as well as the obligation to inform the competent authority on any scientific publication based on the access.[84] Transfer of genetic material is permitted upon the conclusion of a Material Transfer Agreement (MTA) which shall contain a clause prohibiting the transfer of the genetic material to third parties.[85] Transfer of endemic species is only possible for taxonomic research.[86]

---

[78] See https://www.ec.undp.org/content/ecuador/es/home/presscenter/articles/2019/con-204-depositos-voluntarios%2D%2Decuador-consolida-la-proteccion-d.html.

[79] Ministerial Agreement No. 34 of 2015, Art. 2.

[80] Ibid, Art. 3.

[81] Ibid, Art. 4.

[82] Ibid, Art. 14.

[83] Ibid, Art. 4.

[84] Ibid.

[85] Ibid, Art. 16.

[86] Ibid.

By 2018, more than 70 framework agreements for non-commercial purposes had been signed[87] which may tell that facilitation of the procedure by the Ministerial Agreement has helped. In most cases the users are national institutions which integrate international researchers as part of their research staff.[88]

4.9 Sanctions

The Organic Criminal Code of 2014, introduced the crime of misappropriation of the genetic heritage. The Code provides for sanctions of 3 to 5 years of imprisonment any person who accesses GR property of the State or the related intangible component without authorization from the CNA.

5 The Peruvian ABS Regulation

5.1 Legal Context

Like Ecuador, Peru has a civil law system. By virtue of article 55 of the Constitution, ratified international treaties are part of the Peruvian legal system. Article 44 of that instrument proclaims the constitutional obligation of the State to promote integration in Latin America.

Peru ratified the Protocol in 2014. Its first ABS national regulation dates back to 2009 and is contained in the Supreme Decree No. 003-2009-MINAM. Like Ecuador, the purpose of this regulation was to support implementation of Decision 391. It is important to note that, during the debates that preceded ratification of the Protocol, Decision 391 was not mentioned, whereas the existing ABS national regulation was considered to be in line with the Protocol.[89] Notwithstanding, a new ABS regulation has been drafted following ratification of the Protocol, which is based on both the Andean Decision and the NP.[90] The draft was made public in 2019 by the Ministry of the Environment and opened for contributions from civil society.

Over the years, Peru has taken important steps in combating misappropriation, becoming a reference model for the region. At the World Intellectual Property

[87] Cabrera Medaglia (2018), p. 16.

[88] See Frame agreement signed on 19 July 2016 between the National Institute of Biodiversity and the Ministry of the Environment for the program "Genetic Biodiversity in Ecuador". A copy of the agreement was provided by the National Institute of Biodiversity upon formal request made by the author.

[89] See Report of the Committee of International Relations on the draft resolution No. 3092/2013-PE that proposed the ratification of the Nagoya Protocol, Congress of Peru, 12 May 2014, especially p. 36.

[90] Ministerial Agreement 205-2019-MINAM.

Organization, the Peruvian government has continually stressed the importance of linking the international patent system with the international ABS regime.[91] In 2004, the National Commission Against Biopiracy was created to identify and prevent cases of possible misappropriation of GR and aTK. This Commission has actively objected, before international patent offices, patent registry of products on the grounds that they originate from unauthorized use of the Peruvian biodiversity.[92]

5.2 *Recognition of GR*

The Peruvian Constitutions does not mention GR. However, the General Environmental Act adopted in 2005 establishes that the national policy on biodiversity shall be guided by: the recognition of Peru as country of origin of GR and place for the diversification of GR; prevention of misappropriation of GR; and the development of effective mechanisms for the distribution of benefits arising out of the use of GR.[93]

5.3 *Relevant Definitions*

The 2009 ABS regulation does not add new definitions to those established by the Decision 391.

The draft of the new ABS regulation introduces new definitions and reformulates some of those contained in Decision 391. While access is still defined as the process of "obtaining and utilizing GR", the draft regulation ambiguously defines 'obtaining' as the process of "extracting the genetic material and/or their derivatives from biological resources or any other source".[94] This understanding may open the possibility to include, under the scope of the Peruvian ABS regulation, the genetic information understood as the "nucleotides sequence obtained from GR, including sequences digitally stored".[95] The draft regulation also reproduces the definition of 'utilization of genetic resources' contained in article 2 of the NP, which encompasses the conduction of "research and development activities" on GR and/or their derivatives, including through application in biotechnology.[96]

[91] WIPO/GRTKF/IC/11-13 (2007), p. 3.

[92] National Commission Against Biopiracy (31 January 2019).

[93] General Environmental Act, Art. 97 (d).

[94] Ibid, Art. 3.

[95] Ibid.

[96] Ibid.

5.4 *Competent National Authorities*

Under the 2009 regulation, the ABS regulatory function is, at least in principle, separated from the executive function. The Ministry of the Environment is the CNA for the purpose of regulation and monitoring of GR.[97] The task of examining and approving requests for access has been left to the so-called "administering and executing institutions".[98] These entities are: the National Forestry Service (SERFOR) for applications concerning wildlife; the National Institute of Agrarian Innovation (INIA) for cultivated species; and the Ministry of Production (PRODUCE) for marine species.[99]

Despite the above-mentioned separation of functions, one year after Peru's ratification of the Protocol, the Ministry of Agriculture adopted a regulation on forest management which has norms regarding access to GR of wildlife. Such regulation is contained in the Supreme Decree 018-2015 issued by the Ministry of Agriculture (MINAGRI) and makes reference to both Decision 391 and the NP. According to such Decree, the SERFOR shall represent the interest of the State in the negotiation of benefits arising out of the use of GR of wildlife.

5.5 *Access Procedure*

Supreme Decree 003-2009 does not add new elements to the access procedure contained in Decision 391. Notably, it fails to provide timelines to the different steps of the access procedure, thus leaving them to the discretion of the executing institutions.

The Decree elaborates on the content of the access agreement. It indicates that such agreement shall include the following clauses: The prohibition to claim property over the genetic material or its derivatives; prohibition of transfer of GR without the approval of the competent authority; recognition of the origin of the GR; involvement of local researchers in the collection and analysis of GR; and the obligation of the users to transfer to the provider country knowledge and technology related to the access.[100]

---

[97] Ibid, Art. 13.

[98] Ibid, Art. 14.

[99] Supreme Decree No. 003-2009, Art. 15.

[100] Supreme Decree No. 003-2009, Art. 23.

5.6 Procedure for Non-commercial Research

In line with Decision 391, Supreme Decree 003-2009 provides that users can make use of framework agreements for non-commercial research. Framework agreements shall include clauses concerning the involvement of local researchers, reporting obligations on the progress of the research, intellectual property rights and the obligation not to transfer genetic material to parties that are not involved in the research project.[101]

As regard non-commercial research, the draft regulation only requires the authorization from the competent authority to access GR, eliminating the requirement to conclude a framework agreement.[102] To obtain this authorization, the applicant should identify at least three non-monetary benefits for the country of origin, including the "presentation and dissemination of the research outcomes", and the "collaboration, cooperation and contribution to education and capacity building".[103]

5.7 Access to Associated Traditional Knowledge

Like Ecuador, Peru is bound by the obligations set forth in ILO Convention No. 169 on Indigenous and Tribal Peoples. The obligation to consult indigenous peoples on measures affecting their rights, which is contained in that Convention, is further regulated by Act 29785 of 2011. As mentioned early, regulation of traditional knowledge might be subject of consultation.

In 2002, Peru adopted Act No. 27.811, which establishes a regime for the protection of the collective knowledge of indigenous peoples related to biological resources. The Act is based on the principle, according to which, indigenous peoples have the right to decide on the use of their traditional knowledge, which is reflected in Decision 391 as well as in the NP. The Act defines collective knowledge as "the accumulated and transgenerational knowledge developed by indigenous peoples and communities regarding the properties, use and characteristics of biological diversity".[104] However, it fails to define when traditional knowledge can be considered as associated to a biological resource.

In addition, Act No. 27.811 establishes that access to traditional knowledge for scientific, commercial and industrial purposes shall be subject to the PIC of the representatives of the community holder of the knowledge.[105] The determination of

---

[101] Ibid, Art. 25. According to information of the Ministry of the Environment, between 2008 and 2015, SERFOR issued 41 permissions for access for non-commercial research, whereas INIA issued 39.

[102] Supreme Decree No. 003-2009, Art. 35.

[103] Ibid, Art. 49.

[104] Act No. 27.811, 2002, Art. 2.

[105] Ibid, Art. 6.

the community representative for grating the consent is left to the decision of the concerned community.[106] To obtain consent, the user shall provide the community with the relevant information on the purposes, risks and consequences of the access activities, including the potential uses of their collective knowledge.[107] When the collective knowledge is used for commercial or industrial application, the user shall negotiate and conclude with the concerned indigenous community a license agreement stipulating the conditions of the use and the distribution of benefits.[108] According to the Act, 10% of the gross sales resulting from the utilization of GR, before tax deductions, shall be allocated to a national Fund for Indigenous Peoples Development.[109] The Act does not address the situation when traditional knowledge is shared among various communities.

Recently, in 2016, the Ministry of Culture adopted a Decree that establishes a Multisectoral Commission for the safeguarding and revalorization of the traditional knowledge and practices of indigenous peoples.[110] The mandate of this Commission is to propose legislation aimed at strengthening protection of traditional knowledge and includes, among its members, representatives of indigenous peoples. Potentially, this Commission could revise Act No. 27.811 and fill some of its gaps.

5.8 Monitoring and Sanctions

Article 35 of the 2009 ABS regulation provides for sanctions for users that violate the terms of the access. These sanctions include suspension or termination of the authorization for access; confiscation of the genetic material and prohibition to submit new requests for access, in addition to a financial penalty.

It is important to note that the 2009 regulation stresses the supervisory role of the national support institution by including, as one of its tasks, ensuring compliance by the users of the terms of the authorization for access.[111]

Finally, the draft of the new ABS regulation elaborates on the role of checkpoints, in line with article 17 of the NP. A checkpoint is defined as a national entity that "collects and receives information relating to the prior and informed consent, the source of the genetic resource and the establishment of mutually agreed terms relating to the utilization of genetic resources".[112] The draft regulation designates

[106] Ibid, Art. 14.

[107] Ibid, Art. 2 (c).

[108] Ibid, Art. 17.

[109] Ibid, Art. 8. Notably, the Draft Regulation on Access to Genetic Resources and their Derivatives (2019) states that the Ministry of the Environment will prepare a guide for ABS negotiation with indigenous communities in line with Decision 391, the Nagoya Protocol and other relevant norms.

[110] Supreme Decree No. 006-2016-MC of 2016.

[111] Supreme Decree No. 003-2009, Art. 19.

[112] Ibid, Art. 3 (u).

three entities to carry out this task: the National Institute for the Defense of Competence and Protection of Intellectual Property (INDECOPI); the National Council for Science, Technology and Technological Innovation (CONCYTEC); and the National Commission Against Biopiracy.[113]

6 Assessment

In Ecuador and Peru, Decision 391 has been the basis for domestic ABS legislation. It seems that this Decision will continue to play a key role given the importance that the Andean law and regional integration have in the legal system of these two countries. Not surprisingly, the draft ABS regulation developed by Peru after the ratification of the NP continues to refer to Decision 391.

The requirements of PIC and MAT that are provided for in the NP had already been addressed and regulated by Decision 391, as well as the issue of aTK. From such a perspective, the Protocol and the Andean Decision might be regarded as mutually supportive. However, the original objectives pursued by the Andean regime seem to be somehow different from the objectives set forth in the Protocol. Decision 391 was largely conceived as an instrument whose main aim was to prevent misappropriation of GR and to reinforce the regulatory power of the Andean countries. Though the NP also reflects this objective, it seems to put a stronger emphasis on the promotion of international cooperation for scientific research,[114] by calling for facilitated access. Due to the apparent failure of CAN member States to reinforce integration, it is unlikely that Decision 391 will be revised in the near future. It is therefore for States, at the domestic level, to take a proactive approach to reconcile these two instruments. By drafting a new ABS regulation, Peru has attempted to adapt its domestic regulation to the NP's requirement and terminology. As explained earlier in this chapter, such proposal tries to simplify the procedure for non-commercial research on GR seeking to give effect to article 8 of the Protocol. The adoption of a special regulation for framework agreements for non-commercial research by Ecuador also reflects its efforts to comply with that provision of the Protocol.

While it is true that Decision 391 has failed to attain the economic benefits expected by the Andean countries in 1996, it could still serve as legal basis for sub-regional cooperation to address situations of transboundary use of GR and aTK. In this regard, it should be recalled that the Meeting of the Parties to the NP has encouraged parties, non-parties and regional organizations to consider regional approaches in the implementation of the Protocol.[115]

---

[113] Ibid, Art. 61.

[114] Broggiato et al. (2015), p. 1.

[115] CBD/NP/MOP/DEC/3/1 (30 November 2018), para. 7 (D).

At the domestic level, ABS rules have been adopted by Ecuador and Peru but their implementation is still problematic, at least as it concerns access to GR for commercial purposes. Neither of these countries has concluded an access agreement.[116] In the particular case of Ecuador, the constitutional rule that imposes a 50/50 model of benefit-sharing between the provider country and the user may discourage the latter from engaging into commercial research on GR. Rights of users are absent from the ABS legislation in both countries. Timelines of the access procedure are not defined in the Peruvian regulation.

The first review of the implementation of the Protocol indicates that the application of provisions concerning indigenous peoples remains a challenge.[117] While Ecuador and Peru have taken important steps by adopting norms that recognize indigenous peoples as holders of traditional knowledge, the question of what constitutes 'traditional knowledge associated with GR' in the ABS context remains unanswered. Also, it remains unclear who acts as representative of indigenous communities for granting consent.

Acknowledgements The author wants to specially thank María Consuelo Velasco and Dayahina Vera for their support in the collection of legal sources, and to Deyanira Camacho and Lily Rodríguez for their advice in the preparation of this chapter. The views expressed in this chapter are the author's own and do not reflect the position of the ILO or its constituents.

References

Articles

Broggiato A, Dedeurwaerdere T, Batur F, Coolsaet B (2015) Access benefit-sharing and the Nagoya Protocol: the confluence of abiding legal doctrines. In: Coolsaet B et al (eds) Implementing the Nagoya Protocol: comparing access and benefit-sharing regimes in Europe. Brill/Nijhoff, Leiden, Boston

Cabrera Medaglia J (2018) Implementación del ABS y del Protocolo de Nagoya en LAC: Situación actual y Retos Legislativos y de Política 3 años después de su vigencia, Serie Técnica sobre Acceso y Participación en los Beneficios (ABS) No. 4

Caillaux J, Ruiz M, Tobin B (1999) El Régimen Andino de Acceso a los Recursos Genéticos: Lecciones y Experiencias, Sociedad Peruana de Derecho Ambiental

Deplazes-Zemp A (2018) Genetic resources: an analysis of a multifaceted concept. Biological Conservation 222

Gómez-lee M (2012) La Comunidad Andina frente al Reto del Acceso a los Recursos Genéticos y la Distribución de Beneficios, Oasis No. 17

Megiddo T (2018) Beyond fragmentation: on international law integrationist's forces. Yale J Int Law 44

Pavoni R (2010) Mutual supportiveness as a principle of interpretation and law-making: a watershed for the 'WTO-and-competing-regimes' debate? Eur J Int Law 21(3)

[116] Cabrera Medaglia (2018), p. 16.

[117] CBD/NP/MOP/DEC/3/1 (30 November 2018), para. 29.

Ribadeneira M (2017) Veinte años del régimen Andino de recursos genéticos. Opera No. 20
Ruiz M (2008a) Guía Explicativa de la Decisión 391 y una propuesta alternativa para regular el acceso a los recursos genéticos en la subregión andina. Sociedad Peruana de Derecho Ambiental
Ruiz M (2008b) Una lectura crítica de la Decisión 391 de la Comunidad Andina y su puesta en práctica en relación con el Tratado Internacional, Recursos Naturales y Ambiente No. 53
Silvestri L (2016) Conservación de la diversidad genética en el Perú: desafíos en la implementación del régimen de acceso a recursos genéticos y distribución de beneficios. Revista Peruana de Biología 23(1)

Legislations, Draft Legislation and Records of Meetings

Act No. 27.811, 2002 establishing the regime for the protection of collective knowledge of indigenous peoples associated to biological resources (Peru)
Act of the Third Meeting of the Ad-Hoc Andean Committee on Intellectual Property of the Andean Community of Nations, 9 September 2016
Andean Parliament, Normative Framework to Safeguard Genetic Resources and Associated Traditional Knowledge, adopted in 2017. Agreement for the Integration of the Andean Sub-region (Cartagena Agreement) of 1969
Constitution of Ecuador, adopted in 2008
Constitution of Peru, adopted in 1993
Decision 391 is: Decision 391 - Common Regime on Access to Genetic Resources, adopted by the Commission of the Cartagena Agreement in 1996
Executive Decree No. 905 (Ecuador) National Regulation on the Common Regime concerning Access to Genetic Resources in application of Decision 391 of the Andean Community of Nations, adopted in 2011
Executive Decree No. 245 of 2014
General Environmental Act adopted in 2005
Implementing Regulation to the Organic Code on the Social Economy of Knowledge, adopted in 2017
Ministerial Agreement 205-2019-MINAM (Peru) containing the Draft Regulation on Access to Genetic Resources and their Derivatives
Ministerial Agreement No. 34 (Ecuador), adopted in 2015
Organic Administrative Code (Ecuador), adopted in 2017
Organic Code on the Social Economy of Knowledge (Ecuador), adopted in 2016
Organic Criminal Code (Ecuador), adopted in 2014
Organic Environmental Code (Ecuador), adopted in 2017
Supreme Decree 018-2015-MINAGRI (Peru), adopted in 2015
Supreme Decree No. 003-2009 (Peru), adopted in 2009
Supreme Decree No. 006-2016-MC, adopted in 2016
Treaty creating the Tribunal of the Andean Community, 1996

Judicial Decisions

Constitutional Court of Ecuador, Decision No. 011-13-DTI-CC, Case No. 0023-11-T1, adopted on 25 April 2013

Studies/Reports/Reviews

CBD/SBI/2/INF/17, Study into criteria to identify a specialized international access and benefit-sharing instrument, and as possible process for its recognition', prepared by Prof. E. Morgera, Dr. S. Switzer and Dr. E. Tsioumani, 2018

CBD Country profiles at: https://www.cbd.int/countries/ (accessed on 25 November 2020) Report of the Committee of International Relations on the draft resolution No. 3092/2013-PE that proposed ratification of the NP, Congress of Peru, 12 May 2014

CBD/NP/MOP/DEC/3/1, Assessment and review of the effectiveness of the Protocol, 30 November 2018

A/CN.4/L.682 13 April 2006, Fragmentation of international law: Difficulties arising from the diversification and expansion of international law, Report of the study group of the international law commission finalized by Martti Koskenniemi, 2006

CBD/DSI/AHTEG/2020/1/5 29, Fact-finding study on how domestic measures address benefit-sharing arising from commercial and non-commercial use of Digital Sequence Information on genetic resources and address the use of Digital Sequence Information on genetic resources for research and development, 29 January 2020

WIPO/GRTKF/IC/11-13, Combating biopiracy: The Peruvian experience, 2007

Power Point Presentations

Beck E, Rodríguez L, Experience with non-commercial research and access to biological material, (power point slides), presented at the University of Bremen in April 2017. PPT, https://www.uni-bremen.de/fileadmin/user_upload/fachbereiche/fb6/feu/FEU/ABS/ABS_PDFs/Experiences_of_non-commercial_research_with_ABS.pdf (accessed 26 November 2020)

Website Links

https://www.ec.undp.org/content/ecuador/es/home/presscenter/articles/2019/con-204-depositos-voluntarios%2D%2Decuador-consolida-la-proteccion-d.html (accessed 13 April 2020)

National Commission against biopiracy, 'Comisión Nacional contra la Biopiratería ganó 45 casos a nivel internacional por uso indebido de conocimientos tradicionales vinculados con recursos biológicos del Perú' (press release), 31 January 2019., https://repositorio.indecopi.gob.pe/handle/11724/6644 (accessed 3 December 2020)

Maria Victoria Cabrera Ormaza PhD in law, University of Göttingen. LL.M. University of Heidelberg and University of Chile. Legal Officer at the International Labour Organization, specialized on issues concerning indigenous and tribal peoples. Invited lecturer for international law at the Universidad Espíritu Santo, Ecuador. marvic.co@gmail.com.

New ABS Legislation and Practice in Compliance with the Nagoya Protocol: Current Situation and Perspectives in Costa Rica

Jorge Cabrera Medaglia

Abstract This chapter analyzes the background and history of the implementation of Access and Benefit Sharing (ABS) regimen under the Convention on Biological Diversity and other international related instruments in Costa Rica. The chapter describes the administrative and contractual practices of over 650 ABS permits, including two granted for the category of commercialization of genetic resources. It also responds to a set of innovative questions and concerns related to the scope, concept and meaning of utilization of genetic resources, provider and user rights in accordance to the legal framework and permits granted, monitoring of the utilization of genetic resources and associated traditional knowledge (aTK) and incentives (a recent labeling scheme) linked to the commercialization of products in compliance with ABS legislation. This research also addresses new and emerging issues such as treatment of genetic information (digital sequence information, DSI) and the implementation of an on line system for the application, decision making and reporting of ABS activities. Lessons learnt from this practical experience are also highlighted.

Keywords Implementation of the CBD · ABS practice and contract · DSI treatment

---

J. Cabrera Medaglia (✉)
University of Costa Rica, Law Faculty and Lead Counsel on Biodiversity Law, CISDL, San Pedro Montes de Oca, Costa Rica
e-mail: jorge.cabreramedaglia@ucr.ac.cr

E. Chege Kamau (ed.), *Global Transformations in the Use of Biodiversity for Research and Development*, Ius Gentium: Comparative Perspectives on Law and Justice 95, https://doi.org/10.1007/978-3-030-88711-7_5

1 Historical Background and Development of the ABS Legislation: The Biodiversity Law of 1998

1.1 Background and Influence: The Development of the Biodiversity Law

Since the 1970s Costa Rica has implemented a series of political tools for safeguarding the richness of ecosystems and biodiversity within it. It is a country renowned for its leadership in innovative environmental mechanisms being one of the pioneering nations in creating legislation for biodiversity on the basis of the Convention on Biological Diversity (CBD). It also creates a specialized institutional framework to operationalize the legislation and norms.

The first diagnose of the state of the knowledge on biodiversity for Costa Rica was performed in 1992 and it was the basis for the development of the first National Biodiversity Strategy and Action Plan (NBSAP), which was published in 2000. There were established three key elements regarding biodiversity: saving through protected areas and other means, understanding (inventories, ecological studies) and use (for different sectors, including tourism, the cosmetic and pharmaceutical industry).[1] This diagnose has shown that Costa Rica is one of the countries with more biodiversity worldwide, due its wide variety of ecosystems, which have approximately 5% of the species described globally 10% of which are considered endemic.[2]

The CBD became effective in Costa Rica in 1994, giving rise to the need to draft a national law that would implement this international agreement in a clear, simple and precise manner. The process of drafting and approving the proposed Biodiversity Law, took several years to be completed. There were several draft proposals since 1996, but they were not well received by different social sectors, especially the research community because of the requirements set forth for carry out research on biodiversity. Finally, the Legislative Assembly created a Special Joint Commission, which submitted a text that was approved as Biodiversity Law No. 7788 of April 30, 1998 and was published in the Official Gazette No. 101 of May 27, 1998.[3]

The processes used to draft the Law and specially the provisions on the issues of access to genetic resources and benefit sharing, protection of TK and IPR are particularly relevant. In 1996 a biodiversity bill was submitted to the Parliament, but it was considered very restrictive and an obstacle to scientific research on biodiversity. Several negative comments were made. Numerous observations were sent to the Legislative Assembly (Parliament) including a complete bill prepared by

---

[1]MINAE (2000).

[2]Ibid at 22.

[3]It should be mentioned that Costa Rican BL was awarded in 2010 the gold price for the BL as an exemplary law implementing the Convention on Biological Diversity by the World Future Council (see http://www.worldfuturecouncil.org/fileadmin/user_upload/PDF/Survey_of_Future_Just__Biodiversity_Policies_and_Laws.pdf).

the Advisory Commission on Biodiversity which was never formally submitted as a bill.[4]

Considering this initial reaction, in January 1997 a second version of the bill was presented which—despite of considering some of the main objections made—repeated several of the concepts and provisions of the former bill and therefore met with the same opposition. The impasse resulting from the opposing point of views led to the creation of a Special Mixed Commission of the Legislative Assembly to draft a bill based on the existing draft text after having obtained the Assembly's promise to respect the agreement reached by the Commission. The members of the Commission included the main political parties, ILCs representatives, the Union of National Chambers (productive sector), the main public universities, INBio, Minae, and others. At the end of 1997 a new draft was completed and sent to the Congress for approval. With some small changes, the text was finally passed and published as a Law in May 1998.[5]

1.2 Aims and Intentions of the Legal Framework

The general goal of the Biodiversity Law (BL) is to promote the conservation and sustainable use of biodiversity and to ensure the fair and equitable sharing of benefits derived from it (Art. 1). The entire BL responds to this goal as put forth by the Convention on Biological Diversity (CBD. For example, the BL establishes the environmental function of the land (Art. 8), general principles of the law (Art. 9), objectives (Art. 10), criteria for applying the law (Art. 11); the administrative structure of the National Conservation Areas System (SINAC) (including the administration of the national wild protected areas, Articles 22 to 43), the guarantee of environmental safety (biosafety and exotic organisms, Arts 44–48), the conservation and the sustainable use of the ecosystems and species (Arts 49–61), the regulations on access to genetic resources (Arts 62–76), intellectual property rights (Arts 77–85), education and public awareness and research and transfer of technology (Arts 86–91), environmental impact assessment (Arts 92–97), incentives (Arts 98–104) and procedures and sanctions (Arts 105–113). All of these elements are in accordance with the three objectives of the CBD.

In summary the main intention of the legislation was to implement the CBD in its entirety, including the ABS component. The law gives special emphasis to access to genetic resources including a special chapter that defines the use and management of them (Chapter V).

[4] Sittenfeld et al. (2003), p. 197.

[5] Ibid.

2 ABS Legal and Policy Framework

2.1 Current Status of the ABS Legal Framework: The Biodiversity Law and Other Regulations, Instruments and Policies

In addition to the Biodiversity Law No. 7788 of April 30, 1998 (BL), there is a 'General Access Procedure' (GAP) Decree No. 31514-MINAE of December 15 2005. Further, the regulations for access to genetic resources found in *ex-situ* conditions were approved by Decree No. 33677-MINAE of 27 April 2007. These two decrees (No. 31514 and 33677) were recently amended by Decree No 41591-MINAE of May 2019. Furthermore, Decree 39341-MINAE establishing the procedures for the imposition of sanctions for illegal access was approved in 2016. A Memorandum of Understanding (MoU) was signed between the International Treaty on Plant Genetic Resources for Food and Agriculture (ITPGRFA) Focal Point (National Seed Office/OFINASE), the National Commission on Plant Genetic Resources and the National Commission for Biodiversity Management (CONAGEBIO) in 2014 clarifying the implementation of the ABS regimen under the BL and the CBD and the International Treaty. The MoU addresses grey areas and achieves a common understanding on key issues relating to the implementation of the ABS Multilateral System of the IT and the Biodiversity Law and decrees. Finally, the Nagoya Protocol was signed but has not been ratified yet[6] due to opposition by few members of the Parliament. This was based on the premises (in my view incorrectly) that the Protocol was not strong enough in some issues like the protection of traditional knowledge and reduces some of the access requirements.

According to Article 117 BL, the law has been fully in force since its publication in April 1998. However, an action to declare this law unconstitutional was brought by the Attorney General's Office. This claim was admitted for study by the Constitutional Chamber (Unconstitutionality Action Number 98-006524-007-CO-M, admitted by October 1998 Resolution). The Ministry of the Environment and Energy considered these legal competencies unconstitutional;[7] thus the Ministry requested the Attorney General to submit a constitutional challenge.

According to Articles 81 and 82 of the Law of the Constitutional Jurisdiction No. 7135, the suit does not suspend the execution of the LB. However, from the political point of view it has definitely delayed CONAGEBIO's implementation. It had the effect of slowing down many of the necessary decisions to make the law operational. For example, the CONAGEBIO was not put into effect until January

[6] The Commission of International Affairs of the Parliament in August 2019 issued a negative report (Dictamen Negativo) for the ratification of the Protocol. However, the Government submitted the Protocol for approval again. This time, a different Committee of the Parliament, by consensus issued a positive (Dictamen Positivo) report recommending the ratification of the Protocol.

[7] Among the legal arguments submitted were the transfer of legal powers to the civil society and the creation of different bodies within the MINAE which were considered against the Constitution.

2000, almost 2 years later than initially foreseen by the law. The action was decided by the Constitutional Court declaring the Law constitutional.[8]

Since 2004 to 2020, Costa Rica has granted access to genetic resources through more than 650 permits, and several ABS agreements have been negotiated between private companies, universities, farmers and national and international research centers. Two commercial use permits have been granted by CONAGEBIO. One in 2016, for the cosmetics company Chanel, which developed "Blue Serum" using biochemical components of Costa Rican green coffee; and the second one in 2018 for Lisanatura, a Costa Rican company, which produced a cough syrup using organic plants and is obliged to share benefits with a rural cooperative.

Many of these agreements have been concluded by the National Biodiversity Institute (INBio a non-governmental organization) which has over 25 years of experience targeting the systematic search for secondary metabolites and products of commercial interest. INBio has implemented numerous projects involving processes for the extraction, isolation, fermentation, and characterization of compounds of interest in the pharmaceutical, agrochemical and biotechnological industries.

2.2 Scope

2.2.1 Geographical Scope

The Legislation (Art. 3) is applied "...on the elements of the biodiversity under the State's sovereignty, as well as on the processes and the activities carried out under its jurisdiction or control, independently of whether the effects of the actions are manifested inside or outside the national jurisdiction". This law will regulate specifically the use, management, associated knowledge, distribution of benefits and costs derived from the utilization of the elements of the biodiversity (Art. 3). In the same way, article 6 (public domain meaning public property) establishes that "... the biochemical and genetic properties of the elements of the wild or domesticated biodiversity are of public domain. The State will authorize the exploration, research, bioprospecting, use and utilization of the elements of biodiversity that constitute goods of public domain, as well as the utilization of all genetic and biochemical resources, by means of the procedure of access established in Chapter V of this Law". Also, in conformity with articles 62 and 69, every research program or bioprospecting on genetic material carried out in Costa Rican territory requires an access permit, unless covered by one of the exceptions foreseen by the law.

An important starting point in the legal biodiversity framework in Costa Rica is that the Law establishes the State has the exclusive sovereignty and control over the elements of biodiversity and that genetic and biochemical properties of biodiversity are of public domain (public property).

[8]Constitutional Court, Voto (Ruling) No. 01113 of June 2006.

In Summary, the access regime is applied to the components of the biodiversity that are under the sovereignty of the State, as well as to the processes and activities carried out under its jurisdiction or control, independently from those effects manifested inside or outside of the national jurisdiction. In this regard the access regulations are applied to genetic resources on public or private land, terrestrial or marine environments (including migratory species), under *ex-situ* or *in-situ* conditions and in indigenous territories (but additionally the indigenous or customary rules of these peoples should be taken into account, as well as the community *sui generis* intellectual rights when developed).

In principle CONAGEBIO's opinion is that any research, including with foreign (exotic)[9] genetic resources that takes place within the borders of Costa Rica as provided in the Constitution (Art. 6) will be covered by the scope of the access regime of the Country.

2.2.2 Material Scope: Including Treatment of Genetic Information/DSI

The BL covers all the genetic and biochemical resources. Access is defined in the Law (Art. 7.1) as: "obtaining samples (or associated knowledge) of the elements of the wild or domesticated biodiversity found *in situ or ex situ* for basic research, bioprospecting (commercial) or commercial use (aprovechamientoeconómico)". Utilization is also now included in the amended regulations following the definition of the Nagoya Protocol (Art. 6 (y) of the Decree 31514 amended by Decree 41591). In principle it means that R&D is required for any activity to be considered "access for utilization".

The law covers both domesticated and wildlife resources and elements to the extent they are utilized within the national territory.

The law covers basic research, like molecular taxonomy. However, the amendment to the Decree No. 31514 made by Decree 41591 of 2019 under the definition of basic research (Art. 6 n)) determines that "the teaching and routine diagnostic activities which utilize biochemical or molecular biology techniques from which not scientific publishable information is derived will not be considered basic research". Therefore these activities are excluded from the obligation to obtain an access permit. Is not clear the purpose to the qualification "unless scientific publishable information can be derived". The amendment introduced by Decree No. 41591. MINAE of 2018 incorporates in article 2 of the GAP a concept of "utilization" affirming that "the norms of access will apply to the genetic and biochemical elements and **their utilization....**" (emphasis added) which was not in the original text of Decree 31514. Therefore the concept of utilization (as defined in the amended regulations) derived from the Nagoya Protocol (NP) is included in the Regulations. This incorporation should have the purpose of clarifying the scope of access by

[9]The BL already covers migratory species. This interpretation of CONAGEBIO seems to expand the scope, including research done in the country with foreign material (e.g. imported bacteria).

introducing the R&D component on the genetic or biochemical activities. The law is applied equally to genetic agricultural resources (Art. 3). The legislation foresees specifically that in the case of duly registered *ex-situ* collections, the regulation of the law will set the authorization procedure for access permits (Art. 69). This type of access (to genetic resources in *ex-situ* conditions) is now regulated by Decree 33677-MINAE of 2007.

A Manual developed by CONAGEBIO guides the users of the "on line platform for access procedures" providing examples of what activities constitute access to genetic and biochemical resources.[10]

There is no particular reference to access to bulk materials that are later on utilized/accessed for their biochemical or genetic properties. However, the amendment made by Decree 41591 in 2019 of the GAP incorporates in the definition of providers the following statement: "It will not be considered provider any legal or natural person which regular activity is only the distribution or sale of the organisms or their products in ex situ conditions". This means that in case of acquisition of bulk materials or commodities from a market the access procedure may be required to be followed later on if the interested person is seeking to conduct R&D on the genetic or biochemical components, meaning that he/she must negotiate PIC and MAT with a provider and get an access permit from CONAGEBIO.

In relation to access and benefit sharing from the use of genetic information (DSI), a response submitted by CONAGEBIO[11] to a questionnaire sent by the Secretariat of the CBD CONAGEBIO interprets that DSI are covered under the term access to genetic resources in the Biodiversity Law. However, it is not clear whether the legal principles of PIC and MAT will be applicable in the case of access only to genetic information while access to the physical samples is not taking place at the same time. CONAGEBIO[12] seems to indicate that benefit-sharing principles must apply and not necessarily the PIC.

Box 1 DSI Under Costa Rica's Legislation and Practice

Regarding DSI the response submitted by the Government of Costa Rica to the questionnaire prepared by the Secretariat of the CBD on DSI and national measures in summary indicates: that DSI are covered under the definition of access to genetic resources of the BL; but in practice for non-commercial research is not regulated (no PIC and MAT are required) but for commercial research benefit sharing should be established probably through the Global Multilateral Benefit Sharing Mechanism. The legal criteria for the differentiation between commercial and non-commercial use is not clear. Until now no

(continued)

[10] https://www.conagebio.go.cr/Conagebio/public/documentos/manualProcedimientos/Manual_de_Procedimientos_CONAGEBIO_version1.0.pdf.

[11] Gobierno de Costa Rica (agosto del 2019), p. 3.

[12] Ibid, p. 4.

Box 1 (continued)

access permit has been granted for the commercial use of DSI/genetic information per se not involving access to the physical material (genetic or biochemical compound) due to the lack of any application.

The response questionnaire also concludes that on a case by case basis the Technical Office TO of CONAGEBIO (Competent National Authority on ABS) has the power/authority to impose restrictions and prohibitions for the further dissemination/deposit in public data bases of genetic information to avoid losing control on the DIS resulting from an authorized access to genetic/biochemical resources by a permit. The response does not provide any concrete examples.

However, some cases have been identified which have used this restriction. For instance, in the permit **No R-CM-089.2010-OT of January 9 2010**, the following restriction was imposed in the permit granted:

"For the DNA (genetic material) extracted from the requested genetic resources the Technical Office of CONAGEBIO **restricts the publication of complete/full genomic information on the national and international data bases**, meaning that the **entire genomes cannot become public, only the information related with molecular markers.** Likewise, before publishing the sequences of DNA of the molecular markers developed or used for the project purposes, the applicant shall inform in advance the TO and later on submit the number of accession of the sequences" Unofficial translation.

Finally, it is possible that other restrictions/conditions related to the dissemination/deposit/publication of genomes/ gene sequences could have been imposed in the access permit, which exact terms could vary on a case by case basis (response to the questionnaire by Melania Muñoz, OT of the CONAGEBIO, Costa Rica). There is no information on these other cases. SEE DOCUMENT.

INBio practices[13] on ABS contracts can also highlight how DSI related matters can be integrated and regulated under MAT.

For instance in the ABS agreement between INBio, the University of Michigan (U-M) and the University of Harvard (one of the International Cooperative Biodiversity Groups) the following clauses were included in the RCA negotiated for the project "Discovery of Natural-product based Drugs and Bio-energetic Materials from Costa Rica Biota of September 2009":

- INBio will manage the data related to Samples, Isolates, Extracts, Fractions, and DNA pursuant to its activities under the Statement of Work using its databases; however, each of Harvard and U-M shall be permitted to

(continued)

[13] Cabrera Medaglia (2009), pp. 243–270.

Box 1 (continued)

maintain, in parallel with INBio, data sets that wholly or partially overlap the body of data that is managed by INBio.

- Harvard shall manage the information related to the Research in its databases and shall coordinate with U-M any information that needs to be transferred to NAPIS (a data base). Additionally, Harvard shall maintain information updated as long as there is work performed with the Materials.
- Data generated by the Parties in performance of Screens, such as structures and activities of Chemical Entities, will be deposited in **ChemBank by U-M with prior notification to-, and written consent by-, INBio and Harvard. In case a third party has a commercial interest in such information, Harvard, U-M and INBio will require them to negotiate and enter into agreements with Harvard, U-M and INBio** (emphasis added).

In such situations, INBio is the user of the genetic resources (acting here as an intermediary to Harvard and U-M) and the TO of CONAGEBIO as the CNA grants or denies the applications submitted by INBio. In accordance with the national legislation (BL and in situ decree of 2007) all access contracts like this one must be approved by CONAGEBIO. Moreover, the RCA was actually endorsed/approved as part of the permit granted by the TO of CONAGEBIO to INBio. The process of revision of the contracts is covered by a confidentiality agreement with the staff of CONAGEBIO.

Source: *Author's elaboration*

Recently, a bill amending several provisions of the BL was submitted to the Parliament,[14] excluding most of the basic research from the access permitting system and creating a notification procedure, among other changes mainly targeted to facilitate access for basic research for different stakeholders, including public universities. The proposal is at the initial stages of the legislative process therefore the final content and the potential approval are not known at the date of writing this chapter.

2.2.3 Personal Scope

The Law applies equally to foreigners and nationals. There are no specific restrictions for foreign entities seeking access or any condition of obligation to carry out research with the support of a local or national institution. The only specific requirement is the obligation to have a national representative which can be notified of any decisions made by the Technical Office (Art. 63.5). There is no further

[14] Bill No 21807 of February 2020.

reference in the law to any responsibility or liability of this representative (see Art. 63.5 BL and Art. 6.w Decree No. 31514).

In principle, foreign entities or multinational companies can constitute a local/ national enterprise under the laws of Costa Rica (Code of Commerce and related legislation) and be considered nationals.

2.2.4 Temporal Scope

The BL does not apply retroactively in accordance with the Constitutional principles found in article 34. In practice the ABS regime began to apply since 2003 with the entering into force of the first ABS Regulations (Decree31514). Permits granted before for ABS activities were done following more general rules on research on wildlife found in the Wildlife Law of 1992 by SINAC.

2.2.5 Exemptions

All research or bioprospecting programs on the genetic or biochemical material of biodiversity that are to be carried out in Costa Rican territory require an access permit (Arts 62 and 69 of the BL), unless they fall into one of the exceptions provided by the Law. **These exceptions include (Art. 4)**: access to human genetic resources; the non-profit exchange of genetic and biochemical resources and associated traditional knowledge of indigenous peoples and local communities; and research by Costa Rican public universities, which had 1 year (until 7 May 1999) to establish their own controls and regulations for research that implies non-profit access to biodiversity.[15] If none of these exceptions apply, all sectors (pharmaceuticals, agriculture, plant protection, biotechnology, ornamental, herbal etc.) that wish to access genetic components are subject to the BL and must follow its access procedures.

In addition, the resources included in the Multilateral System of the International Treaty are excluded from the scope of the ABS legal framework in accordance with the interpretation provided in the Memorandum of Understanding that was concluded between CONABEBIO and the ITPGRFA Focal Point (2014. This is based on the idea of the nature and characteristics of the self-executing provisions of treaties. What is required is an assessment of the fulfillment of the 3 requirements set by the ITPGRFA for the resources included in the Multilateral System (i.e. management, control and public domain) for a particular plant genetic resource

---

[15] Only one university, the University of Costa Rica has developed its own ABS regulations: See Reglamento de Acceso a la BiodiversidadenActividades de Docencia, Acción Social e Investigación Sin Fines de Lucro de la Universidad de Costa Rica published in La GacetaUniversitaria, No 13-99 of 1999.

for food and agriculture (*ex-situ* or *in-situ*) that is included in Annex 1of the Treaty. This has been the legal interpretation accepted in the case of Costa Rica.

The Biodiversity Commission of the University is the entity responsible for granting access permits for research carried out by University members in accordance to the exception established in article 4 of the BL. For instance, permits were granted in 2018 for 96 projects for taxonomy or inventory, 55 for basic research and in 2019, 19 for taxonomy, 44 for basic research and one for applied research.[16]

This is not an exemption to the obligation to have an access permit, however in practice in this case the National Competent Authority is the University of Costa Rica through the Institutional Biodiversity Commission of the University. There is a divergence of criteria between CONAGEBIO and the UCR Institutional Biodiversity Commission concerning the extent of the exclusion of the UCR access activities: for CONAGEBIO only basic research is under the scope of the UCR regulations and not bioprospecting or commercial use, while for the UCR the exception of article 4 of the BL covers any type of access (basic, applied, bioprospecting, commercial use, etc.) carried out by University members (professors, students or researchers). This disagreement has not been solved so far and in practice UCR continues to give permits for all kind of research done by members of the University.

3 Institutional Arrangements and Functions

3.1 National Focal Point

Decree No. 31514 (Art. 5) provides that the Director of the Technical Office or the person designated by the Director, with the approval of the Ministry of Environment and Energy, will be the National Focal Point on ABS before the Convention on Biological Diversity.

3.2 National Competent Authorities: The National Commission for Biodiversity Management and the Biodiversity Commission of the University of Costa Rica

The Biodiversity Law created the National Commission for the Management of Biodiversity (CONAGEBIO) as the Competent National Authority in Costa Rica. The CONAGEBIO is responsible for proposing policies regarding access to genetic and biochemical elements of biodiversity and related traditional knowledge that

[16]*Pers. comms*, Dr. Adam Karremans, Coordinator, Institutional Biodiversity Commission, 19 February 2020.

ensure proper scientific and technology transfer and the fair and equitable sharing of benefits arising from access (Art. 14 of the BL). The Commission reports to the Ministry of the Environment and Energy and it is the National Focal Point on ABS under the CBD. It acts through a Technical Office as the entity that processes, approves or rejects and monitors access-related activities.

The BL establishes the National Commission for Management of Biodiversity as a decentralized organ of the Ministry of Environment and Energy (Art. 14). Among its responsibilities are the following:

- develop policies and responsibilities set up in Chapters IV, V (Access to genetic and biochemical elements and protection of associated knowledge) and VI of the Law and coordinate them with several responsible organisms on this subjects (subsection 2)
- formulate and coordinate policies for access to biodiversity elements and associated knowledge
- ensure adequate scientific-technical transfer and fair sharing of benefits, which under Chapter V are denominated as General Access Norms (subsection 3)
- implement its agreements and resolutions and institute its procedures through the Executive Director of the Technical Office (Art. 16). This entity is formed by government entities such as Ministry of Environment (President), Ministry of Foreign Trade, Ministry of Health, Ministry of Agriculture, Costa Rican Institute of Fishing and Aquaculture, National Commission of Rectors; Indigenous Peoples, Farmers Organizations, National Union of Chambers (private sector), Costa Rican Federation for Conservation of the Environment (FECON), and the National System of Conservation Areas Director (Art. 15). This Commission shall formulate policies on access and distribution of benefits and can revoke Technical Office resolutions on this subject (Art. 14)

The Commission has a Technical Office for support, which will be integrated by an Executive Director and the staff indicated in the Regulation of Law. The Technical Office will have as a main attribution to process, reject and control the access applications to resources of biodiversity (Art. 17 a); coordinate all about access with Conservation Areas, private sector, indigenous peoples and local communities (Art. 17 b); organize and maintain an updated register of access applications to biodiversity elements, *ex-situ* collections and those from natural and juridical person dedicated to genetic manipulation (subsection c); and gather and to update regulation relative to compliance of agreements and guidelines on biodiversity (subsection d).

In addition, the Commission will execute its agreements and resolutions and will design its internal procedures through the TO's Executive Director (Art. 16). Article 16 allows CONAGEBIO to create *ad hoc* expert committees in complex cases.

The Commission activities have been regulated by means of MEE's decree No. 29680, published in The Gazette of 7 August 2001 and its modifications. Its members are designated for a two-year period.

The legal framework and the BL acknowledge the need for funding sources to manage and operationalize ABS system. By Law CONAGEBIO and the Technical Office will be provided with consignments, legacies and donations, contributions from registration and transaction, benefits from permits and concession, fines and 10% of the tax (Timbre) fee of National Parks (Art. 19).

Finally, in practice the Institutional Biodiversity Commission of the University of Costa Rica can also be conceptualized as another Competent National Authority to the extent the Commission fulfills all the functions of a Competent Authority under the NP, even if it has not been officially notified to the CBD yet.

4 Regulatory Tools and Instruments for Access to In Situ and Ex Situ Genetic Resources and Associated Traditional Knowledge

4.1 Access to Genetic Resources: Procedures

The access procedure is regulated in detail in the Law and the regulations (decrees) of 2003 and 2007. The competent body to grant the access is in first instance the Technical Office of the CONAGEBIO.

The BL regulates the basic requirements for access, including the need of PIC, transfer of technology, equitable sharing of benefits, the protection of associated knowledge, and the definition of the ways in which the above mentioned activities will contribute to the conservation of species and ecosystems. It also mandates the designation of a legal representative in the country, when the person or organization requesting access is domiciled abroad (Art. 63). The procedure to follow is clearly outlined in article 64. It includes proof of the PIC of the owner of the property where the activity will be developed, whether it is an indigenous community, a private owner or public entity. Other interesting provisions incorporate the right of cultural objection (Art. 66), the registry of access applications and the protection of confidential information, except in the case of biosafety concerns (Art. 67).

The BL also regulates in detail commercial and non-commercial bioprospecting permits (Art. 69). These are valid for 3 years and can be renewed. They are given to specific persons or entities and are therefore not transferable (Art. 70). The permits are limited to the genetic and biochemical elements expressly authorized for specific areas or territories (Art. 70). The permits will contain a certificate of origin, permission or prohibition to extract samples, periodic reporting obligation, monitoring and control and any another applicable condition deemed relevant by the TO (Art. 71).

The access request requirements are name and identification of the interested party, name and identification of the responsible researcher, exact location of the place when the genetic resources will be accessed, and the elements of biodiversity that will be the subject of the investigation, indicating the owner and manager or holder of the premises (Art. 72).The applicant will also have to submit a descriptive chronology of activities, aims, and purposes as well as place for legal notifications (Art. 72). The application must be accompanied by the PIC (Art. 72) and a record of individuals or legal entities that are to conduct the bioprospecting (Art. 73).

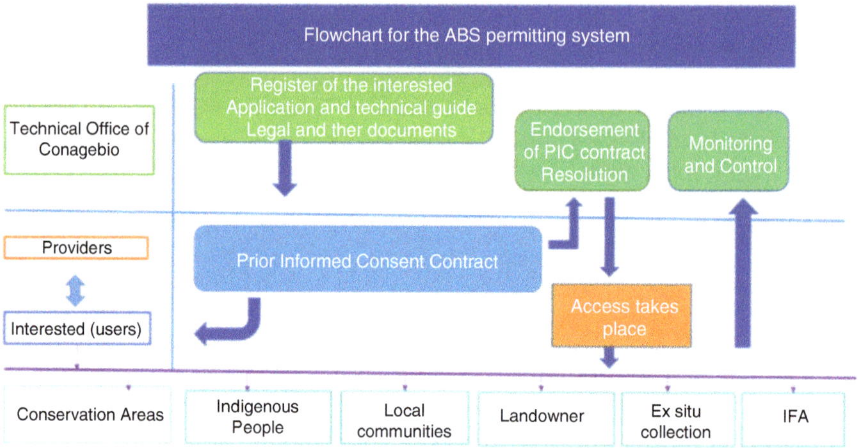

Fig. 1 Flowchart for the ABS permitting system. *Source*: CONAGEBIO's flow chart is included in several power point presentations and is freely accessible

The TO must also authorize those agreements contemplating access to genetic and biochemical elements (Art. 74) signed between individuals, natives, or foreigners, or between them and the institutions registered for such purposes. There is also a possibility to establish framework agreements with universities and other duly authorized centers (Art. 74). It is established that up to 10% of the research budget and 50% of royalties will have to go to the conservation area, the private owner, or indigenous community (Art. 76). In cases in which the TO authorizes the continuing use of genetic material or of biochemical extracts for commercial purposes, applicants are required to obtain a separate concession from the MEE (Art. 75). Further guidance is provided in Decrees No. 31514 and 33697 mentioned before.

A separate PIC should be obtained from other entities such as conservation areas, indigenous territories or public authorities who are owners of lands or in the case of marine resources other authorities such as the National Institute for Fisheries and Aquaculture (Art. 63).

The following diagram presents the current access procedure. The CONAGEBIO has developed and put in practice a full on-line system for the different steps of the ABS procedure (Fig. 1).[17]

In the Annex I of this paper the main steps for the ABS procedures are presented in detail.

[17] https://www.conagebio.go.cr/Conagebio/login.xhtml.

The first step is the fulfillment of the registration. The requirements can be found in article 8 of the Decree No. 31514. The Annex to the Decree includes a form for the registration of the interested[18] (the applicant).

The requirements for access are as follows:

(a) Name and identification of the interested/applicant including information for notifications; in case of a foreign interested the designation of a legal representative residing in Costa Rica is required;
(b) Type of permit who wants to request initially: basic research, bioprospecting or commercial use.

The second stage (Art. 9.1) is the Request Form and Technical Guide with the following information (a form is provided in the Annex of the Decree No 31514):

(a) Name and identification of the interested person, including means and place for notifications;
(b) If the applicant is a legal o juridical person located in a foreign country, it is mandatory to designate a legal representative who is a resident of Costa Rica;
(c) Name and identification of the main researcher or coordinator of the project if different to the applicant;
(d) Name and identification of the scientific collaborator when he or she exists;[19]
(e) Name and identification of the researchers, assistants and other authorized persons to enter into the place where the materials will be collected;
(f) Type of permit requested: basic research, bioprospecting or commercial use
(g) Title of the Project;
(h) Specific and general objectives of the Project;
(i) Type and quantity of materials to be collected and in the case of ex situ collections data of the materials;
(j) Methodology, experimental techniques, data analysis to be used, etc.;
(k) Location of the place where the access activities will take place (private lands, protected areas, ex situ collections, indigenous territory, etc.);
(l) Estimated duration of the Project and number of foreseen visits to the site;
(m) Estimated budget and source of funding;
(n) Destination of the materials;
(o) Indication of the utilization of traditional knowledge if applicable;
(p) Indication of the prior research on the genetic resources or traditional knowledge;
(q) How the access activities will contribute to the conservation of species and ecosystems;
(r) Possible environmental or cultural risk associated to the access activities;

[18] The BL and regulations use the term 'interesado' in Spanish or interested in English. This can be considered as equivalent to the word 'user' or 'applicant'. See Decree 31514, article 6 p (definitions).

[19] Defined as: natural or legal person, foreign or national, who jointly with the interested and the main researcher, participates in the research and development of the project (Art. 6 e Decree 31514).

(s) Timeline of the activities;
(t) Declaration that everything has been submitted under oath.

In addition, the following documentation must be included:

1. Copy of the Project;
2. Certification of legal personality if applicable;
3. Copy of the identity card or passport;
4. Legal powers when applicable;
5. The contract between the applicant and a third party in accordance to article 22 of Decree 31514 and 74 of the BL;
6. Receipt with the payment of the fees (not in established yet);
7. A sworn statement when the access requested is aimed at the research, development and production of living modified organisms in contained use that the activities will not be carry out with the aim of cause adverse effects on the human health, species and ecosystems nor for military or terrorism purposes;
8. A document with the biosafety measures when the access is intended for research and develop of living modify organisms of contained use.

GAP's article 14 establishes the following "Criteria for the evaluation or approval of the request":

- Development options for future generations;
- Food safcty and sovereignty;
- Conservation of ecosystems;
- Protection of human health;
- Improvement of citizens' quality of life;
- Gender issues;
- The conservation, sustainable use and benefit sharing objectives;
- Biosafety measures.

Also, GAP's article 24 allows the imposition of total or partial restrictions on access to the resources to ensure their conservation and sustainable use. These restrictions are issued by the TO in the resolution approving access. In this way, it can prohibit access, set limits, and regulate the methods of collection, in application of the precautionary principle mentioned in LB's article 11.2. To establish complete or partial restrictions some of the aspects that will be considered are (Art. 14):

- The danger of extinction of the species, subspecies, races and varieties.
- Reasons of scarcity and endemic conditions.
- Vulnerability or fragility conditions in the structure or function of the ecosystems.
- Adverse effects on human health, the species, and the ecosystems or on essential elements of the autonomy or cultural identity of peoples and communities.
- Strategic genetic resources or geographical areas qualified as such; and
- The prohibition of access for military purposes or for denaturalization of the resources.

Once access is authorized, the monitoring and control phase begins (article 20 of the GAP). This will be explained later on in this chapter. Applicants will have to follow applicable sanitary and phytosanitary rules for the exportation of the materials and obtain any permit required (article 18 of the GAP).

An environmental impact assessment (EIA) can be requested by the TO based on some general provisions of the BL related to EIA, but not specific to bioprospecting activities (Art. 92 BL). The evaluation is the responsibility of the National Technical Secretariat (a body of MEE) which will coordinate with the CONAGEBIO on the parameters for the assessment (article 26 of the GAP).

In order to facilitate access to genetic and biochemical resources the BL and the GAP foreseen the establishment of "framework agreements" (for commercial and non- commercial access). However, individual permits are required for each access project to be developed. In principle these framework agreements have just simplified some administrative requirements for research institutions like universities and similar centers (Art. 74 BL and Art. 21 GAP).

The procedure for access to the *ex-situ* materials is similar in terms of requirements and type of permits (Art. 7 Decree 33697). The regulation also provides for a registration of the ex situ collections and for that purpose there is a detailed form in the Annex of Decree 33697.

Under this hypothesis the ex situ collection acts as a provider and PIC must be negotiated between the interested person (applicant) and the manager or responsible person of the collection (Art. 7). However, for access to ex situ genetic resources, when the origin and provenance of the material is known and if the collection was established before the entry into force of the Decree (2007) the benefits can be negotiated with the original providers, but the negotiation with the original provider is not mandatory.

When the access to the ex situ materials is for those materials that were collected after the entry into force of the Decree 33697 the benefits must be shared with the providers in accordance with the PIC (which should exist in order to allow the ex situ collection to have the materials included in its banks), and the interested person (applicant) is obliged to submit the PIC with the provider which was necessary for the *ex-situ* collection to legally acquire the material in the first place (Art. 7).

Additionally, from the date of registration and for any type of access permit the interested person must commit under oath to observe a Code of Conduct Annexed to the Regulation. The general objective of the Code of Conduct is to promote the conservation and sustainable use of the genetic or biochemical resources *(both in-situ and ex-situ),* reduce the risk of genetic erosion and protect the interest of the users and providers of the resources. The Code of Conduct requires the user to respect the national law, follow and comply with the terms of the permit and PIC, cooperate, collaborate and facilitate information to the national authorities, communicate any application for IPR, negotiate in good faith the possible monetary and non-monetary benefits derived from the access, respect and comply with the measures for the protection of associated TK, carry out access activities under best practices and mitigate the environmental impact, among others.

This obligation to adhere to the Code of Conduct must be included in the resolution of the TO. In practice other permits for access to *in-situ* genetic resources have also included the obligation of the interested (user) to commit to adhere to the Code of Conduct.

Article 10 of the Decree No. 33697 indicates that the access permit does not exonerate the interested (applicant) to comply and obtain other permits related to the export of materials such as plants, animals, microorganisms, seeds, etc. obtained as a result of the access activities. The TO can also upon request of the interested person issue a "certificate of legal provenance"[20] in the cases of export of materials to accompany their transfer and export to other countries. It seems that in the case of *ex-situ* materials accessed and of a possible export to foreign entities it is to the benefit for the user to get a certificate of legal provenance as a proof of the legality of the access and transfer.

4.2 PIC

The third stage is the negotiation of PIC agreement or contract with the provider. Under Costa Rica's law the contract providing for BS is negotiated between the provider and the user (interested person), but not with the Competent National Authority (see especially Art. 9 GAP).

The PIC is supposed to contain mutually agreed terms that represent the fair and equitable distribution of benefits. Once obtained, this agreement must be endorsed by the TO (Art. 63 BL and Art. 12 GAP). The endorsement is included in the Resolution approving the access request and not in a separated one. The TO limits itself to endorsing the contract rather than negotiating it.

Article 9.2 of Decree 31514 regulates the PIC. PIC could be obtained and negotiated by means of the subscription of contracts, agreements or similar means. The public institutions (as providers) may have a maximum of 3 months to determine whether or not to grant consent.

Paragraph 9. 2.1 recommends in all of these instruments to include the following **common clauses**:

(a) Specific and general objectives;
(b) Place or places where access will take place;
(c) Name and identification of the researchers, assistants and authorized persons. In case of that guide or attendance by persons of local communities or indigenous peoples be required these should be duly hired and paid for;
(d) Type and quantity of material and in the case of *ex-situ* collections the data of the collected materials;

[20]This is not the Internationally Recognized Certificate of Compliance mentioned by the Nagoya Protocol (Art. 17) but may also provide legal certainty about the rightful acquisition and transfer of the materials.

(e) Methodology for the collection, lab techniques, data analysis, etc.;
(f) Estimated time for the development of the Project;
(g) Mutually agreed terms on the exchange of knowledge associated to the elements and resources of the biodiversity;
(h) Agreed terms on any other condition resulting from the participatory process regulated in article 83 of the BL for the protection of the TK;[21]
(i) Express manifestation by the interested party that the protection measures of TK existing in the country will be respected;
(j) Agreed terms of the possible study of the cultural impact of the access if applicable;
(k) Mutually Agreed Terms on the type of transfer of technology or generation of information derived from the access towards the national counterparts, ILC or the providers of the resources;
(l) Mutually agreed terms on the equitable distribution of monetary and non-monetary benefits at short, mid and long term, including possible commercial profits or earnings of any product or sub-product derived from the material accessed. The provider and the interested party will set an amount of money (between 0 to 10%) of the research budget (in practice because it can be zero this is not an obligation for the user) and up to 50% of the royalty obtained. Royalties are defined in article 6.v) as a monetary benefit, payment or economic compensation calculated on the basis of the accrued earnings after taxes and other deductions established in the Law.

    To agree on the percentage the parties can consider the number and price of the samples requested among others. However, in the case of commercial use there is an obligation to negotiate and deposit up to 50% of the royalties. The payment could also be made in tracts. If the interested person is the owner of the land or legal responsible of an *ex-situ* collection (for which no PIC exists) he or she has the obligation to pay to the CONAGEBIO the same percentage;
(m) Formal commitment to provide evidence of the origin of the resources and associated TK in any publication, procedure or further use;
(n) Signature or finger print of the provider and applicant.

In addition, **for bioprospecting and commercial use** is required to incorporate a clause on the final destination of the materials (Art. 9.2). The user and provider can negotiate the way in which the transfer of the materials will take place, which will be notified to the CONAGEBIO. This covers the cases where new places or locations are identified during the process of research and development. To avoid the potential need to renegotiate and amend the PIC contract, for instance a simple communication by the user (interested party) and the consent by the provider will be sufficient and no formal amendment of the PIC contract will be required. The new destination must be also approved by the CONAGEBIO.

---

[21] This process has not been finalized yet, so in practice is not applicable.

For basic research or bioprospecting it is in addition required to include a formal commitment in the sense that if there is a change in the type of access/permit from basic research to bioprospecting or from bioprospecting to commercial use, the user will inform the Technical Office and fulfill all the requirements for the new type of access as appropriate (Art. 9.3 Decree 31514).

For **commercial use** there are three particular requirements (Art. 9.4):

(a) Description of the commercial use of genetic or biochemical resources or associated TK;
(b) An economic feasibility study under oath and certified by a public accountant. The regulation provides specific details on the content of this study including its timeframe for product development;
(c) Include in the PIC the commitment to deposit until 50% of the royalties. In this case a royalty rate is mandatory.[22]

4.3 Traditional Knowledge (TK) Protection in the National Legal System[23]

The Costa Rican protection system for traditional knowledge is based on several premises:

The legal system of access assures prior informed consent and the sharing of benefits related to TK.

(a) From this point of view, it is a combination between access mechanisms, contracts or licenses and a *sui generis* scheme based on registers.
(b) The existence and validity of different forms of knowledge and innovation and the need of protecting them using appropriate mechanisms (Art. 77), such as patents, trade secrets, copyright, plant breeder rights, community *sui generis* intellectual rights, *inter alia* (Art. 78).
(c) Legislation is aimed at protecting knowledge through a registration system, an approach that has been supported by comparable foreign law. Thus, it shall proceed to inventorize community *sui generis* intellectual rights that the communities request to be protected (Art. 84).
(d) For the definition of the extent, nature and requirements of these rights, a participative process of consultation should begin with indigenous peoples and farmer communities (Art. 83 of Law of Biodiversity). This process will determine how community *sui generis* intellectual rights will be used, who will exercise ownership and how the beneficiaries are determined (Art. 85).
(e) Before assigning rights and obligations the following specific issues need to be considered:

---

[22]However, at least in one case the TO did not require a royalty to be established for a commercial use permit. For the CNA a provision included in the PIC to later negotiate a royalty was accepted.

[23]For a detailed study of the legal system, see Cabrera Medaglia (2003).

- the material and associated knowledge to be protected;
- the procedures for protection/registration;
- any granted rights and against whom these will be enforced;
- relevant enforcement mechanisms.

In summary the success of the expected system will depend greatly on the results of the participatory process envisaged in the BL. Unfortunately, this process has not been developed yet, except for pilot cases. Progress in implementing this system has been slow especially because of the lack of appropriate procedures for the consultation with the indigenous peoples and the human and financial resources required. After some unfinished efforts and partial consultations, in 2015, the CONAGEBIO developed the project named "Pilot Plan concerning the Sui Generis Intellectual Community Rights in two Indigenous territories", whose general goal was to develop a pilot plan in two indigenous territories that can define the rules and requirements for a national participatory process for the establishment of the Sui Generis System. A draft "Regulation proposal regarding the sui generis communal intellectual rights related to the protection of traditional knowledge and the intangible elements associated" was developed in the past and was also consulted with the two indigenous peoples in order to receive their feedback and views. This process was developed in coordination with the Costa Rican Indigenous National Table (Mesa Nacional Indígena). The results from this Pilot Plan will serve as the foundation to establish in the future, at a national level, the participatory process with indigenous and rural/local communities to determine the nature, scope and requirements of the Sui Generis Intellectual Community Rights (as foreseen by the BL) and to consequently develop all the legal instruments to be enacted to regulate the system of protection.

Finally, it is expected that this participatory process will also fulfill the provisions of the decree No 40932 of March 2018 which sets out the consultation process for initiatives affecting indigenous peoples in compliance with the ILO Convention 169 (not only for access to TK or genetic resources located in indigenous peoples territories but applicable to other development projects and activities). However a similar system for the development of the participatory process for farmers' *sui generis* rights (campesinos) is not yet available.

4.4 Type of Access Permits

There are three permit categories under the ABS system of Costa Rica (Art. 7 Decree 31514):

Basic research: activity to investigate, examine, classify or increase the existing knowledge about the biological elements in general or their genetic or biochemical characteristics in particular, without **any interest in commercializing its results**. Teaching and routine diagnostic activities that include molecular biology or biochemical techniques (which produces non publishable information) are not considered basic research.

Bioprospecting: systematic research, classification and research for commercial purposes of new sources of chemical compounds, genes, proteins, microorganisms and other products with current or potential economic valued found in biodiversity.

Commercial use: use of genetic or biochemical resources for exploitation or commercialization purposes, which carries out processes of technological-industrial development.

According to article 71 of the BL (characteristics and conditions of access permits), the access requirements will be determined differently depending on whether the research has or does not have a commercial purpose. In the latter case, the non-commercial purpose will have to be verified (there are no further details about how this verification may take place). Nevertheless, the GAP in practice does not contemplate different requirements for non-commercial versus commercial projects.

There are no established criteria for differentiation between commercial and non-commercial access (except for the definitions) and this is done on a case by case basis. In practice there are no mayor differences in the treatment of basic research and bioprospecting (commercial), and thus no simplified procedures or requirements for basic research can be identified. In the case of commercial use (aprovechamiento económico) royalties are mandatory as well as the obligation to submit very studies on the product to be developed.

Under the original text of Decree 31514 basic research was defined by the lack of "direct interest" to commercialize the results. The amended regulations (Decree 41591) eliminates the word "direct" making even more difficult to draw the line between the two concepts (although this is not of particular interest due the similar treatment provided as explained above).

With respect to the change of intent (see Art. 7 Decree 31514): ABS regulations, permits and PIC contracts require to go back and fulfil the appropriate requirements of the applicable permit in the cases of change of intent, but there are no defined milestones developed by CONAGEBIO (when change of intent occurs).

Other Permits (Phytosanitary, Export)

The GAP requires the holder of a permit to fulfil any other stipulations regarding the exportation of plants, animals, seeds, microorganisms or parts thereof obtained from the access (Art. 18). This may include CITES and phytosanitary permits among others. However, it is not necessary to request additional permits from the CONAGEBIO for the export or transfer of the material. The original permit usually authorizes any future export and transfer to a particular destination and only requires complying with other applicable laws.

4.5 Content and Conditions of the Permit

The Resolution granting access of the Technical Office shall be motivated and clearly has to indicate the technical, social or environmental justification for the decision taken (Art. 13 GAP).

(a) Term of the permit;
(b) Obligation to deposit up to 10% of the research budget and up to 50% of the royalties when applicable. The calculation of the percentage of royalties depends of the negotiation between the provider and the user. Likewise any other benefit or transfer of technology agreed in the PIC.
(c) The submission of reports and its periodicity as well as other articles and publications derived from the permits. A recognition of the contribution of country and of the genetic resources or associated TK will be provided in any publications or articles. Copies will be delivered to the provider. If the original language is English a summary in Spanish will be presented.
(d) Biosafety measures when the access takes place with the aim of research, development and production of living modified organisms of contained use.
(e) Any other condition or restriction deemed necessary by the Technical Office.

Permit conditions typically include: conduct research only for the authorized purposes and field; do not transfer to third parties without prior authorization; come back in cases of change of intent; places where GR research will be carried out (no need to further authorization to transfer after the permit was granted).

However, future uses/ ownership of the data is not regulated in the permit. The PIC contract could regulate uses of research results in accordance with the needs and criteria of the user and provider involved depending on each case. However, some permits have restricted the publication of entire genomes in data bases (unless authorized (see Box 1 on DSI)).

It seems from the review of number of permits that the right to utilize ends with the finalization of the term of the ABS permit (3 years with the option for the extension), after that the user must seek for a new permit.

Box 2 Typical Conditions Found in a Basic Research or Bioprospecting Permit[24]

- Authorized the PIC contract with the providers
- Identification of the methodology to be used in conformity with the PIC contract
- Copies, disposals, derived or isolated compounds, of all the materials cannot be utilized for other methodological research different to the one stipulated in the technical guide.
- Description of the elements to be accessed that are authorized, including their location
- Right of the TO to conduct visits, inspections, verification and control where the research activities take place

(continued)

[24]See R-CM-FN-001-2018-OT of February 2018, R-CM-INBio-03-2006-OT March 2006 and R-CM-FN-002-2018 of September 2018.

Box 2 (continued)

- Research reports obligations, timeline and beneficiaries of the reporting and information exchange
- The fair and equitable benefit sharing must be done in accordance with the PIC contract and if applicable with the contract with a third party
- Commitment to provide and state the origin of the resources in future publications and articles or further uses
- Authorized destination and transfer of the materials (places)
- The permit holder cannot transfer the permit and the permit is limited to the materials and resources authorized
- In case of export of sample the obligation to comply with the national legislation in accordance with article 18 of the Regulations
- The duty to submit for approval any future contracts or material transfer agreements with third parties. Lack of presentation may result in the cancellation of the permit
- The materials as well as their derivatives or duplicates (no mention to information) cannot be transferred to other destinations without the prior authorization of the TO.
- Name of the authorized principal researcher
- Respect the Code of Conduct annexed to decree No. 33697-MINAE (also for access to genetic resources *in-situ* conditions)
- Right to cancel the permit without any responsibility if there is evidence of non-compliance with the permit terms or any of the criteria established in article 27 of the Decree 31514. In addition when there is any other non-compliance with the legal framework of the country
- Term of the permit (up to 3 years)

5 Model Agreements

Current legislation provides for a Model Material Transfer Agreement as an annex to the *ex situ* regulations (Decree 33697) but in practice the template can also be used for *in-situ* transfer if appropriate. The Model Agreement is not mandatory but recommended (Art. 7 No. 33697). However, in practices this model is not widely used.

The Model Agreement refers to: Parties; authorized uses; description of the materials; main obligations of the interested party, including reporting requirements; compliance with national and international legislation; further transfer of materials; declaration of the origin and provenance of the resources; benefit sharing terms; dispute resolution; applicable law (Costa Rica); notifications; term; authorization from the TO of the MTA. Other suggested clauses are guarantees, definitions, confidentiality, minimizing environmental impacts, etc.

The Manual (not the regulations) provides for a PIC Model Contract including drafting suggestions for the main issues to be included in accordance to article 9 of Decree 31514.[25]

6 Benefit-Sharing

As it was mentioned before in Costa Rica benefit sharing is left to an individual negotiation between users and providers of the PIC contract.

6.1 Types of Benefits

Article 63 mentions the transfer of technology and equitable sharing of benefits, the protection of associated knowledge, and the definition of the ways in which the project's activities will contribute to the conservation of species and ecosystems. Not all monetary and non-monetary benefits to be distributed are listed but a generic rule is set forth with some specific indications regarding royalties, research budget (Art. 76) and technology transfer (Arts 63 and 76). This was addressed before in this chapter. Unfortunately there is no systematized information on the benefits received by different users and providers in the form of studies, reports, etc.

Other benefits can be negotiated between the users and the providers. Likewise, the user can at the same time reach an agreement with other parties and this agreement can also include benefit sharing provisions.

In the case of INBio[26] contracts with third parties have provided benefits to Costa Rica including:

- Monetary benefits through direct payments.
- Payment for supplied samples.
- Covering research budgets.
- Transfer of important technology which has enabled the development of the infrastructure at the Institute (biotechnology lab, etc.), which can be used for the investigation and generation of their own products.
- Training of the scientists and experts in state-of-the-art technology.
- Negotiation experience and knowledge of the market and the probabilities of searching for intellectual uses for biodiversity resources.
- Supporting of conservation through payments made to the Ministry of the Environment for the strengthening of the National System of Conservation Areas.

---

[25] https://www.conagebio.go.cr/Conagebio/public/documentos/manualProcedimientos/Manual_de_Procedimientos_CONAGEBIO_version1.0.pdf.

[26] Cabrera Medaglia (2009), p. 243.

- Transfer of equipment to other institutions, such as to the University of Costa Rica.
- Future royalties and milestone payments to be shared 50:50 with the Ministry of the Environment.
- Establishment of national capabilities for assessing value of biodiversity resources.

6.2 Terms, Conditions and Content of Benefit-Sharing Agreement

There are no other conditions on BS except those mentioned above- to be included in the PIC agreement or in an agreement between the user and a third party in accordance to article 74 of the BL and 22 of the GAP.

One example of benefit-sharing is the following (see Table 1):

6.3 Commitment of Benefits to Biodiversity Conservation and Sustainable Use Purposes

It must be pointed out that there are no binding requirements that benefits must go towards the conservation of the resources.[27] It is perfectly possible that a private owner, public institution, or indigenous territory could grant the PIC without allocating benefits towards conservation since the legal authority of the TO is limited to endorsing the PIC. In these circumstances, it is valid to ask whether the TO would have the legal authority not to endorse a PIC in the light of the lack of benefits flowing towards conservation (Art. 63). In those cases in which a Conservation Area grants the permits, it is assumed (but not stated in the BL) that the benefit will go in its entirety towards biodiversity conservation.

7 Provider and User Compliance Measures

7.1 Reporting Measures

One of the most important changes made by Decree 41591 is on the content of the reporting measures.

[27] The BL provides as one of requirement of access the definition and ways how access activities can contribute to the ecosystems and species, but there are not further references to BS components to be targeted to conservation (Art. 63).

Table 1 Selected examples of benefit-sharing: The Chanel and Flora Nueva commercialization permit (Blue Serum product)

Biological/genetic resources	Green Coffee and oil extracts from Green Coffee
Associated traditional knowledge	No existing/reported associated TK. The Green Coffee has proved to have special properties found after the research done by the local company Flora Nueva with the support of Chanel
Existing uses of biological/ genetic resources	Green coffee does not have a history of cosmetic use in the country. Flora Nueva in cooperation with Chanel began a process of research on the anti-ageing properties of the coffee. This began because Flora Nueva identified in the green coffee some properties which make them special for a cosmetic formulation. The coffee is grown (cultivated?) in an internationally recognized "blue zone" (area with high longevity).
ABS permits or agreements	In this particular case 2 different permits were granted and agreements were negotiated • A bioprospecting permit (for commercial research) between Flora Nueva (user) and Cerro Azul as a provider of the coffee grains. Flora Nueva did the process of extraction of the oil from the green coffee. • An agreement between Chanel and Flora Nueva was negotiated setting the terms for the collaboration between them for the research activities. The research was made with the support of Chanel to Flora Nueva. • A permit for "aprovechamientoeconómico" or commercialization with Chanel acting as the applicant and Flora Nueva (*ex-situ*) and Cerro Azul (*in-situ*) acting as 2 different providers. • An agreement between Chanel, Flora Nueva and Cerro Azul was negotiated for the commercialization phase determining, among others, the benefits to be shared with the providers Flora Nueva and Cerro Azul and the responsibilities of each of the parties. A product "blue serum" reached the market as an anti-aging cream.
Types and mechanisms of benefit-sharing	For the commercialization phase the following benefits were provided: To Cerro Azul: • Fair prices for coffee supply above the market rate • Financial support for the buying of a coffee roaster machine/ (Tostadora) and other equipment • Minimum amount of coffee to be bought by Chanel during the contract period and exclusivity as provider of the coffee • Support with the branding, marketing and technical assistance to Cerro Azul • Royalties to be defined at a later stage For Flora Nueva: • Fair prices for the oil extracted • Minimum amount of oil to be bought during the period of the contract • Support for the development of value chains for Cerro Azul The permits granted included conditions for the transfer to third parties, reporting and confidentiality among others.

(continued)

Table 1 (continued)

Biological/genetic resources	Green Coffee and oil extracts from Green Coffee
Key lessons	- R&D on genetic resources is a large process involving different stages (research and commercial use) and multiple partners with Flora Nueva acting as intermediary and user. - A web of agreements is required to regulate the roles and responsibilities of the different parties in the different stages of R&D and commercialization. - Understanding of the legal requirements for each type of development/permit, especially for commercialization ("aprovechamiento economico") is key. This was the first permit granted for commercial use in the country, and there was no clarity on key requirements: Such as the feasibility study, royalty rates to be paid for a cosmetic, etc. - The case led to the development of detailed provisions for the issuance of a commercialization permit including reporting requirements, feasibility study, etc. - The practicality of some of the legally determined requirements (royalty rates) versus business practices or models in a particular sector (cosmetic industry) were questioned in this case. - The value of the ingredient (oil extracts for green coffee) versus the total investment and marketing expenses for the development of a cosmetic product was also an issue. What is the value of GR in a cosmetic product and the estimation of the benefit sharing component? - Understanding of the company R and D issues and commercial sensitiveness (confidentiality) by the regulators is necessary. - Constant dialogue with the authorities throughout the process was also important for the success of the project. - The model could be replicated in the future in order to develop value chains with local producers utilizing different GR and for different products.

Decree 31514 includes now very detailed content on the reporting requirements. The aim of the reports is to verify and document the fair and equitable sharing of the benefits derived from the access (Art. 29).

There is detailed content for the **reports of the collection** including: general information, and information on the accessed samples (Art. 30). Article 31 includes also a set of minimum data including: (a) general information, information on the associated samples, results by objective, general analysis of the data, conclusions, recommendations; (b) information on the progress made of any product and its opportunities to be commercialized; (c) information of workshops, events and other type of capacity building activities done as part of the benefit sharing, if there are monetary benefits a copy of the deposit made; (d) indication of any pending activities to be carried out such as payments, trainings, publications and bibliography.

In the case of reporting duties for the activity of commercial use the following information is required: data on the net and gross incomes and documentation on the monetary and non-monetary benefits received by the provider.

7.2 *Contractual Clauses and Permit Conditions Used*

Several PIC contract clauses and permit conditions focus on the compliance with the terms of the PIC contract or the permit. They include:

- Reporting requirements (both in PIC contract and in the Permit).
- Sanctions in cases of non-compliance with the Permit (which in many cases incorporates the relevant PIC provisions on transfer to third parties, benefit sharing and others).
- Prohibitions for the user to transfer of accessed resources to other destinations or to use the genetic or biochemical resources for different purposes without prior approval of the provider and of the TO.

7.3 *Patents/Intellectual Property (IP) Rights Requirements*

Costa Rica has developed some check points (IPR) but limited to the use of national genetic resources and therefore not in line with the Nagoya Protocol (Art. 17).

During the process of drafting the BL the issue of intellectual property rights and their relationship with biodiversity inevitably arose. Thus, the BL establishes that intellectual property rights shall be congruent with the objectives of the Patent Law by virtue of the principle of integration (Art. 79). The Law originally excluded the following from patenting: DNA sequences; plants and animals; unmodified micro-organisms; essential biological processes for plant and animal production; the processes of nature or natural cycles; inventions essentially derived from the knowledge of biological traditional practices or in the public domain; inventions that are produced monopolistically that may affect the processes or basic agricultural products used for food and health purposes (Art. 78).

However, this article was modified by an amendment of an IPR Law[28] which was enacted to comply with the IPR commitments of the CAFTA-DR (Free Trade Agreement between Central American Countries, Dominican Republic and United States). The amendment indicates that (a) DNA and RNA sequences are excluded from patent protection to the extent they don't fulfill the regular patent requirements; (b) microorganisms as they are found in nature are not patentable; (c) it was clarified that non biological and microbiological processes can be protected; (d) it was added

[28] Law No. 8686 published in La Gaceta November, 26 2008.

that patenting is excluded for those inventions commercial exploitation of which would be against the ordre public, morality, the health or life of human beings, animals and plants, or cause serious damages to the environment.

Authorities should consult the TO before granting protection of intellectual or industrial property-related innovations that involve biodiversity elements. The submission of a certificate of origin and prior informed consent is required in order to gain these IP protections. A well-grounded opposition by the TO shall prevent protection from being granted (Art. 80).[29]

The BL provides that beneficiaries enjoying protection of intellectual or industrial property rights are subject to a compulsory license in case of duly justified emergency. In the event of a justified emergency, this license will allow the use of such rights for the benefit of the community. This provision is aimed at solving an emergency, without involving compensation or royalty payment (Art. 81).

The Biodiversity Law recognizes evidential value of the certificate of origin in the case where national genetic resources are accessed. It requires the presentation of this certificate at the competent office if IP rights are applied for. Similarly, a consultation is required with the Technical Office of CONABEGIO in the cases of innovations based on biodiversity elements of Costa Rica. It should be noted that the presentation of the certificate guarantees that the access procedure was followed. This includes the negotiation of prior informed consent, the establishment of mutually agreed terms and the sharing of benefits from the utilization of genetic and biochemical resources. Article 80 BL determines that both the National Seeds Office and the Intellectual Property and Industrial Property Registers shall consult the Technical Office of the Commission before granting protection of intellectual or industrial property to innovations involving components of biodiversity. The applicant shall provide the certificate of origin issued by the Technical Office and the Prior Informed Consent. Justified opposition from the Technical Office will prevent the registration of a patent.

In contrast to patent law, neither the Plant Variety Protection Law No 8631 -nor its regulations- expressly requires the National Seed Office (the competent authority to grant plant breeders rights) to consult with the TO before a plant breeder's rights is issued.[30] Protection exceptions are made for wild plants which are not modified (Art. 2). Likewise, within the concept of "notoriously known variety" ("variedad notoriamente conocida" in Spanish) will be included all the varieties that are

[29] Decree No. 34958-MINAET-COMEX, regulation to the article 80 of the BL limits the opposition grounds to the lack of compliance with the patent requirements. Likewise, in the case of non-compliance (with what? the submission of the certificate) a fine is established. However, the processing of the applications seems to be allowed even if no certificate of compliance was produced. A constitutional action was brought against this regulation and the decree was also annulled by the Constitutional Court (Constitutional Court resolution No. 17058-2012 of December, 5, 2012).

[30] A legal action was brought before the Tribunal Contencioso Administrativo by a NGO (FECON) claiming the illegality of the lack of a mechanism to guarantee compliance with the article 80 of the BL, File (Expediente) No.16-008622-1027-CA October 2916not yet resolved.

protected by community sui generis intellectual rights, whether those rights have been registered or not, in accordance with what is established in articles 82 and 84 of the Biodiversity Law N.° 7788, to the extent that the variety is adequately described and it is possible to verify its existence (Art. 4).

To the present date, no patent applications have been identified that have made use of national genetic resources. There was only one patent application for which the TO was consulted, i.e. a patent application (file N° 7415) named "SUBSTITUTED PIRIDINONAS AS MODULATORIES OF P38 MAP QUINASE". The summary is below:[31]

> Formula 1 components and its pharmaceutically acceptable salts are described in which R1, R2, R3, R4 and R5 are described in this application. These components are useful for the treatment of illnesses and ailments caused by or exacerbated by p38 MAP unregulated quinase activity and/or FNT. Also described are pharmaceutical compositions that contain compounds, methods to prepare compounds and treatment methods that use the compounds.[32]

However, the TO concluded that the strains of *N. meningitidis,* were originated in the United States, Germany and Holland. It also considered that the claims do not include any information that refers to the patenting of microorganisms, DNA sequences, nucleotoids, and/or amino acids as are found in nature. Therefore, this patent application does not infringe upon what is stipulated in BL.

The challenges for the due implementation of the article 80 consultation relates to the fact that the Registry is made up of specialized intellectual property lawyers that are not familiar with the Biodiversity Law. Another challenge is that this consultation is carried out at the stage of substantive review of the application which delays the completion of the patent processing.

7.4 Statistic Tools and Monitoring for Domestic GR's Users

So far the Platform of the CONAGEBIO can facilitate the monitoring, including the compilation of reports from the users. Other actions for monitoring users (domestic or foreigners) are addressed below (under the heading of "following compliance").

7.5 Administrative Sanctions

Article 112 of the Biodiversity Law establishes that "Whoever carries out exploration, bioprospecting or has access to biodiversity without authorization from the

[31] *Pers. comms* with Karen Quesada, head of the Patent Office of Costa Rica.

[32] Through the resolution of 8.07 am on July 30, 2009, the application was rejected due to the lack of an invention unit; lack of clarity and because the claims are related to treatment methods which are excluded from patentability according to article 1 of Law N° 6867.

Technical Office of the Commission, when it is necessary in terms of this law or because of breach of the terms granted in the permit, a fine will be imposed varying between the equivalent of one and twelve salaries, as established in article 2 of Law No. 7337".

Article 106 of the BL indicates that for that purpose the ordinary administrative procedure foreseen in the General Public Administration Law No. 6677 will be applicable. It was considered necessary to create a procedure duly regulated to provide legal certainty to the competent authorities and the stakeholders regarding the mechanism to handle and resolve the potential cases of unauthorized access to the genetic and biochemical resources, complementing the relevant provisions of the General Public Administration Law.

If applicable, the Technical Office will set the fine and other measures established in article 99 of the Organic Environmental Law. The Technical Office must consider the principles of reasonability and proportionality, legality and due process of law. The following are other circumstances to be considered (article 28):

- The kind of intention or negligence of the defendant.
- Expected or actual benefit.
- Previous offenses.
- Non-compliance with previous warnings.
- Transfer of genetic or biochemical resources to third parties in Costa Rica or abroad.
- Publication and dissemination of information about the genetic or biochemical resources.
- Possible loss of opportunities to obtain fair and equitable sharing of benefits.
- Environmental impact, especially in the case of threatened species, in danger of extinction or with reduced populations.

For other cases, the fine can range between one and six basic salaries, according to the general principles established in this article.

Final decisions will be uploaded to the Technical Office's Web Page and in other related WebPages, including the Clearing House Mechanism of the Convention of Biological Diversity. If applicable, final decisions will also be notified to intellectual property rights protection authorities, according to articles 79 and 80 of the Biodiversity Law (Art. 32 Decree).

7.6 Following Compliance

Article 20 of the GAP, "verification and control" provides for following compliance. The Technical Office, in coordination with the authorized representatives of the site where the access to the elements or biochemical and genetic resources takes place, shall perform the pertinent tasks of verification and control. To this effect, the officials shall be able to do inspections in the property or place where the access is materialized, at any moment that the respective permit remains valid or once the activities contemplated in the permit have been concluded. The officials shall prepare acts (Actas) of their control visits. The Technical Office shall also attend to any complaints and investigate the possible violation of the terms of the prior informed consent or the terms of the access permit.

Non-compliance with the agreements and commitments shall give rise to the permit's cancellation as stipulated in article 27.

The certificate of origin is also a form of monitoring the use of genetic resources (see Art. 80 of the Biodiversity Law and 19 of the Regulations). This certificate includes details regarding: place and date of access activity; owner of the biodiversity elements or resources; the type of material obtained; and the person, communities or others who have contributed with their knowledge to the access activity. The certificate will also indicate whether the applicant complied with prior informed consent provisions in the law and its regulation and whether mutually agreed terms where achieved.

The commitment to comply with the Code of Conduct found in the Annex to the Decree 33967 is also a mechanism to promote compliance.

In addition, the control and monitoring is carried out by the use of reports. For that reason, in the approval, among other conditions, 'the obligation of the interested party to present reports and of their periodicity' shall be established (Art. 13 of the Regulations).

Besides, the contract (to obtain PIC) may establish rights to verify compliance, including audits.

Finally, recently a label was created to certify that a product has obtained a permit (limited to the commercial use category) and it is in compliance with the national regulatory framework. It is not necessary to demonstrate any other best practices or to obtain an additional third party verification.

Article 33 of the amended regulations points out that the decision on the commercial use permit, will include the authorization for the use of a distinctive label on the fair and equitable sharing of benefits, which could be included by the interested on the packages, advertising and documentation of the produces related to the permit. Upon request the TO will grant a license for the use of the label. So far the national enterprise Lisan has received the right to use the logo for a natural product derived from the use of biochemical resources.[33]

8 Regime's Strengths and Weaknesses

- The institutional framework is a key element of the ABS legislation allowing the operation and establishment of procedures, general rules and mechanisms for monitoring and control.

[33] https://www.conagebio.go.cr/Conagebio/public/distintivoABS.html.

- Costa Rica's system has evolved over time to address emerging issues and incorporate technical and conceptual developments in international and comparative law. The system has had a capacity to relatively smoothly adapt to new requirements and scientific developments.
- The existence of a national scientific capacities and, consequently, the possibilities of adding value to biodiversity elements have increased the negotiating strengths and benefit sharing which are to be stipulated in contracts or PIC agreements.
- In order to implement the law, it has been critical to count on institutional know-how and experience in key matters regarding the negotiation of agreements and contracts, the determination of what exactly is "fair and equitable", the knowledge of the markets and the existing parameters so as to be able to distribute benefits among the various sectors (pharmaceutical, agricultural, personal care, etc.).
- The scope of access and the conceptualization of what is meant by accessing and utilizing genetic and biochemical resources in practice, as opposed to using organic or natural resources that do not involve access and are therefore not regulated by the applicable legislation is not fully clear despite the amendments made by Decree No. 4591 and seems to be decided and handled on a case by case basis. The cases of access, especially as regards access to medicinal plants, nutraceuticals, taxonomic research, etc. is still giving cause for concern in various sectors. Likewise, the scope of the exceptions granted to public universities, what 'non-profit' stands for, etc. are all conceptual barriers to the adequate functioning of the access system.
- Some critical aspects call for a more meticulous study, such as access to the collections *ex-situ*, the link between access and the conservation of biodiversity, the role of the State (through CONAGEBIO) in the negotiation/approval of access applications, the establishment of efficient monitoring and control mechanisms, and the content and the implications of framework agreements.

9 Assessment and Suggestions

Some suggestions are as follows:

In order to fully implement the NP the Costa Rican Biodiversity Law, other Laws and their regulations should be amended concerning the following issues established by NP:

- Art. 15 NP. Compliance with domestic legislation or regulatory requirements on access and benefit-sharing of foreign countries.
- Art. 16 NP. Compliance with domestic legislation or regulatory requirements on access and benefit-sharing for traditional knowledge associated with genetic resources of foreign countries.

- Art. 17 NP. Establishment of check points in full compliance with the requirements found in article 17 of the NP. A decision by the Court (Tribunal Contencioso) on the legal action brought for the implementation of the Plant Variety Protection Law and the process for the consultation to the CONAGEBIO before a new plant variety protection title is issued will facilitate the further understanding on this topic.
- Simplified measures on access for non-commercial research by and procedures for access and benefit-sharing for pathogens.
- The lack of ratification of the NP and incorporation of some of its innovative provisions are a major disadvantage for a country with a functioning ABS System. By becoming a Party to the Protocol, Costa Rica could benefit from user measures to be provided by other countries of the Protocol.
- -The differentiation between commercial and non-commercial research remains unclear and depends on a decision made on a case by case basis because of the lack of specific or written criteria to provide more certainty to the applicants and the administrators about the nature of the access and when a change of intent may occur requiring the applicant to request for a new type of permit.
- Scope of ABS versus the broader use of biological resources is still not clear even with the incorporation in the amended regulations of the term "utilization". More clarity would benefit the potential users of genetic/biochemical and biological resources and create a smoother environment for the management of biodiversity /bioeconomy in Costa Rica. So far no uniform interpretation exists.
- Exclusion of basic research activities (which probably would require an amendment of the BL) could be advisable to leave out some very basic activities (like molecular taxonomy) and reduce the transaction cost of ABS for these particular cases allowing the authorities to focus on other types of applications.
- A final agreement on the scope of the exclusion of the universities from the BL and related procedures (whether if covers all type of access categories or not) will provide more certainty for the institutions and university researchers.
- The appropriate systematization of the benefits received so far would have the effect to demonstrate that the ABS regime is a tool to promote biodiversity conservation and sustainable use of biodiversity rather than to just controlling (permitting mechanisms) and increasing the acceptability of the legal regime among the different stakeholders.

10 Conclusion

The biodiversity legal framework of Costa Rica is one of the most complete. The elaboration of the law was carried out following the provisions of CBD, Bonn Guidelines and with broad participation of civil society. The starting point of this law has been the recognition of the importance of prior informed consent, traditional knowledge and mutually agreed terms.

Costa Rica is an example of a nation that decided to move ahead on ABS instead of continuing to discuss the difficulties involved. The practical experiences are embodied in permits, contracts and collaborations. It has also proven that legal regimes can adapt to the evolving ABS developments (legal, technical and economic) and build functional ABS systems. While there are still some technical matters unresolved and pending to be addressed properly, the country can provide relevant lessons learnt on how to implement ABS in practice.

References

Book Chapter

Cabrera Medaglia J (2009) The role of the National Biodiversity Institute in the use of biodiversity for sustainable development –forming bioprospecting partnership. In: Kamau E, Winter G (eds) Genetic resources, traditional knowledge and the law, UK and USA. Earthscan, pp 243–270

Reports/Studies

Cabrera Medaglia J (2003) El sistema sui generis para la protección de losconocimientostradicionales: un estudio de la Ley de Biodiversidad de Costa Rica y normativarelacionada", Document submitted to Informal Panel of WIPO on national experiences about sui generis systems of protection, Geneva, 9 July 2003

Gobierno de Costa Rica (agosto del 2019) Encuestasobremedidasnacionales que abordan la distribución de beneficiosderivadas del uso de DSI [Questionnaire sent by the CONAGEBIO in response to the process of preparation of an expert report on DSI]

MINAE (2000) Estrategia Nacional de Conservación y UsoSostenible de la Biodiversidad Obando Vet al (ed.) SINAC-INBio, San José, http://www2.inbio.ac.cr/es/biod/estrategia/Paginas/PDF/Conservaci%BEn/ENBCRfinal.pdf. Accessed 23 Nov 2020

Sittenfeld A, Cabrera J and Marielos M (2003) Bioprospecting frameworks: Policy issues for island countries. Insula Internacional Journal of Island Affairs, February, Year 12. No 1

Legal Documents

Biodiversity Law No. 7788 published in the Official Gazette (La Gaceta) No. 101 of May 27, 1998

Law No. 8686 amendment to the Patent, trademark and industrial design Law, published in La Gaceta November, 26 2008

Plant Variety Protection Law, No. 8631 published in La Gaceta March, 19 2008

Decree No. 31514-MINAE of December 15, 2004 General Access Procedure

Decree No. 33677-MINAE of 27 April 2007 on access to genetic resources found in ex situ conditions

Decree 39341-MINAE of 2016 on the procedures for the imposition of sanctions for illegal access

Decree No. 34958-MINAET-COMEX, of 2008 annulled by the Constitutional Court

Jorge Cabrera Medaglia Professor of Environmental Law, University of Costa Rica; Lead Counsel of CISDL (Montreal); strongly involved in the negotiation process in relation to CBD ABS issues and the Nagoya Protocol including as co-chair and member of several Experts Groups; member of ABS Expert Group established by FAO Commission on Genetic Resources to advice the Commission on ABS issues (2013–2017); Legal Expert Team to provide advice to Working Group for Enhancement of the Multilateral System of the International Treaty from 2017 to 2019; International Consultant on ABS working on capacity building, contract negotiations, design and implementation of ABS legal frameworks in LAC, Asia and Africa. jorgecmedaglia@hotmail.com.

The South African ABS Regime: New Wine in Old Wine Skins?

Evanson Chege Kamau

Abstract Access to biological and genetic resources and associated traditional knowledge in South Africa for bioprospecting and export for research other than bioprospecting is regulated by a number of legislations. The key among them is the National Environmental Management: Biodiversity Act No. 10 of 2004 referred to as NEMBA and the regulations thereto. Significant revisions for access and benefit-sharing (ABS) were done in NEMBA in 2009 and 2013. Besides, the regulations thereto of 2008 were repealed with the coming into force of the new amended version of 2015. Thus, the pre-Nagoya Protocol ABS legislations have been revised after the adoption as well as entry into force of the Protocol. Although South African experts say the revisions were not an action of compliance with the Protocol, several of the revisions nevertheless produce the necessary effect for compliance e.g. by creating legal certainty and transparency and facilitating access. On the other hand, a number of them create new challenges, e.g., the too many scattered amendments can easily go unnoticed by a party interested in accessing resources and/or associated traditional knowledge. In addition, this makes it hard even for a lawyer to understand the regime without difficulty. The guidelines of 2012 have long been overtaken by the new changes and at times are contradictory to these changes. In describing the ABS instruments, approaches and exploring their innovativeness, this chapter examines how the amendments fit into the old regime in view of checking whether conformity with the Nagoya Protocol is achieved and how the different unresolved issues in the Protocol are dealt with either by law or practice. Further, it is examined what new challenges the regime presents and antidotes suggested.

Keywords South African ABS regime · Bioprospecting · NEMBA · BABS regulations · Access and benefit-sharing

E. C. Kamau (✉)
Faculty of Law, University of Bremen, Bremen, Germany
e-mail: echege@uni-bremen.de

E. Chege Kamau (ed.), *Global Transformations in the Use of Biodiversity for Research and Development*, Ius Gentium: Comparative Perspectives on Law and Justice 95, https://doi.org/10.1007/978-3-030-88711-7_6

1 Background of the Development of ABS Legislation

1.1 Status of Biodiversity

South Africa is a megadiverse country and ranks third among the most biologically diverse States of the world.[1] Its richness in terrestrial biodiversity is facilitated by a mix of tropical and temperate climates and habitats, and in marine biodiversity partly due to the extreme contrast between the water masses on the East and West coast.[2] That makes it home to between 250,000 and 1,000,000 species.[3] Over 18,000 of these are vascular plant species of which 80% occur only in South Africa.[4] South Africa is the only country on earth to have within its territory an entire plant kingdom, the Cape Floral Kingdom, which is one of the global "hottest hotspot"[5] of conservation concern.[6] Although it is the smallest plant kingdom among the six recognized in the world, it is the most diverse floristic kingdom known,[7] but the most threatened with 1850 of its plant species (over 20%) facing extinction.[8] There are also other biomes (or ecological units) of global conservation significance with one third of world's succulent plant species, for instance, being in South Africa.[9] Besides the rich and varied plant life, the country has a wealth of animal life both in numbers and variety. It hosts approximately 5.8% of mammal species, 8% of bird species, 4.6% of global diversity of reptile species, 16% of total number of marine fish species in the world, and 5.5% of the world's described insect species making it the fifth richest in Africa in the number of endemic mammal, bird, reptile and amphibian species, and 24th in the world.[10] In the same way is the marine life diverse with its waters hosting over 10,000 plant and animal species comprising almost 15% of the coastal species globally known.[11] An estimated 12% of these occur only in South Africa.[12] These figures only give a picture of some of the described species and do not consider other groups such as fungi and different types of microorganisms.[13] According to a report on status and trends of South African biodiversity

---

[1] The White Paper (1997).

[2] Ibid.

[3] Lewis (2010), p. 71; ibid.

[4] The White Paper (1997).

[5] Biodiversity hotspots are areas characterized by the presence of at least 1500 endemic species of vascular plants and a loss of at least 70% of their original habitat (Lewis, above fn 3, at fn 15) and hence the high levels of species richness, endemism and threat coincide (The White Paper).

[6] The White Paper (1997).

[7] Crouch et al. (2008), p. 355.

[8] See https://www.cbd.int/countries/profile/?country=za#facts (accessed 26 February 2020).

[9] The White Paper (1997).

[10] Ibid.

[11] Ibid.

[12] Ibid.

[13] Ibid.

published on the website of the Convention on Biological Diversity (CBD) based on National Red List shows that, in addition to the threat facing the Cape Floral Kingdom mentioned above, 10% of birds and frogs, 20% of mammals and 13% of plants are threatened.[14] Besides, in terms of natural ecosystems, the 2004 National Spatial Biodiversity Assessment (NSBA)[15] revealed, *inter alia*, that 82% of the main river ecosystems are threatened, with 44% critically endangered, 27% endangered, and 11% vulnerable. Further, of the country's 440 vegetation types, 5% are critically endangered, 13% are endangered and 16% are vulnerable, 65% of the 34 marine biozones are threatened with 12% critically endangered, 15% endangered and 38% vulnerable. In addition, concerning freshwater ecosystems, the assessment revealed that only 29% of the country's main rivers were unmodified, or largely unmodified, and an estimated 50% of wetlands have been destroyed.

1.2 Historical Context, Influence and Aims

Due to the rich biodiversity with a high potential for new commercial product development from the South African biological resources the country has always been a significant interest for bioprospectors.[16] The demand on the biodiversity from local people for their livelihoods has also been always high.[17] Besides, other factors such as rapid agricultural and industrial development, afforestation, mining, urban developments and dams, introduction of exotic species, pollution or toxification of the soil, water and atmosphere have led to habitat loss and degradation, deterioration of ecosystems, impairment of ecological processes and overuse of some species.[18] These factors and the stresses mentioned above and, in addition, the unconstrained access to biological resources under the notion of common heritage for humankind prior to the CBD for use in the botanical medicines, pharmaceuticals, cosmetics and personal care sectors also contributed to the depletion of biodiversity and encroachment on the knowledge of the indigenous peoples on biodiversity-related local uses as witnessed by the cases of Hoodia, Freesias and Pelargonium.[19]

Such tendencies called for the development of a comprehensive biodiversity policy. Following the ratification of the CBD in 1995[20] a meeting was called to discuss civil society involvement in the development of such a policy as the existing

[14] See https://www.cbd.int/countries/profile/?country=za#facts (accessed 26 February 2020).

[15] See also Driver et al. (2005), pp. xiff., 15, 23 and 33.

[16] Crouch et al. (2008), p. 355.

[17] The majority of South Africans are highly dependent on natural resources for their livelihoods, well-being and health care with an estimated 70% dependent on traditional medicine from plants. See https://www.cbd.int/countries/profile/?country=za#facts (accessed 26 February 2020).

[18] The White Paper (1997).

[19] See also Lewis (2010), pp. 72f., 75 and fn 51; ibid, p. 355.

[20] See www.cbd.int.

structure by then did not provide for non-governmental representation.[21] As a result a steering committee keen to take care of this concern was constituted with mandate to manage the policy process, including ensuring the preparation of necessary policy documents, and a reference group of central and provincial governments, statutory boards and non-governmental organizations with different tasks, *inter alia*, to ensure adequate representation of the concerns and interests of different constituencies.[22] After a broad public consultation and participation which provided the opportunity for views to be expressed in different South African languages a national consultative conference was held in Pretoria in May 1996. The conference's purpose was to discuss the issues raised in the submitted documents, which, together with the conference's input, were the basis of the Green Paper.[23] In addition, the conference had to explore the range of policy options available to achieve certain goals.[24] The Green Paper was released for public comment on 28 October 1996 and comments received until 13 January 1997 were incorporated in the document. Thereafter the draft White Paper was submitted to the reference group in February 1997 with a recommendation that the policy document with minor modifications be submitted urgently to the Minister of Environmental Affairs and Tourism, and to MinMEC (Ministers and Members of Executive Councils).[25] The steering committee and the editorial committee were mandated with pursuing the policy process until the adoption of the White Paper by Parliament as formal policy.[26]

The White Paper had several goals among which were the conservation of biodiversity, its sustainable use and benefit-sharing in line with the objectives of the CBD. Shortly after the White Paper a biodiversity bill was drawn and, having passed through the full legislative process including public participation, was signed into law as the National Environmental Management: Biodiversity Act No. 10 in June 2004.[27] The Act adopted the policy goals mentioned above under objectives (s. 2 (a) (i)-(iii)) and transposed the CBD obligations relating to its third objective, i.e. the fair and equitable benefit-sharing following appropriate access to genetic resources and/or associated traditional knowledge (aTK) and their utilization, under chapters 6 and 7. Subsequently the regulations thereto were drafted, commented by the public and finally gazetted in February 2008 and came into force in April of the same year.[28]

[21] The White Paper (1997).

[22] Ibid.

[23] Ibid.

[24] Ibid.

[25] Ibid.

[26] Ibid.

[27] Crouch et al. (2008), p. 357.

[28] Ibid.

1.3 Legal Background

The Constitution (Act 108 of 1996) of South Africa (SA) does not address the issue of GR and their ownership explicitly. Indeed, it does not even use the term 'genetic resources'. However, a few answers can be obtained concerning these in Ch. 2 on 'Bill of Rights' under environmental rights (s. 24, 'Environment') and property rights (s. 25, 'Property').

The Constitution accords concurrent legislative competence to national and provincial governments for most biodiversity-related functions. This include, for instance, the administration of indigenous forests and nature conservation, except in areas demarcated as being of either exclusive national competence, e.g. national parks, botanical gardens, marine resources,[29] or of exclusive provincial jurisdiction such as museums (other than national museums).[30] It also grants executive authority for the administration of some functions at the local government level, e.g. in regard to municipal parks.[31] The national and provincial governments can also assign the local government any other biodiversity-relevant matter in line with the powers provided in s. 156 of the Constitution, in particular if it relates to the local government, or if it "would most effectively be administered locally" as long as the municipality has the capacity to administer it.[32]

The Constitution likewise protects each one's right over property. Accordingly, no law may permit arbitrary deprivation of property "except in terms of law of general application for a public purpose or in the public interest and subject to compensation"; the amount, time and manner of payment must either be "agreed to by those affected or decided or approved by a court" and must be just and equitable. This is of great relevance in the context of SA because under its common law approach,[33] it recognizes the right of ownership by the landowner of everything beneath and above the earth, except wild animals which are by law *res nullius*.[34] Although the State is the trustee of all biodiversity in SA, only over biodiversity that occurs on State land does it have ownership. In SA, much of the land, which is a form of property, falls under private ownership and consequently the biodiversity.[35] Land ownership may be categorized in general under freehold and customary tenures[36] with

---

[29]Constitution of the Republic of South Africa No 108 of 1996 (hereinafter "Constitution"), Schedule 4, Part A. See also UNEP/CBD/WG-ABS/5/5 (30 August 2007).

[30]Constitution, Schedule 5, Part A.

[31]Constitution, Schedule 5, Part B.

[32]Constitution, s. 156 (1) (b), (4) (a) & (b).

[33]SA has a mixed Roman Dutch/common law legal system (having first been colonized by the Dutch and then the British.

[34]UNEP/CBD/WG-ABS/5/5 (30 August 2007).

[35]Ibid.

[36]Wynberg (2004), p. 12.

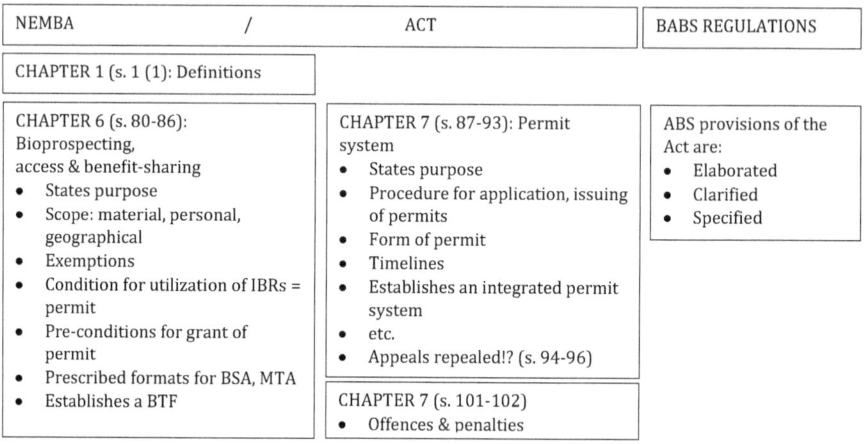

Fig. 1 Structure of the regime with the key provisions (see Addendum 1 and 2 for details). *Source:* Own illustration

most of the State and commercial agricultural land falling under freehold.[37] Consequently, biodiversity falls either under the ownership of the State, the community or an individual, except wild animals which cannot be owned by a community or individual and would be in full trusteeship of the State. To both freehold and customary land tenures, statutory law applies and to the latter (in communal areas), a layer of customary law.[38] However, none of these laws makes a distinction between genetic and natural resources besides giving them different levels of protection.[39]

More clarity is gained from the National Environmental Management: Biodiversity Act of 2004 (NEMBA) which implements the environmental rights embedded in the Constitution under s. 24. The Act says in s. 3 that in fulfilling those rights the State through its organs that are mandated to implement legislation related to biodiversity must "(a) manage, conserve and sustain South Africa's biodiversity and its components and genetic resources; and (b) implement this Act to achieve the progressive realization of those rights". It defines 'genetic resources' as "(a) any genetic material; or (b) the genetic potential, characteristics or information of any species", and 'genetic material' as "any material of animal, plant, microbial or other biological origin containing functional units of heredity" (s. 1). In line with this management mandate of the State the Act includes Ch. 6 which regulates access to, export and use of genetic and biological resources for bioprospecting or any other research and the sharing of benefits from their use, and Ch. 7 on permitting (see Fig. 1 and Addendum 1). Accordingly, the State cannot grant a permit for these purposes unless the person(s) whose interests are protected in regard to such genetic and biological resources have given their consent and material transfer and benefit-

[37] UNEP/CBD/WG-ABS/5/5 (30 August 2007).

[38] Ibid.

[39] Ibid.

sharing agreements have been concluded with them (s. 82). These persons are referred to as stakeholders and are identified under the same section as any organ of State,[40] an indigenous community or a specific individual.[41] Thus any of the stakeholders owns not only the biological resources on their land, but also has the right to control access and use of the genetic components of such resources and to share in benefits accruing from their use.

2 The ABS Framework

2.1 Legislations

The South African ABS framework is influenced by a number of legislations *albeit* two of them are key (see Fig. 1):

- National Environmental Management: Biodiversity Act, No. 10 of 2004 (NEMBA or Act) is the legislative instrument which implements the objectives of the Convention on Biological Diversity in South Africa as well as establishes and governs the ABS regime. It also legitimizes the White Paper on the Conservation and Sustainable Use of South Africa's Biological Diversity,[42] which, *inter alia*, set out the vision, mission and principles guiding a biodiversity policy and strategy in S. Africa and formulated goals for the biodiversity policy and strategy.[43] These goals include conservation of biodiversity, sustainable use of biological resources, and sharing of benefits from the use of such resources.[44] The Act was subsequently amended in 2009 and 2013.
- Bioprospecting, Access and Benefit-Sharing Regulations, 2008 (BABS Regulations or Regulations) elaborate the permit system of the NEMBA in regard to bioprospecting and export of indigenous biological resources. They were repealed and substituted by the BABS Regulations of 2015 which will be used in this study.

Other legislations of relevance to ABS are:

---

[40] An 'organ of State' is defined under Art. 239 of the Constitution as "(a) any department of State or administration in the national, provincial or local sphere of government; or (b) any other functionary or institution (i) exercising a power or performing a function in terms of the Constitution or a provincial Constitution; or (ii) exercising a public power or performing a public function in terms of any legislation but does not include a court or a judicial office".

[41] s. 82 (1) (a), (b)). S. 82 (1) (b) was substituted by s. 40 of Act 14 of 2009 by adding "or a specific individual" thus recognizing individuals' right to provide access to IBRs or aTK for purposes of bioprospecting.

[42] The White Paper (1997).

[43] Ibid.

[44] Ibid.

- Patent Amendment Act, 2005 amended the Patent Act of 1974 creating a disclosure requirement obliging patent applicants to provide any information on the role of genetic resources and/or aTK in the invention. Among others, this amendment helped to close protection gaps identified in the Indigenous Knowledge System Policy of 2004.
- Indigenous Knowledge System Policy, 2004 proposes the integration of Indigenous Knowledge Systems (IKS) in other areas of knowledge and research[45] and highlights the gaps in IKS protection.[46]
- Intellectual Property Laws Amendment Bill, 2011 was introduced to provide for the recognition and protection of certain manifestation of TK as well as amend relevant laws—*viz.* Performer's Protection Act, 1967; Copyright Act, 1978; Trade Marks Act, 1993 and Designs Act, 1993—for the same.[47]
- Provincial Ordinances draw their importance from the governmental structure of the country which comprises nine provinces and the procedure for access to genetic resources as laid down in the NEMBA and the BABS Regulations requiring that permits for export of IBRs collected, gathered or curated in the province for research purposes other than bioprospecting shall be issued by the Member of the Executive Council (MEC) of the Province in accordance to provincial statutes,[48] unless the MEC has delegated this power to the Minister according to sect. 87A (3).[49]

This study will focus on the key ABS statutes that form the South African ABS framework, i.e. The National Environmental Management: Biodiversity Act, No. 10 of 2004 and the Bioprospecting, Access and Benefit-Sharing Regulations, 2015. These legislations were promulgated by the Department of Environmental Affairs (DEA) in order to address the challenge of biodiversity loss in South Africa.[50] The fact that the BABS Regulations were promulgated to elaborate the Act's permit system in regard to bioprospecting and export of IBRs (s. 2 (a) BABS Regulations) makes it expedient for both to be examined in tandem as far as possible. This study hence will focus on the selected issues mainly from the perspective of NEMBA and the BABS Regulations the gist being the implementation of the objective on fair and equitable sharing of benefits arising from bioprospecting involving indigenous biological resources in South Africa. To be in a position to appraise the framework's approach to non-commercial research it is expedient, however, that the topic of its relationship to basic scientific research is, in

[45] Lesley (2007), p. 131f.

[46] National Study (April 2014), p. 7.

[47] Intellectual Property Laws Amendment Bill, 2011.

[48] According to section 87A of the Act as amended by section 22 of Act 14 of 2013 and BABS Reg. 6 (1) and (2), respectively.

[49] Accordingly, the Minister and the MEC may in writing delegate power to each other allowing either to take decision concerning application for a permit or type of permit that by law falls under the others' mandate.

[50] Wynberg et al. (2012).

addition, taken along. To better understand this it is helpful to look at how these statutes are structured (Fig. 1; Addendum 1 and 2).

2.2 Scope

It can be said *ex ante* that none of the definitions or criteria indicate or declare the temporal scope in terms of the validity or limitation of providers' rights, but an attempt is made to identify the activities to which these rights will continue to apply. Also the issue of later use of bulk resources in research and development (R&D) is not addressed. The scopes which are clearly defined in the regime are geographical, material and personal.

The ABS regime covers a geographical scope which is defined in s. 4 (1) (a) of the Act as the Republic. This should be understood as South Africa, including its territorial waters, exclusive economic zone and continental shelf as described in the Maritime Zones Act No. 15 of 1994, and the Prince Edward Islands referred to in the Prince Edward Islands Act No. 43 of 1948. In other words, any terrestrial or marine territory under the jurisdiction of SA. Further it states in ss. (b) that it stretches "to human activity affecting South Africa's biological diversity and its components". It is not clear whether this likewise means activities out of SA. Concerning material scope the regime applies to *indigenous genetic and biological resources* when they are accessed for purposes of *bioprospecting* or export for *any other kind of research*, and to access to aTK. The material scope is only clear enough with the understanding of these terms (shown in italics). It is therefore imperative to examine their interpretations. The terms 'indigenous genetic and biological resources' and 'bioprospecting' are defined in the Act whilst 'any other kind of research' was defined in BABS Regulations 2008 but was neither taken up in the amendments of the Act—*albeit* it is still used in the Act, nor in the BABS Regulations 2015. However, the latter refers to it as 'research other than bioprospecting' in Part 4 and annexure 3 with a similar meaning as its definition in the former. The term 'traditional use/knowledge' is defined in the BABS Regulations 2015.

Indigenous Biological Resources The term 'indigenous biological resources' in connection to bioprospecting is to be understood based on s. 1 on definitions in conjunction with s. 80 (2) on types of biological resources falling under the ABS regime. Accordingly, it means any living or dead animal, plant or other organism of an indigenous species[51] and any derivative[52] or genetic material of such animal,

[51] Section 1 NEMBA defines the term 'indigenous species' as "a species that occurs, or has historically occurred, naturally in a free state in nature within the borders of the Republic, but excludes a species that has been introduced in the Republic as a result of human activity".

[52] In relation to an animal, plant or other organism, 'derivative' "means any part, tissue or extract[,] of an animal, plant or other organism, whether fresh, preserved or processed, and includes any

plant or organism. That would include: any such resource, its derivative and genetic material whether gathered from the wild or accessed from any other source, cultivated bred or kept in captivity or cultivated or altered in any way by means of biotechnology; any cultivar, variety, strain, derivative, hybrid or fertile version of these; and any exotic animals, plants or other organisms, whether gathered from the wild or accessed from any other source which, through the use of biotechnology, have been altered with any of their genetic material or chemical compound (s. 80 (2) (a) (i)-(iii)).

As mentioned in Addendum 1, by virtue of the expansion of s. 80 (1) of the Act, the term 'indigenous biological resources' is enlarged to 'indigenous genetic and biological resources' such that whenever the former is used it should likewise contemplate 'genetic resources'. The definition of the term 'genetic resources' was amended in s. 1 (i) NEMLA Act 14 2013 and means "(a) any genetic material;[53] or (b) the genetic potential, characteristics or information of any species." By virtue of these amendments it implies that while talking about 'indigenous genetic and biological resources' information is implied.

Bioprospecting According to s. 1 of the Act 'bioprospecting' in relation to IBRs "means any research on, or development or application of, indigenous biological resources for commercial or industrial exploitation".[54] Following the revision of s. 81 (1) (a) of the Act a differentiation of the term 'bioprospecting' was made.[55] Accordingly, two categories of bioprospecting are foreseen: 'commercialization phase of bioprospecting' and 'discovery phase of bioprospecting'. The former refers to R&D "... where the nature and extent of any actual or potential commercial or industrial exploitation in relation to the project is sufficiently established to begin the process" and the latter R&D "... where the nature and extent of any actual or potential commercial or industrial exploitation in relation to the project is not sufficiently clear or known to begin the process of commercialization".[56] As the

genetic material or chemical compound derived from such part, tissue or extract" (s. 1 NEMBA. The underlined part was added by NEMLA Act No. 14 of 2013).

[53] 'Genetic material' means any material of animal, plant, microbial or other biological origin containing functional units of heredity.

[54] The definition of the term 'commercial exploitation' was inserted by s. 1(c) of Act 14 of 2013 and "means the engaging in any bioprospecting activity with the intention of making a profit". 'Industrial exploitation' is not defined in any of the instruments but in its usage we presume it means any industrial activity based on bioprospecting.

[55] S. 81. (1) (a) says that "No person may, without a permit issued in terms of Chapter 7, engage in bioprospecting involving any indigenous biological resources". The amendment of this provision by s. 38 of NEMLA Act 14 of 2009 changed ss. (a) to read "engage in the commercialization phase ...". S. 29 of the amendments in turn inserted the terms 'commercial phase of bioprospecting' and 'discovery phase of bioprospecting' in the definitions to differentiate the two types of bioprospecting.

[56] The definition of 'commercial exploitation'/'commercialization' and activities falling thereunder, and the consequences of this differentiation in regard to procedures and permit conditions are discussed below.

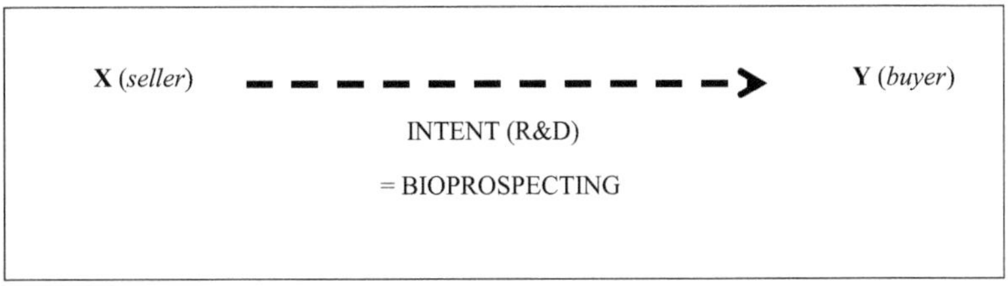

Fig. 2 Trade in and export of biological resources for bioprospecting. Deciding factor in access to IBRs. *Source:* Own illustration

definition of bioprospecting is tightly connected to the meaning given to the term 'commercial exploitation', it makes sense to consider the term under this space. According to s. 1 of the Act it "means the engaging in any bioprospecting activity with the intention of making a profit"[57] and itself comprises commercialization as the definition of the term 'commercialization' will show below. Thus, the intent behind R&D or application is critical in qualifying an act as bioprospecting or not. If there is a commercial intent, then the act qualifies as bioprospecting. For acts listed as bioprospecting, the intent is already decided as being a commercial one.

The activities regarded as bioprospecting are numbered under the definition of the term in the Act of 2004 from (a)-(c), and through the amendment of the definition in s. 1 (b) of Act 14 of 2013 up to (d). According they include:

(a) the systematic search, collection or gathering of such resources or making extractions from such resources for purposes of such research, development or application;
(b) the utilization for purposes of such research or development of any information regarding any traditional uses of indigenous biological resources by indigenous communities;
(c) research on, or the application, development or modification of, any such traditional uses, for commercial or industrial exploitation; or
(d) the trading in and exporting of indigenous biological resources in order to develop and produce products, such as drugs, industrial enzymes, food flavors, fragrances, cosmetics, emulsifiers, oleoresins, colors, extracts and essential oils.

The addition of ss. (d)[58] is, in our opinion, significant to differentiating an activity as either trade in bulk goods or trade with an intention of carrying out R&D on biological resources. In this case any activity involving trade in and export of IBRs "in order" to conduct R&D will be categorized as bioprospecting as this indicates the intent *ab initio* (see Fig. 2) to do so and would therefore fall under the ABS regime.

[57]This definition was inserted by s. 1(c) of NEMLA Act 14 of 2013.

[58]The same content is included under ss. (e) of the definition of 'commercialization'.

Indeed c. 1 of the MTA clearly indicates that the agreement can be entered into purely for trade in IGBRs for bioprospecting purposes.[59] DEA has also notified a permit issued to Zuplex (Pty) Ltd., a private sector (business and industry), to the ABS CH in the form of an internationally recognized certificate of compliance (IRCC) (valid 02.07.18–01.07.23), the subject matter of which is the production and marketing of plant extracts to local and international markets[60] (i.e. bioprospecting) the intention of which is commercial exploitation. However, the use of "trading in and exporting" in ss. (d) makes it slightly unclear whether this is implied only when export is involved, *albeit* read together with s. 1 of the MTA. Based on the definition of the subject matter of the permit referred to, it should be applicable likewise when trading in such resources for R&D and application are taking place in SA.

The definition of the term 'bioprospecting' gives a very wide scope to the SA legal framework. It covers research on, or development or application of any biological material of animal, plant or other organism native to SA and their derivatives and genetic material, and information. In addition, it also covers collection, trade and export of IBRs for commercial or industrial exploitation. This is far beyond the definition of 'utilization' in the Nagoya Protocol[61] (see Table 1). However, it makes the regime clear concerning when benefit-sharing is triggered, i.e. when any bioprospecting activity takes place. Besides, it does not limit benefit-sharing to material alone, but extends to its outputs, e.g. to information, which seems to have been the original mindset and intents of the negotiators of the CBD as reflected in article 15.7.[62]

Any Other Kind of Research In the repealed BABS Regulations of 2008 'any other kind of research' was defined as "research other than bioprospecting and (a) includes the systematic collection, study or investigation of indigenous biological resources, conducted under the auspices of a *bona fide* research institute or organization to generate scientific knowledge; but (b) excludes incidental surveys and searches". This definition might have been an attempt to lay down the ground for

---

[59] "The RECIPIENT/PERMIT APPLICANT will receive [. . .] of indigenous genetic and biological resources from the ACCESS PROVIDER as raw material for [. . . resale to other organizations or further commercial exploitation]".

[60] https://absch.cbd.int/countries/ZA (accessed 11 April 2019).

[61] 'Utilization of genetic resources' is defined in article 2 of the Nagoya Protocol and "means to conduct research and development on the genetic and/or biochemical composition of genetic resources, including through the application of biotechnology". 'Biotechnology' as defined in article 2 of the Nagoya Protocol "means any technological application that uses biological systems, living organisms, or derivatives thereof, to make or modify products or processes for specific use", and 'Derivative' "means a naturally occurring biochemical compound resulting from the genetic expression or metabolism of biological or genetic resources, even if it does not contain functional units of heredity" (Art. 2 (c), (d), (e) NP resp.).

[62] ". . . with the aim of sharing in a fair and equitable way the results of research and development and the benefits arising from the commercial and other utilization of genetic resources . . .". This is also in line with articles 1 and 5 of the Nagoya Protocol.

Table 1 Definitions of 'bioprospecting' and 'utilization'

	NEMBA	NP
	Bioprospecting	Utilization
Activity	• research on . . . • . . . or development or application of . . .	• research and development on . . .
Material scope	• any biological material of animal, plant or other organism • and their derivatives • and genetic resources • and information • aTK	• the genetic composition of genetic resources • the biochemical composition of genetic resources • derivatives • aTK
Method	•??	• . . . application of biotechnology . . .
Purpose	Bioprospecting = commercialization and industrial application with an intention for monetary gain	ALL = commercial and non-commercial

Source: *Own illustration*

differentiation of research of a non-commercial purpose based on types of research,[63] motive/intent[64] and host or proprietor of the research.[65] The intention seems to have been the exemption of what sounds like basic scientific research on biodiversity from similar requirements as those of research for commercial/bioprospecting purposes. However, it was not thorough enough in particular concerning the kinds of research which would normally fall under non-commercial/non-bioprospecting research purposes—its list is shallow and likewise very general in regard to "study or investigation of IBRs". The 'catalogue' of the South African National Biodiversity Institute (SANBI)[66] on non-bioprospecting research activities enlarges it by including:

- Taxonomy, conservation, phylogenetics, systematics, behavior ecology and export of indigenous *ex-situ* indigenous biological resources (must have an export agreement) and include collections outside their natural habitats (in herbaria, museums or in cultivations fields/gardens); and
- The keeping, breeding, cultivation, moving, trading and use of wildlife not directed at bioprospecting or the development of new plant varieties and products.

In addition, the Minister may in line with s. 86 (1) (a) of NEMBA exempt other activities from the purview of Ch. 6 through the publication of a notice based on the

[63] Systematic collection, study or investigation of IBRs as long as these do not include incidental surveys and searches.

[64] Research with none-bioprospecting motive/intent; Research intended to generate scientific knowledge.

[65] Research conducted under the auspices of a *bona fide* research institute or organization.

[66] https://www.sanbi.org/resources/infobases/biodiversity-collection-permits-in-south-africa/ (accessed 26.06.2018).

mandate accorded him under s. 80 (iv). Through Government Notice R149[67] more activities have hence been exempted some of which are, however, not research oriented but are for bulk purposes (see Box 1).

Box 1 Further Exemptions by the Minister

- Research other than bioprospecting, provided that the research is conducted within the borders of South Africa and the research is not conducted for the purposes of commercial or industrial exploitation;
- The export of *ex-situ* indigenous biological resources for purposes of research other than bioprospecting, provided the exporter has entered into an export agreement and notified the issuing authority thereof;
- The trade of commercial products purchased from a bioprospector, provided that the bioprospector has complied with the regulations on bioprospecting, access and benefit-sharing;
- The keeping, breeding, cultivation, moving, trading and use of wildlife not directed at the development and production of:
  - Products such as drugs, industrial enzymes, food flavors, fragrance, cosmetics, emulsifiers, oleoresins, colors and extracts; or
  - New plant varieties and products;
- The collection, use, propagation cultivation or trade of indigenous biolog ical resources for domestic use or subsistence purposes;
- The artificial propagation, multiplication or cultivation of flora species for the local and international cut flower and existing ornamental plant markets;
- Aquaculture or mariculture activities involving fresh water and marine species producing specimens for consumption purposes.

Source: *Government Notice R149*

Traditional Use/Knowledge "'Traditional use or knowledge' (r. 1)[68] refers to the customary utilization or knowledge of indigenous genetic and biological resources by an indigenous community or specific individual, in accordance with written or unwritten rules, usages, customs or practices traditionally observed, accepted and recognized by them, and include discoveries about the relevant indigenous genetic and biological resources by that community or individual".

Thus, the material scope covers any biological material with or without genetic potential/functional units of heredity and their derivatives and information and traditional knowledge on indigenous genetic and biological resources.

---

[67] Government Gazette 30739. Commencement date: 8 February 2008.

[68] Note that the term does not refer to the word "associated" though its definition attests to the fact that the traditional knowledge being referred to relates to genetic or biological resources.

Regarding personal scope the Act does not define the persons that fall under its scope but simply states in s. 81 (1) that "no person may engage in bioprospecting involving any IBRs or export such resources from the Republic for bioprospecting purposes or any other kind of research without a permit". This is done in r. 12 of the BABS Regulations which in addition defines the qualification criteria for such persons to be eligible to apply for a permit. Accordingly, the following fall under the Act's personal scope:

- Juristic[69] person registered in terms of SA law;
- Natural person, who is a SA citizen;
- Natural person, who is permanently resident in SA; and
- Juristic person that is not registered according to SA law or a natural person who is not a citizen of SA if these apply jointly with a registered legal person in terms of SA law or a SA citizen.

Excluded from its scope are:

(i) genetic material of human origin (sect. 80 (2)(b));
(ii) any exotic animals, plants or other organisms, other than exotic animals, plants or other organisms referred to in Sect. 80 (2) (a) (iii)[70] (sect. 80 (2) (b)); and
(iii) indigenous biological resources listed in terms of the ITPGRFA (sect. 80 (2) (b)).

3 Institutional Framework, Access Procedures and Requirements for Access to *In-Situ* and *Ex-Situ* GR and aTK

3.1 Institutional Arrangements and Procedures

The Act establishes a few institutions with a role in the ABS process and defines their mandates. These institutions are the Minister and the Member of the Executive Council (MEC), who are referred to as issuing authorities[71] (s. 87A Act), and the Director-General[72] (s. 85 (3)). Their mandates are elaborated in the BABS Regulations (r. 4 (1) and (2) and r. 40, resp.). In addition, there are other institutions which

[69]The term "juristic" in place of "legal" is used.

[70]I.e. any exotic animals, plants or other organisms, whether gathered from the wild or accessed from any other source which, through the use of biotechnology, have been altered with any genetic material or chemical compound found in any indigenous species or any animals, plants or other organisms.

[71]As defined in s. 1 an 'issuing authority' "means the Minister or organ of State in the national, provincial or local sphere of government designated by regulation ... as an issuing authority for permits of the kind in question".

[72]Defined under s. 1 as "... the Director-General of the Department".

are not mentioned in the Act and the Regulations, but which have relevance for articles 13 and 17 of the NP. These are the National Focal Point (NFP), the Competent National Authority (CNA) and checkpoint. The National Department for Environmental Affairs (DEA) plays the role of the three institutions according to the information submitted to the ABS CH.[73] These roles which concern *inter alia* facilitation and compliance are discussed below under Sect. 3.1.3 ("Department of Environmental Affairs") and Sect. 5 ("Provider and user compliance measures and sanctions for non-compliance").

3.1.1 The Minister and MEC

Accordingly, the Minister is responsible for issuing permits for commercialization phase of bioprospecting involving any IBRs or for their export from SA for the purpose of any type of bioprospecting (commercialization or discovery phase), for biotrade[74] and for the latter two integrated (r. 4 (1)). This would include any bioprospecting targeting threatened, protected or endangered species (s. 87A (1) (a) & (b)), hence permits under the Convention on International Trade in Endangered Species (CITES) Regulations, 2010 and Threatened or Protected Species (TOPS) Regulations, 2007 (No. R. 152 of 2007). For bioprospecting abroad, the Minister can issue an integrated permit for bioprospecting and export (r. 17 (2)). The MEC is responsible for issuing permits for export of IBRs collected, gathered or curated in the province if such IBRs are being exported for research purposes other than bioprospecting (r. 4 (2)). Notwithstanding this division of powers, the Minister and the MEC may in writing delegate power to each other allowing either of them to take decision concerning application for a permit or type of permit that by law falls under the others' mandate.

Section 81 (1) of the Act stipulates clearly that no person may engage in the commercialization phase of bioprospecting involving any IBRs or export from SA of such resources for the purpose of bioprospecting or any other kind of research without a permit issued in terms of Ch. 7. Parties interested in accessing IGBRs for any of these purposes must submit a permit application either to the Minister or the MEC in the prescribed forms (see Table 2). Applications for permits under the Minister's mandate must be submitted to DEA (r. 5 (1), (2)), unless the Minister has assigned or delegated powers and duties to the MEC. Applications for permits under the MEC's mandate as well as those assigned or delegated by the Minister to the MEC must be submitted to the relevant provincial department responsible for environmental affairs (r. 5 (4) & 5 (3), resp.).

[73] https://absch.cbd.int/countries/ZA (accessed 3 April 2019).

[74] According to r. 1 on definitions "biotrade means the buying and selling of milled, powdered, dried, sliced or extract of indigenous genetic and biological resources for further commercial exploitation".

Table 2 Activity, bioprospecting and export permit requirement and issuing authorities

Permit requirement			Issuing authority
Type of research/activity	In South Africa	Abroad	
Commercial phase bioprospecting	BP	BP, EP (can be integrated into one, r. 17 (2) BABS Reg.)	Minister/ DEA
Discovery phase bioprospecting	No (only notification to the Minister needed prior to commencement)	EP	Minister/ DEA
Other kind of research (Research other than bioprospecting	No	EP	MEC

BP Bioprospecting permit, *EP* Export permit, *DEA* Department of Environmental Affairs, *MEC* Member of Executive Council
Source: *Own illustration*

Discovery phase of bioprospecting and non-bioprospecting research undertaken in SA do not require a permit, *albeit* the former must be notified to the Minister prior to the commencement of the activity (s. 81A (1)).

Applications must be submitted to relevant issuing authority *via* the prescribed procedure for submission of permit applications (r. 20) which is either by e-mail, registered mail, ordinary mail, hand delivery, fax; or electronically on-line. The applicant must pay attention to the prescribed procedure for submission for the specific purpose of application.

3.1.2 The Director-General

The DG is empowered to manage the Bioprospecting Trust Fund (BTF) according to the Treasury Regulations issued in terms of the Public Finance Management Act (Act No. 1 of 1999), being "... accountable for the money in the Fund" (s. 85 (3) (a); r. 40 (1)). Following the revision of s. 85 (3) (a) through s. 41 of NEMLA Act 14 of 2009 the DG "... may appoint a trustee in terms of the Trust Property Control Act, 1988 (Act No. 57 of 1988), to administer the fund on the Director-General's behalf in the prescribed manner and under such terms as the Director-General may consider necessary". The DG must ensure that any money shared through the BTF is safe and properly used in accordance with the purpose detailed in the BSA (r. 40 (2)). He must notify the bank details of the BTF to permit holders; such details must be handed to permit holders on issuance of permits (r. 40 (3) (a)). In addition, he must inform the parties to the BSA of money received and amount due to each stakeholder or stakeholders in terms of the agreement and distribute all monies received in accordance with the benefit-sharing agreement (r. 40 (3) (b)-(c)). These powers/ obligations may be discharged annually or according to timelines stipulated in or agreed with parties to the BSA (r. 40 (4)). However, if there is any money in the BTF that is not due to any party in terms of the BSA the DG may use the funds for conservation of IGBRs, to support further research on such resources and aTK, capacity building amongst indigenous communities concerning their rights in terms of the Act, to raise their capacity to negotiate for fair and equitable BSA, to enhance scientific knowledge and technical capacity to conserve, use and develop IGBRs or any other activity that promotes the conservation, sustainable use and development of IGBRs for the benefit of SA (r. 40 (5)). But if there is money belonging to some parties in terms of the BTF but for whatever reason it is not possible to pay it to them the DG must ascertain whether there is any entity to which s/he can legally pay the funds and if none, consult any other party to the relevant BSA and if there are no other stakeholders, distribute the money in accordance with r. 40 (5).

3.1.3 Department for Environmental Affairs

DEA is responsible for advising on applicable procedures and requirements for obtaining PIC and entering into MAT both at the national and federal level. It is also responsible for granting access to all types of GR as well as issuing written evidence that access requirements have been met. In addition, it is mandated to collect or receive relevant information related to PIC, the source of GR, the establishment of MAT and the utilization of GR. Such information is provided to the relevant national authorities, the party providing PIC and to the ABS CH. Thus DEA performs all the functions of an NFP, CNA and checkpoint as foreseen in articles 13 and 17 of the NP.

3.2 Permit Application Requirements and Review

3.2.1 Pre-Conditions for Application of Bioprospecting and Export Permits

Before any of the applications mentioned above can be considered a number of requirements must be met:

(a) The applicant must at the request of the issuing authority disclose to the issuing authority all information concerning the proposed bioprospecting and the IBRs to be used in application in order to facilitate proper consideration of the application (s. 81 (2)).[75]
(b) If the person (stakeholder) providing or giving access to the IBRs and/or aTK has interests, the applicant must pursuant to:

[75] The issuing authority has the right of access to all information that "reasonably has or may have the potential to influence any decision with regard to an application for" the permit "whether or not such information is favourable to the applicant", unless such information is protected by law (r. 8).

- s. 82 (2) (a) and s. 82 (3) (a) disclose all material information relating to the bioprospecting to the stakeholder, obtain the PIC of stakeholder for access and/or use of IBRs and/or aTK.[76]
- s. 82 (2) (b) enter into MTA with the stakeholder that regulates access to IBRs and BSA that provide for the sharing in any future benefits from the proposed bioprospecting. In regard to aTK only a BSA is necessary. The BSA and MTA must have been approved by the Minister[77] (s. 82 (2) (c), 82 (3) (c)).
- rr. 14 (3) (e), 17 (4) (e), 18 (4) (e), 19 (4) (e) pay non-refundable fee for discovery phase bioprospecting, bioprospecting, integrated biotrade and bioprospecting and export permit for research other than bioprospecting respectively, as prescribed by Annexure 4.
- s. 89 provide at own expense independent risk assessment or expert evidence if required by the issuing authority.

(c) Applications must contain all information as well as attached documents prescribed in application forms (r. 20 (3)) which include, according to:

- annex 1 and annex 2 regarding notification and export respectively for discovery phase of bioprospecting, details about the applicant/bioprospector. If it is a juristic person, name of institution or body, whether registered in SA or not and registration number and contact details of contact person (and in regard to export the person's certified copy of ID or PP). If the applicant/bioprospector is a natural person, name of applicant, ID No. and contact details, whether s/he is affiliated to a juristic body and if yes name of the juristic body, contact person and his/her contact details. Details of bioprospecting project leader including name, identity or PP No., citizenship and contact details. Information about the bioprospecting project must be attached to the notification indicating project title, objective, and proposed methodology and timeframe. Classification of all IBRs for which the notification is submitted including genus and species, projected quantity resources to be collected or obtained and the specific area or source from which each resource is to be collected or obtained, prior permits obtained or applied for, project sponsors and their location (in SA or international) and contact details, whether aTK shall inform the project and if yes, details and source of information. For export, the details of access providers for resources and/or aTK must be indicated and MTA and/or BSA concluded with them. A commitment to comply with permit requirements for commercialization phase of bioprospecting when the project enters into commercialization

[76]The issuing authority is required to ascertain that PIC for use of the stakeholder's knowledge of or discoveries about the IBRs to be used in the bioprospecting has been obtained (s. 81 in comb. with 82 (1), (2) (a), (3) (a) NEMBA and r. 8 BABS Regulations).

[77]The intervention of the issuing authority might be sought for the purposes of facilitating the negotiations for such agreement if it has not been possible to conclude the agreement(s) (s. 82 (4) (b); r.16 (4) (d)). It is not clear whether the process of application review can go on without the MTA/BSA as the Minister facilitates their conclusion.

phase signed by the project leader and, if applicable, endorsed by the juristic body.

- annex 3 regarding export permit for research other than bioprospecting and renewal of such a permit, name of institution or body if applicant is a juristic person, whether registered in SA or not and registration number and contact details of contact person and the person's certified copy of ID or PP. If the applicant is a natural person, name of applicant, his/her certified copy of ID or PP and contact details, whether s/he is affiliated to a juristic body and if yes name of the juristic body, contact person and his/her contact details. Names of collaborators and their contact details and certified copy of their ID or PP. Names and contact details of individuals who will conduct research, their citizenship, contact details and ID or PP No., names of research project sponsors and whether they are SA or international, their contact details, names of recipients/importer of IBRs, their contacts and certified copies of their ID or PP, type of indigenous biological resources for which a permit is sought, the family, genus and species, the part of the organism to be collected, the quantity of the resources to be collected or obtained and the specific area or source from which each resource is to be collected or obtained, purpose for which the IBRs are to be exported, any other permit application previously or simultaneously submitted in respect to the same IBRs and if yes whether the application was approved, refused or is pending, if application was approved provide details of permit (No., issuing authority, date of issue) and if pending, the reference No. of issuing authority, state the purpose for which the resources are exported and attach the project proposal showing the research objectives, proposed methodology and timeframe, relevant environmental considerations as well as impacts of collection and antidotes, reporting processes, anticipated outcomes, fate of discarded/waste specimens following the conclusion of the research and possible benefits. If the application is for renewal of a permit, it must be accompanied by project annual and biannual reports. If it is for an amendment of a permit, it must be accompanied by a covering letter with a motivation and only sections of the application form concerning amendment should be completed.
- annex 5 regarding commercialization phase of bioprospecting, details of applicant including name, capacity, nationality, certified copy of ID or PP and contact details, if the applicant is a juristic body or affiliated to a juristic body provide registration of the juristic body, country of registration and No., contact person and his/her contact details, collaborators and their details, whether in respect to the IBRs to be collected/obtained a notification for discovery phase of bioprospecting has been previously made and if yes the contact person, notification reference No. and a summary of results, IBRs to be collected/obtained including their scientific and common names, part of resource to be utilized, physical state of resource (i.e. format in which IBR is commercialized), quantity, locality/source of resource and access provider and aTK holder if applicable, whether any other application for permit has been previously or simultaneously submitted in respect to the same IBRs

(e.g. CITES, TOPS in terms of the Act, or harvesting, phytosanitation in terms of provincial ordinances) and if yes permits granted, their reference No. and issuing authority, access providers for resource and aTK holders and MTA and/or BSA concluded with them, description/nature of aTK, business plan or project summary indicating project title, objectives, desired outcomes, proposed methodology and timeframes, possible benefits for the conservation of SA biodiversity, economic development and other benefits that are in the interest of the public, likely environmental impacts and proposed steps to minimize or remedy them, how unused materials will be disposed, recipient (s) of IBRs and their location and MTA and/or BSA concluded with them.

(d) Applications must be signed by the applicant/project leader and sometimes be initialized in all the pages of the application and, where applicable, be endorsed by the duly authorized person of the juristic body.
(e) Completed documents must be submitted to the relevant issuing authority *via* the prescribed submission procedure.
(f) Any information that may be required in line with s. 88 (2) (a) and r. 8 which gives the issuing authority leeway to request for additional information must be furnished, "unless access to that information is protected by law".

3.2.2 Review of Notifications and Permits

Procedure of Review

The review process is carried out by a National Bioprospecting Advisory Committee established by the Minister, which provides technical input during the evaluation of permit applications.[78] The Committee makes recommendations for the Minister to issue or reject a permit application.[79] It meets four times per year (or when four applications are available[80]) and considers four applications per meeting. However, *ad hoc* meetings are convened when necessary.[81]

Criteria to Be Taken into Account and Requirements to Be Fulfilled by Issuing Authority

The issuing authority can only approve/grant notifications and permits subject to the fulfillment of certain conditions laid down in the Act (s. 82 (1)–(3)) and Regulations (r. 9) as well as other conditions that it may impose, and provision of additional

[78] Marumo et al. (2013).

[79] Ibid.

[80] https://www.environment.gov.za/sites/default/files/legislations/guide_complete_permitapplication.pdf (accessed 31 March 2020).

[81] Marumo et al. (2013).

information that it may require (s. 88 (2) (a)). Therefore, prior to such approval/grant the issuing authority must consider documentary proof of all factors while reviewing applications to ensure *inter alia*:

- it is complying with the Act (r. 9 (1) (a));
- the relevant stakeholders have been identified by the applicant (s. 82 (1); r. 9 (1) (b) (i));
- relevant information has been disclosed to all the stakeholders (s. 82 (2) (a), (3) (a); r. 9 (1) (b) (ii));
- the applicant has obtained the PIC of the provider of resources and/or aTK (s. 82 (2) (a), (3) (a); r. 9 (1) (b) (iii), (iv));
- MTA/BSA have been concluded (s. 82 (2) (b), (3) (b); r. 9 (1) (b) (iii)) and approved by the Minister (s. 82 (2) (c), (3) (c));
- any comments received from relevant stakeholders or affected ILCs or individuals during consultation meetings between them and the issuing authority have been considered (r. 9 (1) (b) (v));
- the activity to be undertaken will contribute to the conservation of biodiversity in and economic development and enhancement of the scientific knowledge and technical capacity of SA (r. 9 (1) (b) (vi), (vii));
- any guidelines and departmental policies relevant to the application are being considered (r. 9 (1) (b) (ix));
- the applicant has not violated any provisions of the Act (r. 9 (1) (b) (x)); and
- the affected ILC has adopted a community resolution authorizing a representative to give PIC, provide access and enter into a BSA on behalf of the community (r. 9 (1) (b) (xii)).

In addition to these criteria the issuing authority must:

- ensure that the applicant and the stakeholder are on an equal footing and benefit-sharing arrangements are fair and equitable. For this the issuing authority may facilitate negotiations (s. 82 (4) (a)-(c)); and
- assess the potential impact of the activity on the environment and ensure that such impact will be negligible and would not jeopardize the integrity of the biological resource (rr. 22–27).

Timeframe for Deciding Notifications and Permit Applications

The timeframe for deciding notifications and permit applications are regulated under r. 21 as follows:

- once the issuing authority has received a complete notification or permit application such a notification or application must be decided within 120 working days from the date of receipt.
- if additional information is required the issuing authority must request the applicant in writing within 10 working days after the date of receipt to submit

such information within 20 working days from the date of receipt of such a request.

- if the additional information requested is not submitted to the issuing authority within 30 working days after such a request, such an application must be deemed withdrawn.
- if the information submitted by the applicant to the issuing authority is sufficient, but the issuing authority is unable to comply with the timeframe of 120 working days from the date of receipt,[82] the issuing authority may extend the timeframe by 30 working days on condition that s/he informs the applicant accordingly before the initial timeframe (120 working days) lapses and provides reasons why it is impossible to comply with the initial timeframe.
- upon the expiry of the extended timeframe[83] the provisions of s. 6 (2) (g) and (3) of the Promotion of Administrative Justice Act, 2000 will apply.[84]

Decision to Issue, Issue Subject to Conditions or Reject

Once the issuing authority has reviewed the application as required by law it will decide to either approve it unconditionally, approve it subject to certain conditions or reject it (s. 88). This decision must be consistent with law (s. 88 (3)) and be communicated to the applicant and relevant stakeholders in writing and, on written request by these, reasons for the decision within 15 working days (r. 10). The issuing authority may approve an application and register a notification as well as issue a permit if it complies with the requirements for applications as summarized above (i.e. in line with rr. 9 and 22 (r. 29 (1) and 9, 23–27 (r. 31 (1) concerning notifications and permits respectively, *albeit* without retrospective effect (rr. 29 (5), 31 (5)).

The issuing authority may refuse to consider as well as register an application that does not comply with the requirements for applications and thus reject it (rr. 28, 29 (1), (4), 31 (1), (4)). An application may also be issued subject to conditions. In regard to both cases of rejection or issuance of the application subject to conditions the issuing authority must, at the request of the applicant, provide written reasons for the decision (s. 88 (5); rr. 29 (3), 31 (3)). A decision to subject the application to conditions or reject it can be appealed by the applicant (rr. 29 (4), 31 (4)).

[82]From a reading of the provision this should be the date of (the first) submission/receipt of the notification or permit application.

[83]That would mean 150 workings days from the date of (the first) receipt.

[84]Accordingly, the action of the issuing authority will be considered as failure to take a decision before the expiration of the period prescribed by law giving the applicant the right to institute proceedings in a court or tribunal for judicial review on the ground that the administrator has a duty to take the decision notwithstanding the expiration of that period.

4 Nature of Regulatory Tools for Access to *In-Situ* and *Ex-Situ* GR and aTK

4.1 What Is Access?

Prior to discussing the tools used to regulate access to IGBRs and aTK it is crucial to understand how the term 'access' is used by the ABS regime. Apparently, none of the instruments define the term 'access'. However, as used by the NEMBA it portrays the act of taking/collecting, e.g.: "... in the areas where the indigenous genetic or biological resources and associated traditional knowledge are **accessed**" (s. 80 (1) (d)); "any indigenous biological resources... whether gathered from the wild or **accessed** from any other source, including any animals, plants or other organisms ..." (s. 80 (2) (a) (i)); "any exotic animals, plants or other organisms, whether gathered from the wild or **accessed** from any other source ..." (s. 80 (2) (a) (iii)); "A person ... providing or giving **access** to the indigenous biological resources ..." (s. 82 (1) (a)); "the applicant ... has obtained the prior consent of the stakeholder for the provision of or **access** to such resources" (s. 82 (2) (a)); "a material transfer agreement that regulates the provision of or **access** to such resources" (s. 82 (2) (b) (i)); and "the type of indigenous biological resources to be provided or to be given **access** to" (s. 84 (1) (b) (ii)) (*bold added*). The same is reflected in the BABS Regulations. Taking/collecting is contrasted well by the term 'provision'. The latter seems to exclude the active participation of the recipient from the initial activity of alienation of the resources from their *in-situ* conditions. Therefore, this term seems not to cover other activities such as R&D, but due to the definition of the term 'bioprospecting',[85] any (systematic) search, collection or gathering of IBRs (which is access) would comprise bioprospecting if the intent is to conduct research for commercial or industrial development.

4.2 PIC for Access

As already mentioned, in order to ensure that the interests of stakeholders are protected their consent must be obtained prior to provision or access and use of GR and aTK. Prior to any of these activities, the PIC of the 'stakeholder', i.e. a person with interest, or whose interests must be protected, must be obtained. Again, under this term it is determined which persons have the right to provide GR and/or aTK and hence to grant PIC. According to s. 1 on definitions a 'stakeholder' means any entity contemplated in s. 82 (1), i.e., in regard to provision of GR, a person (which should mean a natural or juristic person), including any organ of State or a community, and in regard to traditional use or knowledge associated with the GR, an

[85] See text Sect. 2.2, "Bioprospecting".

indigenous community or a specific individual. It can therefore be concluded that a natural or juristic person will give the PIC for access to IGBR (except wild animals) if the resource is located on private/corporate land and/or individual TK associated with the resource; the (indigenous or local) community the PIC to IGBR (except animals) if the resource is located on communal land and/or aTK held by the community; and the State the PIC to IGBR (including animals) if the resource is located on State land.

In the following we shall only focus on the PIC of the State and the community. It is not clear what the common content of a PIC is as there is no model PIC except forms of notification and different permits as well as MTA and BSA.

4.2.1 PIC of State

Following logic or trends of most ABS regimes, an organ of State mandated with biodiversity conservation functions or managing certain ecosystems would normally have the responsibility of granting the State PIC for access in areas under their mandate. Although the SA regime also requires that the applicant of a permit proves that the PIC of the organ of State (if the State was the stakeholder) was obtained as one of the pre-conditions of being granted with a permit, it does not seem to necessarily conform to the same pattern. One, the legislation is not definite concerning which organs must be contacted in regard to PIC, but rather to permit. Second, the organs/institutions mandated to issue permits, i.e. DEA at the national and provincial levels and different organs at the provincial level (referred to either as department, agency or board, *see* Table 3), seem to likewise be responsible for giving consent creating the impression that PIC is subsumed in the permit. However, the instructions of SANBI (South African National Biodiversity Institute) explicitly require that specific institutions are contacted if collection of genetic resources/ biological material/specimens will take place in any of the national parks or forestry areas under their jurisdiction.[86] This suggests that their consent should be sought. These institutions are the Department of Agriculture, Forestry and Fisheries (DAFF), Department of Water and Sanitation (DWS), Eastern Cape Parks and Tourism Agency (ECPTA), North West Parks and Tourism Board, and South African National Parks (SANParks).[87] This, nonetheless, does not seem to be based on the type of genetic resource (permit, purposes (use), intention, or user).

[86] SANBI Website, https://www.sanbi.org/resources/infobases/biodiversity-collection-permits-in-south-africa/ (accessed 22 March 2019).

[87] Ibid.

Table 3 State organs mandated to issue permits

Level	State Organ	
National	Department of Environmental Affairs (DEA)	
Provincial	**Province**	
	Eastern Cape Province	The Department of Finance and Economic Development, Environmental Affairs and Tourism
	Free State Province	The Department of Economic, Small Business Development, Tourism and Environmental Affairs
	Gauteng Province	The Gauteng Department of Agriculture and Rural Development (GDARD)
	KwaZulu-Natal Province	KwaZulu-Natal Nature Conservation Board trading as Ezemvelo KZN Wildlife
	Limpopo Province	Department of Economic Development, Environment and Tourism (LEDET)
	Mpumalanga Province	Mpumalanga Tourism and Parks Agency (MTPA)
	Northern Cape Province	Department of Environment and Nature Conservation (DENC)
	North West Province	Department of Rural, Environment and Agricultural Development (READ)
	Western Cape Province	The Western Cape Nature Conservation Board trading as CapeNature

Source of information: *SANBI Website, https://www.sanbi.org/resources/infobases/biodiversity-collection-permits-in-south-africa/ (accessed 22 March 2019)*

4.2.2 PIC of Community

As mentioned above, the community will give PIC for access to genetic and biological resources located on communal land and to aTK held by it. The process of obtaining the community PIC can be arduous. In order to ease it, a system of representation is established. Accordingly, the granting of PIC, provision of access and conclusion of the BSA are done by the representative(s) of the community. For this a template of community resolution has been developed and is annexed to the Regulations.

The resolution acknowledges the community as the owners of the IGBRs and holders of the aTK to be accessed, records the date the community and its representatives met to discuss the bioprospecting project including the activities foreseen; contains the community's declaration of its awareness of the potential user's interests and planned activities, its consent to share the resources and the associated knowledge; specifies by name the community representative and the number of persons authorized to sign the resolution on behalf of the community and their names and ID numbers. As an alternative the community resolution can be signed by an executive committee. The resolution is categorical concerning authorizations accorded as being valid exclusively and only for the applicant and purposes foreseen therein and not in relation to any other agreement.

4.3 Research Permit

4.3.1 Bioprospecting Permit

A bioprospecting permit is only required for bioprospecting purposes. As already mentioned, there are two types of bioprospecting: commercialization phase of bioprospecting and discovery phase of bioprospecting[88] both of which aim at commercialization. For the former a bioprospecting permit must be obtained whether the activity is to be conducted in SA or abroad. The form of an access permit for commercialization phase of bioprospecting is fixed to the Regulations under Annexure 9 which shows the contents and standard conditions of the permit. For the latter an export permit must be obtained when it is conducted abroad. The form of an export permit for discovery phase of discovery is contained in Annexure 6 and is very similar to the permit for commercial phase of bioprospecting except for a few differences. In regard to general information the former has no option for renewal permit number. Under standard conditions it is explicit that the relevant IGBRs cannot be used for commercialization phase of bioprospecting without a bioprospecting permit issued under BABS Regulations. Besides, concerning third party transfer, no consent shall be granted if the intent of the third party is to use such resources without a bioprospecting permit. In addition, a progress report can either be submitted annually or by a date fixed by the issuing authority. Finally, no benefit-sharing and MTA related issues are addressed, *albeit*, according to r. 14 (c) and (d) signed MTA and BSA must be attached to the application for an export permit.

The term ‘commercialization’ is only elaborated through enumeration of acts that it includes. From its common usage it entails the process of bringing a product or a method of production into the market, thus engaging in the profit-making act, whilst ‘commercial exploitation’ as defined above (see ‘bioprospecting’) is the process through which such a product or method is achieved, but again, with the clear intention of making a profit. The activities that comprise commercialization in relation to IBRs are listed in s. 1[89] and include:

(a) the filing of any complete intellectual property application, whether in South Africa or elsewhere;
(b) obtaining or transferring any intellectual property rights or other rights;
(c) commencing product development, including the conducting of market research and seeking pre-market approval for the sale of resulting products;
(d) the multiplication of indigenous biological resources through cultivation, propagation, cloning or other means to develop and produce products, such as drugs, industrial enzymes, food flavors, fragrances, cosmetics, emulsifiers, oleoresins, colors, extracts and essential oils;

[88] See differences between commercialization phase of bioprospecting and discovery phase of bioprospecting above.

[89] The definition of “commercialization” was inserted by s. 29 of NEMLA Act 14 of 2009 and later amended by s. 1(d) of NEMLA Act 14 of 2013.

(e) trading in and exporting of indigenous biological resources to develop and produce products, such as drugs, industry enzymes, food flavors, fragrances, cosmetics, emulsifiers, oleoresins, colors, extracts and essential oils; and
(f) commercial exploitation.

4.3.2 Contents

The contents of a bioprospecting permit will be discussed only based on the form of a permit for commercialization phase of bioprospecting according to Annexure 9 of the Regulations.

General Information

Accordingly, the permit contains the following information:

1. The name of the issuing authority and contact details;
2. The details of the permit holder and other permit holder if a joint permit application has been made (i.e. names of the applicants, their ID or PP No. of which a certified copy of the first holder must be attached and their detailed contacts including telephone (landline and mobile), fax, email and postal/physical addresses); details of the importer/recipient (i.e. same as permit holder except a copy of the ID/PP);
3. The nature of the permit specifying the IGBRs the permit holder is allowed to access/use (type of organism, family, genus or species—scientific and common names) as well as the quantities and areas of collection, points of export and import if applicable;
4. Permit validation showing when the duration of validity commences and ends;
5. An entry of the MTAs and BSA entered under the permit and together with which the relevant permit must be read; and
6. At the bottom end, the permit number or permit renew number if applicable, the signature of the issuing officer, date of signing, and the stamp of the issuing authority, and the signature of the permit holder, date of signing, and his/her full names.

Standard Conditions

In addition, it contains typical conditions as indicated on a non-exhaustive list, *viz.*:

1. The bioprospecting purpose for which the IGBRs to which the permit relates may only be used;
2. The obligation of the permit holder(s) to comply with all other legislative requirements for the collection of the IGBRs (if any);

3. The permit holder's(s') liability relating to any costs of mitigating or remedying the impact of the bioprospecting on the environment, in accordance with s. 28 of the National Environmental Management Act, 1998 (Act No. 107 of 1998);
4. The obligation to pay all money due to stakeholders in terms of the approved BSA into the BTF, as required by s. 85 (1) of the Act;
5. The obligation of the permit holder to notify the Department when money due to stakeholders as specified in the approved BSA will be transferred to or paid into the BTF;
6. The obligation of the permit holder to notify stakeholder or stakeholders entitled to a monetary benefit in terms of the approved BSA that money was transferred or paid into the BTF;
7. The obligation of the permit holder to submit to the issuing authority and in a form determined by it a status report on an annual basis or timeframes determined by the issuing authority;
8. The obligation of the permit holder not to transfer the IBRs to which the permit relates to any third party without the prior written consent of the issuing authority and then only under a written agreement containing terms no less restrictive than those which apply to the permit holder in terms of the relevant permit and any agreements related thereto;
9. The obligation of the permit holder to notify the issuing authority in writing if new collaborators join the bioprospecting project for which the permit issued relates;
10. The MTA concluded with the subsequent user of traded IGBR should include a clause stating that the use of SA IGBRs for bioprospecting is subject to the provisions of the NEMBA 2004 (Act No. 10 of 2004) and that a bioprospecting permit is required; and
11. Failure to comply with any of the permit conditions renders the permit invalid and may result in criminal proceedings, cancellation of the permit/s and seizure of the consignment/s.

This study pays little focus to biotrade, but since activities related thereto are regulated under the ABS regime it might be helpful to mention that access permits for biotrade and bioprospecting barely differ. The only differences are that for a biotrade permit reporting is on biannual/annual basis and the permit holder is not permitted to engage in R&D or production of final products utilizing the IGBRs to which the permit relates without a bioprospecting permit or an integrated biotrade and bioprospecting permit. In addition, the event of collaborators joining a biotrade undertaking and the consequences are not addressed in the biotrade permit. However, similar to the bioprospecting permit, there is leeway for an issuing authority to include more/additional conditions.

4.3.3 Export Permit for Research Other than Bioprospecting

Nothing in the legislation suggests any access permit requirement for any other kind of research/research other than bioprospecting conducted in SA or abroad. If it is conducted abroad, however, an export permit must be obtained. The form of the export permit is contained in Annexure 7 the general information of which is similar to that of a commercial phase of bioprospecting permit except a few differences under standard conditions. The most critical conditions are that it categorically forbids the use of IGBRs related to the permit for bioprospecting purposes and does not offer a window for a third party to obtain consent for bioprospecting purposes in relation to such resources. Hence, a change of intent from basic/scientific (non-commercial) research to commercial is impermissible. No benefit-sharing issues are addressed in the permit, but the issuing authority has leeway to add more/additional conditions. It is unclear whether such conditions may be of a benefit-sharing nature.

4.3.4 Other Permits

As already mentioned, access to endangered or protected species is subject to CITES and TOPS permits as the case may be.

4.4 Model Agreements

SA operates a MTA and a BSA which are included in the Regulations. These are post-NP agreements being part and parcel of the new Regulations of 2015. The agreements must be entered into by an applicant for a permit and any stakeholders identified in terms of the Act (s. 82 (1)) and the Regulations (r. 38 (1)) for access to any IGBRs and aTK. If the stakeholder is a community, a representative of the community identified in the community resolution will enter into the agreement on behalf of the community (r. 38 (4)). If there are different stakeholders a separate MTA must be entered into with each one of them. Parties can add annexes to the form provided if the space is insufficient or use their own forms as long as the format of the model MTA is followed. There is no indication in the MTA that parties can add other conditions not considered in the provided form but parties to the BSA may include other matters concerning the GR and/or aTK (c. 10). None of the agreements seems to allow parties to expunge, replace or rectify some of the existing clauses. In addition, it is not possible to use other models e.g. availed by the recipient/permit applicant.[90]

[90]The Act states clearly that the BSA and MTA must be in the prescribed format (s. 83 (1) (a) and s. 84 (1) (a) resp.). The Regulations concretize this rule by stating that the prescribed formats must be those set out in Annexures 11 (for MTA, r. 38 (2)) and 12 (for BSA, r. 39 (4)).

Table 4 Types of benefits

Non-monetary, monetary or any kind of benefits	
Support for conservation	Access to international collections by South Africans
Species inventories	Recognition and promotion of traditional knowledge/use
Student training and support	Community development projects
Scientific capacity development	Environmental education
Technology transfer	Fees
Joint research	Royalties
Information	Upfront payments
Equipment and infrastructure	Milestone payments
Other (specify)	Other financial benefits (specify)
Other (specify)	Other (specify)

Source: *c. 5 BSA*

The agreements must be signed by both parties in the space indicated and initialized in every page of the agreement including annexures, and must be approved by the Minister.

4.4.1 Model Transfer Agreement

The MTA is attached to the Regulations under Annexure 11. The typical contractual provisions used to enforce compliance with provider requirements are found under clauses on 'rights and duties' and 'third party transfer'.

Intellectual Property Rights The MTA provides for the right of the recipient/ permit applicant to claim patents or other IPRs connected or referring to the IGBRs for new methods of utilization of such resources or new process of their preparation, production or manufacture subject to compliance with the Act and the Regulations as well as prior expression and written permission from the issuing authority (c. 6).

Change of Intent/Use The recipient/permit applicant must not utilize the IGBRs for other purposes other than those agreed upon and stipulated under c. 5, in particular R&D linked to new and useful properties of the IGBRs, without a (new) PIC and conclusion of a BSA with regards to the proposed utilization changes (c.7). In addition, s/he must not utilize the IGBRs "for germination, propagation, breeding, tissue breeding, cloning or in any way seek to capture genetic material for the purpose of reproduction without PIC from and MAT with the issuing authority and access provider (c. 8).

Third Party Transfer The recipient/permit applicant is only allowed to sell, transfer or make available the IGBRs to third parties under condition that the subsequent recipient and other third parties are equally bound by the terms and conditions of the first agreement—with the initial recipient/permit applicant (c. 9). Hence the MTA embraces the viral clause approach.[91]

Besides, the MTA has clauses on breach and termination of the agreement (c. 14). Accordingly, in case of violation against any material provision of the agreement, the aggrieved party shall serve the breaching party with a written notice requiring the former to rectify that breach within 30 days of receipt (of the notice) failure to which or insufficient rectification of which the aggrieved party may "sue for immediate specific performance of any of the defaulting party's obligations under this agreement, whether or not such obligation is then due". During this time the suing/ aggrieved party must not derogate from any of its other rights or remedies under the agreement or at law. If the remedy of specific performance or damages would not adequately prevent the aggrieved party from being prejudiced, it may cancel the agreement by serving the defaulting party with a written notice of cancellation.

The MTA has no clauses on publication, storage of material or its handling after conclusion of the research and applicable law. The details of the parties and the IGBRs being received or transferred, including inadvertently with such resources, and "any genetic material, biological molecules and biochemical compounds naturally occurring" in such IGBRs and the purpose(s) of access are also inserted in the MTA (cc. 1–2 and 3–4; 10 and 11 respectively).

4.4.2 Benefit-Sharing Agreement

The BSA is attached to the Regulations under Annexure 12 with two sections: one on IGBRs and the other on aTK. If access involves community GR and aTK then both sections must be completed. The difference between the two sections, however, is not much, except that the parties entering into the agreement are different depending on the object of access. Therefore, only the first section of BSA for indigenous genetic and biological resources is examined.

The typical contractual provisions used to enforce benefit-sharing obligations are found under clauses on 'sharing of benefits', 'payment of benefits' and 'third party transfer'. The clause on 'review of agreement' will also be examined.

Sharing of Benefits The BSA contains a non-exhaustive list of non-monetary and monetary benefits that can be shared from bioprospecting projects (Table 4) which is not verbatim to the Bonn Guidelines/NP list. These benefits will vary depending on the case and nature of the project.

Payment of Benefits According to c. 6 any monetary benefits due to the access provider under the agreement except direct payments of fees/charges to the beneficiaries must be paid into the BTF. The BSA serves as a deed of trust in terms of the Public Finance Management Act of money due to the access provider of the IGBRs.

[91] See Kamau EC, Winter G, Stoll P-T (eds) (2015), pp. 19, 282, 346f. Cf. DFG (Ed.) (2020). See also Winter in this volume.

According to information notified to the ABS CH *de facto* benefit-sharing has taken place with the payment of more than two million Rands to the stakeholders providing access to genetic resources and various community development projects implemented as non-monetary benefits.[92]

Third Party Transfer The rule for transfer of IGBRs to a third party under the BSA seems to differ with the rule under the MTA. The recipient/permit applicant may not transfer the resources to a third party without the written authorization of and under a legally binding written agreement with the access provider based on the agreement (c. 8) whilst under the MTA such transfer can be executed under a viral clause. It is not regulated which clause will prevail if the recipient/permit applicant and the access provider have entered into both agreements (MTA and BSA) causing confusion. This matter should be addressed by the guidelines.

Review of Agreement The agreement may be reviewed at an interval not determined in the agreement but, according to a reading of the clause ("fill in agreed timeframe"), the interval must be fixed mutually by the access provider and the recipient/permit applicant (c. 7). The permit holder must disclose any new material information concerning the bioprospecting activity to all stakeholders to enable them to participate in the review from an informed basis. The review can lead to an amendment. This is important because such a review can act as a monitoring tool as well as ensure fairness and equitability in benefit-sharing, but it can also create uncertainty in particular if the access provider insists on very regular reviews.

5 Provider and User Compliance Measures and Sanctions for Non-compliance

5.1 Provider Compliance Measures

Besides, the Patents Amendment Act of 2005 was reviewed to include a disclosure requirement under sections 3A and 3B as a mechanism of checking patent applications based on GR and aTK. Accordingly, every applicant who lodges an application for a patent accompanied by a complete specification must, prior to the acceptance of the application, "lodge with the registrar a statement in the prescribed manner stating whether or not the invention for which protection is claimed is based on or derived from an indigenous genetic or biological resource, or traditional knowledge or use. The registrar shall call upon the applicant to furnish proof in the prescribed manner as to his or her title or authority to make use of the indigenous biological resource, genetic resource, or of the traditional knowledge or use if an applicant lodges a statement that acknowledges that the invention for which protection is claimed is

[92]https://absch.cbd.int/countries/ZA (accessed 3 April 2019).

based on or derived from an indigenous biological resource, genetic resource, or traditional knowledge or use".[93]

Further, users exporting GR to other countries are required to show their permit at the ports of entry and exit as proof of compliance with the national ABS measures. The patent office and the ports of entry and exit hence are critical checkpoints for controlling compliance with national measures.

Any information related to PIC, to the source of the genetic resource, to the establishment of MAT (conclusion of MTA/BSA) and/or utilization of GR, which is usually condensed in an internationally recognized certificate of compliance (IRCC) is submitted to the ABS CH as required by the NP (Art. 17 (1) (a) (iii)). This procedure is likewise critical in effecting compliance after the GR have left national borders.

5.2 *User Compliance Measures*

SA has not taken measures according to articles 15 and 16 NP to ensure compliance with providers' regulations when foreign GR and aTK are utilized in SA. However, practice has some mechanisms to monitor utilization.[94] As mentioned above, the port of entry/exit and the patent office have been established as checkpoints. Users are required to provide the necessary information at these checkpoints. In addition, such information, including from IRCC, which is evidence that the GR and/or aTK was accessed in accordance to PIC and that MAT were established, is made available to the ABS CH without prejudice to the protection of confidential information. Besides, if the issuing authority becomes aware of any utilization of foreign GR and/or aTK by a SA institution within its jurisdiction, the SA NFP notifies the NFP of the provider country about the resources that are being imported into SA. In addition, the user is requested to provide documentation of compliance with the provider country's regulatory requirements, in particular the existence of PIC and MAT. If the user does not provide proof of compliance, the SA NFP informs the NFP of the provider country for it to take the necessary action. Also the disclosure procedure described in Sect. 5.1 above should have a positive impact on checking whether the user is in possession of all necessary documentation. The Act is currently undergoing through an amendment process to include concrete user legislative measures.

---

[93] Patents Amendment Act No. 20 of 2005.

[94] https://absch.cbd.int/countries/ZA (accessed 8 April 2019).

5.3 *Sanctions for Non-compliance*

As already shown in the chapter overviews of the main legislative instruments offences consisting of violation under the Act and the penalties thereof are spelt out in ss. 101 and 102 and elaborated under rr. 41 and 42 of the Regulations.

5.3.1 Offences

A person is guilty of an offence if that person:

(a) contravenes or fails to comply with a condition of a permit issued in terms of these Regulations;
(b) contravenes any provision of these Regulations;
(c) permits or allows any person to contravene any of the provisions of these Regulations;
(d) performs an activity for which the permit was issued otherwise than in accordance with any conditions subject to which the permit was issued;
(e) passes, uses, alters or has in his or her possession any altered or false document purporting to be a permit;
(f) knowingly makes any false statement or report for the purpose of obtaining a permit; or
(g) permits, or allows any other person to do, or to omit to do, anything which is an offence in terms of these Regulations.

5.3.2 Penalties

A person convicted of the above offences is liable to:

(a) imprisonment for a period not exceeding 10 (ten) years;
(b) a fine not exceeding 10 (ten) million rand; or
(c) both such a fine and such imprisonment.

However, "if a person is convicted of an offence involving bioprospecting or biotrading activities without a permit issued in terms of Chapter 7 of the Act, a fine may be determined, either in terms of..." the above penalties "...or equal to three times the commercial value of the activity in respect of which the offence was committed, whichever, is the greater".

6 Compliance with the Nagoya Protocol

PIC or approval and involvement of ILCs for access to GR and/or aTK as foreseen in articles 6.2 and 7 NP is provided for in the legislation. The ownership of GR and aTK by communities in SA is anchored in law. Accordingly, such communities are recognized in the Act as stakeholders whose rights must be protected by the permit issuing authority in regard to their PIC, negotiations for BSA and/or MTA and fairness and equitability. Besides, criteria and processes for obtaining PIC of ILCs according to article 6.3 (f) NP are available. According to the Act PIC must be obtained where the proposed bioprospecting project touches on the interests of ILCs, i.e. where their traditional uses of the IBRs to which the application relates have initiated, or will contribute to or form part of the proposed bioprospecting; or their knowledge of or discoveries about the IBRs to which the application relates, are to be used for the proposed bioprospecting. The interested party must disclose all material information relating to the relevant bioprospecting to the ILCs. The parties must enter into and conclude MTA and BSA. In addition, these processes must proceed with the observation of the community's customary laws, protocols and procedures. To this effect some institutions such as the Department of Science and Technology (DST) and Cooperative Governance and Traditional Affairs (COGTA) are involved in ensuring that these are taken into consideration.[95] Besides, DEA provides support for the development of community protocols through awareness-raising workshops and one-on-one engagements with communities when required.[96] Minimum requirements for MAT and model contractual clauses are included in the model MTA and BSA and the formats for permit application forms, respectively, which are annexed to the Regulations.[97] Hence, conformity of law and practice with article 12 NP is also ensured.

There is an ABS law in place consisting of an Act and Regulations thereto. These legal instruments have been revised severally to adapt them to new insights and developments. The Act was amended in 2009 and 2013 and the Regulations elaborating the relevant provisions of the Act in 2015. Hence, revision work has taken place after the adoption as well as entry into force of the Nagoya Protocol thus bringing numerous issues into conformity with the NP. In addition, there are guidelines of 2012 explaining the instruments. These instruments lay down legal rules upon which rights and obligations are based and actions, omissions are interpreted and judged.

The laws, guidelines and kits forming the regime are easily accessible, except provincial statutes. They have also spelt out the requirements and procedures for ABS, set timeframes for issuing permits, made complaint channels available for aggrieved parties,[98] clearly shown which acts consist of offences, established

[95] https://absch.cbd.int/countries/ZA/PRO (accessed 1 April 2020).

[96] Ibid.

[97] See Annexures 11, 2, 5. Also, ibid.

[98] For critique of the same see text section on "Regimes strengths and weaknesses".

penalties for violations etc. Besides, SA has an independent judiciary.[99] No evidence has been obtained concerning the application of these laws in courts but, essentially, legal transparency is sufficiently provided for.

Although most relevant issues pertaining to legal certainty, clarity and transparency have been addressed, there are still a number of hiccups, which can cause the same. For instance, some requirements are not sufficiently precise and at times seem contradictory, e.g. in regard to PIC and permits for any other kind of research. From the perception one gets from the regime no permit is required when the research is taking place in SA and only an export permit when occurring abroad (Part 4 and Annexure 3 Regulations). But according to practice, collection and research permits[100] and authorization of the owner of the resources are also required.[101] That would imply collection/research permit and PIC for research taking place in SA, and collection/research permit, PIC and export permit for research taking place abroad. This might still imply a significant workload for such research.[102] Likewise, according to the Regulations a permit for export of IBRs for any other kind of research (i.e. research other than bioprospecting) is to be granted by a MEC according to relevant provincial statutes (r. 5 (4)), but these statutes are not easily available. The few accessed while this study was being conducted are pure environmental ordinances dealing e.g. with rules on harvesting but no clear connection to ABS. This raises some questions, e.g.: Which ABS rules apply at the province? Would the MEC strictly apply the national rules of the ABS regime, or are there possibilities of diverting from or conflicting with them? If there are additional requirements, what would they be? This situation might create a vacuum which can be used in an arbitrary manner by the provincial issuing authority.[103]

Besides, these issues are handled in separate instruments. The guidelines which should harmonize, condense and explain the relevant issues into one document are outdated and therefore at times contradict the amended instruments.[104] It takes a great effort to make references to numerous instruments in order to get the right understanding or interpretation of a rule. For this it is necessary at times to refer also to the original version of the Act, the amendments and the Regulations. A non-lawyer applicant might find it hard to understand the regime and to transact with certainty without legal assistance. If an applicant cannot foresee to a reasonable degree the consequences of his/her actions, the regime might be judged as not meeting the standard of legal certainty, clarity and transparency as required by article 6.3 (a) NP.

[99] See s. 165 of the Constitution. See also website of South African Government, https://www.gov.za/about-government/judicial-system (accessed 26 November 2020); Siyo and Mubangizi (30 November 2015).

[100] See also BABS Guidelines (by Wynberg et al.) (2012), p. 44.

[101] Silvestri et al. (2020), p. 133.

[102] Ibid.

[103] Cf. The African Centre for Biosafety (ACB) (2009), p. 27f.

[104] See section on "Regimes strengths and weaknesses".

The laws establish a solid basis for sufficient fairness, non-arbitrariness in access rules and procedures in conformity with article 6.3 (b) NP. The regime applies to all persons equally. However, its personal scope which qualifies permit applicants can be viewed differently. The qualification rule foresees that foreigners qualify *quasi* only in a joint application with South Africans. Depending on how it is judged, this rule can be seen as having the capacity to produce effects that are equivalent to unfairness and arbitrariness.

Chapters 6 and 7 of the Act and the BABS Regulations provide sufficient information on how to apply for PIC/permit as seen above and in Addendum 1. In addition SANBI website contains a list of issuing authorities and their contact details including in the provinces. Specifically concerning PIC, s. 82 of the Act and r. 17 of the Regulations require that the applicant must disclose all material information relating to the bioprospecting to the relevant stakeholders and on that basis obtain their PIC. Further, the parties must negotiate and conclude MTA and BSA. Thus, article 6.3 (c) NP standard is met.

The timeframes for an issuing authority to determine an application and provide a written decision on the same are discussed above. The rules under r. 21 are clear on how long the review process should last and how the decision should be communicated, namely in a written form. If the issuing authority is not able to comply with the set timeframes it must explain to the applicant in writing why it is not able and must do so within the stipulated additional time. Complaint channels are likewise provided in case of delay or a negative decision.[105] The review process is hence sufficiently transparent. It also seems cost-effective, e.g. a communication procedure that does not require the applicant to appear in person (this applies also to applications) for following up purposes, which would *inter alia* involve travelling costs, is provided. Whether the decision is made within a reasonable period of time is debatable. The shortest period provided for is 120 days which translates to 4 months. This can still be considered long if compared to some other regimes (e.g. Vietnam, Costa Rica, Australia).[106] If additional information is needed then the period is extended with 30 days more and if the issuing authority is not able to comply within that period the review can go on for another 30 days. Therefore, the review process can last between 4 and 6 months. Although this period of time might be too long especially for non-commercial research, it is likely that the system has offered what it can best based on its capacity to handle applications. This is also implied in the national interim report on the ABS CH.[107] In addition, the permit review committee[108] meets only 4 times a year and considers 4 applications per meeting, *albeit ad*

---

[105] For critique of complaint channels see ibid.

[106] See Tran et al., Cabrera Ormaza and Burton all in this volume.

[107] "There is ... a human resource capacity issue because there is still so much to be done but the officials are not enough".

[108] The so-called National Bioprospecting Advisory Committee provides technical input during the evaluation of permit applications and also makes recommendations for the Minister to issue or reject the permit application.

hoc meetings may be convened if necessary.[109] We are of the view that the timeframes are still manageable as long as the regime is functioning within the framework provided by the law and can be relied upon by applicants. The requirements of article 6.3 (d) NP can be said to be met at times fairly, sometimes moderately and other times they can only be judged based on how well the system functions.

Sect. 81 of the Act requires that anyone wishing to access IBRs for commercialization phase bioprospecting and for export must obtain a permit. Such permits are provided for by the relevant issuing authorities at the time of access. SA has notified several IRCCs to the ABS CH which are evidence that permits are issued as evidence of PIC and MAT as required by article 6.3 (e) NP and of a functioning permit-issuing procedure.

Clear rules and procedures for establishing MAT as foreseen by article 6.3 (g) NP are existent. The Act and the Regulations state clearly when MAT shall be entered into, i.e. in regard to bioprospecting activity, and at what point it should be concluded, i.e. prior to application for permit. Likewise they state who shall enter into a MTA, i.e. the applicant and the stakeholder or stakeholders contemplated in s. 82(1) (a) of the Act for the provision of access to the IBRs to which the application for any of the permits relates (r. 38 (1)). They have also prescribed the MTA format (s. 84 Act; Annexure 11 BABS Regulations). If a community is the stakeholder and hence party to the MTA, a community resolution authorizing representation must be in the format prescribed in Annexure 13 of the Regulations (r. 38 (4)). For each stakeholder (if there are more than one) a separate MTA must be entered into with each stakeholder and both the parties (applicant and the stakeholder) must sign it in the space indicated and must insert initial(s) on every page of the agreement, including any annexures (Annexure 11, Notes).

Non-commercial research is facilitated in line with article 8 (a) NP. Accordingly, no permit is required for access to IBRs for research other than bioprospecting. If the research shall take place outside SA/abroad only an export permit is required. Conditions on change of intent are provided for in the prescribed MTA format annexed to the BABS Regulations. Accordingly, the user shall only utilize IBRs for other purposes, in particular R&D linked to new and useful properties of the IBRs, after having received prior consent from, and entered into BSA with the access provider with regards to the proposed changes (Annexure 11, c. 7). In addition, many activities have been exempted from permit requirements and fall under the category of 'any other kind of research'.[110]

When considering a permit application, the issuing authority is required to consider the role of the project in conservation of biodiversity. In addition, the Act establishes a BTF under s. 85 where all monies shared must be paid. The DG is responsible for the management of the funds. Regulation 40 elaborates how the DG shall distribute the funds. Accordingly, such funds shall be paid to the stakeholders

[109] Marumo et al. (2013).

[110] See section on definition of the term 'any other kind of research'.

according to how much is due to each and based on the BSA. Funds shall be distributed for other purposes only when there are monies in the BTF which for whatever reasons are not due to any stakeholder in terms of the BSA, or when for whatever reason it is not possible to pay such funds due to the stakeholders in terms of the BSA. Such purposes include conservation of IBRs; enhancement of scientific knowledge and technical capacity to conserve, use and develop IBRs; or any other activity that promotes the conservation, sustainable use and development of indigenous genetic biological resources for the benefit of SA.

Although the framework foresees the possibility of channeling benefits from the utilization of IBRs to conservation and sustainable use, the criterion for actualizing this is evasive. In addition, it is one of several purposes to which the benefits can be allocated even when that criterion is met. Thus, there is no specific provision in law to ensure that a specific sum of money paid in the BTF is committed to conservation and sustainable use each time benefits are shared and thus conformity with article 9 NP is lacking.

As already mentioned above, SA does not have measures which conform to articles 15 and 16 NP, except that the Act is being amended to put the necessary measures in place.[111] Concerning monitoring measures according to article 17 NP checkpoints have been designated which are mandated to collect or receive relevant information on PIC, MAT, the source as well as utilization of GR. However, there are no measures to encourage users and providers to include provisions in MAT in order to share information on the implementation of such terms as provided in article 17.1 (b) NP or the use of cost-effective communication tools and systems as provided in article 17.1 (c) NP and no use of checkpoint communique has been made.[112] However, further amendments of the Act and the Regulations hope to achieve this.[113]

SA has likewise not yet included provisions in MAT to cover dispute resolution as provided in article 18.1 (a) (b) and (c) NP and the utilization of mechanisms regarding mutual recognition and enforcement of foreign judgements and arbitral rewards in line with article 18 (3) (b) NP.[114] However, through the amendment of the Regulations it is hoped that a provision to cover dispute resolution as provided under article 18.1 NP would be included.[115] Although information published on the ABS-CH indicates that the opportunity to seek recourse in case of disputes arising from MAT and access to justice is available under SA legal system as required by articles 18.2 and 18.3 (a) NP respectively,[116] the amended Act gives no clue regarding this following the repeal of the relevant provisions in the old Act (see Addendum 1). The scope of offences provided in the Regulations includes disputes

[111] https://absch.cbd.int/countries/ZA/PRO (accessed 1 April 2020).

[112] Ibid.

[113] Ibid.

[114] Ibid.

[115] Ibid.

[116] Ibid.

arising from MAT as provided in article 18.2 NP which is also supported by access to justice at national level.[117] However, a more effective legal system is under development through the amendment of the Act and the Regulations to deal with issues of dispute resolution, access to justice and arbitral awards.[118]

National authorities (NFP and CNA) have been designated according to article 13 NP and their functions defined,[119] including regarding responsibility for granting access to GR, as seen above. Information on procedures and guidance for obtaining PIC and establishing MAT for access to both IGBRs and aTK has been made available. This can be obtained, for instance, on the Bioprospecting, Access and Benefit Sharing (BABS) CH,[120] or on the ABS CH,[121] *albeit* only for commercial and discovery phases of bioprospecting. Besides, assistance can be sought from DEFF (Department of Environment, Forestry and Fisheries) officials in completing the application forms.[122] There are numerous permits issued, which are written evidence of the decision of the CNA to grant access and a *de facto* proof of its functioning. According to the list on BABS CH 95 permits have been issued from September 2019 to date most of which are for the purpose of bioprospecting (or integrated with biotrade) with only a few expired.[123] Between March 2016 and October 2019, 29 IRCC were notified to the ABS CH.[124] Information on designated authorities, their functions and the contacts has been notified to the ABS CH.[125] There is no list though of relevant indigenous and local communities. However, the Regulations require that a representative acts on behalf of a community whereby the person chosen must get the approval of the entire community and be authorized through a community resolution (r. 9 (1) (xii)). In addition, the point of entry to such communities is traditional councils comprising of the community's elders and chief.[126] These factors definitely ease the identification of genuine owners of GR and aTK among communities and defer deception by unscrupulous entities. There could be cases, however, when the rightful community cannot be identified *ex ante*. In such cases DEA makes provision by including a condition in the permit for *ex post* conclusion of BSA once the relevant aTK holders are identified.[127] In addition, it encourages communities that have not yet got formal structures to organize

[117] Ibid.

[118] Ibid.

[119] https://absch.cbd.int/countries/ZA (accessed 30 March 2020).

[120] https://www.environment.gov.za/projectsprogrammes/bioprospectingaccess_benefitsharing_babs_clearinghouse (accessed 31 March 2020).

[121] https://absch.cbd.int/countries/ZA/PRO (accessed 1 April 2020).

[122] Ibid.

[123] https://www.environment.gov.za/projectsprogrammes/bioprospectingaccess_benefitsharing_babs_clearinghouse (accessed 1 April 2020).

[124] https://absch.cbd.int/countries/ZA (accessed 1 April 2020).

[125] Ibid.

[126] Ibid.

[127] https://absch.cbd.int/countries/ZA (accessed 1 April 2020).

themselves.[128] Obviously the other relevant stakeholder is the State, which is represented nationally by DEA and in the provinces by MECs.

7 Regime's Strengths and Weaknesses

The regime demonstrates a number of strengths. The Act and the regulations thereto are in place. These statutes have been revised severally including after the adoption and entry into force of the NP. Though not done in reaction to the obligations of the NP, these revisions have brought the regime to a certain level of compliance with it. They spell out clearly for what activities a permit must be obtained for access to IGBRs and aTK and the requirements to be fulfilled before such a permit can be issued. Exemptions of activities which do not fall under the regime as well as those that fall thereunder but do not require an access permit are also made. Likewise, they indicate the issuing authorities and show which authority is responsible for which permit and which GR. Besides, their contacts are provided. The stakeholders whose interests must be protected and PIC sought are also clearly defined. Very important is the representation of the ILCs which is established under rr. 9 (1) (xii), 38 (4) and 39 (7).

Major terms such as 'indigenous biological resources' and 'bioprospecting' are defined. Concerning IBRs the regime implies that only bioprospecting on or export of such resources for any other kind of research where SA is the country of origin is regulated. The regime is also evident on the regulation of information on GR, which it incorporates through the definition of 'genetic resources'. Likewise, it makes it explicit that the trigger for benefit-sharing is bioprospecting.

The procedures for permit application, pre-requisites and timeframes are established and are clear. Differentiation of types of research is made with an intention of facilitating non-bioprospecting research, which is referred to as any other kind of research/research other than bioprospecting.

Third party transfer and change of use or intent are precisely regulated. In addition, model transfer and benefit-sharing agreements have been provided. Likewise, a BTF has been established with clear rules on management and distribution of benefits. Offences and penalties have also been defined.

Other strong qualities of the SA regime are that it has communication procedures that can be used by applicants without need to appear in person, which is quite useful for interested persons located abroad; the Minister can assist parties in negotiating MTA/BSA; and permits for bioprospecting and export, or bioprospecting and biotrade can be combined into a single, i.e. integrated, permit.

In spite of the said strengths the regime still suffers a number of weaknesses. *Inter alia*, there are many scattered bits of statutes which make it burdensome for one/an applicant to search, analyze, understand and make a conclusion with certainty of the

[128] Ibid.

right approach to take. The guidelines which should ease the situation are outdated and have been overtaken by revisions of the Act of 2013 and the BABS Regulations of 2015. In addition, they either contradict the legislation or the latter is obscure concerning permit requirements for research other than bioprospecting. Whilst the guidelines state that research other than bioprospecting must obtain research and collection permits[129] there is no hint in the legislation that it needs such permits.

Further, the rules meant to differentiate requirements and procedures for non-commercial research and thus facilitate it are a bit obscure and confusing.

Further, the permits to be issued by the MEC[130] e.g. for export for non-bioprospecting purposes should be issued according to provincial ordinances. These ordinances are fast inaccessible online, save being pure environmental statutes void of ABS regulations. It is therefore unclear how permits at the provinces are issued, *viz.* upon which laws the MEC relies and whether this might be a ground for arbitrariness.

Further, upon repeal of the appeals ss. 94, 95 and 96 of the Act of 2004 by ss. 27, 28, 29 of the NEMLA Act 14, 2013 respectively no clue is given in the repealing Act concerning where to find remedies for an aggrieved party. The changes are arcane. It is only if one bumps on s. 88 (3) (f)[131] and r. 21[132] that one might get a clue on where to search for relevant remedy, namely from the Promotion of Administrative Justice Act (No. 3), 2000.[133] That takes quite a great effort.

Besides, it is good that the regime provides MTA and BSA, but no possibility to substitute or remove any clause is provided. The agreements must be entered into as prescribed. As they are also supposed to contain the terms mutually agreed by the parties, it is hard to judge the terms thereof as mutually agreed and thus always fair because they are pre-determined.

Likewise, it fails to create a strong link between the three objectives of the CBD in terms of benefits shared from the utilization of GR. This is because r. 40 (5) is weakened by making such funds available for conservation and sustainable use purposes only if the funds are not assigned to any party. Besides, it says the DG "may" use such funds for conservation and sustainable use purposes and not "must", which means that even if the funds are not claimed by any party, they can be used in other competing areas. There is therefore a likelihood that no or insufficient funds will be earmarked for conservation and sustainable use purposes.

---

[129] BABS Guidelines (by Wynberg et al.) (2012), p. 44.

[130] Not for bioprospecting purposes issued by the MEC on behalf and with empowerment by the Minister.

[131] Says that the decision of an issuing authority must be consistent with the Promotion of Administrative Justice Act (No. 3), 2000.

[132] On timeframe for deciding notifications and permit applications by issuing authorities, which says that upon the lapsing of the time given to an issuing authority to decide an application, s. 6 (2) (g) and (3) of the Promotion of Administrative Justice Act (No. 3), 2000 will apply.

[133] See fn 84 above for a summary of the content of s. 6 (2) (g) and (3) of the Act. The Act is available online at https://www.justice.gov.za/legislation/acts/2000-003.pdf (accessed 2 April 2020).

In addition, although the revisions undertaken also help in bringing compliance with the NP in general, user compliance measures are still lacking.

Other authors have criticized the regime for poorly publicizing the regime among relevant stakeholders, e.g. landowners and researchers.[134] Accordingly, while landowners are unaware concerning their obligation to grant written permissions for access to GR, researchers are not aware of the process and often fail to comply.[135] Even some regulatory authorities are said not to be well versed with the regime.[136,137]

8 Suggestions for Improvement and Conclusion

The main critiques are summed up in one: the scattered bits of statutes make it very hard to understand with certainty the regime, its requirements and procedures. As mentioned in the text this could be seen as a cause for legal uncertainty and unclarity. To this effect, our suggestion is that a simplified kit is developed which brings together the rules, requirements and procedures into one document with references to the different relevant laws. While this will substantially be useful it might also avert the said impression.

Besides, the guidelines need to be revised to reconcile them with the revisions which have been executed and the ones being undertaken post NP to eliminate contradictions.

Further, it should be made clear how the regime operates when the biological resources are being accessed from the provinces and which ABS laws are applied by the MEC. Besides, to facilitate its smooth running, the awareness of relevant stakeholders and building of the capacities of relevant State institutions would need to be stepped-up.

In addition to these critical points, the regime should create a clear-cut rule that sets aside a fixed amount of benefits for conservation and sustainable use of biodiversity. There is an indication that this is being considered in the new revisions

---

134 Silvestri et al. (2020), p. 133.

135 Ibid.

136 Ibid.

137 Silvestri et al. (2020) illustrate this situation on p. 133 with a recent case of access to herbivorous ladybird beetle Cleta eckloni Mulsant (Coleoptera: Coccinellidae). The beetle was exported by the Centre for Biological Control, Rhodes University, South Africa, to CSIRO (Commonwealth Scientific and Industrial Research Organization), Australia, for testing as a potential biological control agent against the South African plant Lycium ferocissimum Miers (Solanales: Solanaceae), which has become invasive there. The researchers were required to spend considerable time educating landowners who were happy to allow the collection on their land, but were reluctant to give written permission as they felt that they may be held liable if anything went wrong. The provincial authorities were also not sure which directorate was responsible for signing permission and instead sought for explanation from the researchers concerning biological control procedures. Besides, they had no clue concerning the science of classical biological control of weeds.

of the Act,[138] which, if it happens, will make the SA regime one of the few regimes that practically connects the three objectives of the CBD. Also, being one of the most industrialized African countries with evidently extensive biotechnological R&D on biological resources, it is essential that it puts into place user measures to comply with the legislations of other countries. To that effect it will also be important to urgently include provisions in MAT to cover dispute resolution and the utilization of mechanisms regarding mutual recognition and enforcement of foreign judgements and arbitral rewards.

Having said that, it is true that the regime would greatly improve when these and other amendments are done, but the remaining question is: how well does and would the new wine, i.e. accomplished and coming revisions for compliance with the NP, fit in the old wine skins, the old law? In our view, a newly developed legislation might be better placed to examine the obligations of the Nagoya Protocol more holistically and to come up with more plausible law. The new legislations of Brazil, Malaysia, Ethiopia, Cameroon, Spain and France are a good example.[139] We do not claim that these are perfect, but they are relatively straightforward.

Addendum 1

Structure of NEMBA

NEMBA's objectives are *inter alia* the conservation of biodiversity, the sustainable use of indigenous biological resources, and the fair and equitable sharing of benefits arising from bioprospecting involving indigenous biological resources (Ch. 1, s. 2 (a)). The focus of this study is the third objective of the CBD, hence only relevant chapters and provisions are presented. The NEMBA chapters containing the main provisions which form the ABS regime are chapter 6 on bioprospecting, access and benefit-sharing (s. 80–86); chapter 7 on permits (s. 87–93); chapter 1 s. 1 on definitions; and chapter 9 on offences and penalties (s. 101–102) as shown on Fig. 1.

Chapter 6 states its purpose and defines its breadth of application in section 80 (1), i.e.:

(a) "to regulate bioprospecting involving indigenous genetic and biological resources;

(continued)

[138] https://absch.cbd.int/countries/ZA (accessed 1 April 2020).

[139] See Mozini, Kamau (Malaysia), Kamau and Hailu, Mahop (Cameroon and France) and Silvestri (Spain) all in this volume. As at 27 October 2021 the new Ethiopian legislation is still in draft form.

(b) to regulate the export from the Republic of indigenous genetic and biological resources for the purpose of bioprospecting or any other kind of research;
(c) to provide for a fair and equitable sharing by stakeholders in benefits arising from bioprospecting involving indigenous genetic and biological resources; and
(d) to ensure that the nation's indigenous genetic and biological resources are developed and utilized in an ecologically sustainable manner while promoting social and economic development, in particular in the areas where the indigenous genetic or biological resources and associated traditional knowledge is accessed".[140]

It should be noted that the old version of the Act (of 2004) neither used the term 'genetic resources' under paragraph (a) nor contained paragraph (d). These were introduced through the substitution of section 80 (1) by s. 19 of NEMLA (National Environmental Management Laws Amendment) Act 14 of 2013. In particular it is notable that the augmentation of the term 'genetic resources' and its definition under s. 1 of the amended Act expands the breadth of application of Ch. 6 and generally its material scope. Thus the terms indigenous biological resources (IBRs) and indigenous genetic and biological resources (IGBRs) are used synonymously in this study. Besides defining its scope Ch. 6 establishes exemptions to the scope (s. 80 (2) (b)). Further, it conditions commercialization phase of bioprospecting involving any IBRs or export of such resources from SA for the purpose of bioprospecting or any other kind of research to a compulsory permit issued in terms of Ch. 7 (s. 81 (1)) and states the pre-condition for grant of such a permit (s. 81 (2); 82 (1)–(3)). Section 81A[141] establishes different permit conditions for discovery phase of bioprospecting exempting such type of bioprospecting from a compulsory bioprospecting permit and creates a separate procedure for the same.[142] Under ss. 83 and 84 Ch. 6 prescribes formats for benefit-sharing and material transfer agreements (BSA, MTA), respectively, and establishes a

(continued)

[140] Revisions of the Act according to section 19 of NEMLA Act 14 of 2013 (*Commencement date of section 80 (revisions): 1 January 2006 [Proc. No. 47, Gazette No. 26887]*).

[141] Section 81A inserted by section 39 of NEMLA Act 14 of 2009 *(Commencement date of section 81A: 1 April 2011 [Proc. No. 17, Gazette 34072])*.

[142] For definition of the terms 'commercial phase of bioprospecting' and 'discovery phase of bioprospecting' see chapter Sect. 2.2 (under definition of the term 'bioprospecting').

bioprospecting trust fund (BTF) under section 85. Finally, the chapter gives the minister powers under section 86[143] with a *caveat*[144] to exempt by notice indigenous biological resources or an activity relating to such resources, categories of research or commercial exploitation, activity of collection, use, propagation cultivation or trade of indigenous biological resources for domestic use or subsistence purposes from the application of Ch. 6 (ss. 1 (a)-(bA). In addition, the minister may declare that the BSAs and MTAs contemplated in sections 83 and 84 do not apply to certain categories of commercial or industrial exploitation of indigenous biological resources (ss. 1 (bB).

Chapter 7 states its purpose, which is to regulate the issuing of permits, *inter alia*, for bioprospecting and export of IBRs in terms of section 81 (1) (s. 87 (c), (d)). Through an amendment of 2013 the authorities for deciding an application and issuing permits are established under section 87A.[145] Further, Ch. 7 creates the procedure for application and issuing of permits (s. 88), prescribes the form of a permit (s. 90), establishes a procedure for issuing of integrated permits (s. 92) and lists grounds on which an issuing authority may deny a permit (s. 92A (1)).[146] It also provides grounds for cancellation (s. 93), renewal and amendment (s. 93A),[147] and suspension (s. 93B[148]) of permits. Sections 94–96 on appeals were repealed[149] and the amended Act does not give a clue concerning which channels exist for an aggrieved party to seek recourse.

Chapter 1 contains vital definitions of the ABS system. The ones which have been considered in this study are 'genetic resource', 'indigenous biological resource', 'indigenous species', 'bioprospecting', 'commercial exploitation', 'commercialization', 'commercialization phase of bioprospecting', 'discovery phase of bioprospecting' and 'issuing authority'. The term 'any other kind of research' is not defined in NEMBA but is very important in distinguishing the types of research as well as delineating permit conditions. This term was defined in the BABS Regulations 2008 (s. 1) and is also considered in this study in line with the definition of the repealed version of the Regulations.

(continued)

[143] This section was substituted as well as amended by NEMLA Act 14 of 2009 (s. 42) and NEMLA Act 14 of 2013 (s. 20(a) and (b)).

[144] According to ss. (2) the Minister must follow a consultative process in accordance with sections 99 and 100 before publishing a notice in terms of ss. (1), or amending or repealing such a notice in terms of ss. (1A).

[145] Section 87A was inserted by section 22 of NEMLA Act 14 of 2013.

[146] Section 92A was inserted by section 24 of NEMAL Act 14 of 2013.

[147] Section 93A was inserted by section 44 of NEMLA Act 14 of 2009.

[148] Section 93B was inserted by section 26 of NEMLA Act 14 of 2013.

[149] NEMLA Act 2013, ss. 27-29 [*Proc. No. 47, Gazette No. 26887*].

Finally, chapter 9 provides a list of offences for which, if committed, a person is guilty (s. 101), and penalties to which a person is liable if convicted of an offence (s. 102).

Source: *by Author*

Addendum 2

Structure of BABS Regulations

Chapter 1 contains definitions which were either not used (e.g. 'biotrade') or interpreted (e.g. 'applicant') in the NEMBA 2004 and its revisions thereto, or substituted ones (e.g. 'permit') (r. 1). Neither of the definitions of the (new) Regulations are examined in this study. Further, it establishes the purpose of the Regulations, i.e. to prescribe the notification process for the discovery phase of bioprospecting contemplated in s. 81A (2) of the Act; prescribe the permit system set out in Ch. 7 of the Act insofar as it applies to the scope of Ch. 6 as defined in s. 80 (1); set out the form and content of, and requirements and criteria for BSA and MTA; and set out the administration process of the BTF (r. 2). Finally, it defines the scope or breadth of application of the Regulations, which is commercial or industrial sectors that utilize any IGBRs and/or aTK "... for research, application or development of drugs, complementary medicines, neutraceuticals, industry enzymes, food flavors, fragrances, cosmetics, emulsifiers, oleoresins, colors, extracts, and essential oils, and non-commercial sectors that export ..." IGBRs from SA for basic scientific research – research to generate scientific data (r. 3).

Chapter 2 identifies the issuing authorities, i.e. the Minister and the Member of Executive Council (MEC), and establishes their mandates in terms of which permits they issue (r.4) as well as indicates where notifications and applications must be submitted (r.5). Further, it creates a mechanism of assistance to the applicant or stakeholder and consultation between relevant stakeholders and the issuing authority (r. 6 & 7). It also gives the issuing authority the right of access to all information of relevance to deciding the application and the obligation of the applicant to disclose it to the former unless such information is protected by law (r. 8). Besides it establishes the criteria the issuing authority must comply with as well as take into account when considering notifications and permit applications (r. 9), how to communicate the decision to the applicant and within which timeframe (r. 10). Finally, it establishes the obligation of the issuing authority to keep a register of permit applications and notifications received in terms of the regulations and decisions made in respect of the permit applications and BSAs and MTAs (r. 11).

(continued)

Chapter 3 on permit system defines the personal scope (r. 12) and establishes the procedure for notification and application of access permits (rr. 13–19) as well as the methods for submission of these (r. 20). Further, it sets the timeframes for the issuing authority to consider permit applications (r. 21) and factors to be taken into consideration while considering notifications (r. 22), discovery phase export permits (r. 23), biotrade permits (r. 24), bioprospecting permits (r. 24), integrated biotrade and bioprospecting permits (r. 25), integrated biotrade and bioprospecting permits (r. 26) and export permits for research other than bioprospecting (r. 27). It also lists circumstances under which the issuing authority may refuse to register a notification or deny a permit (r. 28) and lays down conditions under which the issuing authority may register a notification, issue a permit or subject them to certain conditions, and consequences thereof, including the obligation of the issuing authority to inform the applicant of his/her right of appeal (rr. 29, 31). The permits must be in the permit form prescribed in the Annexures (r. 32). Regulation 33 prescribes compulsory conditions applicable to permit holders. The period of validity of permits, timeframes for renewal and the form of the application and fee for renewal are fixed in rr. 34–35. The considerations that an issuing authority must make while deciding the renewal or amendment application and the timeframes for informing the applicant of the decision as well as the obligation to give (written) reasons to the applicant for refusal and to inform him/her of his/her right to appeal are also provided in Ch. 3 (rr. 36–37).

Chapter 4 defines the parties to MTAs and BSAs, their (prescribed) forms, the authority empowered to approve them, the conditions a BSA must achieve as well as conditions it must fulfil before it is approved (rr. 38–39). Further, the Ch. establishes the administrator for the BTF and defines how its administration must be carried out (r. 40).

Chapter 5 on general matters defines the offences (r. 41) and penalties (r. 42) and repeals the BABS Regulations 2008.

Chapter 6 contains transitional provisions (rr. 44–45), provides the title of the new Regulations (2015) and defines the date they take effect (r. 46).

Finally, Annexures 1–13 contain forms for: Discovery phase notification form; Application form for a discovery phase export permit; Application for an export permit for research other than bioprospecting; Non-refundable fees for permits application; Application form for a bioprospecting permit or biotrade permit or integrated biotrade and bioprospecting permit; Discovery phase export permit; Export permit for research other than bioprospecting; Biotrade permit; Bioprospecting permit; Integrated biotrade and bioprospecting permit; Material transfer agreement; Benefit-sharing agreement; and community resolution.

Source: *by Author*

References

Books

Kamau EC, Winter G (eds) (2009) Genetic resources, traditional knowledge and the law. Solutions for access and benefit sharing. Earthscan, London

Kamau EC, Winter G, Stoll P-T (eds) (2015) Research and development on genetic resources. Public domain approaches in implementing the Nagoya protocol. Routledge, London and New York

Articles

Crouch NR et al. (2008) South Africa's bioprospecting, access and benefit-sharing legislation: current realities, future complications, and a proposed alternative. SAJS 104 (September/October): 355–366

Driver A et al (2005) National Spatial Biodiversity Assessment 2004: priorities for biodiversity conservation in South Africa. Strelitzia 17. South African National Biodiversity Institute, Pretoria

Lesley JFG (2007) The indigenous knowledge systems policy of 2004: challenges for south African universities. Soc Dyn 33(1):130–154. https://doi.org/10.1080/02533950708628746

Lewis M (2010) Access and benefit-sharing in the wake of CBD COP10: the Nagoya protocol and its potential implications for South Africa. SAJELP 17:69–96

Silvestri L et al (2020) Implementation of access and benefit-sharing measures has consequences for classical biological control of weeds. BioControl 65:125–141. https://doi.org/10.1007/s10526-019-09988-4

Siyo L, Mubangizi J (30 November 2015) The independence of south African judges: a constitutional and legislative perspective. Potchefstroom Electr Law J, Vol. 18, No. 4, 2015, https://doi.org/10.4314/pelj.v18i4.03. Accessed 26 Nov 2020

The African Centre for Biosafety (ACB) (2009) Critical overview of South Africa's bioprospecting laws. PressPrint, South Africa

Legal Documents

Bioprospecting, Access and Benefit-Sharing Regulations 2015

Constitution of the Republic of South Africa No 108 of 1996, https://www.gov.za/sites/default/files/images/a108-96.pdf (accessed 15 April 2020)

Convention on Biological Diversity, 1992

Convention on International Trade in Endangered Species (CITES) Regulations, 2010

Government Notice R149, Government Gazette 30739, 2008

Government Notice No.R.152/Government Gazette No.29657, 23 February 2007, https://www.environment.gov.za/sites/default/files/legislations/nemba_threatenedspecies_regulations_g29657rg8638gon152.pdf (accessed 15 April 2020)

Government Notice No.R.173/Government Gazette No.33002, 5 March 2010, https://www.environment.gov.za/sites/default/files/legislations/nemba_wildfaunaandflora_g33002rg9240gon173_0.pdf (accessed 15 April 2020)

Intellectual Property Laws Amendment Bill, 2011, http://www.ip-watch.org/weblog/wp-content/uploads/2011/10/So-Africa-TK-Bill-Sept-2011.pdf (accessed 15 April 2020)

Nagoya Protocol on Access to Genetic Resources and the Fair and Equitable Sharing of Benefits Arising from their Utilization to the Convention on Biological Diversity, 2010
National Environmental Management Act No. 107, 1998
National Environmental Management Laws Amendment Act 14, 2009
National Environmental Management Laws Amendment Act 14, 2013
National Environmental Management: Biodiversity Act, No. 10 of 2004
Patents Amendment Act No. 20 of 2005, https://wipolex.wipo.int/en/text/179614 or http://www.saflii.org/za/legis/num_act/paa2005189.pdf (accessed 15 April 2020)
Promotion of Administrative Justice Act (No. 3), 2000, https://www.justice.gov.za/legislation/acts/2000-003.pdf (accessed 15 April 2020)
The White Paper on the Conservation of and Sustainable Use of South Africa's Biological Diversity. Government Gazette No. 18163 of July 1997. https://www.environment.gov.za/sites/default/files/legislations/biodiversity_whitepaper_18163_gen1095_0.pdf (accessed 15 April 2020)
Threatened or Protected Species (TOPS) Regulations, 2007 (No. R. 152 of 2007)
Trust Property Control Act No. 57, 1988

Model Clauses and Guidelines

DFG (Ed.), Model clauses for mutually agreed terms on access to genetic resources and benefit sharing, Deutsche Forschungsgemeinschaft 2020, https://www.dfg.de/download/pdf/dfg_im_profil/gremien/senat/biologische_vielfalt/191212_model_clauses.pdf (accessed 15 April 2020)
South Africa's Bioprospecting, Access and Benefit-Sharing Regulatory Framework. Guidelines for Providers, Users and Regulators (by Wynberg R et al (2012). DEA, South Africa, https://www.environment.gov.za/sites/default/files/legislations/bioprospecting_regulatory_framework_guideline.pdf (accessed 15 April 2020)

Studies, Reports and Presentations of State Officials

Marumo M et al. (March 2013) Framework on bioprospecting, access and benefit-sharing. Presentation at the 7th Pan-African ABS Workshop, Hans Merensky Hotel, Phalaborwa, 25 February - 1 March 2013, http://www.abs-initiative.info/uploads/media/Lactitia_Tshitwamulomoni_-_DEA_-_South_Africa_s_legislative_framework.pdf (accessed 15 April 2020)
National study on ABS implementation in South Africa (April 2014) Commissioned by the ABS Capacity Development Initiative in collaboration with the Government of South Africa, http://www.abs-initiative.info/fileadmin/media/Knowledge_Center/Pulications/ABS_Dialogue_042014/National_study_on_ABS_implementation_in_South_Africa_20140716.pdf (accessed 15 April 2020)
UNEP/CBD/WG-ABS/5/5 (30 August 2007) Report on the legal status of genetic resources in national law, including property law, where applicable, in a selection of countries, https://www.cbd.int/doc/meetings/abs/abswg-06/official/abswg-06-abswg-05-05-en.pdf (accessed 15 April 2020)
Wynberg R (2004) Bioprospecting, access and benefit-sharing in South Africa: Towards a strategic assessment. A paper prepared for the National Botanical Institute as a contribution towards the Southern African Biodiversity Support Programme and National Biodiversity Strategy and Action Plan, https://www.researchgate.net/publication/229004887_Bioprospecting_Access_and_Benefit-sharing_in_South_Africa_Towards_a_Strategic_Assessment (accessed 15 April 2020)

Important Links

www.cbd.int (accessed 15 April 2020)
https://absch.cbd.int/countries/ZA (accessed 15 April 2020)
https://www.sanbi.org (accessed 15 April 2020)

Evanson Chege Kamau Dr. iur., is Senior Research Fellow at the Research Centre for European Environmental Law (FEU), University of Bremen. He is also an Adjunct Lecturer at the CvO University of Oldenburg. His area of expertise is public international law, European law and intellectual property law. His research focuses on genetic resources, environment, development and traditional rights on biological resources and traditional knowledge. Kamau has over 17 years of working experience in the area of ABS in which he has published widely. For many years he has also worked as a freelance legal consultant for governments, African Union (hired by IUCN and Capacity Building Initiative resp.), private industries and research organizations. echege@uni-bremen.de.

The Post Nagoya Protocol ABS Regime in Cameroon: Exploring the Extent to Which Ongoing Policy, Regulatory Developments and ABS Practices Uphold the Obligations of the Protocol

Marcelin Tonye Mahop

Abstract In 1995 and 2017 respectively, Cameroon ratified the Convention on Biological Diversity (CBD) and its Nagoya Protocol (NP) on Access and Benefit Sharing (ABS). Cameroon's ratification of the NP on ABS indicates the country's commitment to translate and operationalize its provisions in a domestic ABS regulatory framework. A process was initiated in 2013 by the Government of Cameroon (GoC) that was aimed at: (1) Adopting a ministerial order on ABS that would govern ABS negotiations on the specific case of *Echinops giganteus* and other emerging cases and; (2) Constructing a comprehensive NP compliant ABS policy, administrative and legal framework. This chapter looks at the progress so far in Cameroon in establishing a NP compliant ABS regulatory framework. It also discusses the newly adopted ABS law of July 2021 and the draft implementing regulations around the following issues: the scope of the regulatory framework and the proposed institutional framework; the nature of the key regulatory tools such as the prior informed consent (PIC), mutually agreed terms (MAT/ABS contracts) and benefit-sharing schemes; the approach taken by the emerging instruments to address the different facets and levels of compliance. Likewise, it discusses the extent to which the ABS instruments are compliant with the NP, highlights their strengths and weaknesses and suggests way forward in view of delivering an effective and NP compliant ABS regime.

Keywords Cameroon · Access and benefit-sharing · Nagoya protocol · Regulatory framework · ABS compliance

M. T. Mahop (✉)
Sustainability Research Institute (SRI), School of Earth and Environment (SEE), Faculty of Environment, University of Leeds, Leeds, UK
e-mail: m.tonyemahop@leeds.ac.uk

E. Chege Kamau (ed.), *Global Transformations in the Use of Biodiversity for Research and Development*, Ius Gentium: Comparative Perspectives on Law and Justice 95, https://doi.org/10.1007/978-3-030-88711-7_7

1 Introduction

Cameroon is located in the Congo Basin, which is the second largest and richest ecosystems and biodiversity area in the world after the Amazonian Basin. Cameroon is a biodiversity rich country, ranking immediately after South Africa as the most biological diverse country in Africa.[1] The biological and ecosystems diversity of Cameroon comprise different types of habitats, including the moist tropical forest dominating the South and South-East covering 54% of the country, the mountain forest and savannah in the highlands and the sub-Sahelian savannah and near desert in the far north.[2] These diverse habitats host more than 9000 species of plants, of which 160 species are endemic.[3] The majority of the endemic taxa are concentrated around Mount Cameroon and other highland areas.[4] These habitats and their incredibly rich biodiversity have been subjected to various threats, casting doubt on their sustainability and the many research, scientific, environmental and livelihoods benefits they provide.[5]

In becoming party to the Convention on Biological Diversity (CBD) in 1995 and its Nagoya Protocol on Access to Genetic Resources and the Fair and Equitable Sharing of Benefits Arising from their Utilization in 2017, Cameroon demonstrates its recognition of the values of biological and genetic resources and the necessity for their conservation and sustainable use. Specifically, Cameroon's ratification of the Nagoya Protocol (NP) on access and benefit-sharing (ABS) is a clear indication of the country's commitment to translate the provisions of this international environmental agreement to domestic ABS regulations. The ultimate target is to ensure that access to and utilization of the country's genetic resources (GR) will lead to tangible (monetary or non-monetary) benefits shared between Cameroon and users of GR. The proceeds of these benefits earned by Cameroon can be reinvested in domestic conservation, sustainability, socio-economic, and livelihoods improvement efforts.

Despite being party to the CBD and the NP since 2017, Cameroon has just adopted Law No2021/014 of 09 July 2021 on Access to genetic Resources, derivatives, associated traditional knowledge and the fair and equitable sharing of the benefits arising from their utilization (hereinafter the July 2021 ABS law, or ABS law). The law is a step forward in respect of the country's commitment to fulfil its obligations to the NP, which will be followed by the adoption of relevant implementing regulations. The fact that the country had yet to enact a domestic ABS law until recently does not mean that there was inactivity on the ABS front from a regulatory perspective, quite the contrary. Cameroon has been active in ABS

[1] Sunderland et al. (2003), pp. 9–17.

[2] Ibid.

[3] Ibid.

[4] Fonge et al. (2013).

[5] Duguma et al. (2018). For the different threats on Cameroonian habitats and biodiversity see Tchouto (2004).

discussions and framing of ABS regulatory frameworks at all levels: international, continental (at the African Union level), regional (within the Congo Basin in the context of the Central African Forestry Commission-COMIFAC-ABS processes) and national. For example, Cameroon participated actively in international negotiations leading to the adoption of the NP in 2010. Furthermore, Cameroon played a leading role in the development of the COMIFAC regional ABS strategy in 2011[6] and participated effectively in the process of formulation of the African Union Strategic and Practical Guidelines for a Coordinated Implementation of the Nagoya Protocol on ABS in 2015.[7] At the national level, following the adoption of the COMIFAC regional ABS strategy in 2011, Cameroon adopted its national ABS strategy in 2012 and established a National ABS Committee in 2013. The role of the committee as provided in a 2013 draft ABS ministerial order (which never came into force), was to be the advisory arm to the Ministry of Environment and Nature Protection in the process of development of Nagoya Protocol compliant ABS regulations and to assist the ministry in its handling of a new case of interest in genetic resources -*Echinops giganteus* - by a French company.[8]

Although progress on the domestic ABS regulatory front in Cameroon can be described to have been slow[9] despite national awareness and recognition of the value of its biological/genetic resources and associated traditional knowledge (aTK), two key occurrences were instrumental in changing the mood and boosted domestic actions. The first occurrence was the ongoing interest by a French company in access and utilization of a plant called *Echinops giganteus*. This plant grows in the Western and Southwestern regions of Cameroon, largely in mountains habitats at altitudes ranging from 950 to 2400 m above sea levels and is endemic to Cameroon.[10] Although the rhizomes of this plant are traditionally used by the local populations as ingredients in various culinary preparations and for the treatment of various diseases, it is its use in the cosmetic industry, based on its peculiar fragrance discovered by the French company Mane et Fils S.A. that was the main bioprospecting interest in this plant. Based on this commercial potential, *Echinops giganteus* emerged as a case of interest that can be used to test the functionality of ABS value chains in Cameroon and their role in delivering on the objectives of the NP. Many actors came together to make an Echinops value chain happen including the Government of Cameroon through the Ministry of Environment and Nature Protection supported by the ABS Capacity Development Initiative, a local NGO, the Environment and Rural Development Foundation (ERUDEF) and the United Nations Development Programme (UNDP) small grants scheme through the National UNDP office in Cameroon. Critically, the creation of a value chain was

[6]COMIFAC (2011).

[7]Africa Union (2015a, b).

[8]IUCN (2015).

[9]Some of the reasons slowing domestic process could be possibly Cameroon's involvement in international negotiations leading to the Nagoya Protocol, continental and regional ABS processes.

[10]Tchetcha (2012).

dependent on a credible, legal and sustainable supply of Echinops raw materials to the French company for its research and development activities. The problem is that there is an ABS regulatory vacuum, which prompted the Government of Cameroon (GoC) to accelerate the formulation of specific Nagoya Protocol compliant ABS regulations.

The second significant occurrence was the findings of a study carried out by Oldham et al. (2013) that looked at biodiversity in the patent system in the case of Cameroon.[11] The aim of the study was to investigate patent activity involving genetic resources and traditional knowledge by answering the question: what can patent information teach the ABS community of actors about the range of potential economic uses of African genetic resources and traditional knowledge? Using data sourced from the Global Biodiversity Information Facility (GBIF), which at the time recorded 15,988 species names from Cameroon, 135,776 patent documents were found containing species names known to have Cameroon distribution.[12] Of these documents, 184 made some form of reference to Cameroon and contained 2425 species. Critically, some 21 species, which are clearly recorded in the GBIF as only distributed in Cameroon (that is endemic to Cameroon), appeared in 97 patent documents where their source/origin as Cameroon was not mentioned.[13] Additional findings of this research revealed that the species of plants, animals, bacteria and viruses that were linked to Cameroon were found in patent documents linked to different research and development sectors including genetic engineering, pharmaceuticals, biocides, new agricultural plants, foodstuffs and detergent.[14]

A process was initiated in 2013 by the GoC with the support of international donors that was meant to: (1) in the interim, adopt a ministerial order on ABS that would govern the ongoing case of *Echinops giganteus* and any other emerging cases and; (2) construct a comprehensive Nagoya Protocol compliant ABS policy, administrative and legal framework. Although a draft ministerial order was elaborated and went through some degree of consultations, it was never finalized. An assessment of the draft ABS order conducted by the legal department of the office of the Prime Minister considered that this instrument should be elevated to the level of a Prime Ministerial decree instead of a ministerial order. The ABS decree on ABS in turn never came to fruition, as by 2015 the domestic ABS regulatory process in Cameroon took another turn with the decision to go for a complete ABS law and implementing regulations.

In the following, the chapter discusses progress so far in Cameroon in domesticating the Nagoya Protocol on ABS through the recently adopted July 2021 ABS law and the current drafts implementing regulations that are under development and going through the national consultations and validation process. Subsequently, it

[11] Oldham et al. (2013). The other cases studied in this series included Kenya, Madagascar, Mozambique, Senegal and South Africa.

[12] Ibid.

[13] Ibid.

[14] Ibid.

addresses how key principles of ABS are considered in the Cameroon case in the way that could or not be viewed to be Nagoya Protocol compliant.

2 The ABS Regulatory Framework in Cameroon

2.1 Current Situation and ABS Instruments in Cameroon

Domestic process for the development and eventual adoption of a Nagoya Protocol compliant ABS regulatory framework in Cameroon has been ongoing. While this process is still underway, Cameroon has thus far not just only marginally included some aspects of key ABS principles enshrined in the CBD in some sectoral laws but has adopted the July 2021 ABS law and has drafts implementing regulations. With regards to the inclusion of the ABS principles of the CBD in sectoral laws, we can note the principle of State's sovereignty[15] over its resources and therefore power to authorize their access, exploitation and use in scientific and commercial undertakings. Indeed, now and until the recent adoption of the domestic ABS law, some national sectoral laws are interpreted in view to their applicability in ABS cases. Central to these laws is the 1994 forest law[16] which asserts State's sovereignty over all forestry and genetic resources found in the national territory and stresses the State's sole prerogative to authorize access to and exploitation of these resources. Furthermore, pursuant to article 12.2 of the 1994 forest law, should access to and use for scientific and commercial purposes generate any benefits, such benefits should be shared fairly and equitably.

Another piece of law generally referred to in ABS cases is the 1996 Environmental Framework Law.[17] This law reinforces the principle of State's sovereignty over its resources and is prerogative to authorize exploitation and adds a critical perspective to ABS processes by stressing the need to always involve local communities when exploitation of biological resources occurs in their localities.[18]

Alongside these key instruments, specific implementing regulations in the form of ministerial or joint ministerial orders were issued. These include the 1995 forestry decree,[19] the 2006 ministerial order pertaining to research authorizations for

[15] See the Convention on Biological Diversity, Art. 3 Principle; and Art. 16 of the CBD on access to and transfer of technology.

[16] Art. 12.1 of La loi No 94-01 du 20 janvier 1994 portant Régime des Forêts, de la Faune et la Pêche.

[17] La loi No 96/12 du 5 août 1996 portant loi-cadre relative à la gestion de l'Environnement.

[18] Ibid, Art. 65.1.

[19] Décret No 95-531 du 23 août 1995 fixant les modalités d'application du régime des forêts.

scientific research,[20] the 2003 phytosanitary law[21] and the joint ministerial order between the ministries of territorial administration, finance and forestry pertaining to the management of the revenues earned for forest related activities.[22] Pursuant to these implementing regulations and ministerial orders, the ministry in charge of forestry in consultation with the ministry of scientific research, can deliver the collection permits for GR for scientific and culture use.[23] The ministry of scientific and technical research is the agency responsible for the issuance of the research permits to foreign researchers who are in partnership with a domestic research entity.[24] Furthermore, the importation and export of GR for scientific research or cultural purposes is subject to the issuance of certificates of origins and of the importation and export permits by the ministry in charge of forestry in consultation with the ministry in charge of scientific and technical research.[25] To ensure the safe handling of all the GR involved in the importation or exportation operations, all materials involved should be accompanied with a phytosanitary certificate which is delivered by the ministry of agriculture.[26] Lastly, management of the revenues earned through forest related activities, especially the sharing of the revenues in terms of what goes to the State in terms of taxes and what is entitled to local communities and for what purpose these earnings can be used, is governed by the joint ministerial order.[27] Although these revenues are not generated from the utilization of GR from the utilization perspective of the NP, this sharing of forest related revenues has been until to date, the benefit-sharing mechanism that is seen in the Cameroon regulatory context to be closer to the ABS framework.

Although these instruments are generally characterized by national actors as very relevant to ABS matters, in reality they do not considerably uphold several other key ABS principles of the CBD and the NP. Indeed, current sectoral laws fail to properly uphold the prior informed consent (PIC) and mutually agreed terms (MAT) principles, address benefit-sharing from the utilization of GR as defined by the Protocol, and their monitoring and surveillance of utilization as well as compliance approaches are not aligned with the spirit of the NP.

In the absence therefore of a Nagoya Protocol compliant ABS regulatory framework, an examination of the access and benefit-sharing regulations and possibly

---

[20] L'arrêté No 00002/MINRESI/B00/C00 du 18 mai 2006 fixant les conditions d'octroi d'une Autorisation de Recherche.

[21] Loi No 2003/003 du 21 avril 2003 portant protection Phytosanitaire fixant les principes et règles qui régissent la protection phytosanitaire au Cameroun.

[22] Arrêté conjoint N° 0000076 MINATD/MINFI/MINFOF du 26 Juillet 2012 fixant les modalités de planification, d'emploi et de suivi de la gestion des revenus provenant de l'exploitation des ressources forestières et fauniques destinés aux communes et aux communautés villageoises riveraines.

[23] Ibid, above at 21, Arts 13.1 and 13.2.

[24] Ibid, above at 21.

[25] Ibid, above at 21, Art. 13.2.

[26] Ibid, above at 22.

[27] Ibid, above at 23.

practice in Cameroon can only be based on the recently adopted July 2021 ABS law and the associated 2018 version of the drafts implementing regulations.[28] These instruments are:

- The July 2021 ABS law;[29]
- Draft implementing decree setting the terms of access to genetic resources and associated traditional knowledge and the fair and equitable sharing of the benefits arising from their utilization;[30]
- Draft model ABS permit;[31]
- Draft model PIC;[32] and
- Draft model MAT[33]

When all these instruments are eventually adopted, they will hopefully form the post Nagoya Protocol ABS regime of Cameroon.

2.2 *Scope of Cameroon Draft ABS Instruments*

The scope of the emerging ABS regulatory regime of Cameroon is set out in the July 2021 ABS law. Although the law has not provided clear indications of the different dimensions of the scope, these are indeed embedded in some of the provisions as we briefly outline below. Hence, reading through the current version of the law, it can be depicted that the following aspects are taken into consideration: (1) the geographical scope, (2) the material scope and the activities that are regulated; (3) the exemptions from the scope and (4) an inclusion of some key definitions which have significant bearing on the scope, despite as it is evident, the obvious confusion emanating from these definitions. Below, we address these aspects in turn and touch upon some missing and confusing aspects of the law and other issues left untouched.

---

[28]It must be stressed that with financial support provided by the National UNDP office in Cameroon, work was commissioned to an international consultant who delivered drafts ABS law to the National ABS committee of Cameroon.

[29]Loi No2021/014 du 09 Juillet 2021 Régissant l'Accès aux Ressources Génétiques, à leurs dérivés, aux Connaissances Traditionnelles associées et le Partage Juste et Equitable des Avantages issus de leur Utilisation.

[30]Projet de Décret fixant les Modalités d'Accès aux Ressources Génétiques aux Connaissances Traditionnelles associées et de Partage Juste et Equitable des Avantages découlant de leur Utilisation.

[31]Model Permis APA.

[32]Projet de Décision de Consentement préalable donné en connaissance de cause en vue de l'accès aux ressources génétiques / biologiques et / ou aux connaissances traditionnelles associées d'origine Camerounaises.

[33]Projet de: Conditions Convenues d'un Commun Accord (CCCA) en vue de l'Accès aux Ressources Génétiques, Biologiques et/ou aux Connaissances Traditionnelles Associées.

Geographical Scope The July 2021 ABS law stresses that the State of Cameroon has sovereign rights over the biological and genetic resources that originate from the country's national territory and those rights apply to *in-situ* and *ex-situ* GR, derivatives of GR and information associated with Cameroon GR and to the sharing of benefits derived from Cameroon's GR, which are regulated by international instruments.[34] Article 5 of the ABS law therefore appears to indicate that the geographical scope of the law extends to the national territorial jurisdiction of Cameroon. However, by the failing to stress the fact of access to and collection of biological and GR from within the national territory but instead indicating the element of 'origin' of the biological and GR including the derivatives from these resources, it is reasonable to stress that the spirit of this instrument is that it can apply to GR, aTK and derivatives accessed outside the geographical borders of Cameroon as long as the biological and GR are proven to have their origins from Cameroon.

Material Scope and Activities Regulated The July 2021 ABS law applies to access and utilization of the genetic resources, derivatives and aTK including: access to plant, animal and microbial genetic resources in the national territory, conservation of GR, application for and acquisition of intellectual property rights (IPRs) from the use of GR and aTK, transfer to third party of GR and aTK for research and commercial purposes, transboundary cooperation; current use of GR and/or previously acquired.[35]

Exemptions from the Scope The July 2021 law however provides for some exemptions from its very broad scope. Pursuant to article 4, it does not apply to biological resources that are accessed from Cameroon but are not utilized as genetic resources in line with the draft its definition of the term 'utilization'. In addition, the law does not apply to the exchange of GR and aTK among rural communities for their livelihood.[36] There are obvious grounds for confusion raising questions about the ABS law's clarity when it comes to implementation of article 4, especially in respect to the exclusion from its scope of 'biological resources' which are not used as 'genetic resources'. This confusion comes from the fact that, biological resources and genetic materials have the same definition, an approach that is not helpful for its operationalization.

Important Definitions In the July 2021 ABS law, the definitions of the terms 'biological resources', 'genetic resources' and 'genetic material' are drawn from the exact definitions provided by the CBD.[37] With the inclusion of genetic resources in the title of the ABS law as one of its core regulatory targets, there is a signal that the domestic approach is to closely align national ABS regime to the NP for which genetic resources are one of the subject matters of the regulations. Although the NP

[34] Art. 5 of the July 2021 ABS law.

[35] Art. 3, ibid.

[36] Art. 4, ibid.

[37] See definitions, Art. 7, ibid.

does not define the term 'access', a number of countries that have developed or are in the process of elaborating their ABS legislations have chosen to define the term, in a way that in their view can begin to clarify what it entails.[38] Cameroon has followed the same path. According to the July 2021 ABS law, access means "the possibility for a person or a group to reach some knowledge, to own it and master it. This is connected with the collection or acquisition including any transaction involving the genetic resources, derivatives or associated traditional knowledge by a user".[39] As currently framed, Cameroon's definition of access appears very broad and complex because it encapsulates critical ABS issues such as the possession and mastery of knowledge a concept that not only is not defined by the CBD and the NP, but which is likewise very difficult to practically implement in the operationalization of the NP. Furthermore, the definition of 'access' in the Cameroon law includes the notion of derivatives, that although defined by the NP, consensus on their scientific and operational implications in domestic ABS regulatory framework is yet to be reached among parties. The complexity of the definition of access in the law is exacerbated by the fact that it includes a loose notion of '*any transaction*' associated with the genetic resources, derivatives and traditional knowledge. While the definition of access does not directly point to access for utilization, it is tempting to assume that the inclusion of 'any transaction' include utilization as defined by the NP. Furthermore, 'any transaction' could also be interpreted to accommodate utilization of genetic information and DSI linked to Cameroon's materials in this definition. Therefore, it does not matter whether the user collected or acquired the materials themselves directly from Cameroon or through intermediary, or acquired them by any means overseas, utilization of such materials trigger the ABS rules and should earn benefits to Cameroon.[40] Also included in the law is its approach of defining 'utilization of genetic resources and associated traditional knowledge'. Pursuant to its article 7 utilization of GR and aTK refers to research on the properties of plants, animals, micro-organisms and to associated traditional knowledge and derivatives in view to enhancing scientific knowledge or to developing commercial products. This definition encapsulates important considerations of two broad

[38] E.g. section 2 of the 2011 draft ABS law of Liberia defines access as: collecting, obtaining, or using genetic resources or traditional knowledge associated with genetic resources, for the purposes of bioprospecting, non-commercial research, or export; Article 2 of the 2016 draft Ministerial Order of Rwanda Governing the Access to Genetic Resources and the Fair and Equitable Sharing of Benefits Arising from their Utilization in Rwanda defines access as: 'obtaining and possessing the right to utilize genetic resources conserved, whether derived products and, where applicable, intangible components, for purposes of research, bio prospecting, conservation, industrial application or commercial use. Algeria defines access to biological resources as: any prospection, collection or sampling of biological resources. Moreover, Prospection is defined as: exploration of biological diversity in view to isolate its elements or components whose value is yet to be established. See Loi n 14-07 du 13 Chaoual 1435 correspondant au 9 aout 2014 relative aux ressources biologiques.

[39] Art. 7 of the July 2021 ABS law on definitions.

[40] I must stress that this is an extreme view shared with me by an anonymous informant. It is not an official interpretation of the definition of access.

objectives for such utilization: (1) enhancement of scientific knowledge and, (2) development of commercial products from the resources, associated traditional knowledge or their derivatives. This definition therefore signals that Cameroon is embracing both commercial and non-commercial utilization of the GR and aTK upon which the country can lay claims of sovereign rights, thus the authority to determine the terms of access.

Some Missing and Some Confusing Issues The July 2021 ABS law is not clear on the principal trigger of benefit-sharing provisions in the sense that it fails to provide whether benefit-sharing is triggered by access to the resources, or at a specific stage in the utilization process. Furthermore, although the law covers current use of previously accessed resources in its material scope, there is no temporal scope for the law's applicability. In other words, it is not clear whether when the law is operationalized including through the implementing regulations that are expected to be adopted, it will apply to the resources accessed before or after the entry into force of Nagoya Protocol and/or of the CBD. Although the Nagoya Protocol itself has not directly addressed both the main trigger of benefit-sharing and the temporal scope, some of its provisions have been interpreted to support utilization as the most logical trigger of the ABS regulations.[41] On its part, in not addressing these issues, Cameroon's ABS regime generates some degree of confusion on the country's approach in handling cases of collections and access to its biological and genetic resources that happened before the entry into force of the NP and of the law and its implementing instruments. This confusion will especially be more acute when it comes to setting out the basis of claims for compensation for or benefit-sharing from scientific and/or commercial use associated with GR accessed before the entry into force of the NP and the adoption of the law. However, it can be argued that the approach Cameroon has taken to define the temporal scope of its ABS regulatory framework in a way that can be broadly interpreted (to express the State sovereign rights over the resources and to define utilization of genetic resources) signal an intent to pursue claims of benefit-sharing entitlements from the utilization of any GR that have been collected from Cameroon at any point in time.

There is no mention of Digital Sequence Information (DSI) anywhere in the law or implementing regulations. In effect, from the perspective of the Cameroon ABS law, products derived from synthetic biological based on or relating to Cameroon's biological and genetic resources are part of Cameroon's heritage and under the country's sovereign rights. It will be interesting to see how Cameroon monitors synthetic biology activities associated with the resources that are part of its national heritage and to what extent the country may be able to effectively assert its sovereign rights over its products. In a similar vein, the law provides that the utilization of 'genetic information' is a legally recognized type of utilization, amongst different

[41] Lassen (2016): This report provides an interpretation of a joint reading of article 2, article 3 and article 5(1) of the NP which leads to the interpretation that utilization should be the logical trigger for benefit-sharing.

types of utilization of GR and aTK.[42] On its part, the draft decree provides that genetic information, as information that is produced through basic and fundamental research can be used freely if such utilization is for non-commercial and non-profit making purposes.[43] However, utilization of genetic information obtained through any other means should be subject to the normal access authorization procedure set out by the ABS law, meaning that the access permit must be issued by the competent national authority (CNA), which is the ministry in charge of the environment and nature protection.[44]

2.3 The Proposed Institutional Framework of the Cameroon ABS Regulatory Framework

A national focal point (NFP) and a CNA have been designated to perform their responsibilities in accordance with Article 13 NP. The Ministry in charge of the Environment and Nature Protection is the CNA and is assuming its responsibilities with technical and advisory support of the GoC backed by the National ABS committee established in 2013. The ABS committee comprises representatives from various government departments including the ministries in charge of the environment and nature protection, fauna and flora, agriculture, research and technical innovations as well as representative from the academia, non-governmental organizations (NGOs) and intergovernmental organizations, and from COMIFAC.

The July 2021 ABS law formalizes the establishment of the three key agencies mentioned above as part of the ABS institutional framework in Cameroon.[45] The ABS law designates the ministry in charge of the environment and nature protection as the CNA and assigns it the CNA responsibilities in respect of the grant of access permit, supervision of the National ABS committee, elaboration of the conditions for PIC and MAT negotiations.[46] In addition, the ABS law formalizes the establishment of the National ABS committee, which is placed under the supervision of the Minister in Charge of the Environment and Nature Protection. The committee is set up as an advisory agency to the CNA, with the ability to create internal technical and scientific committees that can enhance its performance and strengthen its

[42] Art. 8 (2), July 2021 ABS law.

[43] Art. 31 (1) of the draft implementing decree setting the terms of access to genetic resources and associated traditional knowledge and the fair and equitable sharing of the benefits arising from their utilization.

[44] Art. 31 (2), ibid.

[45] Article 33 of the ABS law establishes the CNA and outlines its roles and prerogatives, article 34 establishes the national ABS committee and outlines its roles and prerogatives and article 10 establishes the national ABS focal point and outlines its role and prerogatives.

[46] Art.33 (2) of the July 2021 ABS law.

recommendations on ABS cases to the CNA.[47] It studies applications for access to genetic resources, assists national stakeholders such as local communities, researchers/research institutions and the government in decision making pertaining to the granting of access permits. In practice, it is the ministry (CNA) that makes the final decision on the grant or refusal of the ABS permit. The ABS also formalizes the designation of the NFP on ABS. In addition to article 13 NP roles, the NFP is responsible for the overall coordination of the secretariat and serves as the technical secretary of the ABS committee. Interestingly, the NFP is also responsible for monitoring the utilization and ensuring traceability of Cameroon's genetic resources that are exported and used overseas and of Cameroon's biological resources that are held in international *ex-situ* collections.

3 Nature of the Regulatory Tools for Access to *In-Situ* and *Ex-Situ* Materials in Cameroon

Like an increasingly growing number of countries that situate themselves more on the supply side of the ABS equation, even though in practice many countries are both suppliers and users of at least their own resources, Cameroon has chosen to regulate access to its genetic resources by including the requirements of PIC and MAT in its emerging ABS regulatory instruments. Drawing from the definition of access by the July 2021 ABS law of Cameroon as discussed above which provides no direct link to in situ or *ex-situ* GR, it seems to be the case that the conditions for access which are based on PIC and MAT apply equally to *in-situ* and *ex-situ* GR materials.

3.1 Prior Informed Consent in the Cameroon Draft ABS Regulatory Framework

Cameroon is following the path of many other parties to the NP to require in its ABS legislation that access to GR and aTK for their utilization is subject to the PIC of the CNA in accordance with the Nagoya Protocol.[48] The 2021 ABS law defines PIC as "an authorization issued to an applicant to access the genetic resources and/or

[47] Art. 34, ibid.

[48] Indeed, in accordance with article 6 of the NP, unless a country decides otherwise which is its absolute right, access to its genetic resources for their utilization should be subject to the PIC of the party providing access. In practice, many countries are choosing to require PIC in their ABS legislations. The Africa Union strategic guidelines for the coordinated implementation of the Nagoya Protocol on ABS encourage African countries to include the requirements for PIC in their domestic administration, policy and regulatory instruments on ABS.

associated traditional knowledge for continued utilization".[49] This is preceded by authorization to negotiate, a procedure described by a former key actor in Cameroon's ABS landscape as some form of pre-PIC agreement between the CNA and the applicant. 'Authorization to negotiate' is defined as a consent given by the CNA to an applicant to take steps to secure permits and agreements from other stakeholders to access GR and/or aTK in the national territory. This paves the way for the application to embark on proper ABS negotiations and to secure any additional and relevant certificates such as the phytosanitary certificate, the research and/or export permits as appropriate.[50]

With regards to the PIC process, the ABS law provides that PIC is given by the CNA under recommendation of the national ABS committee,[51] as already said above. In a move that mimics a simplified access, it exempts from the PIC requirement any fundamental research carried out by the national research system within the national territory.[52] Instead, it requires a declaration to collect the required GR and the negotiation and signing of MAT with local communities. However, the caveat for this exemption is that the national research system cannot benefit from this exemption if research is carried out in partnership with an overseas partner leading to the exportation of genetic resources or their derivatives.[53]

In addition, the ABS law has provisions for access in situations of emergencies that threaten humans, plants and animals' health as defined under international law.[54] Pursuant to article 15 of the draft ABS law of Cameroon, the conditions for access in these circumstances will be set out in specific implementing regulations. In operationalizing both the simplified approach and access in situations of emergencies (declared by the CNA), the draft decree clarifies that access in these cases should be free to national/public research institutions if there is proof that such research contributes to conservation and sustainable use of GR and aTK.[55] These provisions represent the Cameroon's current move in domesticating the special considerations provisions of article 8a and 8b NP.

The draft decree expands on the requirements for the PIC application. Among the elements that the applicant is required to submit to the CNA are a detailed research project, an identification document (national ID for Cameroon nationals and passport for foreigners) and the research permit which is issued by the ministry of scientific and technical research.[56] However, in a twist that seems to contradict the law's definition of PIC, the draft decree which is meant to be one of the implementing instruments of the ABS law, contends that the grant of PIC by the CNA is to enable

[49] July 2021 ABS law, Art. 7, definitions.

[50] See Art. 3 of the draft model prior informed consent.

[51] Art. 9 (1) of the July 2021 ABS law.

[52] Art. 9 (3), ibid.

[53] Art. 9 (4), ibid.

[54] Art. 15, ibid.

[55] Art. 45 draft ABS decree of Cameroon.

[56] Arts 5 and 11 of the draft ABS decree.

the applicant to commence negotiations leading to the signing of the MAT with the local and indigenous communities.[57] This approach of the draft decree seems to suggest that PIC is not really an authorization given by the CNA to an applicant to access the resources, but is more aligned to the definition of the 'authorization to negotiate' of the ABS law, which is described as a pre-PIC. The draft model PIC, another prospective implementing instrument of the ABS law, is strongly aligned to the PIC approach of the draft decree as it stipulates that PIC is granted to the applicant to commence MAT negotiations and to take steps to obtain all the necessary permits and certificates from the relevant administrations including the research permit, the phytosanitary permit, the export permits where necessary.[58]

3.2 *The ABS Permit Scheme of the Emerging ABS Regime of Cameroon*

The July 2021 ABS law provides that the CNA is the sole agency that delivers the ABS permit upon presentation of the PIC and completion of negotiations and signing of the MAT and the permit is valid for 5 years but is renewable.[59] Once the ABS permit has been issued by the CNA, it must be published with the PIC and MAT on the Access and Benefit-sharing Clearing-House (ABS CH) by the CNA.[60] The ABS permit or certificate of conformity according to the ABS law is defined as "a legal title attesting that access to the genetic resource to which it relates was made in accordance with the requirements of access and benefit-sharing legislation".[61] The access permit can be granted for commercial or non-commercial utilization of GR and aTK including specifically the following types of activities: fundamental/scientific research, bio prospection, research and development, commercialization, utilization of genetic information and the application of intellectual property rights.[62]

The ABS law does not define 'commercial' or 'non-commercial' utilization and makes no difference on the requirements for access to GR for the purposes linked to these categories of utilization. Accordingly, it is only the fact of undertaking fundamental research and research and development in the national territory by an institution of the national research system and provided there is no transfer/export of

[57] Art. 7 (1), ibid.

[58] See draft PIC model, Arts 2, 3 and 4.

[59] Art. 17 of the July 2021 ABS law.

[60] Art. 24 (2) of the draft ABS implementing decree. Cameroon's approach goes a little bit beyond the NP, which only requires publication of the permit to the CHM to become an internationally recognized certificate of compliance. See Art. 17.2 Nagoya Protocol on ABS.

[61] Art. 7, definitions, July 2021 ABS law.

[62] Art. 8 (1) and (2) of the July 2021 ABS law and Art. 30 and 31 of the draft ABS implementing decree.

the materials overseas that leads to avoidance of the rigorous access permit application process with the requirement of PIC and MAT.

With regards to the content of the ABS permit, it is structured in the way that it identifies the issuing authority and date of issuance, the permit applicant/user of the materials that are sought, the type of activity that the materials will be subjected to e.g. commercial or non-commercial use (despite the law and implementing decree not actually defining these types of activities). Furthermore, the draft permit identifies the specific indigenous and local communities that are the custodians of the GR and holders of aTK, it makes specific references of the PIC and MAT that were granted by the CNA and it defines the period of validity of the ABS permit.[63] These features of the ABS permit are aligned with the minimum information expected from the internationally recognized certificate of compliance in accordance with article 17.4 NP.[64]

3.3 *Mutually Agreed Terms in the Emerging ABS Regime*

From the foregoing, it is evident that the CNA is the sole entity with the responsibility to issue the access permit upon submission of PIC and MAT.[65] In practice, MAT can only be negotiated after PIC has been granted.[66] The draft ABS implementing decree clarifies that negotiations of MAT between the applicant/users of materials and indigenous and local communities should be carried out under supervision by the CNA and that the PIC of ILCs for access to the genetic resources in their territories and to the aTK is evidenced by the negotiated MAT.[67]

A very broad outline of key clauses that should be included in MAT captures the following elements: the identities of the contracting parties, the types and quantities of the materials sought, the types of benefits to be shared (monetary or non-monetary benefits), the modalities of payments or retrocession of the benefits and the conditions for the transfer to and use by third parties[68] concerning the application of IPRs over the products of research based on the materials accessed. The draft ABS implementing decree provides that MAT provisions on IPRs need to address

[63] See draft ABS model permit for access to Cameroon's genetic resources and associated traditional knowledge.

[64] Article 17.4 of the NP provides that the internationally recognized certificate of compliance should contain the following minimum information: the issuing authority, date of issuance, the provider, the unique identifier of the certificate, the person or entity to whom prior informed consent was granted, the subject matter or genetic resources covered by the certificate, confirmation that mutually agreed terms were established, confirmation that prior informed consent was obtained and the commercial and/or non-commercial use.

[65] Art. 17 of the July 2021 ABS law.

[66] Art. 10, ibid.

[67] Art. 21 of draft ABS implementing decree.

[68] Art. 20 (1), ibid.

among others, the objectives for applying IPRs and the risks to local and indigenous communities, regular reporting on research progress and possible industrial developments and capacity building aimed at indigenous and local communities.[69]

A very important component of the emerging ABS regulatory framework is the draft model MAT which among other things, covers a very wide range of obligations for the parties involved, the applicant/user and the suppliers, specifically indigenous and local communities concerned. In a nutshell, the range of obligations for users include but are not limited to the realization of the environmental and social impact assessment during the application process which should determine the vulnerability of the resources sought and impact of exploitation and utilization of aTK to local communities; obligation to ensure transparency in the declaration and sharing of monetary and non-monetary benefits arising from commercial use of the GR and aTK; obligation to seek further agreements/authorization from the CNA and local communities for transfer of the resources to third parties; obligation to train local and indigenous communities on the impacts of the application of IPRs and to acknowledge Cameroon and the relevant communities in any publications.[70] The obligations of suppliers of GR, who the draft model MAT narrows down to local communities, include but are not limited to ensuring guaranteed access by the user to the genetic resources as authorized by the CNA and to ensure periodic supply of the required quantities to the user; in the event that there may be delay in supplying the required materials, inform the user in advance of the difficulties or challenges encountered; ensure transparency in the utilization of the monetary and non-monetary benefits received and obligation to guarantee price stability of the materials throughout the lifetime of the MAT.[71] There is no specific mention of or clauses on the use of digital sequence information and not even genetic information in the draft model MAT. The draft model MAT appears to be focused on access and utilization of physical GR and the sharing of monetary and non-monetary benefits.

The draft model MAT provides for a possible three-stage approach to the resolution of conflict that may arise in the course of its implementation. In the first instance, contractual parties are expected to resolve any conflict amicably. Should the amicable approach fail, parties should submit to the arbitration of the CNA. Lastly, if the outcome of arbitration process is not satisfactory to any of the parties, the issue can be submitted to the Cameroon's judiciary as Cameroon's courts are the only one to hear disputes arising from the implementation of the MAT.[72]

---

[69] Art. 24, ibid.

[70] Art. 5 draft model MAT ABS regulatory framework.

[71] Art. 6, ibid.

[72] Arts 17 and 18 of the draft Model MAT.

4 Benefit-Sharing Framework of the Emerging ABS Regime of Cameroon

The emerging ABS regulatory framework of Cameroon deals with the fair and equitable sharing of monetary and non-monetary benefits arising from the utilization of GR and aTK in the July 2021 ABS law,[73] the draft implementing decree[74] and the draft model MAT.[75] Further details on what non-monetary benefits entail are provided by both the ABS law and the draft implementing decree. Non-monetary benefits encompass such elements as technology transfer, training, local development, information exchange, the supply of goods and services and any other elements as may be consensually agreed in the MAT.[76] In turn, the draft implementing decree describes different types of payments of monetary benefits by the user to the beneficiary. Such payments may be made for the purpose of access and collection of the materials. Other types of payments include milestone payments, license payments, salaries of local/community assistants, research funding, payment to trust funds for example those set up for the purpose of conservation and sustainable use of biological diversity.[77] The benefits generated and allocated to the beneficiaries shall be managed by local communities in the interest of community economic, social and cultural development and for the conservation of biodiversity.[78]

One of the three instruments that deserves special attention is the model MAT which seems to have been drafted on the presumption that ABS agreements will always be (1) focused on access to and utilization of physical biological resources, (2) will therefore always involve the communities in localities where the materials are collected and (3) will always be for commercial utilization. Based on such presumption, the Cameroon benefit-sharing approach runs the risk of losing sight of the benefits that can be generated from scientific use of its resources. Furthermore, by focusing on the benefits arising from the commercialization of Cameroon's biological, genetic resources and aTK, the draft model MAT has adopted a very narrow perspective in the implementation of the benefit-sharing provisions of the draft ABS law and the draft decree. Lastly, although the main objective of the draft model MAT is to define the mechanisms to share the benefits derived from the commercialization of the biological, genetic resources and aTK, its content is heavily focused on outlining the detailed obligations of both the users and local communities. The draft model MAT does not define specific benefit-sharing mechanism in terms of e.g. providing the basis for estimating the benefits to be shared

[73] Arts 27 to 29 of the July 2021 ABS law.

[74] Arts 39 to 44 of the draft implementing decree.

[75] Arts 1 and 8 of the draft Model MAT.

[76] Art. 27 (b) of the July 2021 ABS law and Art. 41 of the draft implementing decree.

[77] Art. 40 of the draft implementing decree.

[78] Art. 29 of the July 2021 ABS law and Art. 42 (2) of the draft decree.

such as percentage of profits, or contribution to research and development. Including the specific mechanism for benefit-sharing discussions in the model MAT is important because that provides all parties with clear terms that will guide discussions on the monetary and non-monetary benefit-sharing in the course of MAT negotiations.

The ABS law and implementing instruments therefore include benefit sharing chapters (of the law and the draft implementing decree), an approach which can be interpreted as the basis for compliance of the Cameroon ABS regime with article 5 of the NP on the fair and equitable benefit sharing. However, the extent to which these benefit-sharing chapters are in Cameroon's interest or are very clear to users is questionable. Reshaping the draft model MAT to make it more accommodating of non-commercial research and associated non-monetary benefits while recognizing that other national actors (apart from local communities) are likely to be parties in MAT will deliver wider benefits to Cameroon.

5 Provider and User Compliance in the Emerging ABS Regime of Cameroon

The draft ABS law and implementing decree have specific provisions aimed at ensuring compliance with ABS regulatory framework in relation to access for utilization of the GR and aTK of local and indigenous communities. In effect, those provisions are meant to deliver compliance of the Cameroon ABS framework with Articles 15, 16 and especially article 17 of the NP, all of which are the core articles of the compliance cluster of main obligations of the NP. Chapter VII of the draft ABS law deals broadly with monitoring measures of activities pertaining to access and utilization of GR and aTK in Cameroon. However, the way these measures are drafted does not establish a system/mechanism of monitoring and surveillance of access and utilization of genetic resources and aTK but provides that such a system will be set up by the CNA.[79] The system will be deployed to enable access to a static or mobile equipment of the permit's applicant and for monitoring the use of GR and aTK at all stages, including in the context of bio-prospecting, collection of samples and information, research, development, innovation, pre-marketing and marketing.[80] Concerning access, the system will be designed to address illegal and unscrupulous access and acquisition of GR and aTK.

The way the law envisages the system to monitor illegal activities is through the setting up of checkpoints and through collaborations between the CNA and other sectoral ministries.[81] Furthermore, the law empowers the CNA to establish

[79] Art. 38 of the July 2021 ABS law.

[80] Ibid; Art. 38 of the July 2021 ABS law.

[81] Art. 39, ibid.

partnerships with institutions overseas to ensure that GR and aTK from Cameroon used overseas were acquired legally.[82]

The ABS law also empowers the CNA to set up what can be seen as a system of denunciation[83] of illegal access and utilization of GR in the national territory. Under this system, local and indigenous communities and local authorities will be able to report any illegal activity they are aware of in their localities. In clarifying some of the monitoring and surveillance mechanisms introduced by the ABS law, the draft decree identifies road checkpoints as the point of controls where anybody in possession of biological and genetic material must show the permit that authorized collection of such materials.[84] Parties to MAT are also required to deliver annual reports to the CNA on the implementation of the agreement[85] and users of GR and aTK are required to provide the CNA with annual statements on the monetary and non-monetary benefits from the utilization of GR and aTK.[86] The law has provisions for a range of sanctions for failure to comply with the regulations which vary depending on the severity of the violation and include administrative sanctions,[87] fines and even imprisonments of between 2 to 10 years for using the GR, the derivatives and associated traditional knowledge without PIC and MAT.[88]

6 Compliance with the Nagoya Protocol

As already seen, Cameroon requires the PIC of the CNA as part of the requirements for an access permit. What is remarkable in the PIC approach enshrined in the Cameroon's instruments is the very significant involvement of local and indigenous communities in the access application process whether this is only for GR aTK or even both.[89] Arguably, there is a significant alignment of the draft instruments with NP articles 6.1, 6.2 and 7 which deal with access to GR and aTK. Pursuant to article 8 (1) of the July 2021 ABS law, any individual or institution can seek access to the GR and aTK of Cameroon, as long as access is for the activities outlined in Article 8 (2) of the law, be they for commercial or non-commercial ends and regardless of

[82] Art. 40, ibid.

[83] Art. 41, ibid. and Art. 48 draft ABS implementing decree.

[84] Art. 47 draft ABS implementing decree.

[85] Art. 49, ibid.

[86] Art. 50, ibid.

[87] Arts 49.1 and 49.2 of the July 2021 ABS law. Among the administrative sanctions, the ministry in charge of the environment may decide the withdrawal of the ABS permit, seizure of any materials collected, ban of up to 5 years or more to undertake any ABS related activity in Cameroon etc.

[88] Section III, specifically article 50 of the July 2021 ABS law.

[89] Art. 11 (1) of the draft ABS implementing decree. In particular, the draft decree stresses that applications for access to traditional knowledge associated to genetic resources are transferred by the CNA under the recommendation of the national ABS committee, to the relevant local communities.

where utilization will take place, domestically or overseas. There is therefore no segregation or discrimination in terms of nationalities. In this respect, the Cameroon's approach can be viewed as compliant with Article 6.3 (b) as it portrays fairness and tries to prevent discrimination. However, like many countries, Cameroon has taken the approach to promote biodiversity research that is carried out domestically, by exempting national institutions from the stringent PIC and MAT rules. Institutions undertaking research domestically are subject to simple declaration. This declarative system is however not applicable to research envisaged in collaboration with overseas partners with the intent to export materials. This is a slight but legitimate and understandable deviation from article 6.3 (b), as the country's approach is designed to address a very specific domestic interest, enhancement of biodiversity research capacities through ABS in Cameroon.[90]

As presented in the table below (Table 1), the ABS law and prospective drafts implementing instruments are largely aligned to the relevant provisions of the Nagoya Protocol on ABS, except in relation to compliance with mutually agreed terms, article 18.3 (b) on the utilization of mechanisms regarding mutual recognition and enforcement of foreign judgements and arbitral awards. There is no provision on mutual recognition of foreign judgements or foreign arbitral award in the drafts of Cameroon ABS regime. The draft model MAT asserts that Cameroon law[91] is the applicable law and conflicts arising from MAT implementation may be resolved amicably or through the arbitration of the CNA or failure in both instances the matter should be submitted to the Cameroon's judiciary should that be necessary.[92]

It is however questionable whether the 2021 law and accompanying draft implementing instruments provide enough legal clarity and transparency to actors interested in accessing materials from Cameroon. Arguably, the various instruments of the emerging ABS regime are generating substantial confusion on the role of the PIC given by the CNA. There is a need therefore to recalibrate some of the definitions to make them coherent across the various instruments. This recalibration should clarify that only the access permit granted based on PIC and MAT by the CNA, should be used as an authorization for access/collection of GR and/or aTK for the activities outlined in article 8.2 of the July 2021 ABS law. Furthermore, despite the ABS law's attempt to define the functions of the National ABS committee, the mechanism by which the committee will deliver its recommendations to the CNA is not clearly elucidated in any of the instruments, raising a serious question of transparency. Although members of the ABS committee are drawn from various sectors, its independent judgment may be called into question because the committee is under the authority of the minister in charge of the environment and its secretariat is serviced by the ABS focal point.

Another confusing element of the draft instruments relates to the various fees associated with either the transfer of materials or related commercial information and

[90] See Greiber et al. (2012), explanations on article 6.3.b, at p 103.

[91] Art. 17 draft model MAT.

[92] Art. 18, ibid.

Table 1 Alignment of the domestic ABS regime with the provisions of the Nagoya Protocol

	Provisions of the ABS regulatory framework of Cameroon				
Nagoya Protocol provisions	July 2021 ABS law	draft decree	draft model MAT	draft PIC and permit	Comments
Art 6.3(a)	×	×	×	×	A holistic reading of the various instruments does not portray sufficient legal certainty, clarity and transparency e.g. complexities of PIC and authorization to negotiate; operation of the ABS committee
Art 6.3(b)	Art 8(1)				However, national research institutions submit a simple declaration to the CNA but must sign MAT with local and indigenous communities
Art 6.3(d)	Art 17	Art 5(1), Art 6; Art 24; Art 27	√	√	There is a draft Model PIC and a draft Model permit
Art 7	Art 18	Art 11	√	√	
Art 18.2			Art 18		
Art 18.3			Art 18		There is no provision on mutual recognition and enforcement of foreign judgements and arbitral awards
Art 6.3(c)		Art 5(3); Art 11(3)	√	√	
Art 6.3(e)		Art 24		√	
Art 6.3(f)		Art 5; Art 7	√	√	Negotiations and conclusion of MAT with ILCs is the process through which ILCs are involved in the PIC and permit processes
Art 6.3(g)	Art 9.3(c); Art 10.1 and 10.2;	Art 20 to 23	√		
Art 18.1			√		Arbitration is the only alternative dispute resolution approach considered in the draft MAT. Cameroon NCA is the arbitrator
Art 12	√	√	√	√	
Art 13	Art 33 to 37				Article 33 to 37 of the law outline the roles of the CNA, the national ABS committee and the National ABS focal Point.
Art 15/16	x	x	x	x	A holistic reading of all the ABS law and all the draft implementing instruments does not reveal user measures for foreign GR or aTK used in Cameroon

(continued)

Table 1 (continued)

Nagoya Protocol provisions	Provisions of the ABS regulatory framework of Cameroon				
	July 2021 ABS law	draft decree	draft model MAT	draft PIC and permit	Comments
Art 17	Art 38 to 41	Art 47 to 50			Measures not reflecting utilization as defined by the ABS law or the NP

Table is adapted from the author's interpretation of the compliance dimension of the ABS regulatory framework of Cameroon with the specific provisions of the NP that were identified to be of special interest to this research
Tick = Specific ABS instrument is generally in compliance with the specific article of the NP
No tick = No specific observation to highlight in respect of the specific compliance area
Source: *Own illustration*

the IPRs fees that are outlined in the draft implementing decree. The draft ABS decree requires users of GR and aTK to pay 10,000,000 FCFA (about USD 1700) if they request to transfer their research results for the purpose of commercial use and 500,000 FCFA (about USD 850) if they want to apply for IP over their research results with the intent to use such intellectual property (IP). None of the draft instruments gives an indication of the stage when these fees could be discussed (during the PIC process or in the context of MAT negotiations) and it is even questionable if these fees are practical, pragmatic and whether the CNA is the authority to deal with application fees for IPRs.

7 Strengths and Weaknesses of the Emerging Cameroon ABS Regulatory Framework

7.1 Strengths

The principal strength of the drafts ABS instruments of Cameroon is the way they approach the involvement of local and indigenous communities in the ABS permit application, be it in respect to access to GR, aTK or both. Based on the July 2021 ABS law[93] and the draft decree,[94] it is the CNA that makes the final decision on PIC as a prelude to MAT negotiations. However, in the case of access to aTK only, permit applications are duly notified to the relevant ILCs whose responsibility is to grant PIC, with assistance and supervision of CNA staff, even though eventually the written PIC decision is issued by the CNA. The current versions of the draft implementing instruments put local communities at the heart of the PIC process and MAT negotiations, allowing them to determine and express their benefits-

[93] Arts 7 to 17 of July 2021 ABS law.

[94] Arts 5 to 11 draft ABS implementing decree.

sharing expectations and concerns over the cultural impacts of prospective activities. This is done by making the community bio-cultural protocols the basis for any interactions between local communities and the outside world.[95] This approach is clearly in line with the NP on ABS and may even be viewed as going beyond the NP particularly in relation to the Protocol's approach to PIC for access to genetic resources. The NP providers that PIC for access to GR for their utilization is given by the party providing access that is the country of origin of the resources in question.[96] Therefore, the NP does not obligate parties (through their designated entities) to involve local communities in PIC decision in respect of access to GR for their utilization. However, the protocol provides that PIC and involvement, or approval of indigenous and local communities may be obtained if they have established rights to grant access to such resources" (Art. 6.2 NP).

There are currently no clearly established rights of ILCs over GR in Cameroon and the ABS law has in fact reasserted the State's rights over all the genetic resources in the national territory. Again according to the July 2021 ABS law, all the genetic resources in the national territory belong to the state and the state's rights extend to genetic resources, associated traditional knowledge, genetic resources in the national domain, public domain and private domain, derivatives.[97] These resources cannot be utilized by anyone without the PIC of the State which is delivered by the CNA. In Cameroon the CNA could therefore grant PIC and coordinate or oversee MAT negotiations over access to GR for their utilization between a research entity in Cameroon and an overseas partner, with very minimal to no involvement of local communities. So, even in the absence of clearly established rights of ILCs over GR, an approach to sideline them in PIC and MAT processes would violate the 1996 environmental framework law which requires community involvement in environmental and forestry related issues happening in their localities.

In Cameroon, community involvement in collection and supply of biological resources was operationalized in the case of the PIC and MAT negotiations for access to and utilization of the roots of *Echinops giganteus* by a French company. In the western region of Cameroon, its rhizomes are used as ingredient in different culinary preparations and traditionally as medication against numerous diseases.[98] However, it is neither the therapeutic and culinary properties of the plant nor the local knowledge of the communities living in the localities that the French company is interested in. The company is interested in the rhizomes of the plant for their potential use in the fragrance and cosmetic sector. As already said, an ABS contract (MAT) was signed in April 2015 between the Ministry in charge of the environment, the company and the communities of Magha - Bamunbu, based in the South West Region of Cameroon. It is in the Magha Bamunbu locality where this plant grows

[95] Arts 18 to 21 of the July 2021 ABS law and articles 10 and 11 of the draft ABS implementing decree.

[96] Art. 6.1 of the NP.

[97] Art. 8.1 of the July 2021 ABS law.

[98] Ibid, above at 10.

naturally that domestication efforts have been initiated with the support of the company to ensure long-term and sustainable supply of the roots to the company. The objectives of the MAT are to define benefit-sharing modalities between the local community and the company arising from commercial exploitation of the roots of the plant in conformity with the PIC of the CNA. The MAT signature was the culmination of 2 years of negotiations between the company and local communities, with assistance and facilitation of a local NGO ERUDEF under the supervision of and based on the Pre-PIC previously granted by the ministry in charge of the environment in January 2013. Pre-PIC resulted from consultations between the government, the communities and the company and was in principle an authorization given to the company to engage with local communities, commence negotiations of full MAT while conducting research on Echinops roots. The negotiated MAT comprised obligations for both the communities and the company. The MAT obliged the company to carry out environmental and social impact assessment in conformity with Cameroon's legislation and to ensure rational, conservation and sustainable management of the site; not to hold intellectual property rights on the genetic materials of Echinops and only sell products deriving from the roots of Echinops only as agreed with local communities; to obtain and comply with the terms of any other permits including the research permit, export permit, phytosanitary permit. With regards to local communities, their main obligations are to ensure exclusive supply of Echinops roots to the company during the 3 years MAT validity; to ensure transparency in the management of the benefits earned.

Giving local communities a significant role in this Echinops MAT despite Cameroon's strong assertion of the State's ownership of GR is testament of the strength of the emerging regulatory framework on the issue of community involvement in ABS arrangements and the reaffirmation of the country's commitment to respect the 1996 environmental framework law.

On a general note, the current draft instruments attempt to cover all the key obligations NP parties are expected to fulfil in their domestic legislation, which is also one of the regime's strengths. This shows Cameroon's willingness and efforts in the process that should hopefully deliver a Nagoya Protocol compliant ABS regulatory framework. It is also an evidence of its commitment to designing a domestic ABS regulatory framework which should not only enable the country to fulfil its obligations under the NP but likewise generate development, conservation and sustainability outcomes for the country.

7.2 Weaknesses

The principal weakness of the emerging ABS regime of Cameroon is its approach in monitoring the utilization of GR, specifically the designation of checkpoints. Checkpoints designation is one of the measures proposed by the Nagoya Protocol to monitor and to enhance transparency about the utilization of genetic resources.

Parties are invited to designate checkpoints and assign them among other functions "to collect or receive as appropriate, relevant information related to the PIC, to the source of the genetic resources and to the establishment of mutually agreed terms and/or the utilization of genetic resources as appropriate".[99] Critically, the NP stresses that checkpoints must be effective and should have functions relevant to the utilization of genetic resources, or to the collection of relevant information at *inter alia*, any stage of research, development, innovation, pre-commercialization or commercialization.[100]

The July 2021 ABS law and the current versions of the draft implementing instruments of Cameroon have not operationalized article 17.1 (a) NP in the way that reflects the broad spectrum of utilization of GR. Pursuant to the draft ABS law,[101] the various mode of utilization and activities that can be applied on Cameroon's GR include fundamental research, bioprospection, research and development, utilization of genetic information, commercialization and the application of IPRs to the products of these various modes of utilization. All these forms of utilization reflect the definition of the term utilization of the NP. However, the monitoring and surveillance regime that is being constructed in the emerging ABS regime of Cameroon which focuses on the establishment of checkpoint does not reflect the modes of utilization defined by the law[102] but rather focuses on the control of movement of raw materials (biological and genetic resources) through, for example, access to physical infrastructure where these materials are supposedly transformed. The draft ABS law provides for the establishment of checkpoints by the CNA through collaboration and consultations with other sectoral departments.[103] Looking at the way the draft implementing decree operationalizes these provisions of the ABS law, the only checkpoints that appear to have been considered are the road checkpoints set up by the ministry of forestry to control the movement of forest products including timber and non-timber forest products. Although the NP protocol does not dictate parties how to go about designating checkpoints domestically, well-known possible checkpoints discussed during the NP negotiation process (though not listed in the final negotiated NP text) such as IP offices, market approval offices and research funding offices are examples of checkpoints that are more relevant to GR utilization but are completely missing in the Cameroon regime. The draft ABS regime contends that the CNA may also collaborate with other national and overseas institutions in order to identify cases of utilization of illegally accessed GR and aTK from Cameroon. Although this provision provides an opportunity for collaboration

[99] Art. 17.1.a.i of the NP.

[100] Art. 17.1.a.iv of the NP.

[101] Art. 38 to 41 of the July 2021 ABS law.

[102] Art. 38, ibid. The ABS law provides that the CNA will establish a regime of monitoring of the exploitation of the GR, their derivatives and the associated traditional knowledge to enable access to static and mobile infrastructure of the application for access to have information on the utilization of GR at all stages including in relation to bioprospection, collection of samples and information, research and development, pre-commercialization and commercialization.

[103] Art. 39 of the July 2021 ABS law.

with IP and market registration offices nationally and internationally, the failure to designate these national bodies as checkpoints appears to be a major weakness. The emphasis of the ABS law is on control of physical collection and movement of raw biological materials. This is strengthened by the denunciation system designed to enable local communities to accuse actors who carry out suspicious or illegal plant collection activities happening in their areas.

In addition, the July 2021 ABS law and the current versions of the drafts implementing instruments (1) generate significant confusion in some key areas, and (2) include some unrealistic and unpragmatic provisions the implementation of which may not deliver the intended results. These generate more weaknesses for or points of critique of the regime, including the following:

- The lack of coherence in definitions. It is therefore necessary to instill coherence between the definitions and their operational meaning in the texts of the various instruments and reduce confusion in the way they are intended to be operationalized.
- Unclarity in allocation of mandates to the CNA and the ABS committee. The law assigns the committee an advisory role to the CNA in the PIC process, not a decision-making role but then article 9 (2) of the draft decree creates confusion by providing that PIC of the ABS committee is required for an applicant to export plants of special interest. It is therefore not clear if the CNA is responsible to grant PIC for access and the ABS committee responsible for the grant of PIC in the context of exportation of plants of special interest. This responsibility is anyway not mentioned among the responsibilities earmarked for the ABS committee by the draft ABS law.
- The mechanisms to realize benefits and the expectation set in the draft instruments may well be unrealistic and may never deliver the intended objectives. The ABS law and the draft implementing decree have provisions pertaining to the access applications fees, special taxes on the revenues generated from commercial exploitation of genetic resources and aTK and fees for the application and exploitation of the IPRs. For example, regarding the application for and exploitation of IPRs based on the utilization of GR, the draft implementing decree requires that the user of the GR and aTK to pay 500.000 FCFA (about USD 850) into the National Environmental and Sustainable Development funds respectively for the application and for the exploitation of the IPRs.[104] These fees could be viewed as prohibitively high and may even stand against Cameroon's ambition to promote the creation of ABS value chains, enhance research capacities and benefit from scientific or commercial use of its biodiversity. Broadly speaking, this provision raises several questions including when exactly this money should be paid, e.g. during the PIC or MAT negotiations if at that stage the applicant discloses their intent to eventually pursue IPRs protection. Or is it when the user returns to the CNA to request PIC for commercial use? And most of all, is the

[104] Art. 37 and 38 of the draft ABS decree.

CNA the agency to handle IPRs related fees for IPRs protection? Understandably, this could represent a serious disincentive for users of GR to seek access to Cameroon's resources because at this stage, there is no way the user will know whether his prospective activity will generate an IPR protectable result/derivative. To address these issues, Cameroon needs to get some experts contributions to its current drafts regulations in order to make them more reasonable, technically coherent with the IP system and realistic with regards to benefit expectations.

– Still in relation to the interface between IPRs and the protection of GR and aTK of local and indigenous communities, there is another mismatch created by the law which is unworkable and may never deliver any practical outcome. The law provides that a *sui generis* IP tool will be adopted by special regulation aimed at the protection of the rights of local communities over GR and aTK alongside other existing IP tools which consist of patents, plant breeders' rights (PBRs), geographical indications; trademarks, respect of community rights over GR and associated TK.[105] Pursuant to article 19.2 of the ABS law, a panoply of measures can be considered alongside the classic IP tools for protection of traditional knowledge through a *sui generis* all which may include legislation on anti-competitive practices, confidential information, use of contracts and licenses, use of registers and databases, etc.[106] Furthermore, the ABS law requires the State to ensure that the existing IPRs tools are applied to derivatives of GR and aTK, to the protection of the rights of ILCs and to assist ILCs in their pursuit of IP protection and acquisition through patents, trademarks, PBRs, geographical indications over their aTK.[107] It is very difficult to reconcile these three provisions of the law which at the same time signal recognition of the inappropriateness of the conventional IPRs over GR and aTK and the need for a *sui generis system*, and provides State's support to local communities in their bid to pursue the same conventional IPRs over their aTK. Furthermore, the way the law is drafted indicates minimum understanding of the workings of conventional IPRs because traditional knowledge per say does not fit the eligibility criteria as a subject matter for protection of patents or PBRs. How then the State will assist local communities in acquiring some of these IP forms is highly questionable. The *sui generis* approach contemplated in articles 19.1 and 23 of the law needs careful consideration and expert contributions if it is to deliver a system that works for Cameroon's local communities. As a member of the Organisation Africaine de la Propriete Intellectuelle (OAPI), Cameroon can effectively only deliver this *sui generis* system through the OAPI framework. These are some of the practical issues that must be taken into consideration when dealing with IP in relation GR and aTK in Cameroon.

[105] Arts 19.1 and 23 of the July 2021 ABS law.

[106] Art. 19.2, ibid.

[107] Arts 24 and 25, ibid.

- Unrealistic approach adopted by the ABS law in assigning roles to the NFP in monitoring the use and traceability of GR exported and the GR from Cameroon found in *ex-situ* collections overseas. While the intent of the ABS law is commendable, from a practical point of view, it is hard to see how the NFP would handle this responsibility which is arguably not the prerogative of the NFP pursuant to Article 13.1 NP. One way around this is to assign this role to the National ABS committee, which could establish a special technical task force to focus on monitoring use of materials in exports or the resources in *ex-situ* collections overseas. Success in generating any positive results from monitoring the traceability of GR will highly depend on the financial, technological means given and the individual capacities of members of this task force.

8 Conclusion

This chapter demonstrates that with every good intent, Cameroon is very well on the path of establishing what the country hopes to be a very comprehensive national ABS regulatory framework that upholds the various facets of Nagoya Protocol obligations. The comprehensiveness of this framework is reflected through the development not only of the stand-alone ABS law adopted in July 2021, but also of a panoply of implementing regulations as well as useful support tools such as the model PIC, model MAT and model access permit. However, these instruments if adopted in their current shapes bear significant shortcomings which not only raise the question about their proper compliance with the NP, but may complicate practical implementation therefore failing to deliver their stated objectives.

Acknowledgements The author expresses sincere gratitude to all the experts who agreed to be interviewed in this case study. My heartfelt thanks to Dr. Evanson Chege Kamau, for kindly inviting me to be part of the team he put together for this research project and for reviewing the draft materials. I would also like to thank the other members of team, whose comments shared during all the research meetings held at the University of Bremen between 2017 and 2019, were very instrumental in fine tuning my case studies. I want to thank Dr. Kamau assistants at Bremen, who were always ready to help in any way, especially in relation to all the meetings that were held in Bremen. I want to thank my family, my wife and children for putting up with me through all related travel. All views and mistakes in this chapter are mine.

References

Publications

Duguma LA, Minang PA, Foundjem-Tita D, Makui P, Mandiefe Piabuo S (2018) Prioritizing enablers for effective community forestry in Cameroon. Ecol Soc 23(3):1. https://doi.org/10.5751/ES-10242-230301

Fonge BA, Tchetcha DJ, Kembi L (2013) Diversity, distribution, and abundance of plants in Lewoh-Lebang in the Lebialem highlands of Southwestern Cameroon. Int J Biodiv 2013:13 pp

Greiber T, Pena Moreno S, Ahren M, Carrasco JN, Kamau EC, Medaglia JC, Oliva MJ, Perron-Welch F (2012) An explanatory guide to the Nagoya Protocol on access and benefit sharing, IUCN Environmental Policy and Law Paper No.83. IUCN, Gland, Switzerland

Sunderland TCH, Comiskey JA, Besong S, Mboh H, Fonweban J, Dione MA (2003) Vegetation assessment of Takamanda forest reserve Cameroon. In: Comiskey JA et al (eds) Takamanda: the biodiversity of an African rainforest. Smithsonian Institution, pp 9–17

Tchouto MGP (2004) Plant diversity in a central African rainforest. Implications for biodiversity conservation in Cameroon [Ph.D.thesis], University of Edinburgh/Royal Botanic Garden of Edinburgh

Other Documents and Websites

AU (2015a) Africa Union strategic guidelines for the coordinated implementation of the Nagoya Protocol on access to genetic resources and the fair and equitable sharing of the benefits arising from their utilization, Africa Union Commission, http://www.abs-initiative.info/fileadmin/media/Knowledge_Center/Pulications/African_Union_Guidelines/AU_Strategic_Guidelines_On_ABS_-_20150215.pdf (accessed 17 February 2020)

AU (2015b) Africa Union practical for the coordinated implementation of the Nagoya Protocol in Africa, Africa Union Commission, http://www.abs-initiative.info/fileadmin/media/Knowledge_Center/Pulications/African_Union_Guidelines/AU_Practical_Guidelines_On_ABS_-_20150215.pdf (accessed 17 February 2020)

Cameroon's accession to the Nagoya Protocol., https://www.cbd.int/countries/default.shtml?country=cm (accessed 17 February 2019)

COMIFAC (2011) Stratégie des pays de l'espace COMIFAC relative à l'Accès aux ressources biologiques/génétiques et au Partage juste et equitable des Avantages découlant de leur utilization. Série Politique No 4, Commission des Forest d'Afrique Centrale

IUCN (2015) Mise en Œuvre du Protocole de Nagoya au Niveau National et Local dans les Pays de l'Espace COMIFAC: Etat des Lieux de la Mise en Œuvre du Protocole de Nagoya a la CDB, Contraintes et Besoins d'Appui dans les Pays de l'Espace COMIFAC, IUCN, Gland, Suisse

Lassen B (2016) The two worlds of Nagoya-ABS legislation in the EU and provider countries: discrepancies and how to deal with them. Public Eye & Natural Justice, Zurich/Cape Town

Oldham P, Barnes C, Hall S (2013) Biodiversity in the patent system: Cameroon. A country study of genetic resources and traditional knowledge in the patent system of relevance to Cameroon. Deutsche Gesellschaft für Internationale Zusammenarbeit (GIZ), UNEP and others. https://doi.org/10.13140/RG.2.1.3124.7767., http://tinyurl.com/zqloo7ts (accessed 15 January 2020)

Tchetcha D (2012) Literature review on *Echinops giganteus,* Environment and Rural Development Foundation (ERuDeF). Buea, Cameroon

Legal or Draft Legal Texts

Old Forestry, Environmental, Research and Benefit-Sharing Framework

Cameroon forest Law: La loi No 94-01 du 20 janvier 1994 portant Régime des Forêts, de la Faune et la Pêche., http://www.fao.org/faolex/results/details/en/c/LEX-FAOC004845 (accessed 30 August 2020)

Cameroon environmental Framework Law: La loi No 96/12 du 5 août 1996 portant loi-cadre relative à la gestion de l'Environnement., http://www.fao.org/faolex/results/details/en/c/LEX-FAOC019578 (accessed 30 August 2020)

1994 Forest Law implementing decree: décret No 95-531 du 23 août 1995 fixant les modalités d'application du régime des forêts., http://extwprlegs1.fao.org/docs/pdf/cmr4471.pdf (accessed 30 August 2020)

Research permit regulation: L'arrêté No 00002/MINRESI/B00/C00 du 18 mai 2006 fixant les conditions d'octroi d'une Autorisation de Recherche

Phytosanitary Law: loi No 2003/003 du 21 avril 2003 portant protection Phytosanitaire fixant les principes et règles qui régissent la protection phytosanitaire au Cameroun

Benefit sharing regime on forest/timber and NTFPs exploitation: The joint ministerial order 'arrêté conjoint N° 0000076 MINATD/MINFI/MINFOF du 26 Juillet 2012 fixant les modalités de planification, d'emploi et de suivi de la gestion des revenus provenant de l'exploitation des ressources forestières et fauniques destinés aux communes et aux communautés villageoises riveraines

2018 Drafts/Emerging ABS Regulatory Framework (Not Available Online)

ABS Law: Loi No 2021/014 du 09 Juillet 2021 Regissant l'Accès aux Ressources Génétiques, a Leurs dérivés, aux Connaissances Traditionnelles associées et au Partage Juste et Equitable des Avantages Issus de leur Utilisation

Draft implementing decree setting the terms of access to genetic resources and associated traditional knowledge and the fair and equitable sharing of the benefice arising from their utilization

Draft model ABS permit

Draft model PIC and

Draft model MAT

Algeria 2014 law on biological resources: Loi n 14-07 du 13 Chaoual 1435 correspondant au 9 aout2014 relative aux ressources biologiques

Liberia draft ABS law: The 2011 adopted version of the draft Access and Benefit Sharing law of Liberia

Rwanda draft ABS decree: the 2016 draft Ministerial Order of Rwanda Governing the Access to Genetic Resources and the Fair and Equitable Sharing of Benefits Arising from their Utilization in Rwanda

Marcelin Tonye Mahop PhD, held academic research positions in the UK and the Netherlands, senior programme management position with ECCAS in Central Africa and he is currently a Research Fellow at the School of Earth and Environment (SEE), the University of Leeds. Marcelin was a freelance consultant working on various topics including Access and Benefit Sharing/Genetic resources policies broadly and BioTrade, IPRs, agricultural biotechnology, seed regulations and seed systems, forest law enforcement and trade, projects and programmes evaluations, project formulation. tonye2169@aol.com, or tonye2169@gmail.com.

The Ethiopian Access and Benefit-Sharing Regime: Stringent with a Purpose

Ashenafi Ayenew Hailu and Evanson Chege Kamau

Abstract It is said that the Ethiopian regime regulating access to genetic resources, aTK and benefit-sharing is too stringent and seems to be overly concerned about the loss of GR. Its development was highly influenced by lessons learned from the agreement on access to and use of Teff and the related case of bio-piracy and, therefore, it is not a hidden fact that its intention is to establish ownership and maintain sovereign control over its genetic resources and aTK aimed at tackling misappropriation, act of bio-piracy and unfair exploitation. Thus, it emphasizes tight control on every act on its genetic resources and aTK, and on compliance of users. This can be observed e.g. in its definition of the term access, the requirement of an assurance letter from the competent authority of the user's country guaranteeing that it shall uphold and enforce the access obligations and also the restriction made on the export of genetic resource out of the country. The instruments regulating ABS, the Proclamation No. 482/2006 and Regulation No. 169/2009, hence pursue the third objective of the Convention on Biological Diversity with this mindset. With the understanding that Ethiopia, being a party to the Nagoya Protocol, is obliged to comply with the new obligations, this chapter looks at how the law and practice are coping. However, it is to be appreciated that the regime has been undergoing revision and a draft law has been released. This chapter therefore, in addition, examines the proposed changes and tries to see how conformity with the Nagoya Protocol is balanced against the control approach. Besides, the draft deals with some of the very topical issues, e.g. digital sequence information. Concerning this issue the draft has included it within the scope of regulation and hence, the chapter also takes interest on how such issues are addressed. It is notable that in spite of its severity, this regime can nonetheless offer a number of lessons. For instance, the

A. A. Hailu (✉)
Genetic Resources Access and Benefit Sharing Directorate of the Ethiopian Biodiversity Institute, Addis Ababa, Ethiopia
e-mail: ashenafi.ayenew@ebi.gov.et

E. C. Kamau
Faculty of Law, University of Bremen, Bremen, Germany
e-mail: echege@uni-bremen.de

E. Chege Kamau (ed.), *Global Transformations in the Use of Biodiversity for Research and Development*, Ius Gentium: Comparative Perspectives on Law and Justice 95, https://doi.org/10.1007/978-3-030-88711-7_8

benefit distribution approach and the differentiation of access procedures for access with commercial and non-commercial purposes, which are exemplary for the implementation of articles 9 and 8a of the Nagoya Protocol respectively, are useful to legal transplants. Hence, while exposing the regime's weaknesses as well as proposing changes, this chapter likewise tries to identify its strengths that can be used to fertilize other regimes.

Keywords Proclamation No. 482/2006 · Ethiopian ABS regime · Access and benefit-sharing · Genetic resource · Community knowledge · Bio-piracy · Nagoya protocol

1 Development of Legislation

1.1 Status of Biodiversity

Ethiopia is one of the top 25 biodiversity-rich countries in the world,[1] and hosts two of the world's 34 biodiversity hotspots, namely, the Eastern Afromontane and the Horn of Africa hotspots. It is recognized as one of the world's eight major Vavilovian centers of origin and diversity of crop plants.[2] Many crop plants have their primary center of origin in Ethiopia.[3] Some of these include Coffee (*Coffea arabica L.*), Teff (*Eragrostis tef (Zucc.)Trotte*), Enset (*Ensete ventricosum (Welw.) Cheesman*), Noug (*Guizotia abyssinica (Lf.) Cass.*), Anchote (*Coccinia abyssinica (Lam.) Cogn.*), Ethiopian potato (*Plectranthus edulis* (Vatke) Agnew), Chat (*Catha edulis Forsk.*), Ethiopian mustard (*Brassica carinata A. Braun*), and Taro (*Colocasia esculenta (L.)Schott.*).[4]

The so far known higher plants in Ethiopia are 6000 species of which 10% are endemic.[5] However, since the vegetation description is not yet exhaustive in the country, new native and endemic plants are described every time. Among these higher plant species, woody plants constitute about 1000 species.[6]

Ethiopia is endowed with a large domestic animal population with a huge genetic diversity within species and breeds that is relevant even at a global scale.[7] These are comprised of 28 cattle, 9 sheep, 8 goat, 7 camel, 6 donkey, 8 horse, 2 mule and 7 chicken breeds.[8] Ethiopia is also rich in wild fauna diversity and this includes 284 mammal, 861 bird, 201 reptile, 200 fish, 63 amphibian and 1225 arthropod

[1] WCMC (1994).

[2] Vavilov (1951).

[3] Dalle (2015), p. 125.

[4] IBC (2012b).

[5] Hedberg et al. (2009).

[6] IBC (2012a).

[7] Yimegnuhal and Degefa (2004), pp. xi, 23.

[8] EBI (2016).

(324 butterflies) species. Of these faunal resources, 29 mammal species, 18 bird, 10 reptile, 40 fish, 25 amphibians and seven arthropods (most of them butterfly) species are endemic to the country.[9]

1.2 *Historical Context, Influence and Aims: Stories Behind the Development of the Legislation*

Before the adoption of the Convention on Biological Diversity, genetic resources were considered as the common heritage of mankind and deemed to belong to everyone and as a result they were accessed freely without any regulation. Developed countries with their advanced biotechnology developed products of a commercial value and as a result generated income and accumulated wealth without sharing any benefits with the countries from which the resources were taken. This led to unfair and unequitable distribution of wealth arising from the utilization of genetic resources (GR) and associated traditional knowledge (aTK). Besides, the unregulated access led to the unsustainable utilization of resources. As a result, developing countries, which are rich in biodiversity, raised concerns about this trend and campaigned for the establishment of ownership and sovereign control over their GR and aTK in order to protect them from misappropriation and unfair exploitation.

The ownership and control of GR became an intensely debated issue in the negotiations and the quest to solve the problem of unfairness and unsustainable utilization of resources, which resulted in the adoption of the Convention on Biological Diversity (CBD) in 1992 and its entry into force in 1993. Consequently this resulted in the demise of the notion of the common heritage of mankind and the recognition of national sovereignty over own GR as well as the authority of States to determine access to their GR based on national legislation, which is affirmed in the CBD. The CBD, therefore, requires all its parties to take legislative, administrative or policy measures, as appropriate, in order to regulate access to their GR and ensure fair and equitable benefit-sharing from their utilization.

The issues concerning access and benefit-sharing came into the attention of scholars in Ethiopia since the adoption of the CBD and have been widely discussed since the country ratified the CBD in 1994. These discussions culminated in the establishment of a national committee under the auspices of the Institute of Biodiversity Conservation and Research (now called Ethiopian Biodiversity Institute, EBI) in 1998.[10] The institute was mandated to prepare and submit to the government a draft proclamation on access to GR.[11] The drafting process went through a number of stages, including the convening of a couple of stakeholder workshops to discuss

[9]EBI (2015).

[10]Yifru (2003), p. 107 ff.

[11]Ibid.

the elements and contents of the draft legislation as proposed by the national committee.[12]

In Ethiopia relevant governmental institutions take the lead in the need to regulate access to GR and aTK and to develop access and benefit-sharing (ABS) legislation. Some Non-governmental Organizations (NGOs) were involved in the policy debate on ABS.[13] The Gaia Foundation of the United Kingdom, Third World Network (TWN) of Malaysia, and Genetic Resources Action International (GRAIN) of Spain, have been working closely with the Environmental Protection Authority (EPA) (now called Environment, Forest and Climate Change Commission, EFCCC) of Ethiopia and the Institute of Biodiversity Conservation and Research (now EBI) were involved in the relevant discussions.[14] In general, these organizations advocate for what the CBD provides in relation to access to genetic resources and benefit-sharing and oppose patenting of any life forms.

There has been limited involvement of local communities in the policy debate on ABS issues. Communities did not really appreciate the need for regulation of ABS as the system they were historically used to was based on free exchange and open access to GR.

There were attempts to consult with farmers in areas covered by the Global Environmental Facility (GEF) project, and with communities inhabiting the vicinity of some protected areas.[15] Farmers and local communities were amused to learn about cases where collectors of their local varieties were claiming exclusive rights over these varieties claiming that they have improved them to some extent.[16] In order to increase understanding of the issue by communities, some practical cases have been used. For example, in one case a plant breeder right was obtained over one of the teff varieties, an important and endemic cereal crop of Ethiopia and the major staple food for Ethiopians. A certain American collected and exported a teff variety from Wollo, an administrative region in Northeast Ethiopia, and claiming having made improvements on it, submitted an application for plant variety protection to the US Department of Agriculture (USDA) in November 1988 in the name of "Teff Company".[17] The USDA granted a certificate in 1996 which entitles the company to exclude others, presumably including farmers from Wollo area, from "selling or producing it or exploiting it for propagation or making it for any of the above purposes or using it in producing a hybrid or different variety ... to the extent provided by the plant variety protection act" (extracted from the certificate). This and similar stories have served as illustrations which get the ordinary person involved in the policy debate in a more meaningful manner.

---

[12] Ibid.

[13] Ibid.

[14] Ibid.

[15] Ibid.

[16] Ibid.

[17] Ibid.

After a series of debates, discussions and consultations, Ethiopia issued Access to Genetic Resources and Community Knowledge, and Community Rights Proclamation (No. 482/2006) and Regulation (No.169/2009), which is the main legislation regulating access to genetic resources and associated traditional knowledge in Ethiopia. The major driving force for the country to enact the Proclamation was the urgent need to conserve and sustainably utilize the country's rich biodiversity; recognize the historical contribution the Ethiopian communities have made to conserve, develop and sustainably utilize biological resources; recognize and protect the knowledge of the Ethiopian communities generated and accumulated with respect to the conservation and utilization of genetic resources; and promote the wider application of such knowledge with the approval and sharing of benefits with such communities as well as involve them in the decision-making.[18]

Since Ethiopia is party to both the CBD and the International Treaty on Plant Genetic Resources for Food and Agriculture (ITPGRFA), its ABS system is influenced by these two international ABS instruments and it takes into consideration both their provisions. The Ethiopian ABS legislations focus on prior informed consent (PIC), mutually agreed terms (MAT), the Multilateral System (MLS) of access of the ITPGRFA and how to implement relevant provisions. Furthermore, Ethiopia's ABS law development is inspired by the African Model Law on Community, Farmers' and Plant Breeders' Right and Access to Biological Resources, as well as by the Bonn Guidelines.

The development of Ethiopian ABS laws is highly influenced by lessons learned from Teff ABS agreement and an act of bio-piracy. It is worth mentioning the patent application filed on the processing of teff flour and related products in the Netherlands in July 2003 and which was later filed under the European Patent Office (EPO) in July 2004, before the conclusion of the agreement and without the consent of Ethiopia, was considered as an act of bio-piracy and misappropriation of teff genetic resource and associated traditional knowledge (aTK). It was also noted that the Company came to the negotiation of ABS agreement on teff not in good faith. The lack of good faith during the negotiation of the teff agreement and the act of misappropriation and bio-piracy on teff genetic resource and aTK led to the development of very stringent measures to regulate access and a more prudently regulated legal regime. Therefore it was intentional to establish ownership and maintain sovereign control over GR and community knowledge in order to tackle the above-mentioned vices and unfair exploitation. The law thus emphasizes on tight control on every act on its GR and associated community knowledge and compliance on users. This can be observed e.g. (1) in the definition of the term access,[19] (2) in the requirement that the competent authority of the user's national State or that of his domicile issue an assurance letter to the effect that it shall uphold and enforce

[18] Dalle (2015), p. 126.

[19] "'Access' means the collection, acquisition, transfer or use of genetic resources or community knowledge".

the access obligations,[20] and (3) in the restrictions made on the export of genetic resource out of the country.[21] It aims to ensure that the country and its communities obtain a fair and equitable share of benefits arising from the utilization of GR and aTK (referred in the legislation as and hereinafter as (associated) community knowledge) so as to promote the conservation and sustainable utilization of the country's biodiversity resources.

1.3 Legal Background

Ownership of genetic resource and the right to regulate access in the Ethiopian ABS law is guided by constitutional provisions. The Constitution stipulates the exclusive ownership of land and all natural resources by the State and the peoples of Ethiopia.[22] This is reiterated by the ABS law, i.e. the Proclamation No. 482 of 2006, which, in addition, states that the ownership of community knowledge unlike GR belongs to the concerned local community (Art. 5 Proclamation[23]). This right is therefore well established and the State is committed to protect and enhance it. Article 10 (1) of the Proclamation[24] states that "The rights of local communities over their genetic resources and community knowledge shall be protected as they are enshrined in the customary practices and norms of the concerned communities".

2 The ABS Framework

2.1 Legislation

Access and Benefit Sharing in Ethiopia is regulated by the Access to Genetic Resources and Community Knowledge, and Community Rights Proclamation (No. 482/2006) and Regulation (No.169/2009).[25] The legislation is a stand-alone (self-standing) and pre Nagoya Protocol law, which establishes a bilateral ABS system but provides legal space for multilateralism. The legislation is under review to harmonize it with the Nagoya Protocol.

[20] ABS law (2006), Art.12 (4).

[21] ABS law (2006), Art.12 (6) and Art.14(4).

[22] Constitution of Ethiopia, Art. 40.

[23] Revised draft law, Art. 5.

[24] This is found in article 10 (2) of the revised draft law.

[25] See texts at http://www.ebi.gov.et/regulation/ (accessed 11 September 2020).

2.2 *Scope*

The Geographical scope of the regime applies to all kinds of access to genetic resources and community knowledge found in both *in-situ* and *ex-situ* conditions. In regard to material scope it extends to any genetic material of biological resource containing genetic information and derivatives which are biochemical compounds extracted or developed from biological resource. The draft revised Ethiopian ABS Proclamation extends its scope to digital sequence information (DSI). The draft defines 'genetic resource' as "any genetic material of biological resource containing genetic information having actual or potential value for humanity and including its derivatives and digital sequence information" (Art. 2 (6)). The ABS law also applies to bulk commodities used in research and development as understood from a reading of article 4 (2) (b) of the draft law. Accordingly, the law shall not apply to "the sale of produce of biological resources . . . for direct consumption . . . that do not involve the use of genetic resources . . .". That means if the use (of these biological resources later) involve GR the law will apply.[26] The personal scope extends to all users. Accordingly, "no person shall access genetic resources or community knowledge unless in possession of a written access permit granted by the Ethiopian Biodiversity Institute based on prior informed consent" (Art. 11 (1) Proclamation) ". . .and mutually agreed terms" (Art. 12 (1) draft). There is a requirement to involve Ethiopian nationals designated by the EBI in the research-based access of GR of Ethiopia. This restriction aims to promote technology transfer. The regime applies since February 2006 which is the date establishing its temporal scope. From its scope the customary use and exchange of genetic resources and community knowledge by and among Ethiopian local communities are exempted. Besides, also not included in the scope is the sale of products of biological resources for direct consumption which do not involve the use of the genetic resource thereof.

3 Institutional Framework and Mandates

Ethiopian Biodiversity Institute (EBI) is the Competent National Authority designated according to article 27 of the Proclamation (Art. 37, draft) to issue access permits, negotiate and enter into access agreements in accordance with the access law, and follow up their implementation. EBI is responsible to negotiate and conclude genetic resources and community knowledge access agreements. EBI grants written access permits for GR and community knowledge as per the Ethiopian ABS national law. It is also authorized to grant export permits for GR that involve utilization of genetic material and an exploration permit to the applicant. EBI is also responsible for follow-up and to ensure that access is carried out in accordance with

[26]EBI official, *pers comms*, 30 September 2020.

the Proclamation as well as regulations and directives issued. EBI also acts as a checkpoint.

Apart from the EBI the law gives some administrative responsibilities to local communities and regional bodies concerning regulation of access (Art. 28 and 29 respectively). The former is now under article 39 of the draft law and the latter under article 38 and are renamed as regional administration bodies. According to the draft law their responsibilities shall not touch on the customary use and exchange of GR and CK by and among Ethiopian local communities (Art. 4 (2) (a)). The kebele administration[27] and regional bodies responsible for the protection and conservation of GR at all levels shall ensure that no person accesses GR from their respective jurisdiction without a permit. If they find any person accessing without a permit they shall seize the GR and present such a person to the concerned law enforcement organ and notify EBI together with the detailed particulars of the GR and the person found in possession of the GR. Local communities on the other hand shall require any person collecting or taking GR from their localities, to show his access permit, and if he is without permit immediately notify or present him to the nearest kebele or woreda administration.[28]

4 Application Procedures and Nature of Regulatory Tools for Access to *In-Situ* and *Ex-Situ* GR and aTK

4.1 What Is 'Access'?

The term 'access' is defined broadly in the Ethiopian ABS legislation and it means the collection, acquisition, transfer or use of genetic resources and/or community knowledge. The term access applies to all genetic resources that are found in *in-situ* or *ex-situ* conditions and community knowledge.

4.2 Pre-Conditions for Access: Application, PIC, MAT and Permit

Any person wishing to access genetic resources or community knowledge must be in possession of an access permit granted based on PIC and MAT (Art. 11 (1) Proclamation).[29] To obtain an access permit a written application must be submitted to the

[27]Kebele administration is the smallest administrative unit of Ethiopia.

[28]Woreda administration is the third-level administrative division of Ethiopia which is further subdivided into a number of kebeles.

[29]Revised draft law, Art. 11 (1) and 12 (1).

EBI (Art. 14 (1) Proclamation).[30] An application for access for non-commercial purposes must be presented in accordance with the form specified in Annex 2 of the ABS Regulation and must fulfill the foreseen requirements.

On the other hand, an access application for commercial purposes must be presented in accordance with the form specified in Annex 1 of the ABS Regulation.

The EBI grants PIC for GR and the local community PIC for their CK (Art. 12 Proclamation).[31] EBI helps to procure the PIC for the local community by submitting the access application along with the proposal for the decision and asking for the consent of the concerned local community. MAT are established as well as access agreement concluded with the EBI for both GR and CK based on the PIC (Art. 14 (2), (3) Proclamation).[32] The local communities have the right to demand restriction or withdrawal of the PIC given by EBI for access to their GR if they find out that it is likely to be detrimental to their socio-economic life or their natural or cultural heritages (Art. 7 (1) (b)-(c) Proclamation).[33] There is no difference on type of genetic resource to obtain PIC and it applies to all genetic resources that involve utilization of genetic resources for research and development.

Upon the fulfillment of the requirements of Annex 2 and signing of an access agreement, an access permit will be issued in writing for non-commercial research. An access permit for non-commercial purposes gives permission only to conduct research with no intention of commercialization. If any intention to change the purpose from non-commercial to commercial arises, the permit holder must first seek an access permit for commercialization. An access permit for commercial purposes allows the permit holder to develop a product and commercialize it.

For commercial research, after examining the access application and if there is sufficient ground to deny the intended access the EBI shall communicate the decision to the applicant in writing with reasons for denial (Art. 5 (2) (a) Regulation).[34] If there are no sufficient ground to deny, the access application will be published for public notice in a national newspaper with a wide distribution and a local newspaper of the locality where access shall take place (Art. 5 (2) (b) and Art. 6 (1) Regulation)). After 30 days of its publication and if there is no public objection from any person or any relevant institution (if its opinion is sought) the EBI will determine to grant the access and call upon the access applicant to negotiate an access and benefit-sharing agreement (Art. 6 (3), (4) and Art. 10 (1) (b) Regulation). The permit is only issued after MAT have been negotiated and concluded.

[30]The revised draft law does not say this directly but implies it in article 2 (11) on definition of PIC (PIC of EBI and local community must be based on an access application) and 41 (b) on offences and penalties (provision of false information in the access application is an offence punishable by law).

[31]Revised draft law, Art. 11 (1), (2).

[32]Ibid.

[33]Ibid, Art. 7 (2).

[34]Grounds for denial of an access permit are listed under article 13 of the revised draft law.

Besides these requirements, a foreign access applicant (for either commercial or non-commercial permit) shall present a letter from the competent authority of his/her national state or that of his/her domicile assuring that it shall uphold and enforce the access obligations of the applicant (Art. 12 (4) Proclamation). This requirement proves difficult to enforce from a legal perspective and it has been proposed to leave it out in the revised draft Proclamation. Further, the collection of GR and community knowledge (CK) by a foreigner shall be accompanied by the personnel of EBI or the personnel of the relevant institution designated by EBI (Art. 12 (5) Proclamation).[35] In addition, research-based access to GR shall be carried out in Ethiopia and with the participation of Ethiopian nationals designated by the Institute, unless where it is impossible (Art. 12 (6) Proclamation).[36] When research of GR or CK shall take place abroad, the institution sponsoring and/or hosting the research should give a letter of assurance that it shall observe the access obligations attached thereto (Art. 12 (7) Proclamation).[37]

Export permits are granted by the EBI (Art. 11 (3) Proclamation).[38] In addition, it is responsible for issuing exploration permits without which no person can conduct exploration of GR in Ethiopia (Art. 22 (1) Proclamation).[39] EBI is also empowered to issue import permits for the introduction of biodiversity specimens into Ethiopia.

4.3 *Contents of an Access Permit*

The contents of an access permit and agreement includes the type and quantitative description of the genetic resource permitted to be accessed, the description of the community knowledge permitted to be accessed or associated with the genetic resource to be accessed, the intended use the genetic resource or the community knowledge, type of benefit sharing, the duration of access, dispute settlement mechanisms and the obligations of the access permit holder.

4.4 *Conditions of an Access Permit*

Permit conditions (or obligations of the permit) are included under article 18 of the revised draft law. Accordingly, an access permit will be issued under the conditions that population of farmers' planting stock or wild species are not depleted or significant genetic variation removed from local gene pool during collection.

[35] Revised draft law, Art. 11 (3).

[36] Revised draft law, Art. 12 (3 and 6).

[37] Ibid, Art. 11 (6).

[38] Ibid, Art. 12 (3).

[39] Ibid, Art. 33 (1).

Where GR is to be collected repeatedly, the environmental and socio-economic impact of the access must be followed up and a report thereon submitted. If the GR is to be collected from protected areas, the rules and regulations of the administration of the protected area must be observed. Besides, the type and quantitative limits of the GR permitted to be accessed must be observed.

A permit holder/user while undertaking collection and exploration of Ethiopian GR or community knowledge must deposit a copy of the access permit granted to him with the relevant institutions in the collection area and show the access permit on request. Besides, when seeking to acquire IPRs over the GR accessed or parts thereof, the user shall apply to the Intellectual Property Office based on the relevant laws of Ethiopia and also notify this to the Institute. Further, in the application for IPRs, users must disclose and recognize the origin or source of the GR or community knowledge as well as give information in writing concerning all the findings of the research and development based on Ethiopian GR or community knowledge. The law also establishes mandatory compliance measures to submit to the EBI regular status reports.

A permit holder is forbidden from transferring to third parties the GR and the CK accessed, or the access permit, or the rights and obligations there under, or using the GR and the CK for any purpose other than that originally intended, without first notifying and obtaining written authorization from the EBI. Any unused genetic material must be returned at the end of the planned research or upon termination of the MAT. In the application for commercial property protection of the product developed there from the locality where the GR or CK was accessed from must be recognized as the origin. Any person who obtained an access permit must observe the terms and conditions of the MAT established.

5 ABS Agreements

Ethiopia has model agreements but also several executed agreements on access to genetic/biological resources and benefit-sharing. We present one of the agreements executed after the adoption of the Nagoya Protocol for commercial use and a model material transfer agreement for non-commercial purposes.

5.1 ABS Agreement for Commercial Purposes

The agreement between the EIB (provider) and Docomo Oils Plc (user), whose parent company is American was entered into in 2012 (and added onto in 2013) with a validity of 10 years for the commercial use in Ethiopia of a plant, *Osyris* species, for the development of essential oil without the community knowledge, except prior to the obtainment of the community's consent and establishment of MAT. A number of ABS issues are dealt with in the agreement.

5.1.1 Access

The agreement restricts the regions/territories from where the company is allowed to harvest, collect and use the plant. These are listed under article 3.1.

5.1.2 Use and Change of Intent

The resource is to be accessed only for the purpose of developing essential oil products (Art. 3.2) and must be used in a sustainable manner in order to hinder overexploitation (Art. 3.5). The company cannot use the resource for any other purpose whatsoever unless the EBI gives an explicit written permit (Art. 3.3). In other words the company must first seek the PIC of the provider.

5.1.3 Third Party Transfer

The genetic/biological resource and any component thereof may not be transferred to third parties except with prior and explicit written consent of the provider (Art. 5).

5.1.4 Intellectual Property Rights

The company cannot claim IPRs over the genetic/biological resource or any part thereof (Art. 4.1) and any inventions based thereon shall be jointly owned with the provider (Art. 4.2). Besides, the company shall acknowledge Ethiopia as the geographic origin of the genetic/biological resource in the manufacture of the final product (Art. 4.3).

5.1.5 Benefit-Sharing

The agreement foresees both monetary and non-monetary benefits (Art. 7). The former include an upfront payment of USD 40,000 upon signing of the agreement, a lump sum equal to the amount of 3.5% of net profit after taxes to be made immediately after the publication of the annual account of the company, an annual royalty of 3.5% of the net profit of the company, and an annual license fee of 2000 USD. The latter include the involvement of local scientists in the research undertaking as mutually agreed by parties in the research plan of the company, the sharing of results of research undertaken on the genetic/biological resource including the knowledge or technologies the company might generate using the genetic/biological resource except where these constitute undisclosed information, and training and enhancing skills of experts from relevant institutions and local communities once in a year during the life of the agreement on conservation of genetic/biological resources, evaluation, development, propagation and use to ensure their sustainable utilization.

5.1.6 Sanctions

The agreement foresees a penalty of 50,000 USD for breach by a party of its terms (Art. 11). It is important to mention that this penalty applies likewise to the provider. Although only clauses that are used to enforce compliance with provider rules have been examined, the agreement has clauses which the provider too must observe. These include no grant to other parties of access to the genetic/biological resource in the area where PIC was obtained for the same product with the company unless the company grants consent (Art. 3.4), release of results of joint research without consent of, or disclosure of information identified as confidential by the other party (Art. 8.1), or disclosed to it (Art. 8.3). Besides, in the same way as with the company, the provider may not assign his rights, benefits or obligations under the agreement to any third party without the written consent of the company (Art. 8.2).

Apart from the above clauses the Ethiopian MTA has more model contractual clauses that may be included in an agreement for commercial research between parties to ensure compliance. These include provisions that prevent the user not to collect and use the GR to an amount greater than the sustainability threshold;[40] oblige the user to recognize and declare Ethiopia as a geographic origin[41] of the GR in the manufacture of the final product[42] and to retain the material for the period of the research;[43] and to submit to the provider reports according to determined intervals.[44] Furthermore, it includes a provision that obliges the user to comply with obligations stated under the Ethiopian ABS legislation[45] and also on dispute settlement and applicable laws.[46]

[40]Ethiopian model contractual clauses in MAT, Art. 3(1–3).

[41]This provision does not consider situations when Ethiopia might be only a providing country and not the country of origin.

[42]Ethiopian model contractual clauses in MAT, Art. 4 (1 and 3).

[43]Ibid, Art. 4 (1).

[44]Ibid, Art. 17 (2).

[45]Ibid, Art. 3 (10).

[46]Ibid, Art. 13 and 15.

5.2 *MTA for Non-commercial Purposes*

The model material transfer agreement[47] presented deals with various issues, including use, change of intent, third party transfer, storage, fate of material after research, reporting etc.

5.2.1 Use and Change of Intent

The researcher is allowed to utilize the material only for the non-commercial research purpose and cannot use it for commercial purposes, or obtain IPRs on it. Any change of intent must occur only with prior and explicit consent of the provider.

5.2.2 Third Party Transfer

The material cannot be transferred to any third party except with an explicit written consent of the provider, and if any third party obtains the material without such consent, such third party shall not have any right whatsoever over it and its components.

5.2.3 Storage and Fate of Material After Research

The model agreement does not state whether the material must be stored in the specific institution where research is undertaken but states that the material shall be retained in the country of destination. The question arises as to whether the researcher would hence be permitted/entitled to move the material from one institution to the other for his research as long as the institutions are within the same country. After the research the researcher is obliged to return any unused material left to the provider.

5.2.4 Reporting and Results

The researcher is obliged to make periodic reports of the research progress. The model MTA, however, does not state how often the reports should be. At the end of the research the researcher shall provide the provider with a hard and electronic copy of the research results.

[47] The agreement is accessible at www.ebi.gov.et.

5.2.5 Benefit-Sharing

No benefits are defined in the model agreement except a statement that any benefit that accrues for the use of the material shall be subject to the relevant existing and future national and international laws. Therefore, this must be defined according to Proclamation No. 482/2006 and regulations thereto (Regulation No. 169/2009) as well as the CBD and its Nagoya Protocol.

Other issues regulated by the model MTA for non-commercial research include the purpose of research, the amount of samples to be taken, the destination and justification why the research cannot be undertaken in Ethiopia.

6 Sanctions

The draft law provides administrative fines and criminal sanctions in case of alleged violation of the legislation and the MAT. Such sanctions include: limitation of the amount of GR to be accessed and other restrictions in particular if access to GR is found to pose threat of genetic erosion, degradation of the environment or threat to the cultural values of communities, which cannot be easily averted; suspending or terminating MAT and prohibiting the access to GR or community knowledge, if it is recognized that the access will cause risk of damage to GR or the environment or affects overriding public interest;[48] confiscation of the GR accessed, cancellation of the access permit granted[49] and punishing the civil liability arising thereof, depending on the gravity of the circumstance, with rigorous imprisonment of not less than 3 years and not more than 5 years and a fine of not less than fifty-thousand and not exceeding three hundred-thousand Ethiopian Birr.[50] If the offences committed are in relation to endemic GR the punishment will be, depending on the circumstance, rigorous imprisonment of not less than 5 years and not exceeding 12-years and a fine ranging from two hundred thousand to five hundred thousand Birr.[51] In the event of a conviction for any of the offences under article 41 the court may order the convicted person to repair the damages caused on the environment and/or, depending on the gravity of the offence, impose a heavier fine than indicated thereunder.[52]

---

[48] Revised draft law, Art. 17 (1, 2).

[49] According to article 41 (1a-e) the acts that will attract these penalties are accessing Ethiopian GR or community knowledge without an access permit; providing false information in the access application therefor or in the course of subsequent monitoring of MAT, changing the purpose of access specified in the MAT without obtaining permit; exploring GR without an exploration permit or providing false information in the application therefor; and possessing or using GR or community knowledge obtained from another country in contravention of the ABS laws of such country.

[50] Revised Ethiopian ABS Proclamation (Draft), Art. 41.

[51] Ibid, Art. 41 (2).

[52] Ibid, Art. 41 (3, 4).

7 User Compliance and Monitoring Measures and Sanctions for Non-compliance

The existing Ethiopian ABS law completely lacks user compliance measures as required by articles 15 and 16 of the NP. According to information from the national report notified to the ABS CH,[53] there are a few measures to monitor or support compliance which satisfy article 17.1(b) and (c) NP. They encourage users and providers to include provisions in MAT to share information on the implementation of such terms. There are rules and procedures for sharing information in MAT. According to article 17 (7) and (8) of Proclamation No. 482/2006 users (access permit holder) of genetic resource and/or community knowledge are obliged to submit regular status report of the research and inform the EBI in writing of all the findings of the research and development based on the genetic resource and community knowledge accessed. Second, to encourage the use of cost-effective communication tools and systems a web-based ABS information-sharing system (http://www.ebi.gov.et/gm-access/) has been established. All required documents such as forms, guidelines, laws and other information are posted to this system and regularly updated for easy access for users of GR and community knowledge. Documents and information available online include the legislation (Proclamation No. 482/2006), model MTA, instructions for users on how to process applications for access, sample forms of assurance letter to be produced by the research hosting institution abroad, and list of potential GR publicized for bioprospecting and benefit-sharing purposes. These measures nevertheless do not produce fully article 17 NP compliant rules. The lack of user compliance measures as well as fully compliant measures for monitoring/supporting compliance is hence one of the major reasons for revising the existing law. These measures are included under articles 21–32 of the draft revised law. They aim to ensure that GR and aTK utilized within the jurisdiction of Ethiopia have been accessed in accordance with PIC and MAT as required by the regulatory requirements of other provider countries. The draft law also includes measures to also ensure that local or foreign users comply with national ABS laws when accessing Ethiopian GR and/or CK as well as carrying out utilization either locally or abroad. These measures aim to prevent access and utilization to Ethiopian GR and CK without PIC and MAT.

7.1 Compliance Measures

Compliance measures are included under articles 21 and 22 of the revised draft law. Accordingly, users must declare to the EIB or to the checkpoints designated by the revised draft law that the GR and/or aTK they use "… have been accessed in

[53] https://absch.cbd.int/countries/ET (accessed 19 February 2020).

accordance with the applicable access and benefit-sharing legislation or regulatory requirements of the provider country".[54] This must be done by showing the necessary documents proving that PIC was obtained and MAT established in accordance with the ABS rules of the provider country.[55] Such documents shall clearly indicate that the PIC has been obtained and MAT established in accordance with the access and benefit-sharing rules of the provider country.[56] If the user fails to satisfy these conditions, or if the information in their possession is insufficient, uncertain, or wrong the EBI "... shall issue a written warning to the user and order the user to fulfill those requirements ...and report the measures it took against the user to the concerned institution of the provider country" and the ABS CH.[57] Besides, until the user has fulfilled the said requirements the EBI "... shall hold as exhibit the genetic resources in the hands of the user or of third party and the equipment owned by the user for use of genetic resources, report the measures it took against the user to the concerned institution of the provider country and the Access and Benefit-Sharing Clearing House, and report the defaults of the user, as appropriate, to the media".[58] The user will also be banned from accessing GR or community knowledge in Ethiopia.[59] These measures are applied accumulatively. In addition, users and importers of GR are obliged to specifically indicate the country of origin of the GR in case the country which provided it is not the country of origin.[60]

7.2 Measures to Monitor/Support Compliance

Measures to monitor/support compliance are included in articles 23–32 of the revised draft law. Article 23 deals with measures to be accomplished by the EBI and articles 24–32 by designated checkpoints.

7.2.1 Measures by EBI

Article 23 defines mechanisms that the EIB will use to monitor the execution of MAT of the provider country. These include designation of checkpoints; inspection; periodic progress and status reports by access permit holders; a report by any person (i.e. the public); information from ABS CH and internationally recognized certificates of compliance; information from checkpoints designated to collect or receive

[54] Art. 21 (1) draft law.

[55] Ibid.

[56] Revised Ethiopian ABS Proclamation (draft), Art.21 (1).

[57] Art. 21 (2) draft law.

[58] Art. 21 (3) draft law.

[59] Ibid.

[60] Art. 21 (4) Draft law.

relevant information related to PIC, the establishment of MAT, the source of the GR or community knowledge, or the use of GR or community knowledge; and information obtained from any other mechanism deemed appropriate. Further, the permit holder and the local cooperating institution, which also has the mandate to monitor the implementation of MAT, shall give periodic reports to EBI on collection, progress of research and findings. Besides, the EBI shall inform the concerned local community about the progress of the research, the findings and the accrued benefits as well as notify the information received from the checkpoints and other sources on compliance to the ABS CH.

7.2.2 Measures by Checkpoints

Article 24 designates checkpoints with responsibility to monitor the use of GR in Ethiopia. The government may designate other checkpoints upon nomination by the EBI if necessary. Currently 8 checkpoints have been designated, i.e. the intellectual property right office; research funding or publishing institutions; research and educational institutions; quarantine control institutions; customs authorities; mail service institutions; and food, medicine and health care administration and control authority. Checkpoints have general responsibilities to collect or receive from users relevant information related to PIC, the sources of the GR or community knowledge, establishment of MAT, or the use of GR. Where users fail to provide this information, checkpoints shall seize all documents, GR and equipment and transfer them to the EBI. In addition, checkpoints shall report the details of their own performance to the EBI. Specific functions of each checkpoint are defined in articles 25–32 (Table 1[61]). Generally, they must report any relevant information related to PIC and MAT which they have collected or received from users and any measures they have taken against users who fail to provide such information. The draft law foresees that the EBI shall itself act as a checkpoint.[62]

7.3 *Compliance Checks: Who Is Checked, What Is Checked, at What Stage?*

Users of GR or community knowledge will be checked to ascertain whether they obtained PIC and established MAT or not at any stage of the value-chain.[63] Local permit holders will be checked whether they observe the terms and conditions of

[61] Only three of the authorities/institutions exist in the current law as administrative organs. Their roles have been changed to checkpoints in the draft law and the responsibilities of some of them are now more.

[62] Art. 37 (2) (m) draft law.

[63] Ibid, Art. 21(1) and Art. 24 (3a).

Table 1 Checkpoints and their functions

Checkpoint	Responsibilities	Reference article	
		Current law	Draft law
Intellectual property right office	– Ensure the user has produced the internationally recognized certificate of compliance before issuing IPR protection over a product developed from GR or community knowledge – If an IRCC is not available demand information and relevant documents on access permits, MAT and BS arrangements – Ensure the user has revealed the origin or source of the GR or CK, where utilization will be conducted and how – If public notice for IPR protection is posted on a public notice give details to EBI on the contents of such a notice and how long it remains posted, and – Facilitating conditions for the introduction of the special IPR system of *sui generis* to protect CK	None	Art. 25
Research funding or publishing institutions	– **Research funding institutions:** Ensure that the researcher has accessed the GR or CK according to the ABS laws of the providing country before granting fund or other supports for research works on GR or CK – **Research publishing institutions:** Ensure that the research has complied with ABS laws of the providing country before publishing research results on GR or CK	None	Art. 26
Research and educational institutions	– Ensure that the researcher has complied with ABS laws of the providing country when receiving research topics or proposals related with GR or CK	None	Art. 27
Quarantine control institutions	– Ensure that the quarantine certificate they issue for biological resource contain a statement indicating that the certificate does not constitute a permit to use GR contained in the biological resource and that doing so is prohibited and would constitute an offence	Art. 32	Art. 28
Customs authorities	– Ensure GR being exported or imported is accompanied by an export or import permit issued by EBI – Require any person transporting or possessing GR while leaving the country to produce a permit issued by EBI to that effect – Seize GR being transported out of the country without a permit from EBI and the person transporting it and immediately	Art. 30	Art. 29

(continued)

Table 1 (continued)

Checkpoint	Responsibilities	Reference article	
		Current law	Draft law
	report same to the nearby relevant body such as federal or state police, militia, inland security, anti-terrorism, or immigration forces and the institute – Ensure that a statement is written on the package of a biological resource product to be exported indicating that the use of the GR contained in the product is prohibited and doing so would constitute an offence – Ensure that GR imported into Ethiopia is being imported with permission of the providing country according to ABS laws of the relevant country and seize GR being imported into the country illegally and the person importing it and immediately report the matter to a nearby relevant body such as federal or state police, militia, inland security, anti-terrorism, customs or immigration forces and the institute		
Product or market approval institutions	– Ensure that the applicant has complied with the ABS requirements of the country providing GR or CK before approving production or marketing of the product based on GR or CK	None	Art. 30
Mail service institutions	– Require their clients to produce permit from EBI to export GR out of the country before receiving and transporting it as mail – While EBI is issuing export permit of GR via mail service institutions, the latter shall register the recipient's address with the view to prohibiting delivery to unauthorized third parties	Art. 31	Art. 31
Food, medicine and health care administration and control authority	– Ensure that the GR has been accessed according to the ABS laws of the providing country before granting on issuing permit for any production using GR	None	Art. 32

Source: *Own illustration*

MAT or not at any stage of the value-chain: collection research, development, innovation, pre-commercialization or commercialization.[64]

An applicant for intellectual property right protection of a product developed from GR or community knowledge accessed will be checked whether or not he/she has produced internationally recognized certificate of compliance or relevant

[64] Ibid, Art. 18 (7–8) and Art.18 (17).

documents on access permits and MAT, including benefit-sharing arrangements before intellectual property right protection is granted.[65] In addition, it will be checked whether he/she has disclosed the origin or source of the GR or community knowledge.[66]

An applicant for funds or other supports for research works on GR or community knowledge will be checked to ensure that he/she has accessed the genetic resource or community knowledge according to the ABS laws of the providing country before the funds or other supports are granted.[67] A researcher will be checked to ensure that the research is compliant with the ABS laws of the providing country before publication of research results and approval of research topics or proposals related to the GR or community knowledge.[68]

A user that exports or imports GR will be checked to ensure that the GR exported or imported is accompanied with an export or import permit and that GR imported into Ethiopia has got permission as per ABS laws of the country providing the GR.[69]

An applicant for product and/or market approval will be checked to ensure that the applicant has fulfilled the ABS requirements of the country providing GR or community knowledge before approving production or marketing of the product based on GR or community knowledge.[70]

8 Regime's Strengths, Weaknesses and Assessment

8.1 Concerning Compliance with the Nagoya Protocol

Ethiopia operates a pre Nagoya Protocol ABS law, but the effort to bring it into full conformity with the NP is ongoing. The draft Proclamation 2019 is ready and has been submitted to the Council of Ministers for consideration after which it will be submitted to the parliament for endorsement to become law. The current legislation designates a competent national authority, the EBI, and defines its functions in line with article 13 of the NP. In practice, though not detailed in the current law, EBI also carries out all other functions foreseen under article 13 of the NP and is responsible for all types of GR, for community knowledge and benefit-sharing. The draft law has added all the obligations of article 13 of the NP which were not in the current law to the functions of the EBI.[71] This gives a seemingly heavy burden to the competent national authority it being likewise the national focal point. The draft law, however,

[65] Ibid, Art. 25 (1a, b).

[66] Ibid, Art. 25 (2).

[67] Ibid, Art. 26 (1).

[68] Ibid, Art.26 (2) and Art. 27.

[69] Ibid, Art. 29 (1, 5).

[70] Ibid, Art. 30.

[71] See revised draft law, Art. 37.

also gives it power to delegate some of its duties to other legally established bodies where necessary and convenient to enable it carry its duties in a better way.[72] The contacts of the EBI have been notified to the ABS CH.

The legislation spells out the pre-requisites for access, i.e., access is subject to PIC (Art. 12 (1) Proclamation) and a written access permit from the EBI and if export shall take place, an export permit from the same institute (Art. 11 Proclamation). It makes it clear that a person who wants to obtain an access permit to access GR or community knowledge must submit a written application to the Institute (Art. 14 (1) Proclamation[73]). The procedure according to which the application must be presented and PIC given are also specified in the regulations (Art. 14 (1) Proclamation[74]). The grounds for denial of an access permit (Art. 13 Proclamation) or an export permit (Art. 14 (4); 12. 6/ Proclamation[75]) are clear. The Proclamation also specifies the obligations of the permit holder (Art. 17) and provides a non-exhaustive list of elements an access agreement may contain (Art. 16) as well as reasons why the institute may alter the agreement or reduce the size of GR to be accessed, or put any other limitation (Art. 21). Likewise, it states the offences which are punishable and the consequent sanctions (Art. 35). There seems to be no contradiction in the Proclamation and Regulations thereto. Besides, both are freely and easily available.[76] It can be said that a person interested in accessing GR/CK can foresee to a degree that is reasonable in the circumstances, the consequences of his/her actions.[77]

Like much other jurisprudence, it is difficult to ascertain the level of legal certainty of the ABS legislation and other relevant laws in Ethiopia as required by article 6.3 (a) NP. Following the legal doctrine of qualities of legal certainty, one would have to ask himself whether the ABS legislation and other relevant laws provide those subject to them with the ability to regulate their conduct with certainty as well as protect them from arbitrary use of State power.[78] The opinion regarding this can be very subjective. For example, consider Article 21 (1) of the Proclamation. It gives the EBI power to limit the rights of the permit holder where it is recognized that, *inter alia*, "violation of the cultural values of communities which cannot be easily averted" has occurred. Hence, legal certainty and transparency in regard to this aspect will depend on whether the permit holder has legal channels to defend himself/herself against any allegations of the relevant authority and avert arbitrary limitation of rights as well as try to save his/her undertaking from consequences thereof. Here the law does not give direction. The Civil Procedure Code of Ethiopia and the model contractual clauses on dispute settlement may, on the one hand, be

[72] Ibid, Art. 37 (2) (j).

[73] Art. 22.1 and Art. 23.1 in the revised draft law.

[74] Art. 12 (7), revised draft law.

[75] In the draft law this is regulated under articles 12 (3) and 12 (6).

[76] For critical aspects of legal certainty, clarity and transparency see Greiber et al. (2012), pp. 102–103.

[77] Ibid.

[78] Ibid.

considered as being such a channel. On the other hand, the Regulations are clear concerning confidential information. In article 9 (1) it states that "Confidential information may not be communicated to third parties except upon court order, expiry of the period of confidentiality, or mutual agreement of the parties". That means the aggrieved party can seek the intervention of the court if the relevant authority intends to disclose, or discloses its confidential information unlawfully. Further, it needs to be checked whether these laws change in an undesirably sudden manner and if decisions are made in a lawful way.[79] The Proclamation and its Regulation have not changed since they came into effect which proves a certain level of stability, consistency and predictability. They have also laid down regulations in different provisions based on which decisions are taken.

Further, the Regulation lays down procedures which every applicant wishing to undertake an access for commercial (Art. 3–10) or non-commercial (Art. 11–13) purposes, or to plant genetic resources for food and agriculture (PGRFA) under Annex I of the IT (Art. 14–20) must adhere to.[80] All information that must be included in an application is provided in the Regulations under Annex I for commercial purposes, Annex II for non-commercial purposes, and Annex III for the said PGRFA. Within the same provisions standard rules are also set for the EBI to follow while deciding/examining applications. One might argue concerning the fairness of some rules and their ability to meet the standards of article 6.3 (b) NP, e.g. the period for which confidential information shall be protected against disclosure by the EBI for access with commercial intent. For example, according to article 8 (1) of the Regulations the period for which such information shall be treated as confidential shall not be more than 10 years. Considering that some R&D e.g. in the pharmaceutical sector might take longer than 10 years, especially if it starts from scratch (for instance without prior research or CK), this rule could be seen as aggrieving to the user/applicant. Another example is the rule that where GR is exported for non-commercial use by an Ethiopian institution the foreign institution hosting the research shall be bound in the access agreement to follow up and monitor the observance of the agreement by the Ethiopian institution (Art.13 (3) Regulation). According to the provisions on the procedure for non-commercial research (Art. 11–13 Regulation) there is not the slightest indication that this involves access and export in a joint venture relationship between an Ethiopian and a foreign institution such as implied in article 12 (6)–(7) of the Proclamation.[81] Therefore, to put that degree of responsibility on the foreign institution is excessive. However, the foreign institution has the prerogative to reject the request to host the local institution. According to practice there are more than 800 researchers hosted by foreign

[79] Ibid.

[80] See http://www.ebi.gov.et/wp-content/uploads/2018/01/ABS-Regulation-Ethiopia.pdf (accessed 17 June 2019).

[81] See http://www.ebi.gov.et/wp-content/uploads/2018/01/ABS-Proclamation-Ethiopia.pdf (accessed 17 June 2019).

institutions to date and the EBI has not been faced with any challenges.[82] All foreign institutions have been willing to provide the requested assurance letter. The only challenge experienced so far is having the requested assurance letter provided by the foreign CNA.[83] Hence, in spite of this critique, the overriding consideration might be whether any participant in the ABS interaction has the ability to know the obligations involved and foresee the consequences in order to make an informed decision on whether to participate or not.

Besides, information on conditions and procedures on how to apply for PIC for GR as required by article 6.3 (c) NP are found in the Regulation. According to the national interim report on the ABS CH of the CBD[84] this information is also provided through the national ABS CH (website), workshops, seminars, awareness-raising programs, print media and brochures, emails and other communication tools.

Also, the Regulation requires the EBI to communicate its decision to grant or deny an access permit (Art. 5 and 10)in writing which fulfills article 6.3 (d) and (e).[85] A permit is evidence that PIC was obtained and MAT established.

The legislation also takes measures to ensure that CK of local communities is accessed with their PIC or approval and involvement, a requirement established under article 12(2) of the Proclamation. In addition, article 7 (2) of the Proclamation requires that conditions and procedures for granting PIC by local communities are specified in a regulation,[86] which is done under Part III of the Regulation establishing procedures and criteria for the same. This fulfills the requirements of articles 7, 6.2 and 6.3 (f) NP. Since the EBI shall assist in procuring the PIC of the local community as well as negotiate and conclude access and benefit-sharing agreements on their behalf, the processes involved are wrapped up in the institute's procedures and help to eliminate the usual complications and uncertainties involving the PIC of communities. In conformity with, article 12 (4) of the NP, the ABS law in its article 8(1) states that local communities shall have an inalienable right to use or exchange among themselves their genetic resources or community knowledge in the course of sustaining their livelihood systems in accordance with their customary practices or norms.

Further, the legislation possesses rules and procedures for establishing MAT as required by article 6.3 (g) NP. The elements that an access agreement may include

[82] EBI official, *pers. comms*, 8 September 2020.

[83] Ibid.

[84] https://absch.cbd.int/countries/ET (accessed 14 June 2019).

[85] Actually the Regulation does not say directly that EBI shall grant a **written** evidence to grant but rather it just says that "if it ascertains that there exists no ground sufficient to deny access, it shall determine that the access may be granted . . ." (Art. 10 (1) (b)). It is, however, direct concerning a written decision to deny (Art. 5 (2) (a)). The revised draft law is clear concerning this as it defines clearly under article 37 that issuing of written evidence that access requirements have been met is one of the powers and duties of the EBI.

[86] "The conditions and the procedure in accordance to which local communities shall give prior informed consent for access to their community knowledge shall be specified by a regulation".

are contained under articles 16–19 of the Proclamation. These elements are terms normally mutually agreed by parties to the agreement e.g. those that have been looked at in Sect. 5 above. Furthermore, clear rules and procedures are detailed in the Regulation (Art. 3–13).

The legislation differentiates the types of research and procedures thereto. The Proclamation waives access permit requirement for organs of the state which are empowered by law to conserve GR. Accordingly, such organs are not obliged to obtain access permits for access to GR and/or CK from the EBI in the discharge of their duties (Art. 11 (4)). Their employees must, however, carry with them a letter to this effect when collecting GR and CK (ibid). They may also not transfer to third parties or export the same without an explicit permit from the EBI (ibid). In addition, the legislation provides facilitated access to Ethiopian national public research and higher learning institutions and intergovernmental institutions based in the country for purpose of development and academic research activities which they undertake within the country (Art. 15 (1)). Under the facilitated access the EBI grants such institutions 'specific access permits' without the need to strictly follow the access procedure provided in the Proclamation. However, the EBI determines, as appropriate, the obligations they shall assume while having access under such a permit (ibid). Further, the Proclamation has put measures in place to address issues on third party transfer and change of intent (Art. 17 (9)). These measures fulfil article 8a NP on special considerations for non-commercial research.

Also, the Proclamation takes measures to direct benefits towards conservation and sustainable use as required by article 9 NP. It states clearly that the share of benefits received by the State after deducting the share of local community as determined pursuant to its article 9 (1) "shall be allocated for conservation of biodiversity and the promotion of community knowledge" (Art. 18 (2)). According to article 9 (1) and (2) of the Proclamation the local communities shall have the right to obtain 50% "of the benefit shared by the state in the form of money from the benefits derived out of the utilization of their genetic resources". Hence the rest which comprises 50% shall be used for biodiversity conservation purposes and the promotion of CK (Art. 27 (1) Regulation) in line with the criteria stipulated in article 27 (1)–(4)) of the Regulation. In other words, the State puts a 100% of the benefits accruing to it into conservation.

As already mentioned, the current legislation has no compliance measures for users and monitoring measures are insufficient. However, the revised draft law has put in place the former as well as augmented the latter to comply with articles 15–17 NP. In addition, according to information notified to the ABS CH[87] there are already measures to comply with article 18 of the NP on compliance with MAT. First, concerning measures to encourage the inclusion of provisions in MAT to cover dispute resolution article 16 (12)[88] of Proclamation No. 482/2006 states that MAT

[87] https://absch.cbd.int/countries/ET (accessed 19 February 2020).

[88] Revised draft law, Art. 16. Accordingly, the dispute settlement mechanisms are to ensure the rights of parties to access justice through negotiation, mediation or arbitration as alternative dispute

should include a provision of dispute resolution. Second, in regard to ensuring availability of recourse in cases of disputes arising from MAT the legal system generally grants everyone the right to bring justifiable matter to duly established courts and tribunals and to obtain from them judicial remedies.[89] Suits regarding contracts (including MAT) may be instituted at the discretion of the party claimant in the court of the place where the contract was made or was to be executed, unless the parties agreed otherwise in their contract.[90] Third, concerning access to justice the Constitution of Ethiopia upholds the right of each person to bring justifiable matter to a court of law or any other competent body with judicial power and to obtain judicial remedies.[91] Accordingly, the principle of access to justice on the rule of law emphasizes the right of equal access to justice for all, including members of vulnerable groups. It also reaffirms the commitment to take all necessary steps to provide fair, transparent, effective, non-discriminatory and accountable services that promote access to justice for all. Fourth, in respect to mutual recognition and enforcement of foreign judgements and arbitral awards these can be executed generally in the Ethiopian legal system on conditions stipulated by different laws.[92] These laws set the jurisdiction and conditions of courts before the foreign judgement and arbitral award executed in Ethiopia. The Federal Courts Proclamation (1996) provides that the Federal High Court shall have first instance jurisdiction over cases of enforcement of foreign judgements[93] and hence applications for enforcement of foreign judgements and arbitral awards are to be brought before it.[94] On the other hand, the Civil Procedure Code lays down the requirements and hence the framework for enforcement of foreign judgements and awards in articles 458 and 461.[95] Accordingly, foreign judgements may not be executed in Ethiopia unless certain substantive and formality requirements are fulfilled. According to article 458 permission to execute a foreign judgement shall not be granted unless: (a) the execution of Ethiopian judgements is allowed in the country in which the judgement to be executed was given (which is a reciprocity rule); (b) the judgement was given by the court duly established and constituted; (c) the judgement-debtor was given the opportunity to appear and present his defence; (d) the judgement to be executed is final and enforceable; and (e) the execution is not contrary to public order or

resolution mechanisms, and, by specifying the applicable law and jurisdiction, have their cases adjudicated in a court of law. Where parties have not specified dispute settlement mechanisms, Ethiopian laws shall be applied in Ethiopian courts.

[89] Art. 37 (1) Constitution Proclamation No. 1/1995.

[90] Federal Courts Proclamation No. 25/1996 and Civil Procedure Code of Ethiopia extraordinary issue No. 3 of 1965.

[91] Art. 37 (1) Constitution Proclamation No. 1/1995.

[92] Federal Courts Proclamation No. 25/1996 and Civil Procedure Code of Ethiopia extraordinary issue No. 3 of 1965.

[93] Teshale (2000), p. 569.

[94] Article 11 (2) (c)) Federal Courts Proclamation No. 25/1996.

[95] Teshale (2000).

morals.[96] In the same vein, as stated in article 461, foreign arbitral awards may not be enforced in Ethiopia unless: (a) reciprocity is ensured as provided for by Art. 458 (a); (b) the award has been made following a regular arbitration agreement or other legal act in the country where it was made; (c) the parties have had equal rights in appointing the arbitrators and they have been summoned to attend the proceedings; (d) the arbitration tribunal was regularly constituted; (e) the award does not relate to matters which under the provisions of Ethiopian laws could not be submitted to arbitration or is not contrary to public order or morals; and (f) the award is of such nature as to be enforceable on the condition laid down in Ethiopian laws.[97]

The revised draft law has also put other measures in place to comply with NP articles 10 and 11 on transboundary cooperation.[98] The provision of the revised draft law in this regard creates space for implementation of the multilateral ABS system (Art. 15 (1), (2)) and also for bilateral cooperation with neighboring countries in establishing "common access and benefit-sharing frameworks... with the view to avoiding competing against each other and negotiating common prior informed consent and mutually agreed terms as far as possible and as far as feasible" (Art. 15 (3)). It also establishes facilitated access in cases of emergencies according to article 8 (b) NP[99] and exempts for the first time PGRFA under Annex I of the International Treaty on Plant Genetic Resources for Food and Agriculture from its scope,[100] a measure complying with article 4 NP.

8.2 General Issues

Although Ethiopia currently operates a pre Nagoya Protocol legislation, it has a number of strengths that can be identified. We shall mention only the major ones.

First, only one competent national authority is designated, the EBI, which also serves as the NFP. In other words, this is a kind-of one-window or one-stop shop. This simplifies by far the usually cumbersome and complex/confusing provider-country requirements and procedures. It also eliminates the danger of having conflicting, insufficient and/or inexact information from two different authorities. Some of the countries, for example, have designated high ranking officials in the government who have little knowledge about ABS as NFPs and who are almost never accessible, so the Ethiopian approach is commendable.

The second, which is related to the first, is that the CNA (EBI) assumes the PIC and MAT responsibilities of the local communities. This eases the complex procedures of such communities which are widely bemoaned.

[96] Ibid, 577.

[97] Ibid, 578.

[98] Revised draft law Art. 15.

[99] Ibid, Art. 14 (1) (b).

[100] Ibid, Art.14 (1) (c).

Third, it is one of the rare models of traditional provider states' legislations that are very clear concerning the sharing of benefits to and between the State and the local communities.

Fourth, it fully implements the core thought of the third objective of the CBD which is benefit-sharing for the sake of biodiversity by investing all benefits due to the State to conservation and sustainable use.

Fifth, it differentiates non-commercial and commercial types of research and creates a separate procedure for the former.

However, the current legislation, Proclamation 2006, also possesses several weaknesses. Among the main ones, first, it lacks definitions of key terms used therein e.g. 'access', 'exploration', 'genetic resources', 'community knowledge', 'local community', 'commercial use' and non-commercial use'. These have been defined in the revised draft law (Art. 2).

Second, it lacks user compliance and monitoring measures. These are included now in the revised draft law. In this, the revised draft law must be complimented for implementing articles 15–17 NP properly and not muddling the measures relating to the obligation thereunder like it is the case with a number of new measures of traditional provider countries.

Third, the existing ABS law is not sufficiently clear on how to apply and obtain PIC and establish MAT for **access to community knowledge** associated with GR. The revised draft law now establishes clear rules and procedures for obtaining PIC and establishing MAT for access to such knowledge and GR.

Fourth, the legislation has not addressed in a proper manner how to involve the local communities or seek their approval in PIC procedures for **access to GR**. This can be seen as a limitation of the law, *albeit*, such communities have the right to reject the PIC decisions of the CNA involving GR in their territories.

Fifth, some of its provisions are very protective e.g. those seeking denial of access permit to users from jurisdictions that do not have effective monitoring measures, including effective checkpoints, and requiring foreign institutions and competent authorities to guarantee that users are complying with their obligations as compliance support measures. At least the provision requiring oversight by foreign CNAs has been repealed in the revised draft law. This obligation was difficult to implement anyway.[101]

Sixth, it does not provide an option for IT-based submission of an access application and granting of access permits and thus is void of a cost-effective communication tool. Such tools and mechanisms need to be established to facilitate access and monitoring of the implementation of contractual agreements.

[101] EBI official, *pers. comms*, September 2020.

9 Conclusion

Ethiopia is a biodiversity and associated community knowledge rich country, but its resources have often been misappropriated or utilized without fair and equitable sharing of the benefits arising from utilization of its genetic resources. Therefore, efficient and effective legal and institutional frameworks should be put in place to regulate access, prevent misappropriation and ensure fair and equitable sharing of the benefits arising from utilization of its genetic resources and associated community knowledge.

Ethiopia is party to CBD and ITPGFA, which established a bilateral and multilateral ABS system respectively. The country put in place a national ABS legal framework that ensures mutually supportive implementation of these instruments/frameworks. The existing law addresses the ownership of GR and associated community knowledge and it recognizes the right of local communities to regulate access and share benefits arising from the use of its GR and associated community knowledge. Currently the law regulates any activity on genetic resources and associated community knowledge under its scope, including collection, export, transfer, exploration, acquisition, and use in an attempt to close every window of misappropriation and misuse and thus ensure sovereign rights over its resources. This makes the current law a bit too restrictive and protective. It also lacks the innovative provisions of the Protocol and also has gaps to effectively address ABS issues. It is therefore important for the law to be revised in order to harmonize it with the NP and likewise fill existing gaps. The revision has taken place and the current draft has been submitted to the decision makers for approval.

The draft revised ABS law includes the innovative provisions of the NP *viz.* on user compliance measures and legal certainty, clarity and transparency. It is hoped that it will contribute to the global fight against bio-piracy and ensure fair and equitable benefit-sharing that supports conservation and sustainable utilization of genetic resources. It also contains provisions that provide a facilitated access to Ethiopian national public research and higher learning institutions and intergovernmental institutions based in the country to undertake research and development activities within the country. In addition, facilitated access is created for non-commercial research, PGRFA under Annex I of the International Treaty and emergencies that threaten or damage human, animal, or plant health. This will contribute to the promotion and encouragement of scientific research and ensure food security worldwide. It will also help in the prevention of public health risks as well as in hastening response during such risks.

The approval of the comprehensive revised draft law will further promote access to GR and associated CK and ensure fair and equitable sharing of benefits arising from their use and contribute to the conservation and sustainable use of biological diversity, poverty eradication and thereby contributing to achieving the Sustainable Development Goals.

On a general note, an assessment of the implementation of the existing law indicates that there is limited human and institutional capacity to negotiate and establish effective ABS contractual agreements and enforce the provisions of the law and contractual agreements. There is also limited awareness on ABS law and process among decision- and policy-makers, researchers, higher learning and regulatory institutions, and local communities to effectively implement the ABS law. Even though there is much improvement in collaboration and integration among the different key-stakeholders on ABS, still it is inadequate for effective implementation of the ABS law.

Further, practice in Ethiopia indicates that 882 access permits have been granted to applicants for non-commercial purposes and MATs established on 14 species for commercial purposes. There is limitation in addressing lack of compliance with the provisions of reporting and benefit-sharing obligations of the MTAs and MATs signed so far. The MATs established and signed are not effective to ensure fair and equitable benefit-sharing, mainly due to limited capacity to determine the value of GR to be accessed.

Decision- and policy-makers should give attention and priority for the approval of the revised draft law to ensure fair and equitable sharing of benefits arising from utilization of genetic resources and associated community knowledge and prevent bio-piracy.

There is a need to put in place measures to raise awareness of the importance of genetic resources and traditional knowledge associated with genetic resources, and related access and benefit-sharing issues. Human resources and institutional capacities should be strengthened to negotiate and establish effective ABS contractual agreements and enforce the provisions of the ABS law and contractual agreements.

Documentation of community knowledge, valuation and bioprospecting research should be promoted and valorization strategy developed to maximize benefit-sharing. Collaboration among key stakeholders on ABS should be strengthened to effectively implement the ABS law. Finally, national ABS clearing house should be established and strengthened to ensure effective information exchange on ABS and support compliance with provisions of the ABS law and MAT.

References

Texts

Dalle G (2015) Research on genetic resources and indigenous knowledge in the framework of the Ethiopian ABS law. In: Kamau EC, Winter G, Stoll P-T (eds) Research and development on genetic resources. Public domain approaches in implementing the Nagoya Protocol. Routledge, New York, pp 125–131

EBI (2015) Government of the Federal Democratic Republic of Ethiopia: Ethiopia's National Biodiversity Strategy and action plan 2015–2020. Ethiopian Biodiversity Institute, Addis Ababa

EBI (2016) Farm animal diversity of Ethiopia: breeds and ecotypes catalogue. Ethiopian Biodiversity Institute (EBI), Addis Ababa

Greiber T, Peña Moreno S, Åhrén M, Carrasco JN, Kamau EC, Cabrera Medaglia J, Oliva MJ, Perron-Welch F in cooperation with Ali N and Williams C (2012) An explanatory guide to the nagoya protocol on access and benefit-sharing. IUCN, Gland, Switzerland

Hedberg I, Friis I, Person E (2009) General part and index to Vol. 1–7. Flora of Ethiopia and Eritrea Volume 8. The National Herbarium, Addis Ababa, Ethiopia and Uppsala, Sweden

IBC (2012a) The state of forest genetic resource of Ethiopia. Country report submitted to FAO, Addis Ababa, Ethiopia

IBC (2012b) Ethiopia: third country report on the state of plant genetic resources for food and agriculture to FAO, Addis Ababa, Ethiopia

Teshale S (2000) Reciprocity with respect to enforcement of foreign judgments in Ethiopia: critique of the supreme Court's decision in the Paulos Papassinous case. Afr J Int Comp Law 12(3):569–578

Vavilov I (1951) The origin, variation, immunity and breeding of cultivated plants. Translated from the Russian by Chester KS. Ronald Press Co., New York

WCMC (1994) Biodiversity data source book. World Conservation Monitoring Centre, World Conservation Press, Cambridge

Yifru WD (2003) Access to genetic resources in Ethiopiain. In: Nnadozie K et al (eds) African perspectives on genetic resources: a handbook on laws, policies, and institutions. Environmental Law Institute, Washington DC, pp 107–122

Yimegnuhal A, Degefa T (eds) (2004) Farm animal biodiversity in Ethiopia: Status and prospects. Proceedings of the 11th annual conference of the Ethiopian Society of Animal Production (ESAP) held in Addis Ababa, August 28–30, 2003, ESAP, Addis Ababa, Ethiopia

Legal Texts

Access to Genetic Resources and Community Knowledge and Community Rights Council of Minister Regulation No. 169/2009, Federal Negarit Gazeta Year 15 No. 67, 9 November, 2009, Addis Ababa, http://www.ebi.gov.et/regulation/ (accessed 11 September 2020)

Access to Genetic Resources and Community Knowledge, and Community Rights Proclamation No. 482/2006, Federal Negarit Gazeta Year 13 No. 13, 27 February, 2006, Addis Ababa, http://www.ebi.gov.et/regulation/ (accessed 11 September 2020)

African Model Legislation for the Protection of the Rights of Local Communities, Farmers and Breeders, and for the Regulation of Access to Genetic Resources adopted by OAU in Algiers, Algeria on 2000

Agreement on Access to, and Benefit Sharing from Teff, concluded between the Institute of Biodiversity Conservation and Research (Provider), the Ethiopian Agricultural Research Organization and Health and Performance Food International BV, signed on 5 April 2005

Bonn Guidelines on Access to Genetic Resources and Fair and Equitable Sharing of the Benefits Arising out of their Utilization, Conference of Parties to the Biodiversity Convention, adopted by COP-6 in Bonn, Germany on April 2002

Constitution of the Federal Democratic Republic of Ethiopia Proclamation No. 1/1995

Convention on Biological Diversity (1992) Text of the CBD. Rio de Janeiro, Brazil, United Nations

Ethiopian Biodiversity Institute Establishment Council of Ministers Regulation, No 291/2013, Federal Negarit Gazette No. 57, 24 July, 2013, Addis Ababa, http://www.ebi.gov.et/regulation/ (accessed 11 September 2020)

Federal Courts Proclamation No. 25/1996. http://www.fsc.gov.et/content/Negarit%20Gazeta/Gazeta-1988/Proc%20No.%2025-1996%20Federal%20Courts.pdf (accessed 19 February 2020)

Nagoya Protocol on Access to Genetic Resources and the Fair and Equitable Sharing of Benefits Arising from Their Utilization to the Convention on Biological Diversity, adopted by COP-10, in Nagoya, Japan on October 29, 2010

Revised Ethiopian ABS (Access to Genetic Resources and Community Knowledge and Community Rights) Proclamation (Draft)

Ashenafi Ayenew Hailu a national of Ethiopia, is a researcher at Ethiopian Biodiversity Institute. He served for 6 years as ABS National Focal Point and the Director of Genetic Resource Access and Benefit Sharing Directorate at Ethiopian Biodiversity Institute. He has more than 10 years working experience in ABS process. He was involved in the negotiation and drafting of several ABS agreements and led the team that drafted the revised ABS law. He holds a Master of Science degree in Dry land Biodiversity from Addis Ababa University. ayenewashenafi2007@gmail.com or ashenafi.ayenew@ebi.gov.et

Evanson Chege Kamau Dr. iur., is Senior Research Fellow at the Research Centre for European Environmental Law (FEU), University of Bremen. He is also an Adjunct Lecturer at the CvO University of Oldenburg. His area of expertise is public international law, European law and intellectual property law. His research focuses on genetic resources, environment, development and traditional rights on biological resources and traditional knowledge. Kamau has over 17 years of working experience in the area of ABS in which he has published widely. For many years he has also worked as a freelance legal consultant for governments, African Union (hired by IUCN and Capacity Building Initiative resp.), private industries and research organizations. echege@uni-bremen.de.

Abracadabra! Or When and How Will the Kenyan ABS Law Be Born?

Evanson Chege Kamau

Abstract Kenya operates a pre-Nagoya Protocol ABS law of 2006 that very much replicates the first generation laws in this area, which were marred by post-CBD reactionary approach meant to curtail bio-piracy and capture the anticipated huge benefits from "green gold". As a consequence, the ABS scenario is clouded with requirements and procedures that rather cripple than aid access to genetic resources and benefit-sharing. About 5 years ago, it was undertaken to revise the current legislation in order to deal with some of its shortcomings and align it to the Nagoya Protocol, but the bill, which resulted from a report of a country-wide stakeholder consultation disappeared in thin air after being prepared for tabling at the parliament. Since approximately 3 years ago (as at the time this chapter was authored) a UNDP-GEF program has been going on to try and assist with legislation and bring compliance with the Protocol but still there is no clear resolution whether to revise the old law or produce a stand-alone law. In addition, more independent consultants were hired late 2019 "to support the process of exploring and proposing a legal framework for Kenya to implement the Nagoya Protocol". The time for a comprehensive and balanced ABS law in Kenya is overdue, but the process seems to be going in circles. This chapter therefore asks: What is stopping the birth of a healthy ABS law: lack of or bad midwifery? When is a new law expected? In the meantime, practice has introduced some mechanisms to ameliorate the situation including by trying to eliminate competition between relevant government agencies, abandoning some of the requirements and *quasi* integrating the permit procedure and thus drastically reducing the duration from the time of permit application to grant. After describing the current legal situation and the efforts being made to implement the Nagoya Protocol this chapter critically looks at these mechanisms interrogating how practice can function in *quasi* legal lacuna, and also the legality of such mechanisms in case of a legal suit.

E. Chege Kamau (ed.), *Global Transformations in the Use of Biodiversity for Research and Development*, Ius Gentium: Comparative Perspectives on Law and Justice 95, https://doi.org/10.1007/978-3-030-88711-7_9

Keywords Access and benefit-sharing · Regulations 2006 · Kenya wildlife service · NEMA · Traditional knowledge

1 Development of ABS Legislation

1.1 Status of Biodiversity

Kenya covers a land area of approximately 583,000 km^2 straddling the Equator between approximately 4.5 degrees South and 4.5 degrees North latitude.[1] It has a coastline of approximately 640 km and an Exclusive Economic Zone (EEZ) covering about 230,000 km^2.[2] Of the total ecosystem area 28% is marine and 72 terrestrial.[3] Kenya is rich in diverse and unique flora and fauna[4] making the country one of the richest megadiverse countries in the world.[5] In spite of 80% of the terrestrial land being arid and semi-arid, it has a high species index due to the confluence of five major phytogeographical regions, which include Somali-Maasai, afromontane, Swahilian Zanzibar-Inhambane, the Lake Victoria region and the Guinea-Congolian mosaics.[6] From the described and documented biodiversity there are approximately 7004 species of plants (which comprises 57% of all 12,317 plant species recorded in the East African region[7]), 2071 species of fungi and bacteria (although it is estimated that over 5000 fungal species occur throughout the country[8]) and 30,000 species of animals (with the richest avifauna diversity in Africa of around 1100 bird species[9]).[10]

This richness in biodiversity is attributed to a number of factors, including a long evolutionary history, varied and diverse habitat types and ecosystems, diversity of landscapes and variable climatic conditions.[11] About 70% national biodiversity resources are found outside the protected areas, while 30% are within protected areas that include national parks, reserves, sanctuaries, gazetted forests, and heritage

[1] Republic of Kenya (2015), p. 3.

[2] Ibid.

[3] Ibid.

[4] Ibid.

[5] See Watai et al. (2015), p. 110.

[6] Republic of Kenya (2015), p. 7.

[7] Ibid, p. 4.

[8] Ibid, p. 10.

[9] Ibid, p. 14.

[10] Ibid, pp. 4, 10, 13 respectively.

[11] Ibid, p. 3.

forests.[12] Besides, Kenya is endowed with rich *ex-situ* conservation facilities comprising arboreta/botanical gardens, animal orphanages and animal parks.[13]

1.2 Historical Context, Influence and Aims: Stories Behind the Development of the Legislation

Kenya joined the world community in search for a global approach for environmental protection as early as 1972 by participating in the UN Conference on Human Environment where the United Nations Environment Programme (UNEP) was born.[14] In 1974 the government of Kenya founded the National Environment Secretariat (NES) as the lead environment agency to coordinate and oversee environmental activities in the country.[15] During the Earth Summit in Rio de Janeiro in 1992 Kenya endorsed Agenda 21[16] and signed the Convention on Biological Diversity (CBD), which it ratified in 1994.[17] In the same year it ratified the Convention. NES founded the Interministerial Committee on Environment (IMCE), a multi-sectoral and multi-disciplinary team with membership from the government, private sector, and non-governmental organizations (NGOs), which had a sub-committee responsible for the implementation of the CBD.[18] However, Kenya lacked an integrated national strategy and action plan for biodiversity.[19] Hence, a multi-disciplinary and multisectoral task force was established in 1996 to function as a broad-based consultative and advisory group to steer the development of a National Biodiversity Strategy and Action Plan (NBSAP) with the objective of addressing the national and international undertakings elaborated in article 6 of the Convention.[20] Some of the guiding principles in developing such a NBSAP were

[12] Ibid, p. 3f.

[13] Kamau (February 2009), p. 76.

[14] Ministry of Environment and Natural Resources (March 2000). The conference was held in Stockholm in response to the UN General Assembly Resolution No. 2850 (XXVI) (https://undocs.org/en/A/RES/2850(XXVI)).

[15] Ibid.

[16] Among other things, Agenda 21 specifically calls for the development of national strategies for the conservation of biological diversity and the sustainable use of biological resources.

[17] Ministry of Environment and Natural Resources (March 2000).

[18] Ibid.

[19] Ibid.

[20] Ibid. Article 6 of the CBD requires each contracting party, in accordance with its particular conditions and capabilities, to: "(a) Develop national strategies, plans or programmes for the conservation and sustainable use of biological diversity or adapt for this purpose existing strategies, plans or programmes which shall reflect, inter alia, the measures set out in this Convention relevant to the Contracting Party concerned; and (b) Integrate, as far as possible and as appropriate, the conservation and sustainable use of biological diversity into relevant sectoral or cross-sectoral plans, programmes and policies".

that conservation and sustainable utilization of Kenya's biodiversity would contribute to fair and equitable sharing of benefits arising therefrom and that the utilization of Kenya's biodiversity should be transparent, equitable, and efficient.[21] A number of national goals were also foreseen *inter alia* that the country will have an enabling policy, legislative and constitutional environment for conservation and sustainable utilization of biodiversity in the shortest term and in the long term, will have realized the benefits of sustainable utilization of biodiversity resources and improved living standards of the people.[22] These goals had to contend with a number of challenges identified among them overexploitation of biodiversity for consumptive uses and unregulated bioprospecting.[23] For ABS the latter was a critical issue to consider with so many species having been smuggled out of the country,[24] with false permits,[25] with both local and foreign bioprospectors ignoring requirements concerning depositing of duplicate of collected species,[26] and with lack of knowledge on the importance of genetic resources,[27] among others.

The first NBSAP of 1999 mentioned the need for policy and legislative measures to regulate Kenya's biological resources. The initiatives to develop a regulatory system started in 2000 first by an expert group made up of experts from the NGO sector, relevant government ministries and departments accompanied by a committee to oversee its future implementation.[28] At the same time, although with no statutory mandate, a second initiative was undertaken by the National Museums of Kenya (NMK), and pursuant to its activities of collecting plant and seed varieties with the Royal Botanic Gardens at Kew, a drafting sub-committee including government ministries and the Attorney General developed a preliminary draft of possible regulations.[29] Both initiatives stalled.[30]

On the institutional level NES was faced with many challenges which were mainly a result of the fact that it was never provided with statutory legal status.[31] As a result, it lacked direct influence on the activities of the various lead agencies (LAs) and hence there was no concerted approach and each LA acted according to its interests leading to fragmentation of legislation and sectoralism.[32] NES was also consistently underfunded and was thus unable to respond to the breadth of its

[21] Ministry of Environment and Natural Resources (March 2000).

[22] Ibid.

[23] Ibid.

[24] Zehle (2001); Kwayera (1999).

[25] Mbaria (23 August 2004).

[26] Kamau (February 2009), p. 76.

[27] Ibid.

[28] Angwenyi (2009), p. 177.

[29] Ibid.

[30] Ibid.

[31] Kamau (February 2009), p. 77.

[32] Ibid.

responsibilities.[33] In response to its difficulties the Environment Management and Coordination Act 1999 (EMCA), a framework legislation on environmental management, was enacted.[34] The EMCA transposed the CBD and brought issues concerning the conservation of biological diversity and access to genetic resources under the purview of the Act's general administration.[35] It established the National Environment Management Authority (NEMA) with mandate over environmental issues including legislating on access and benefit-sharing (ABS). In 2006, 3 years after NEMA assumed office, it adopted the so-called Environmental Management and Co-ordination (Conservation of Biological Diversity and Resources, Access to Genetic Resources and Benefit Sharing) Regulations, 2006, which implemented the relevant provisions of the framework legislation, including ABS provisions.[36] This law (Regulations 2006) started badly with a slow operationalization take off[37] due to lack of recognition by the main stakeholders, i.e. the main state agencies (known as LAs), lack of clarity in many issues, and lack of capacity as well as shortage of implementation resources for NEMA.[38] Even today, a number of the initial challenges have not found a legal solution.

1.3 Legal Background

Kenya inherited the Anglo-Saxon common law tradition pursuant to which ownership is said to extend to everything found beneath and above someone's private property except specifically identified assets, e.g. minerals or mineral oils and rights of way (such as through airspace) over which the government retains control for the benefit of the public.[39] Hence, the concept of property is based on the common law tradition, except in instances where this is superseded by the Constitution or statute.[40] It is most likely that based on this the Registered Land Act,[41] which governs individual ownership of land and the nature of interests one gets under these laws, vests in any person registered as the proprietor of land absolute ownership of the same together with all rights and privileges belonging or appurtenance

[33] Manek (April 2001), p. 3.

[34] Ibid; Kamau (February 2009), p. 80. EMCA entered into force on 14 January 2000.

[35] Kamau, ibid., p. 80f.

[36] Angwenyi (2009), p. 173.

[37] According to Angwenyi (2009, p. 181) it was only in January 2008 when the actual operationalization began.

[38] Kamau (February 2009), p. 82.

[39] Lettington (2001).

[40] Angwenyi (2009), p. 174.

[41] The Registered Land Act (1989, Rev. 2010), s. 27 (a) and s. 30.

thereto (s. 27 (a)) subject to overriding interests enumerated under s. 30.[42] This approach was upheld by the High Court in the case of *Abdikadir Sheikh Hassan and 4 Others v. Kenya Wildlife Service*[43] which ruled that "... according to common law and/or customary law of the inhabitants of this country, those entitled to the use of land are also entitled to the fruits thereof which include the fauna and flora unless this has been negated by law". The old Constitution of 1992 provided every citizen with protection of personal property including land, and from deprivation thereof without compensation (s. 70) and referred to the environment in the context of governmental powers for the purposes of conservation.[44] However, together with other laws, it did not solve the question about ownership and control of genetic resources.[45] It is only after the enactment of the EMCA that law began to confront this question first by leanly implementing articles 15 and 8j of the CBD.

The (new) Constitution of 2010 addresses this question by including provisions to protect genetic resources and biological diversity (s. 69 (1) (e)). It expresses the citizens will on how their biodiversity should be governed. Besides, it emphasizes every person's duty to cooperate and ensure ecologically sustainable development and use of natural resources (s. 69 (2)). Further, it recognizes indigenous and local knowledge of communities on biodiversity and their intellectual input and gives guidance for parliament to enact legislation on access and protection of their traditional knowledge as well as use and compensation thereof (s. 11 (3) (a); s. 69 (1) (c))—thus the Constitution commits the State to protect and enhance such rights.[46] In addition, it defines the term 'traditional community' and 'indigenous community' under the term 'marginalized community' in s. 260 (b) and (c). This constitutes the basis for enactment of domestic legislations, policies and institutional arrangements. Accordingly, the EMCA was amended in 2018 to accommodate in particular the provisions on the rights of the indigenous and local communities over their biodiversity, genetic resources and (a)TK.

EMCA establishes the National Environment Management Authority (NEMA or Authority[47]) with a statutory mandate to oversee the coordinated management of the environment[48] and to implement the Act.[49] The main provision on access to genetic resources and benefit-sharing (ABS) is s. 53. It gives the Cabinet Secretary (for the time being responsible for matters relating to environment and natural resources) on

[42] The exceptions to these broad categories of property generally involve specifically identified assets, such as minerals or mineral oils, and rights of way, such as through airspace, over which the government retains control for the benefit of the public.

[43] Civil Case No. 2959 (High Court of Kenya, 1996).

[44] CBD (30 August 2007), p. 16.

[45] Angwenyi (2009), p. 174f.

[46] Same, Stanley Atsali (KIPI), *pers. comms*, 22 September 2017.

[47] The term "Authority" was defined in section 2 ("Interpretation") of EMCA and "means the National Environment Management Authority", in short NEMA, which was established under section 7 of EMCA.

[48] Kamau (February 2009), p. 81; Watai et al. (2015), p. 113.

[49] Kamau, ibid., p. 81f.

the recommendation of NEMA the power to issue guidelines and prescribe measures for the sustainable management and utilization of genetic resources for the benefit of the people of Kenya (s. 53 (1)). This power belonged to NEMA in the old EMCA which had to perform these functions in consultation with relevant lead agencies.[50] That these powers have now been centralized could solve the past sluggishness in ABS process caused by conflicting interests of the LAs. Further, the revised version made additions to the issues to be specified by such guidelines and measures, which will be marked with bold. Therefore the new section 53 requires that the following are specified: appropriate arrangements for access to **biological resources**, genetic resources **and ecological services** by non-citizens, including the issue of licenses and fees to be paid for access (s. 53 (2) (a)); measures for regulating the import or export of germplasm (s. 53 (2) (b)); the sharing of benefits derived from genetic resources (s. 53 (2) (c)); measures necessary to regulate the development, access to and transfer of biotechnology (s. 53 (2) (e)); **measures necessary for the recognition, protection and enhancement of indigenous knowledge and associated practices in the conservation of the environment and natural resources (s. 53 (2) (f)); and measures necessary for the protection of indigenous knowledge of biodiversity and genetic resources of communities (s. 53 (2) (g))**. In addition, it gives the Cabinet Secretary (and not NEMA like in the old version) leeway to issue guidelines or prescribe measures to specify any other matter that it considers necessary for the better management of genetic resources (s. 53 (2) (h)).

Other provisions of the EMCA that are relevant to ABS are sections 50–52. Accordingly the Cabinet Secretary with NEMA's advice/recommendation has power to prescribe measures necessary to ensure the conservation and sustainable use of *in-situ* and *ex-situ* biological diversity. Such measures include: the specification of national strategies, plans and government programs for conservation and sustainable use of biological diversity (s. 50); integration of traditional knowledge with mainstream scientific knowledge (s. 51); and issuing of guidelines for the management of germplasm banks and botanical gardens (s. 52). These are important e.g. in checking whether the government has succeeded in undertaking measures to conform to articles 9, 12 and 20 of the Nagoya Protocol (NP) to connect benefit-sharing to conservation and sustainable use of biological diversity, integrate associated traditional knowledge (aTK) to mainstream scientific knowledge and to develop guidelines on ABS from gene banks and botanical gardens respectively.

As already mentioned, ABS provisions of the EMCA were concretized in Regulations 2006. The Regulations establish an access permit as a condition for access to genetic resources (GR) in Kenya. To receive such a permit one must as a prerequisite seek the prior informed consent (PIC) from interested persons (r. 9). These are

[50] EMCA defines the term 'lead agency' as "any Government ministry, department, parastatal, state corporation or local authority, in which any law vests functions of control or management of any element of the environment or natural resource".

named in the first schedule (Part III 2.0(g)) as relevant LAs, local community[51] or private **owner** of the genetic resources. Therefore, the different biodiversity laws/ Acts and in addition an Act on TK establish a PIC requirement for access to biological/genetic resource depending on where and from whom this is collected, and for access to aTK from the local communities, respectively. Thus together with the Constitution, the EMCA and the Regulations, the main laws which form the ABS framework are: The Wildlife Conservation and Management Act 2013 (WCMA); The Forest Conservation and Management Act No. 34 of 2016 (FCMA); and the Protection of Traditional Knowledge and Cultural Expressions Act No. 33 of 2016 (PTKCEA).[52] Other legislations of relevance to the subject are: The National Museums and Heritage Act No. 6 of 2006; Industrial Property Act, No. 3, of 2001; Science, Technology and Innovation Act No. 28 of 2013; and The Seeds and Plant Varieties Act of 1972.[53]

Based on the above-said the ownership of genetic resources according to the Kenyan law depends on the land tenure and is vested either in the State, local community or private landowner. However, the issue of control can be observed to follow the following pattern:

- For GR that are under the State's ownership, the government has competence in matters of PIC,[54] access (permit)[55] and benefits.
- For GR under the territories of the local communities PIC is either granted corporately by the government and the local community or alone by the community depending on the status of the territory, whereas access (permit) and benefit-sharing are controlled by the government.
- For GR on private land the landowner has power to grant PIC and the government the access (permit). There is nothing in law to show how the issue of benefits is dealt with.

[51] The term 'local community' is neither defined in the Regulations nor in its revised draft version, which is not yet law.

[52] For details on focus of these legislations (except the PTKCEA) see Watai et al. (2015), pp. 113–115.

[53] Ibid.

[54] 'Prior informed consent' is defined in the ABS instruments as "an international operation procedure for exchanging, receiving and handling notification and information by a competent authority" (in Regulations 2006), "a certificate from Service showing prior acceptance of collector by the Service and the concerned community or land owner, to access biological resource, genetic resource or associated knowledge" (in the draft bioprospecting regulations), and "the giving of, by the prospective user, complete and accurate information, and based on that information, the prior acceptance, by the owners, to the use of their traditional knowledge or cultural expressions" (in the PTKCEA).

[55] An 'access permit' is defined in the ABS instruments as "a permit that allows a person to access genetic resources issued under Regulation 4 of the Environmental Management and Coordination (Conservation of Biological Diversity and Resources, Access to Genetic Resources and Benefit Sharing) Regulations, 2006 (See e.g. Regulations 2006 and draft bioprospecting regulations).

This indicates State's competence or oversight over genetic resources even when they are under community territory or private land. For aTK the control over PIC and authorization for access to registered TK is fully in the hands of the local communities as the ownership is, but, as it would be seen below, the government again infringes in the issue of benefit-sharing.

2 Scope of ABS Framework

The geographical scope of the regime applies to the territorial jurisdiction of Kenya, including marine areas, exclusive economic zone and the continental shelf. Concerning material scope it applies to access to biological/genetic resources[56] for research, bioprospecting, conservation, industrial application or commercial use activities and to derivatives and, where applicable, to intangible components[57] the definition of which some administrative officers believe could be extended to cover Digital Sequence Information.[58] It also applies to associated traditional knowledge. The regime does not address the question of personal scope directly. This can be drawn, however, e.g. from r. 9 (1) which states that "Any person who intends to access genetic resources in Kenya shall apply to the Authority for an access permit . . ." (*emphasis (underline) added*). "Any person" can be a natural or legal person. This is in line with the wording of the ABS brochure which defines providers of genetic resources as ". . . any persons or entity in custody of GR that provides genetic resources and/or associated knowledge within the Kenyan territory" and ". . . include public or private land owners, and resource managers, county government, local community and individual traditional knowledge holders".[59] It also defines a user of genetic resources as ". . . anyone pursuing collection activities in Kenya, whether of tangible materials or intangible information" and includes "scientists, academic researchers, collectors or curators, herbal doctors, bio-prospectors, and investors or industry researchers".[60] The academic researchers, collectors or curators, investors

[56] Regulations 2006 adopt the CBD definition of 'genetic material' which elaborates the term 'genetic resources' but does not include the latter or give own definition. Apropos, the definition is also adopted with an error. In section 2 on interpretation the definition reads: "'genetic material' means any **genetic** material of plant, animal, microbial or other origin containing functional units of heredity" (*bold added*). The word 'genetic' in bold is redundant and should not be included as it is what is defined as material with "functional units of heredity".

[57] According to Regulations 2006 and the draft bioprospecting guidelines "'intangible component' means information held by persons, **parties or communities** that is associated with or regarding genetic/**biological** resources within the jurisdiction of Kenya" (the draft bioprospecting guidelines have added the words in bold).

[58] Mukonyi Kavaka Watai (Head of Bioprospecting, KWS), *pers. comms,* 20 September 2017; Jane Nyandika (ABS Desk Officer, NEMA), *pers. comms,* 25 September 2017.

[59] NEMA (2014a), p. 7f.

[60] Ibid, p. 8.

and industry researchers often execute actions on behalf of the institutions, collections and industries and hence also imply the *persona ficta* behind them. This implies that if such entities are acting on behalf of the legal personality incorporating them the latter can also be made responsible. The question of temporal scope for biological/genetic resources is also not clear but an attempt to resolve it can be grounded on the provision of the Regulations in r. 22 which requires any person carrying out any activity involving access to GR "... immediately before the coming into force of these Regulations" to "... take all necessary measures to ensure full compliance with these Regulations" within 6 months from the time of their coming into force. If this is well understood, the legislation has a retrospective resonance; it extends to a period prior to its entry into force not concretely defined. What is "immediately before the coming into force?" How long before its coming into force can one assume that their activities involving access (which presumably based on the definition of 'access' means both collection and research and development (R&D)) do not fall under the scope of the legislation and hence are not affected by the compliance requirement: 1 day, 2 weeks, a month? The other question that arises here involves new R&D on GR accessed prior to its entry into force. New legislations (*viz.* Cameroonian law 2021, Malaysian legislation 2017) suggest that such GR should fall under their regulation. In the Kenyan context, would it mean that new R&D on GR taken/obtained prior to the temporal scope here established is not regulated? We submit that any new R&D on such GR or aTK should be subject to benefit-sharing requirements. A clause could be inserted in the new law similar to s. 63 (3) of the Malaysian Act.

In regard to aTK the transitional provisions of the Act on the protection of traditional knowledge establish a rule on temporal scope, but a confusing rule. In s. 42 (1) it states that "Upon the commencement of this Act, any person who, before the commencement of this Act, was lawfully involved in the exploitation and dissemination of traditional knowledge shall be required to comply with the provisions of this Act within twelve months, subject to equitable treatment of the rights acquired by third parties in good faith" and then in 42 (3) says, "This Act does not affect or apply to contracts, licenses or other agreements entered into by traditional owners before the commencement of this Act in relation to the use of traditional knowledge ...". So what is what? In their present state the two provisions contradict themselves. Presumably there is an error with the term "lawfully" in s. 42 (1) which we guess should be "unlawfully". If our assumption is correct, any activities involving aTK that were not regularized had 12 months from the time of coming into force of the Act to comply whilst activities which were conducted legally were/are not affected by the Act.

According to Regulations 2006 (r. 3) the regime does not apply to: (1) Exchange of genetic resources, their derivative products, or intangible components associated with them, carried out by members of any local Kenyan community amongst themselves and for their own consumption; (2) Access to genetic resources derived from plant breeders in accordance with the Seeds and Plant Varieties Act of 1972 (as amended in 2002); (3) Human genetic resources; and (4) Approved research activities intended for educational purposes within recognized Kenyan academic and research institutions governed by relevant intellectual property laws, except if the biological material is taken and used abroad.[61]

The regime also exempts plant genetic resources for food and agriculture (PGRFA) under Annex I of the International Treaty on Plant Genetic Resources for Food and Agriculture (ITPGRFA) which are subject to the standard material transfer agreement (SMTA) of the same administered in Kenya by the Kenya Agricultural and Livestock Research Organization (KALRO).

3 Institutional Framework and Mandates

The main institutions involved in ABS are the National Environment Management Authority (NEMA), Kenya Wildlife Service (KWS), Kenya Forest Service (KFS) and the National Council for Science and Technology (currently known as National Commission for Science, Technology and Innovation, NACOSTI) for access to genetic resources, and the Cabinet Secretary responsible for matters relating to culture (CS culture) in regard to access to aTK.

3.1 National Environment Management Authority (NEMA)

As said, NEMA[62] is mandated to issue access permits which is a prerequisite for access to genetic resources. According to r. 9 (1) of Regulations 2006 "Any person who intends to access genetic resources in Kenya shall apply to the Authority for an access permit". Accordingly, NEMA is the only government organ with the mandate to grant access permits to all types of GR.

3.2 Kenya Wildlife Service (KWS)

KWS is established under s. 6 of the Wildlife Conservation and Management Act (WCMA) with numerous functions as listed under s. 7 *inter alia* to conserve and manage national parks, wildlife conservation areas, and sanctuaries under its jurisdiction. According to the current practice KWS gives prior informed consent (PIC) for access to wild genetic resources (in and outside protected areas), which includes

[61] Watai et al. (2015), p. 116.

[62] Usually referred to in Regulations 2006 as 'Authority' which is defined in the EMCA as "the National Environment Management Authority established under section 7".

mutually agreed terms (MAT) and material transfer agreement (MTA).[63] According to the WCMA, the KWS/Cabinet Secretary for wildlife[64] (CS wildlife) must ensure that the consent of stakeholders (in territories under the jurisdiction of KWS) has been obtained before a permit is granted (s. 22 (4) (a)), as will be seen below (Sect. 4.2.2.). The WCMA ordains the CS wildlife as the granting authority for permits on bioprospecting[65] and export for bioprospecting and any other kind of research[66] who issues the permits on the advice of the Kenya Wildlife Service (s. 22 (1)).[67] The consent of the KWS is hence expressed through these permits. Accordingly, a person wishing to engage in any of the activities mentioned must obtain a permit from the CS wildlife (see details below, Sect. 4.2.2.). Besides, the CS wildlife may promulgate regulations to operationalize the ABS relevant provisions of the Act (s. 22 (8)). The Act also establishes the Wildlife Research and Training Institute (WRTI) under s. 50 and empowers it to issue research permits (s. 59 (1)).[68] The Service/KWS is responsible for maintaining a register of all permits issued under the Act (s. 87 (1) (a)).[69] These rules apply to both foreign and local persons and are available on the KWS website.[70] Collection of samples/specimens from the field is to be executed in strict adherence to KWS scientific specimen collection and depository procedures.[71]

3.3 *Kenya Forest Service (KFS)*

KFS was established by the Kenya Forest Act of 2005 as amended in 2016 and now named Forest Conservation and Management Act. The KFS is responsible for *inter alia* the conservation, protection and management of all public forests (s. 8 (a)). Unlike the previous Act,[72] the current one barely has provisions which can be interpreted as establishing any ABS role for KFS. However, according to the

---

[63] Mukonyi Kavaka Watai (Head of Bioprospecting, KWS), *pers. comms*, 20 September 2017. The KWS was supposedly responsible for granting access permits to the country's genetic resources prior to the adoption of the ABS subsidiary legislation.

[64] Under the WCMA the term 'Cabinet Secretary refers to "...the Cabinet Secretary for the time being responsible for matters relating to wildlife" (s. 3).

[65] Defined under s. 3 as "... the exploration of biodiversity for commercially valuable genetic and biochemical resources".

[66] The term 'any other kind of research' is not defined but seemingly means any research other than bioprospecting.

[67] See also s. 7 (p) which states that "The functions of the Service shall be to ... grant permits". The term 'Service' is synonymous to 'Kenya Wildlife Service' (s. 6 (1)).

[68] For research (not involving bioprospecting) and bioprospecting undertaking conducted abroad it implies that two permits are required from the KWS, research/bioprospecting and export permit.

[69] KWS also issues all CITES-related permits (s. 48 WCMA).

[70] http://www.kws.go.ke/.

[71] Mukonyi Kavaka Watai (Head of Bioprospecting, KWS), *pers. comms*, 20 September 2017.

[72] See Kamau and Winter (2009), p. 368f.

information provided on the ABS CH website concerning functions of checkpoints (see Table 2), current practice shows that it is responsible for the management of forest GR and establishes MAT with users. This makes KFS a strong stakeholder and is one of the key lead agencies. Besides, it is one of the members of the ABS permitting committee. In addition, a few provisions of the amended Act still indicate important functions which need to be considered in a streamlined ABS system else more unclarity and uncertainty of the regime is created. These include the power of the Cabinet Secretary responsible for matters relating to forestry to declare that a forest produce may not be exported (or imported) (s. 61); to make regulations *inter alia* (selection from s. 71) for formulating guidelines on ... benefit-sharing; specifying the conditions subject to which any license, permit or authorization may be granted or issued under the Act; regulating activities in public forests, provisional forests, nature reserve ...; controlling the harvesting, collection, sale of and disposal of forest produce ...; and ensuring compliance with international obligations (*which would include obligations under the CBD and NP*); and the enforcement powers given to the Chief Conservator of Forests or any authorized officer concerning checking of licenses and possession of forest resources (s. 62).

3.4 National Commission for Science, Technology and Innovation (NACOSTI)

NACOSTI is an advisory institution of the Government of Kenya on matters of national science, technology, innovation and research. It is mandated to issue authorizations/licenses for research, including on genetic/biological resources before any research or access permit can be granted.[73]

3.5 Cabinet Secretary on Matters of Culture (CS Culture)

The CS culture has different roles in regard to aTK as foreseen in the Protection of Traditional Knowledge and Expressions Act of 2016. Accordingly, the CS culture shall prescribe the form or manner in which aTK registered by any county shall be entered in the repository upon collection, documentation and transmission of information[74] thereto (s. 8 (6)); may grant a compulsory license for the exploitation of TK

[73]This is a prerequisite for a research permit to be granted by the KWS (r. 3 of draft Wildlife Regulations 2016) and an access permit by NEMA (r. 9 (2) Regulations 2016).

[74]According to Stanley Atsali (KIPI, *pers. comms*, 22 September 2017) this is quite important for the patent examination process. In his words, a patent examiner can only refuse to grant a patent if there is evidence in documentation that TK has been disclosed either orally, in writing or otherwise. What the patent office always lacks for such purposes is data and documentation.

"Where protected traditional knowledge is not being sufficiently exploited by the owner or rights holder, or where the owner or holder of rights in traditional knowledge refuses to grant licenses for exploitation . . .", *albeit* with the PIC of the owner[75] and subject to article 40 (3) (b) of the Constitution (s. 12 (1)); "shall make regulations prescribing the criteria and conditions for the grant of a compulsory license" (s. 12 (4)); shall consider and determine[76] within 60 days an application for compulsory license, reject or grant the consent and make rules to guide the determination of just compensation (s. 27 (5)–(8)); shall receive a written copy of access grants, authorizations, assignments or licenses[77] issued by the holders of TK in respect of protected traditional knowledge (s. 22 (3);[78] give advice to parties when drawing up a document to a licensing agreement if his/her advice is sought and keep a register of all licenses and assignments granted under section 22 (6) (on assignment and conclusion of licensing agreements). The CS (culture) must ensure that all holders of TK have been identified before its owners grant their consent for exploitation and use under s. 25[79] and then make a written determination of an application for consent containing the details identifying the holders as well as publish a

[75] According to s. 22 (2) the traditional knowledge belonging to a local or traditional community shall not be assigned or licensed without the authorization of the custodian of the local or traditional community. This is confusing and sounds like a contradiction. It is not clear how a compulsory license from the CS and PIC of the TK owner(s) would work simultaneously. The reference to article 40 (3) (b) of the Constitution may suggest that the CS can grant a compulsory license for the utilization of TK in the public interest as long as interested persons have a right of access to a court of law. Article 40 (3) (b) forbids the State from depriving a person of property or any interest in or right over property except for a public purpose or interest as long as prompt, full and just compensation is made and the person has the right of access to a court of law.

[76] Before determining such an application the CS must follow some rules under s. 28. The CS must first, give a copy of the application for consent to the holders of the traditional knowledge to which the application relates, second, publish a notification of the application in a newspaper with nationwide circulation stating the website and other public information centers (these are provided in ss. 95 and 119 of the County Governments Act No. 17 of 2012) where the interested persons may obtain a copy of the application, and third, record in writing the details of any written representation. This gives an opportunity to any person who claims to be a holder of the traditional knowledge to which the application relates make a representation (in writing) to the CS. This should be done within 28 days after the application is published or broadcasted, whichever is the later.

[77] These only have effect only if done in writing (s. 22 (4)).

[78] One copy of the same goes to respective county executive committee member in charge of matters relating to traditional knowledge and culture.

[79] The owners of TK have the right to grant authorization for the exploitation and use of their TK or authorize the national government, county government or any other person to exploit their traditional knowledge prior to appropriate and documented consultations with the members of the communities, in accordance with their traditional processes for decision-making and public affairs management. If the CS is not satisfied that all of the holders have been identified or that there is a dispute about ownership of the rights, the CS shall refer the matter to the parties for resolution in accordance with customary laws and practices or such other means as are agreed to by the parties and the parties shall inform the CS of their resolution once the matter is resolved (s. 30).

notification of the application in a newspaper with nationwide circulation (s. 29).[80] Besides, s/he shall make regulations prescribing the matters that should be included in an authorized user agreement and the procedures to be followed before a user agreement is entered into (s. 33 (2)). The CS (culture) shall keep a register of such agreements[81] and ensure that intellectual property rights shared between the holders and users of TK are not transferred without his/her authorization[82] (s. 35). Further, s/he may also make regulations prescribing the matters that should be included in a benefit-sharing arrangement (s. 24 (3)). The CS (culture) shall likewise receive notification from the owners of TK in writing of every authorization agreements they enter into (s. 25 (2)). In addition, the CS culture (or the person authorized under s. 25 (1) (b))[83] shall transfer the benefits arising from the use of TK directly to the relevant communities—"... after consultations with the person in charge of matters relating to traditional knowledge and culture in the relevant county" (s. 25 (3) (d)). It is not clear to which fund the benefits are paid, or how the said persons should transfer non-monetary benefits.

S. 43 gives the CS culture leeway to make regulations in consultation with the Council of County Governors for the better carrying into effect of the Act. S. 18 (4) likewise asks the national government in consultation with county governments to "... establish mechanisms that enable the communities to prevent the misappropriation, misuse or unlawful access and exploitation of traditional knowledge ... without prior consent". As at the publication of this chapter none of these exist(ed).

3.6 *Others*

Other important institutions are KALRO, which executes the SMTA for PGRFA under Annex I of the ITPGRFA including *ex-situ* genetic resources of the same under the National Gene Bank, and NMK which is mandated with the management of cultural heritage, natural history collections and GR obtained from monuments and heritage sites.[84] Any material collected in any herbarium such as NMK is to be

[80] In the notification the CS must state the website and other public information centres provided in sections 95 and 119 of the County Governments Act No. 17 of 2012 where the interested persons may obtain a copy of the application.

[81] Agreements not registered by the CS are null and void.

[82] Such intentions must be communicated to the CS in writing or in such a manner as the CS may prescribe. The CS may grant approval for transfer of such rights after consultation with the relevant communities subject to the terms and conditions it deems fit, including imposition of charges by way of royalties or may for reasons recorded in writing reject the application.

[83] Accordingly, the holders of TK may authorize the national or county government or any other person to exploit their TK on their behalf.

[84] NEMA (2014b), p. 23.

accessed and/or transferred subject to the execution of a MTA in compliance with national ABS laws of PIC, MAT and MTA.[85]

4 Application Procedures and Nature of Regulatory Tools for Access to *In-Situ* and *Ex-Situ* GR and aTK

4.1 What Is 'Access'?

Before going into this question it is critical to understand what the term 'access' means in the Kenyan regime. It is defined under r. 2 of Regulations 2006 and "means obtaining, possessing and using genetic resources conserved, whether derived products and, where applicable, intangible components, for purposes of research, bio-prospecting, conservation, industrial application or commercial use". This definition covers a wide range of activities from obtaining (which should be understood as first-hand acquisition including through *in-situ* collection or from an *ex-situ* collection or parataxonomists), possessing (maybe through e.g. transfer, or buying) to using (which most likely has a similar meaning as 'utilization') GR, derivatives or intangible components.[86] As long as the GR is meant for the purposes listed these activities are to be considered as 'access' and would fall under the ABS regime. In implication, if one buys an avocado of Kenyan origin from a market in Germany and intends to conduct R&D on it, his activities will be categorized as access. The term 'genetic resources' is not defined but the Regulations adopt the CBD definition of 'genetic material'. Hence, it is assumed that 'genetic resources' means genetic material of plant, animal, microbial or other origin containing functional units of heredity and having actual or potential value. The Regulations therefore do not address Digital Sequence Information (DSI) as the only reference to information is in regard to intangible components, *albeit* the permitting requirements of the new practice for access to biological resources includes DSI under scope.[87]

4.2 Pre-Conditions for Access

For an application of access to succeed at NEMA it must fulfil the following conditions:

[85] Mukonyi Kavaka Watai (Head of Bioprospecting, KWS), *pers. comms*, 20 September 2017.

[86] 'Intangible component' means any information held by persons that is associated with or regarding genetic resources within the jurisdiction of Kenya'.

[87] *Pers. comms*, Watai, 28 July 2020.

1. It must be made according to the form set out in the First Schedule, be submitted in ten hard copies and one soft copy and be accompanied by the fees prescribed in the Second Schedule to the Regulations (r. 9 (1); and
2. It must be accompanied by evidence of PIC from interested persons and relevant LAs, and a research clearance certificate/license from the National Council for Science and Technology (currently known as National Commission for Science, Technology and Innovation, NACOSTI) (r, 9 (2)).[88]

4.2.1 Form of Application

The First Schedule is divided into part I which must be filled by applicants applying as individuals, part II for corporates/organizations and part III for all applicants. Part I and part II contain the contacts of the applicants including their names, identification, postal address, qualifications etc. Part II contains information on financial details (budget of the project and details of corporate or individual sponsors), technical particulars and renewal details (whether the application is for renewal of an access permit and if yes the permit numbers and when granted). We shall mention a few elements under technical particulars. One, the applicant is required to declare the details of previous collection/research (if any) not only conducted in Kenya, but also in any of the East African countries of Uganda and Tanzania. This is vital because it can be a ground for establishing transboundary cooperation in compliance with article 11 NP. Further, the applicant must provide information about the GR sought e.g. the scientific names of taxa, parts to be collected or whether access is sought to derivatives and/or products, and in case the GR is held *ex-situ*, the details of the relevant depository institution(s). Besides, information on any known or proposed uses of the GR and expected results must be provided. Further, the identification of the provider(s) of the GR for which access is sought must be provided and information whether Kenyan nationals or institutions will be involved in the undertaking and/or if assistance to identify and access the GR described in the application would be required, and if yes, details of the assistance required. Finally, the applicant must provide details of any royalties, payments and/or other compensation that s/he offers for access to the GR. The application must be signed (first schedule). Together with the application the applicant must provide research clearance from the National Commission for Science, Technology and Innovation (NACOSTI) and proof of PIC from interested persons and relevant LAs (r. 9 (2), see below)).[89] Finally, the applicant must pay a fee of Kshs 20,000 for individual

[88] According to the draft bioprospecting regulations access (and transfer) to any wildlife genetic resources within and outside the country for bioprospecting purposes can only be done subject to the conclusion of a MTA between the user and the Service (r. 14). That means the application for an access permit at NEMA must be accompanied by proof of a concluded MTA, which, of course, is logical as the existence of a MTA is a prerequisite for the obtainment of a bioprospecting permit from the KWS (s. 22 WCMA).

[89] Same, Dr. Benson Mburu (NACOSTI), *pers. comms*, 26 September 2017.

applicants and Kshs 50,000 for corporate applicants as prescribed in the Second Schedule[90] without differentiation of commercial or non-commercial purposes.

4.2.2 PICs, Permits/Licenses and Clearances

Consents of Lead (State) Agencies

(a) Research and bioprospecting permits for access in nature reserves and parks under the jurisdiction of the Kenya Wildlife Service

According to the WCMA it is understood that a permit granted is an exhibition of the PIC of KWS. Rules on research permits are established in s. 59 of the WCMA. Accordingly, permits for research on wild resources are granted by the Wildlife Research Training Institute (WRTI or Institute) established under s. 50 of the Act. If research is being conducted outside Kenya the applicant must have a sponsoring institution from the home country and a locally recognized collaborating institution which shall guarantee that the researcher complies with the requirements under the Act. It is also a condition to deposit a copy of the research report, thesis or assessment with the Institute in a manner prescribed by the Institute upon completion of research (s. 59 (2)). Currently there are no rules in force to elaborate these provisions e.g. concerning procedures, timelines and application review process but s. 59 (4) mandates the CS on matters of wildlife to make regulations, on his own motion or at the recommendation of the Board of the Institute, to carry out the relevant purposes. According to the draft regulations an application must be submitted to the Institute through the form A prescribed in the first schedule to the regulations (r. 5 (1)).[91] It must be accompanied with the documents shown in the second schedule ((r. 5 (2)) and the fees prescribed in the fourth schedule (r. 6).[92] The Institute has 60 days to review the application and announce the outcome to the applicant (r. 5 (3)). They contain a list of reasons why a permit may be denied (r. 8). If after review the proposed research is considered to be for commercial purposes the application shall be subject to the bioprospecting regulations (r. 5 (4). If the permit is issued it shall remain valid for the period specified in the permit (r. 11 (1)). An application for renewal can be made to the Institute using the same procedure described above and the Institute in reviewing will take into account a number of

[90] According to the stand of the law when this chapter was written. The currency exchange rate as at 18 November 2020 is 1 USD = ~109 KES (1 KES = 0,00914135 USD).

[91] The form contains details on the applicant, species and research project. It also contains a list of other documents to be checked the details of which are listed in the second schedule. Form B of the same schedule contain the form of the permit, which is to be signed by the Director General of KWS.

[92] The current draft erroneously (as at 28 May 2020) refers to the third schedule for prescribed fees. The Institute may waiver the fees on application of the applicant if the permit relates to research that may support public benefit (r. 7 (1)) or grant the applicant time of up to a maximum of 6 months from the date of issue of the permit to pay the whole or any part of the fee for any permit issued pursuant to the Regulations (r. 7 (2)). As at the date mentioned the fees have not been specified.

issues[93] (r. 11 (2), (3)). The Institute can also give direction on variation of the permit (r. 12) and can also revoke it on reasons listed under r. 13.[94] The decision to revoke a permit must be communicated to the applicant within 30 days after which the applicant also has a right to show why the permit should not be revoked (r. 13).

Permits for bioprospecting or any other kind of research on wildlife resources are issued by the CS on matters of wildlife on the advice of the Service (KWS) (s. 22 (1)).[95] Applications must be addressed to the 'Authority'[96] in a prescribed format and on payment of prescribed fees (s. 22 (2)) and must be submitted 3 months before the intended date of the start of research project with the following documents attached: (1) A comprehensive research project proposal; (2) A letter of support from the institution of affiliation/sponsorship; and (3) The applicant's CV. In addition, foreigners must submit: (1) Copies of valid passport; (2) Copies of a visa or resident permit from the Department of Immigration; and (3) Any other relevant documents. The ensuing procedure after the application has been submitted to the Authority is detailed under s. 22 (3)-(6). Accordingly, the Authority must ensure that any interests of stakeholders, i.e. a person[97] or community,[98] are protected. Where a stakeholder has interest the Service/Authority upon approval by the CS may issue a permit only if the applicant has disclosed all material information relating to the relevant bioprospecting to the stakeholder and, on the basis of that disclosure, has obtained PIC of the stakeholder(s) for the provision of access to or use of discoveries about such resources. Besides, the applicant and the stakeholder must have concluded the relevant agreements: MTA and benefit-sharing agreement (BSA) for a person, and BSA for a community. In addition, the Service must be a joint partner in all bioprospecting involving any wildlife resources on behalf of the people of Kenya.

It is not clear according to which timelines the Service and CS wildlife have to review the application and grant a permit or which specific information regarding

---

[93] According to r. 11 (3) the things to be taken into consideration are: (1) the reason necessitating an extension; (2) the likely impact of the extension on the affected species; (3) the benefit of the extension to the applicant, the species, community and human life; and (4) any other factor the Institute may consider necessary.

[94] These include, *inter alia*, (1) fundamental breach of the terms of the permit; (2) use of the animal for unauthorized purposes; (3) if the animals in the permit holders custody are in danger due to neglect and unlawful use; and (4) if the Institute determines that it is in the interest of the long-term conservation of the species that the permit be revoked.

[95] The KWS also issues import and re-export permits on wildlife species (s. 85 (1) WCMA).

[96] According to its usage in s. 22 the term 'Authority' clearly refers to the Service/KWS as it is the one mandated to review applications and grant permits for bioprospecting and export for bioprospecting and any other kind of research. However, it is confusing how the Act uses it interchangeably with 'Service' without defining it or indicating its usage in that effect.

[97] Including a public or private organization or institution or community, providing or giving access to the wildlife resource to which application relates.

[98] I.e. a community (i) whose traditional uses of the wildlife resources to which the application relates have initiated or will contribute to or form part of the proposed bioprospecting; and (ii) whose knowledge of or discoveries about the wildlife resource to which the application relates are to be used for the proposed bioprospecting.

some of the application documents (e.g. details of research project) must be submitted. The Act, however, gives the CS wildlife power, on recommendation of the Service and by notice in the Gazette, to promulgate rules, regulations and guidelines as are necessary and appropriate in order to operationalize the provisions of the Act concerning permits (s. 22 (8)). It is expected that the details and specifics of relevant issues will be dealt with in these rules, regulations and guidelines. Currently only a draft of the regulations exists. Accordingly, an application for a permit must be made to the Authority using the form A in the first schedule (r. 7 (2)).[99] The applicant must include a PIC (in form C of the first schedule) executed with the Service (r. 15). The Service will have 60 days from the day of receipt of the application to review it and communicate its decision in writing to the applicant (r. 9 (1). If the application is rejected, i.e. the permit is denied, the Service shall justify its decision (to reject), state recommendations and give the applicant a further 30 days to submit a revised/new application (r. 9 (2)). If granted the permit shall be in form B of the first schedule (r. 9 (3)).[100] Its validity would be as indicated in the permit and can be renewed upon expiry following the same procedure of application of the initial permit, and upon provision of justifications for renewal,[101] which the Service must consider while reviewing the renewal application (r. 11). It can also be varied with recommendations from the Service to effect compliance with the terms of the permit (r. 12). The permit can be revoked for similar reasons as those used in revoking a research permit[102] upon communication to the applicant by the Service as well as giving the applicant 30 days to defend his permit (r. 13 (1)).

(b) Permit by the Kenya Forest Service for entry into public forests and harvesting/ collection of forest resources

Entry into public forests and harvesting/collecting of forest resources (include flora, fauna and microorganisms, s. 2) is only permitted with a permit/license from KFS (s. 64 (a), (b); Table 2). According to FCMA, KFS shall "receive and consider applications for licenses or permits in relation to forest resources . . ." (s. 8 (c)). The amended Act does not lay down a procedure for application of permits/licenses. As already said, there is no actual ABS procedure under the FMCA. Generally there is confusion regarding the mandates of KWS and KFS concerning wild resources under the jurisdiction of the latter. The WCMA puts all wildlife resources under its purview both concerning access for research and bioprospecting as well as export. On the other hand the FMCA requires a permit from KFS for harvesting/collection of forest produce (includes leaves, moss, flowers, soil, plants, seeds, sap, water etc.,

[99] The contents of form A of the first schedule of the application for a bioprospecting permit in the bioprospecting regulations are similar to the content of form A of the first schedule of the application for a research permit in the wildlife research regulations.

[100] The form of the permit for a bioprospecting (form B) in the first schedule of the bioprospecting regulations is similar to the form of the research (form B) in the first schedule of the wildlife research regulations.

[101] The justifications needed are similar to those of the renewal of a research permit.

[102] See fn 94 above.

s. 2) as seen in this section, and for export of such produce from Kenya (s. 60 (1)). Although the FMCA is so much focused on trade in forest resources, there is no indication that it excludes research and bioprospecting activities. This makes the procedure vague, a situation which needs to be resolved.

(c) Research authorization/license (or clearance) from NACOSTI

A research authorization/license will be issued in the name of the applicant: the project, expedition or institution leader. An appropriate fee must be paid for each research project. An institutional/company authorization/license should not exceed three researchers. The holder of the research authorization/license will be held responsible for making sure that the authorization/license conditions are observed.[103] If an undertaking involves collaboration with a foreign public or private institution the Kenyan counterpart is required to submit to NACOSTI details of the nature of the collaboration.

For application of a research authorization/license or clearance permit the following steps must be taken:[104] Step 1: Register online for free at oris.nacosti.go.ke; Step 2: Follow the instructions; Step 3: Submit a fully completed application together with research proposal (which must outline in detail the nature of the project, also in order for the institution to inform whether other licenses are needed[105]), CV, copy of ID/passport, letter of affiliation to a local institution and others as appropriate (a checklist is provided at the end); and Step 4: Upload the application as one file. After the application has been submitted/uploaded the following procedure ensues: (1) The Research Coordination Division forwards the application to the experts of the relevant science schedule for assessment and advice; (2) The relevant science schedule sends comments to the Research Coordination Division following the review of the application; and (3) The Research Coordination Division issues the research permit or informs the applicant to provide further information. NACOSTI issues the license to local researchers for research taking place in the country within 7 days and 14–30 days for collaborative research involving foreigners and local researchers taking place abroad.[106]

The researcher is obliged to submit two copies of research reports to NACOSTI and upload a pdf copy into its system upon completion and finalization of research work failure to which no authorization/license can be issued for subsequent research work.[107]

(d) Prior informed consent of *ex-situ* collections

These are public *ex-situ* conservation facilities in which prospecting of biological and genetic resources is undertaken, e.g. National Gene Bank, Zoos and Botanical

[103] Dr. Benson Mburu (NACOSTI), *pers. comms*, 26 September 2017.

[104] See Biological Authorization Procedures in Kenya; NEMA (2014b), pp. 22, 47.

[105] Dr. Benson Mburu (NACOSTI), *pers. comms*, 26 September 2017.

[106] NEMA (2014b), p. 47.

[107] Biological Authorization Procedures, above fn 104.

Gardens.[108] However, there are also private collections such as International Agricultural Research Centres (IARCs), Consultative Group of International Agricultural Research (CGIARs) and Non-Governmental Organizations (NGOs) which, like with public collections, the relevant LA will be involved in the PIC process and would issue a letter of support.[109] *Ex-situ* resource providers must aspire to disclose the original source of the material or resource as part of the negotiations.[110]

(e) Others clearances

In addition to the above a researcher hoping to access biological genetic/resources from Kenya must apply for a research license/pass from the Department of Immigration (DoI) prior to the application of a visa.[111] The pass, which is valid for 12 months can only be granted subject to the exhibition of the research license from NACOSTI and affiliation with a local institution.[112] It will also be required for application for an access permit at NEMA.[113]

Prior Informed Consent by Local Community

The PIC of local communities would include the obtainment of the right to access components of biological diversity in territories occupied by such communities, mainly native territories, which, as already said, is given together with the KWS. In regard to aTK the Protection of Traditional Knowledge and Cultural Expressions Act of 2016[114] establishes a PIC requirement prior to access which must be obtained from the owner or holder of the knowledge (s. 36 (1), (4)).[115] Accordingly, every community[116] has the right to: make and adopt its rules prescribing procedures for authorization of exploitation of its TK; authorize exploitation based on appropriate and documented consultations with the members of the communities and according to community rules, procedures and traditional processes for decision-making and

[108] NEMA (2014b), p. 32.

[109] Ibid.

[110] Ibid.

[111] NEMA (2014b), p. 47.

[112] Ibid.

[113] Ibid.

[114] See the Act at http://kenyalaw.org/kl/fileadmin/pdfdownloads/Acts/ProtectionofTraditionalKnowledgeandCulturalExpressionsAct_No33of2016.pdf (accessed 02 December 2017).

[115] See also s. 22 (3), (4) WCMA.

[116] The term 'community' is defined by TK Act under s. 2 and means "a homogeneous and consciously distinct group of the people who share any of the following attributes - (a) common ancestry; (b) similar culture or unique mode of livelihood or language; (c) geographical space; (d) ecological space; or (e) community of interest." Although the adjective 'local' is not used in the definition in other references of the Act it is understood that it refers to such communities. In its definitions of 'owner' and 'traditional knowledge', for example, it uses both the terms 'local' and 'traditional' communities. In the definition 'owner' it also explains that such are communities in whom certain traditional rights are entrusted "in accordance with the customary law and practices of that community".

public affairs management; prevent any person from doing so without their PIC; and, in addition to other remedies and actions available, institute legal proceedings against any person who exploits it without the owner's permission/PIC (ss. 10, 25). The prospective user may obtain PIC directly from the holder(s) of TK and then inform the CS culture accordingly (s. 36 (1), (2)). PIC to the proposed use is considered to have been obtained once the owner/holder of aTK and the prospective user have entered into a so-called 'authorized user agreement' (s. 36 (4)). The user must provide the CS culture with a copy of the proposed authorized user agreement for comment and advice as well as with a signed copy within 30 days of the agreement coming into force for registration to avoid it becoming null and void (ss. 36 (2), (3) (5); 35 (2)). The owners of TK shall also notify the CS culture in writing of every authorization agreements they enter into (s. 25 (2)). An authorized user agreement must provide all the matters provided for under s. 34.[117]

In addition, community protocols and procedures are being developed which must be adhered to once they are in place. In August 2019 the Biocultural Protocol (BCP) of the Endorois community was launched[118] and is published on the community's website.[119] It establishes a PIC requirement for any project taking place within the community's lands whether involving resources or aTK.[120] No agreements (BSAs, MTAs) may be entered into without prior consultation of/with the community and its participation in decision-making with full understanding of all details concerning the project.[121] The relevant LA will normally be involved during negotiations for PIC as a practice and as indicated in the WCMA (s. 22).

Prior Informed Consent of Private Landowners

In the same vein, private landowners participate by giving PIC for entry into or access to GR from their land. Such entities include individuals or corporates owning private land and practicing conservation of wildlife and other forms of biological resources.[122] They are referred to as private resource managers e.g. in private wildlife ranches such as Lewa Conservancy where bioprospecting is undertaken.[123] During negotiation for PIC, the relevant LA is involved to issue a letter of support.[124] There is no information concerning the format of negotiations for MAT or benefit-sharing but it is assumed that this is left to the parties.

---

[117] s. 34 lists the terms and conditions of a user agreement.

[118] See https://naturaljustice.org/the-endorois-community-launch-their-biocultural-community-protocol/ (accessed 15 June 2020).

[119] Endorois Peoples' Biocultural Protocol. According to Cicilia Githaiga (*pers. comms*, 17 June 2020), the BCP has legal force being part of the schedule of Lake Bogoria Management Plan.

[120] Ibid, p. 27f.

[121] Ibid, p. 28.

[122] NEMA (2014b), p. 32.

[123] Ibid.

[124] Ibid.

4.3 Review of Access Application

Once an access application is complete it is submitted to NEMA for review. The application review process is described under rr. 10, 11 and 13 of Regulations 2006. Accordingly, upon receipt of the application, NEMA carries out a formal examination of the application to ascertain that it meets all the formal requirements including the attachments. If the application does not meet formal requirements, the applicant will be notified to remedy the application within a specified period of time. If all the requirements have been met, NEMA publishes the application in the Kenya Gazette and in at least one newspaper with nationwide circulation or in such other manner as it may consider appropriate. This is to invite public comments on the application for the access permit and to encourage the public to participate in the management, protection and conservation of GR. The publication specifies the name and other particulars of the applicant, the activity to be undertaken for which the access permit is required,[125] and the time within which representations or objections in respect of the proposed access permit may be made to NEMA.

In line with r. 11 (1) and (2), once NEMA has received public representations or objections, it "shall review the application and if satisfied that the activity to be carried out shall facilitate the sustainable management and utilization of genetic resources for the benefit of the people of Kenya, issue an access permit to the applicant". Where reasonable grounds exist to deny the permit, NEMA shall inform the applicant of such reasons.

The Regulations do not give a clue concerning the role played by public presentations and/or objections. For example, is public opinion solicited again in case of objections, or NEMA just acts upon its convictions once it is of the opinion that the activity shall facilitate the sustainable management and utilization of genetic resources? Would NEMA's action qualify the undertaking as of benefit to the people of Kenya and therefore legal *ipso facto* even if the Kenyan people have opposed it? Only a slight change to the new practice seems to indulge minimally into these questions, but it does not clear them. Accordingly, NEMA does not review the application alone but with the help of an ABS Technical Committee (ABS TC). The multi-stakeholder committee is supposed to ensure a broad consultation, range of expertise, perspectives and opinions on the proposed project in order to ascertain that decisions made are fairly inclusive and objective.[126] It is charged with the functions to determine applications for access permits; ensure that they are in the national interest; monitor compliance with agreements and permit conditions; and assist in

[125] At the request of an applicant of an access permit the Authority (NEMA) may hold some information relating to access to genetic resources which is the subject of the application as confidential (r. 21 (1)). Such information will remain confidential even after the permit has been granted and cannot be accessible to persons inspecting the register of access permits (r. 21 (2)).

[126] Mukonyi Kavaka Watai (Head of Bioprospecting, KWS), *pers. comms*, 20 September 2017; Jane Nyandika (ABS Desk Officer, NEMA), *pers. comms*, 25 September 2017.

the implementation of the ABS Regulations.[127] The Committee is invited to determine the application once the stipulated publication period has expired. In determining the application it considers the comments from the public and other stakeholders,[128] however, the criterion for approval of the access permit remains the same: 'facilitation of sustainable management and utilization of genetic resources for the benefit of the people of Kenya'. Once more, it is still not clear how public views are assessed and whether they can tilt the decision to grant.

According to r. 11 (3) any person aggrieved with the decision of NEMA (and by practice TC) to deny the permit may appeal to the National Environment Tribunal in accordance with s. 129 of EMCA. However, the Regulations do not provide any complaint mechanism for the public if it is aggrieved by the issuing of the permit by NEMA/TC without proper consideration of its presentation and/or objections. But, of course, s. 70 of the Constitution of 2010 and the standing created by EMCA under s. 3[129] accords anyone aggrieved by this action the right to bring the matter before the court.

The permit is not transferable and is valid for a period of 1 year but may be renewed for a similar period upon payment of the fee prescribed in the second schedule[130] and upon such terms and conditions as NEMA may deem necessary to impose (r. 14). The actual problem is that the time of validity of the permit is very short and besides NEMA might impose unbearable conditions as a prerequisite for renewal, which could be a cause of legal uncertainty.

According to the Regulations, the decision to grant or reject the application for access permit is communicated in writing to the applicant within 60 days from the date of receipt of the application (r. 13). However, the formation of the TC has changed the procedure in practice and thus drastically cut the duration. One, it has "softened" the stringent procedures of the various LAs the main ones of which are part of the TC. Second, it eliminated the requirement of advertising the application in a local newspaper and instead makes notifications by asking for informed comments.[131] This has cut the duration to 3 weeks.[132] Although there is still no study to examine how the latter practice might have impacted on public opinion there is an identifiable progress in terms of the durations of the timelines for deciding applications and issuing access permits.

According to r. 17 of Regulations 2006 NEMA "shall keep, manage and update a register of all access permits which it has granted" which it shall make publicly

[127] Watai, ibid.

[128] Ibid.

[129] For details see Kamau (2005), p. 239f.

[130] The fee for renewal of an access permit is Kshs 10,000 for individual applicants and Kshs 25,000 for corporate applicants (as at 08 May 2019).

[131] Mukonyi Kavaka Watai (Head of Bioprospecting, KWS), *pers. comms*, 20 September 2017.

[132] Ibid. This has also increased the speed of issuing of permits. According to *pers. comms* of 7 January 2021 with Joyce Imende, ABS Desk Officer, NEMA, 113 permits have been issued since the entry into force of the Nagoya Protocol (12 October 2014) compared to 35 issued between 2010 and 2014. All permits are for non-commercial purposes.

accessible in a prescribed manner. Any person wishing to peruse the register must make an application to NEMA and pay a fee of Kshs 1000 for residents and Kshs 5000 for non-residents as prescribed in the second schedule.[133]

4.4 Pre-Export Permits and Clearances

Prior to export of wildlife resources for bioprospecting or any other kind of research an export permit must be sought from the KWS according to the WCMA (s. 22). The procedure for application of an export permit is the same as the one for a bioprospecting permit (see above). To the application for an export permit must be attached a phytosanitary or an animal health certificate.[134] A phytosanitary permit (also for import)[135] is issued by KEPHIS (Kenya Plant Health Inspectorate Service) and animal health clearance certificate by the Department of Veterinary Services (DVS).

From the said it is obvious that the access permitting system under the Kenyan regime is complex and hence we summarize it beneath (Table 1).

5 Common Content of PIC and Permit for Access to GR

5.1 Common Content of PIC

The current Regulations 2006 do not contain forms of a PIC. However, a process to review them, which seems to have stalled, had suggested a draft to be included as a schedule in the version of the proposed legislation. This draft has been included in

[133] As at 08 May 2019.

[134] NEMA (2014b), p. 53f.

[135] This is to ensure that all plant materials and/or plant products for export meet the recommended quality standards and are free from pests and peat or any foreign material. In order to certify this, inspection is usually done during active growth of plants destined for export market and at exit points. Samples are inspected according to the type and volume of the commodity and inspection levels are determined by the assessed risk of the commodity. A phytosanitary certificate is only issued once it has been certified that materials are free from any infections. Inspection involves the following steps: (1) Document verification (i.e. invoice, plant import permit detailing condition for importation, quantity allowed, validity of the permit, plant species allowed for export); (2) Produce presentation for inspection in the ware house; (3) Sampling; and (4) Inspection and documentation (issuance of phytosanitary certificate). At the exit points, the following standards are checked: (1) Freedom from pests/pest damage; (2) Freedom from diseases/disease damage; (3) Maximum Residue Levels compliance; (4) Freedom from physiological disorders; (5) Properly graded (size, color, shape etc.); (6) Freedom from excessive moisture; (7) Whole (no mechanical damage); and (8) Packaging (clearly labelled, clean and well ventilated).

Table 1 Prior informed consents, permits and other clearances and mandates to grant them

Granting authority/ entity		Type of permission	Purpose
DoI		Research pass/ visa	Immigration clearance before travel for research
NACOSTI		Research license	Prior research clearance
KWS	WRTI	Research permit	Entry into and collection of GR from national parks and reserves for research
	CS wildlife	Bioprospecting permit	Entry into and collection of GR from national parks and reserves for bioprospecting
		Export permit	Export of any GR from Kenya for any R&D
KFS		PIC	Entry into and collection of GR from gazetted forests
ILCs	Owners or holders of TK	PIC	Consent for access to TK
	CS culture	Compulsory license	Authorization to access insufficiently utilized TK, or where its owner or holder refuses to grant a license for exploitation
Private persons		PIC	Consent for access to GR on private lands
Public *ex-situ* collections (National Gene Bank, Zoos, Botanical Gardens)		PIC	Consent for access to GR in relevant facilities
Private *ex-situ* collections (IARCs, CGIARs, NGOs)		PIC	Consent for access to GR in relevant facilities
NEMA		Permit	Prior access to any GR
KEPHIS		Permit	Prior export of plant GR
DVS		Permit	Prior export of animal GR

Source: *Own illustration*

the ABS Tool-Kit[136] and hence is in use in practice. It is also likely to be adopted in the revised legislation as is or with minimal changes, if it is ever adopted, or in a stand-alone ABS law.

Part A contains information of the natural or legal person seeking to access GR and/or aTK, and local partner/affiliates, if applicable, including their names and applicant's home organization if the applicant is an individual or person to contact if a corporate, ID/PP, PIN and visa No., country of origin and nationality, research permit No. and contact details. Further, information follows on the project (whether it is educational or involves commercial use, industrial application, biosprospecting, conservation etc.), the GR to be accessed (type, i.e. whether plants, animals, vertebrates, invertebrates or micro-organisms; scientific name; status of conservation), methodologies, sites and duration of collection; location of R&D and duration of analysis, if the applicant is aware of any product made of the material s/he is

[136] NEMA (2014b), Appendix I.

proposing to access or specific uses of the material by the local community where it is found, or any current industrial application of the material and whether the local community will play any role in the project with details of how it will participate. Besides, information of potential impacts of the access on the environment and socio-economics of the provider community and proposed mitigation measures are included. In addition, the potential of third party involvement and commitment of sharing the benefits according to concluded MAT showing the ways in which an audit of accrued benefits will be ensured during and after the project period are included. Finally, it shows the overall project budget and any additional information that the provider should know in order to make an informed decision in giving access to the genetic resource(s).

Part B contains details and contacts of the provider of the GR. According to the draft a provider can either be an individual, corporate, organization or community. For a community the authority to act on behalf of the community duly executed by the community's representatives and/or a resolution to that effect in case of the more organized communities, e.g. duly signed minutes, will have to be attached, and for a corporate/organization the authority/resolution to act on its behalf. Besides, a declaration to be signed by the provider stating that s/he comprehends all matters under Part A ("I/We understand/confirm") and hence consents to the access of the listed GR and/or aTK follows.

Finally, the PIC will have to be signed by the provider and user of the GR/aTK together with their witnesses.

5.2 Common Content of Access Permits

The form of the access permit issued by NEMA is found in the third schedule of the current Regulations 2006. This occupies the first page of the permit (not the one in the schedule, but the issued ones). The rear side of the permit lists the terms and conditions of an access permit. Regulations 15 (2) has listed some conditions that shall be implied in every access permit. However, NEMA has the discretion to impose such terms and conditions as it may deem necessary (r. 15 (1)). Thus, it makes sense to look at examples of such terms and conditions from already issued access permits.[137] One of the permits has 16 conditions and the other 17. The conditions will not be mentioned separately according to each access permit as some are similar but will be synchronized into one list.

[137] For the sake confidentiality the reference numbers of the applications, the registration numbers of the permits and the recipients of the two permits will be kept anonymous.

5.2.1 General Information

The permit starts by providing the reference number of the application and the registration number of the permit. Further, the applicant's details, including the names and ID/PP No. are inserted. Besides, the employer, if that is a corporate, and the address are included. In addition it is stated in accordance with which regulation the permit is granted (i.e. r. 11, see procedure above) and finally, the title of the project and what the research is planning to examine is provided.

5.2.2 Substantive Information

The description of the GR, its derivative product(s) or intangible component(s) as stated in the MTA and the amounts to be collected/obtained as well as stored in duplicates are inserted (e.g. 49 soil eDNA or 10 water eDNA samples). Further, the geographical description of the location of the GR and the place they will be exported is named. Besides it is stated that the "permit is issued subject to the Regulations and all agreements concluded pursuant to its grant, and may be suspended, cancelled or revoked should the holder breach any of those agreements and the conditions of issue and those contained in the Regulations". In addition, it contains a statement binding the holder of the permit and his agents and assignees to abide by the conditions of the permit and to promptly report to NEMA any matter that may prejudice the interests of Kenya and other parties concluded pursuant to the grant of the permit.

Finally, the permit contains the signature of the Director General of NEMA and the date of signature.

5.2.3 Permit Terms and Conditions

The validity of the permit is indicated (i.e. 12 months from the date of issuance) and a statement that it is not transferable. The GR to be accessed, the amounts and where the access shall take place are shown. Further, it requires NEMA to be provided with a list and quantity of all collected species and conditions the access of the GR and its derivatives strictly to "research and innovation action". It also requires the recipient to exercise due diligence in sample collection method (non-destructive), packaging, transport/export, laboratory analysis and disposal. It might also be required that further analysis takes place in local institutions and that the local partner is involved in the identification and preparation of the samples for export. The GR and the derivative should not be transferred to a third party without the consent of the provider of the GR and NEMA, or used for commercial or for purpose of profit without their understanding. Besides, it shows where the deposit and records of the duplicates of GR shall be kept and evidence of deposition provided to NEMA. It is required that reasonable access is ensured to all collected GR and intangible components by all Kenyan citizens locally or abroad. Likewise, the recipient must ensure that all agreements entered into are strictly adhered to with regard to the purpose of the access of the GR as well as the MTA with the providers. Where aTK of a local community has been accessed it would be required that such knowledge and its uses are well safeguarded. Quarterly reports must be provided to NEMA on the status of research, including all discoveries from the research on the GR and/or intangible components thereof and data generated. In addition, the status report on the environmental impacts of collection of the GR must be provided and the recipient must comply with the improvement orders issued by the environmental inspectors of NEMA throughout the project cycle. S/he must also abide by all relevant laws of Kenya. Finally, the recipient is required to return the remaining GR to Kenya upon completion of the intended studies and deposit it with the relevant LA and notify NEMA of the return.

In the event of a breach of the conditions provided in the permit or any contravention of the Environmental Management and Coordination CAP 387 and regulations thereunder NEMA reserves the right to suspend, cancel, revoke or vary the permit.

6 Benefit-Sharing

Regulations 2006 takes monetary and non-monetary benefits that can be shared (Part IV) from the list of the Bonn Guidelines[138] but leaves out a few types of benefits listed in the latter. They do not say, however, to which fund the benefits should be paid or establish criteria for sharing. The WCMA on the other hand gives the CS for wildlife the mandate, on the recommendation of the Service, to require that a particular per centum of the proceeds from the bioprospecting is contributed to the Wildlife Endowment Fund (WEF) (s. 22 (7)). First, the language of the Act is a bit confusing as "contributed" is not an equivalent of "paid into". Contribute can imply donate to/share with (the WEF) from another source. This statement is further weakened by s. 23, which establishes the WEF and lists the sources of monies to be paid into the fund but does not list research and bioprospecting as some of the sources. Anyway, the CS is empowered to promulgate rules, regulations and guidelines as are necessary and appropriate to carry out the purposes of these arrangements (s. 22 (8)), and in the draft bioprospecting regulations, it has been clearly stated that a holder of a bioprospecting permit shall have the duty to disclose the proceeds from the bioprospecting venture and to contribute a certain minimum[139] of the same to the WEF (r. 19). Here it does not only become clear that the term "contribute" is used synonymously with "pay into", but also that the WEF shall be the fund into which benefits from bioprospecting are paid. However, the

[138] See Bonn Guidelines, Appendix II. This list has also been annexed to the Nagoya Protocol.

[139] The draft regulations foresee a minimum of 10%.

measurement of the contribution remains too vague ("a certain minimum") and can cause legal uncertainty. Likewise the draft bioprospecting regulations adopt almost verbatim the monetary and non-monetary benefits that can be shared (r. 21) from the list of Regulations 2006 and propose a benefit-sharing plan in the second schedule with types of benefits[140] and level of contribution, and how much of the contribution goes to the national government, the county government and to the communities. It is strange though that the draft regulations provide a benefit-sharing plan for benefits derived from the utilization of aTK and proposes the sharing of the same not only with communities, but also with the national and county governments.[141] Except where no ownership of aTK by indigenous and local communities was established as foreseen in s. 31 of the TK Act,—a question the (draft) regulations need to be clear with—the WCMA and regulations thereto have no business in the question on how such benefits should be shared. This should be a prerogative of the communities to be dealt with exclusively under the TK Act.

By establishing the WEF the WCMA makes a good step, but a number of questions remain: (1) Since the regulations are not in force how can the current vacuum be filled? (2) Into which benefit-sharing fund are proceeds that do not fall under the ambit of the WCMA to be paid, and how should they be shared? We are of the opinion that these questions could be solved by using The Natural Resources (Benefit Sharing) Bill of 23 October 2018 as the overarching national law for collecting and sharing proceeds from the exploitation and utilization of all natural resources, including biodiversity. The Bill was enacted following numerous discoveries of various mineral resources in Kenya such as oil, and the increased need to explore mineral deposits ranging from traces of natural gases, coal and other rare earth minerals, large aquifer of water etc. With very little or no benefits reaching the local communities, the Bill aimed at, *inter alia*, establishing a system of benefit-sharing in resource exploitation between resource exploiters, the national government, county governments and local communities. The Bill established a Natural Resources Royalties Fund to be managed by the Commission (s. 14. (1)) on Revenue Allocation established under Article 215 (1) of the Constitution of 2010. The Commission is mandated, *inter alia*, to determine and review the allocation of funds (s. 6 (1)), except "Where a written law prescribes the royalty, fees, payments or benefit-sharing in a particular natural resource sector" (s. 6 (3)). Where it prescribes, such a law shall apply with respect to the relevant sector (s. 6 (3)). There was no law prescribing benefit shares from utilization of GR at the time the Bill was gazetted, and, again, none exists so far as the regulations are still in draft form. Therefore, the Bill should have seized the opportunity to regulate this question by bringing the revenue from utilization of GR under its ambit.

Whichever the case, a flow of some benefits has been experienced. At the time the Interim National Report was submitted to the Access and Benefit-Sharing

[140] The types of benefits are listed under five categories: research fund, royalties, milestone payments, taxes and fees.

[141] Second Schedule, draft bioprospecting regulations.

Clearing-House (ABS CH) an amount of USD 171,407.51 in monetary benefits was reported to have been generated through permitting fees, royalties, upfront payments and trainings excluding research grants.[142] Besides, non-monetary benefits had been received in form of technology transfer, e.g. equipment, infrastructure, and training through long-term PhD, masters and diploma programs as well as short-term exchange programs, seminars and workshops.[143] Likewise, benefits from utilization of TK may be monetary and non-monetary as agreed by the parties in the authorized user agreement and may include fees, royalties, sharing of intellectual property rights (IPRs) etc.[144]

7 Model Agreements

7.1 MTA as a Conditio Sine Qua Non *for Access and Transfer*

According to Regulations 2006 GR can only be transferred outside Kenya subject to the execution of a material transfer agreement (r. 18). Kenya operates a model (referred to as "sample") MTA but does not have a model BSA. No executed BSA could also be found. The transfer of GR abroad is only permitted prior to the execution of a MTA.[145] The model/sample MTA can be found attached to the ABS Tool-Kit published in 2014 by NEMA as appendix III.

Unlike the SA MTA the Kenyan MTA must not be entered into as is and may be varied. It is structured as follows: Preamble (s. 1); Parties to the Agreement which contains their names and contact details (s. 2); Purpose and objectives showing the type of the research whether non-commercial or commercial, the type of GR and its source, documentary evidence of PIC, MAT and deposition of duplicate specimen in a designated repository and if aTK is involved, its source (s. 3); Rights and obligations of providers and recipients (s. 4); Right of the Government of Kenya to require the repatriation of the GR as well as the obligation of the recipient to use the GR and/or aTK for the purposes foreseen in the MTA and/or "continue to keep the genetic resource in safe custody in accordance with standard procedures and practice" (s. 5); Termination of Agreement (s. 6); Warranty (s. 7); Applicable laws (s. 8); Force majeure (s. 10); and Notices indicating how any notice should be served, i.e. "delivered by hand or sent by registered mail or by international courier service to be served at the addresses" provided in the MTA (s.11). Finally, the MTA

[142] See https://absch.cbd.int/countries/KE (accessed 9 June 2020).

[143] Ibid.

[144] See s. 34 Protection of Traditional Knowledge and Cultural Expressions Act, 2016.

[145] According to r. 18 of Regulations 2006 ". . .no person shall transfer any genetic resources outside Kenya unless such person has executed a Material Transfer Agreement".

must be signed by the provider, the recipient and their witnesses and the competent authority. Documents to be included in the MTA (e.g. list of GR and GPS, letters of evidence of deposition of the material(s) and relevant permits) must be annexed thereto.

In the following we shall only make reference to a few issues on rights and duties of parties which are interesting.

7.2 Rights and Duties

7.2.1 Intellectual Property Rights

Any IPRs for inventions arising from the utilization of GR accessed according to the MTA shall be jointly owned by the recipient and the provider (s. 4.2.vii.).

7.2.2 Change of Intent/Use

The recipient is permitted to use the GR only for the purpose stated in the agreement (s. 5.2.i.). Thus, no commercialization may take place prior to notification of and negotiation of an agreement with the provider (s. 5.2.iv.). If the recipient violated this rule or any person connected to or acting under him/her (e.g. servants, agent) s/he would be required to pay 50% of the gross income arising from the GR whether the violation is through his/her act or omission (s. 5.2.v.). In the event of commercialization, the provider shall give a separate consent and the provider shall become the duly recognized supplier of the GR (s. 5.2.viii.).

7.2.3 Third Party Transfer

The recipient is not permitted to transfer any GR obtained according to the MTA to a third party except with prior written authorization of the provider and the conclusion of a MTA between him/her and the third party (s. 4.2.ix.). The recipient and the provider must inform the designated national authority of any material transfer to a third party (s. 4 a)).

7.2.4 Non-disclosure

The MTA forbids any disclosure of TK, information and data disclosed and or generated during access to the GR to third parties without the consent of the provider (s. 4 b)). In the same vein both the provider and recipient are prohibited from disclosing confidential or proprietary GR information unless such information is in the public domain or if the disclosure is made in public interest (s. 4 c)).

8 Compliance Measures and Sanctions for Non-Compliance

There are no user compliance measures yet in place. On the provider side, Kenya has a number of measures to enforce compliance. These include the following:

- Participation in the collection of genetic resources by government officials, e.g. wardens, or by local communities;
- Establishment of different checkpoints as can be seen below (Table 2);
- Requirement that a foreign institution where research takes place, or which sponsors the researcher takes over the monitoring function as well as assumes liability (s. 59 WCMA). Also this condition will be included in the PIC conditions and concluded agreements—it deserves special attention. This kind of measure is known from the Ethiopian legislation and is relatively new in the Kenyan legislation, the WCMA;
- Police controls, including road controls, and customs checks at exit points;[146]
- Use of local communities as informers of government agencies as well as enforcers e.g. as foreseen under s. 41 (b) WCMA where one of the functions of wildlife community associations and individuals (registered wildlife managers) is to assist the Service in combating illegal activities. Also according to s. 49 (1) (d) FCMA a forest community association shall "assist the Service or any other relevant authority in enforcing the provisions of this Act including in relation to illegal harvesting of forest products"; and
- Sanctions established by legislations for specific offences. Some legislations e.g. Regulations 2006 are very general and regard a contravention or failure to comply with any of the matters provided in the legislation as an offence (r. 23) while others are more concrete concerning the type of offence (see Addendum 1).

9 Regime's Strengths, Weaknesses and Assessment

9.1 Concerning Compliance with the Nagoya Protocol

First and foremost it is important that there is an ABS regime in place. Under the regime national authorities are established and their mandates defined in line with article 13 NP. The measures and contacts of the National Focal Point and the Competent National Authority have been notified to the ABS CH of the CBD as required by article 14 §2 (a) and (b) NP. It has also been specified which authority is responsible for which GR. The authorities establish pre-conditions for access to GR including PIC and MAT and lay down procedures for the same. These have been published at the ABS CH and can be judged as fair and non-arbitrary because they

[146] NEMA (2014b), p. 60.

Table 2 Checkpoints and functions

Checkpoint	Function
National Commission for Science, Technology and Innovation (NACOSTI)	Issues permits to carry out research based on genetic resources and is relevant in checking scope and conditions attached to the research license
Kenya Copyright Board (KECOBO)	Mandated to establish and maintenance of the repository on traditional knowledge and cultural expressions of communities in Kenya
Kenya Agriculture and Livestock Research Organization (KALRO)	Promote, streamline, co-ordinate and regulate research in crops, livestock, genetic resources and biotechnology in Kenya. Expedite equitable access to research information, resources and technology and promote the application of research findings and technology in the field of agriculture
Kenya Industrial Property Institute (KIPI)	Manages IPRs, disclosure of source of GR and TK and establishment of MAT in patent applications
Kenya Plant Health Inspectorate Service (KEPHIS)	Issuance of phytosanitary certificate and administration of plant breeders' rights
Kenya Forest Service (KFS)	Management of forest genetic resources and associated biodiversity. Obligations of user and provider in MAT and forest product permit
National Museums of Kenya (NMK)	Obligations of user and provider in MAT and collection export permit
Kenya Wildlife Service (KWS)	Management of wildlife genetic resources, and issuance as well as monitoring of PIC, MAT and research, bioprospecting and export permit conditions

Source: *Information on ABS CH website and own research*

apply equally to all applicants including Kenyans.[147] However, the legal language of the instruments and pre-conditions are often very broad (e.g. r. 11 (1) of Regulations 2006) and thus leaves administrators with a very wide discretion which can be used arbitrary[148] and lead to violation of article 6 §3 (b) NP. Besides, the regime is very complex and obscure and barely fulfils the requirements of article 6.3 (a) NP on legal certainty and clarity.

At least information, criteria/processes for obtaining PIC and procedures for establishing MAT in line with article 6 §3 (c), (f) and (g) NP are provided. A sample MAT has been developed and is attached to the ABS-Tool-Kit which is also accessible at the ABS CH.[149] The sample gives examples of terms that may be

[147] Except, of course, to traditional exchange of GR and uses by the local communities and to approved research activities intended for educational purposes within recognized Kenyan academic and research institutions, which are governed by relevant intellectual property laws (r. 3 (a), (d) Regulations 2006). These are exempted.

[148] For details see Kamau and Winter (2009), p. 376f.

[149] NEMA (2014b).

included in the MAT *inter alia*, terms on use(s) of GR, reporting, third party transfer, change of intent, termination of agreement and handling of GR thereafter, IPRs and benefit-sharing. The ABS Tool-Kit state how a PIC for access to GR should be applied for which includes filling in a form modelled along Appendix I.[150] These rules do not differentiate between procedures for commercial and non-commercial research and hence no facilitation measures are available for the latter as required by article 8 (a) NP although two types of researches are identified. Since access to GR might involve local communities, which might be hard to identify and with which one cannot negotiate with as a whole, the relevant lead agency (mainly KWS) will get involved in the process. This eases such issues as identification of representatives, heightens legal certainty (in that the outcome is also acknowledged by the relevant state agency) and accelerates the process. The draft revised version of Regulations 2006 proposes county environment committees as advisors of the competent authority in matters involving access to resources and knowledge held by local communities. The committees are supposed to give advice based on consultation with local communities in the counties on procedures and terms for such access. Hence, the challenge of researchers searching for or making inquiry of the rules of local communities by themselves should be surmounted if this becomes law in future. Besides, in regard to access to TK the Protection of Traditional Knowledge and Cultural Expressions Act gives guidance on how to apply for PIC of (and establish MAT with) the local communities,[151] which likewise is in line with article 7 of the Nagoya Protocol. What is lacking is information on how the PIC of individuals, e.g. (private) landowners, should be applied. As already seen the law states that such PIC may also be required for access to GR.[152]

Further, the law requires the CNA to provide a written decision to approve or reject an application. This decision must be clear indicating the grounds for rejection if a decision is made to reject. Besides, the law promotes transparency as the public have a right to inspect the register of permits and, in addition, the applicant has a right to appeal to a body/tribunal provided in law.[153] The period of time defined by the current law of 60 days for the CNA to reach a decision is reasonable. However, the procedures it creates yield unreasonable timelines. This, as already said, is being dealt with by the new practice such that the time needed to make such a decision is shorter than the one described by the law. Upon successful review of an application by NEMA/ABS TC as well as its approval, an access permit is issued. This permit comprises the evidence that PIC was obtained from and MAT established with the persons granting access to or providing GR and/aTK. Usually, NEMA would notify such permits to the ABS CH as evidenced by a number of notifications. Thus, the requirements of article 6 §3 (d) and (e) and article 14 §2 (c) NP are fulfilled.

[150] Ibid, p. 40. For common content of the model PIC in the kit see Sect. 5.1 above.

[151] See above, Sect. 3.5 and section "Prior informed consent by local community".

[152] Regulations 2006, r. 9 (1) together with Second Schedule 2.0 (g).

[153] See review of an application for access, above.

The rights of local communities over their biodiversity, genetic resources and traditional knowledge as well as intellectual property thereof are anchored in the Constitution and the state committed to protect and enhance them. Hence, they are implemented through domestic laws which establish PIC, approval and involvement, MAT, BSA and MTA requirements for access to GR and aTK from such communities.[154] These measures are in compliance with articles 7 and 6 §2 of the NP. Besides, community customary laws, protocol and procedures are recognized in the Constitution and reflected in different legislations e.g. in provisions concerning conservation and sustainable use of biodiversity or access to genetic resources and benefit-sharing. For instance, the FCMA under s. 48 recognizes community forest associations in conservation and sustainable use of forest resources. Likewise, the WCMA under s. 40 approves the establishment of community wildlife associations with functions listed under s. 41 which include *inter alia* the protection, conservation and management of wildlife conservancies and sanctuaries under their jurisdictions. Likewise, the WCMA makes a bioprospecting permit conditional on PIC of the community as well as establishment of MAT and conclusion of BSA and MTA if the community's traditional uses of the *in-situ* GR being accessed from a native territory initiated the proposed bioprospecting or will contribute thereto, or their TK will be used in the bioprospecting project (s. 22 (3) (b), (4)). The processes of obtaining PIC, establishing MAT and concluding BSA and MTA prior to grant of a bioprospecting permit in this case, or obtaining an access permit to GR held by the community as required by Regulations 2006[155] as well as resolving disputes will normally take place according to customary laws, protocols and procedures which are recognized in the PTKCEA (s. 25 (3) (c); s. 30; s. 40 (c)). None of the instruments restrict the exchange of GR and aTK amongst the local communities.[156] The relevant instruments and an ABS Tool-Kit containing templates of minimum requirements for MAT and MTA are published on the ABS CH and are therefore accessible to users. In addition, several initiatives and projects between the government, development partners and communities are going on to establish community platforms and protocols for improved resource management and to develop model clauses for

---

[154] How communities should be involved is usually defined in the MAT and MTA.

[155] First Schedule 2.0 (g) Regulations 2006.

[156] Regulations 2006 state in r. 3 (a) that "These Regulations shall not apply to - (a) the exchange of genetic resources, their derivative products, or the intangible components associated with them, carried out by members of any local Kenyan community amongst themselves and for their own consumption". The FCMA in s. 52 gives the right to take and use forest produce to any member of a forest community "as it has been the custom of that community . . . otherwise than for the purpose of sale". The WCMA is not articulate in this issue but probably that could be covered under s. 40 that states an application for registration of a community wildlife association shall show the "type of wildlife user rights being proposed", albeit if these will enhance conservation and survival of wildlife in their area. The PTKCEA also says in s. 19 (1) (a) that "the protection of traditional knowledge or cultural expressions shall not restrict or hinder the normal usage, development, exchange, dissemination and transmission of traditional knowledge or cultural expressions by members of a particular community within the traditional and in accordance with the customary law and practices of that community".

equitable benefit-sharing.[157] One of the first results is the finalization of the BCP of the Endorois community which *inter alia* talks about procedures for obtaining PIC and establishing MAT for access to resources in Lake Bogoria and the surrounding territory and the community's aTK.[158] There are not many practical examples of implementation of the rights of communities in Kenya though,[159] but existing legal measures are in line with article 12 NP.

However, there is no legislation currently implementing article 11 NP. Transboundary cooperation is important for Kenya due to the existence of enormous *in-situ* cross-border resources and indigenous and local communities who move freely across the neighboring countries and share freely the resources as well as traditions. Currently only Fist Schedule, 2.0 (a) of Regulations 2016 can be interpreted as striving towards such cooperation in East Africa. It requires an applicant for an access permit to GR to give details of his/her "… previous collection/research (if any) conducted in any of the East African country (Kenya, Uganda and/or Tanzania)". This can be a good beginning to build such cooperation. There are also other regional agreements which can be critical in this question e.g. the African Regional Intellectual Property Organization (ARIPO) and East Africa Community (EAC) agreements. According to information on the ABS CH there are efforts before the regional cooperation platforms to develop appropriate protocols on resource governance and utilization taking into considerations the requirements of Nagoya Protocol on article 11.[160] Such efforts include the Draft Policy Biodiversity Mainstreaming of the Busia county.[161]

Likewise, there are no clear rules on directing benefits to conservation and sustainable use of biodiversity according to article 9 NP. Regulations 2006 are silent on the issue of distribution of benefits from utilization of GR, including for such purposes. In this regard even the Benefit Sharing Bill 2018 fails as it does not consider benefit allocation for conservation and sustainable use purposes. Although it gives the Commission the function of identifying "… benefits to promote the conservation of natural resources" (s. 5. (1) (m)), it does not recall conservation of natural resources and their sustainable use when it talks about the revenue sharing ratios (s. 8. (1): "The revenue collected under this Act shall be shared as follows—"). It could have been wise to state here explicitly the percentage to be allocated for this purpose. In this question the wildlife Act is more innovative as it explicitly lists the functions of the WEF, which are mainly in the area of conservation and sustainable

[157] https://absch.cbd.int/countries/KE (accessed 15 June 2020).

[158] Gino Cocchiaro (Natural Justice), *pers. comms*, 6 April 2018; Cicilia W. Githaiga *pers. comms*, 17 June 2020 (both were part of the Natural Justice team which gave technical and legal support to the drafting and reviewing process of the BCP). See Endorois Peoples' Biocultural Protocol.

[159] Gino Cocchiaro (Natural Justice), *pers. comms*, 6 April 2018.

[160] https://absch.cbd.int/countries/KE (accessed 15 June 2020).

[161] Ibid.

use (s. 23 (3)). Besides, the draft bioprospecting regulations thereto try to propose a percentage of benefits to be directed to such purposes.[162]

Also, there are no compliance measures to ensure that GR and/or aTK from another party utilized within the jurisdiction of Kenya have been accessed in accordance with PIC and that MAT have been established as required by articles 15 and 16 NP. There could be measures to address non-compliance with PIC and MAT requirements of another party, e.g. criminal law measures, but one, they are not explicitly directed toward such situations and two, there are no mechanisms to detect such violation unless the authorities find out accidentally. Likewise, there have not been any instances or experience of cooperation in cases of alleged violation of domestic ABS legislation or regulatory requirements of another party. The lack of appropriate measures is due to the fact that the ABS regulations were enacted prior to the NP. There is an ongoing process to develop a comprehensive ABS legislative framework though, which hopefully will take these aspects into consideration.

Some checkpoints have been established. The ABS Tool-Kit which was published in 2014 listed 17 likely checkpoints (CPs)[163] but the list of 2016//2017 that has been notified to the ABS CH has eight in total.[164] We shall consider the latter as the true list. As can be established from their functions (Table 2), none is mandated with the functions of monitoring the utilization of genetic resources including collection of relevant information in the value chain as required by article 17 NP. Although they are relevant for the implementation of the Protocol in regard to the provider State requirements, they do not fulfil the NP standards of checkpoints.

It must be said that although the above CPs have been named (not legally established) they are not yet apt for monitoring compliance with other parties' measures when foreign GR and/or aTK is being locally utilized but only with own (Kenyan) ABS measures. In other words, Kenya does not have any *de jure* monitoring mechanisms at the moment; *albeit*, as mentioned above, some of the existing measures can also serve as such, e.g. disclosure requirement during patent application.[165] No checkpoint communiqué has been reported to show whether practice is

[162] Second Schedule, draft bioprospecting regulations.

[163] NEMA (2014b), p. 61. Also available at the CBD website at https://absch.cbd.int/countries/KE (accessed 10 June 2020).

[164] https://absch.cbd.int/countries/KE/NDB (accessed 10 June 2020).

[165] According to Stanley Atsali (KIPI, *pers. comms*, 22 September 2017), for the patent examination process to be of assistance the current Intellectual Property Act of 2001 (No. 3 of 2001) needs to be revised to include a *proviso* that if one is seeking patent protection for an invention based on TK and associated genetic resources he/she must disclosure the origin of such knowledge or resources. The only provision that one can relate to GR at the moment with difficulties is s. 29 (1) which refers to inventions concerning "…micro-biological process or the product thereof". However, Kenya supports the proposal that all patent and breeders applications utilizing GR and aTK be disclosed in patent, utility model, and breeders applications (Stanley Atsali (KIPI), *pers. comms*, 22 December 2020).

sealing this gap. The challenges in the implementation of some of the requirements of article 17, *viz* taking measures to address situations of non-compliance, again, boils down to the fact that there is no compliant post-NP regime yet.[166] What practice can afford, e.g. developing MAT and PIC which take into account the elements/terms recommended by the Protocol and encouraging users to include such in their MAT, is being done.[167] As mentioned already, an online IT permitting system is also being developed which should serve as a cost-effective communication mechanism as required by the Protocol.[168]

Finally, the question about compliance with MAT in line with article 18 NP is critical for users of GR and aTK and in achieving legal certainty. As said above, a model MTA has already been developed with relevant contents including clauses on applicable law, the jurisdiction to which the parties should subject disputes as well as alternative dispute resolution mechanisms.[169] However, if justice can be accessed with ease is a different issue. This question can be examined by looking at the past record and the changes which have taken place so far. In spite of Kenya being in possession of a number of laws meant to facilitate access to justice for all persons,[170] the functioning of the legal system in the past is said to have been below standards to the point that the general public lost confidence in the judiciary[171] and preferred/chose alternative systems of justice e.g. mediation and community-based traditional justice systems.[172] Some of the issues quoted are poor case management including slow delivery of justice,[173] judgements that people could not relate with, human rights ignorance of those at the forefront of justice delivery, including judges,[174] corruption,[175] high costs that limited accessibility to justice, understaffed and overworked courts,[176] instruments which were drafted in a manner that is too complicated for an ordinary person to understand, hardship in obtaining copies of

[166] Also implied by the Interim National Report.

[167] Ibid. See also e.g. cl. 7 of the model MTA in the ABS Tool-Kit, NEMA (2014b), p. 75.

[168] According to the Interim National Report.

[169] See cl. 15 and 16, model MTA, fn 167, p. 79f.

[170] According to articles 48, 49 and 50 of the Constitution of Kenya the provisions of laws on access to justice apply not only to Kenyans but to all persons.

[171] The Danish Institute for Human Rights (DIHR) (2011), p. 28; Kameri-Mbote and Akech (March 2011), p. 17; Waruhiu and Odhiambo Otieno (2014), p. 185.

[172] DIHR, ibid., p. 30; Wakonyo and Odhiambo Otieno, ibid.

[173] Kameri-Mbote and Akech, p. 18.

[174] DIHR, p. 28f.

[175] Fombad (2020), p. 38.

[176] DIHR, p. 29; Kameri-Mbote and Akech, p. 18. According to Kameri-Mbote and Akech, from the 4681 staff establishment for the judiciary, 554 for magistrates (judicial officials), 14 for Court of Appeal judges and 70 for High Court judges, only 1456, 277, 11 and 46 people respectively are in posts.

legislative acts, a poorly resourced judiciary,[177] interference with the judiciary by the executives,[178] court structures that do not facilitate equal access to justice, and lack of regulation to ensure that the proceedings before the courts conform to international law and constitutional standards of due process.[179] These challenges have been blamed on a lack of political will to provide quality legal aid.[180]

A number of past policy initiatives on poverty reduction, elimination and wealth creation[181] recommended that access to justice must be responsive, affordable, accessible and speedy in promoting sustainable economic development.[182] In addition, the government has made international and regional commitments[183] to enhance access to justice.[184] Although this shows the government's recognition of access to justice as a critical pillar for economic development and poverty reduction,

[177] DIHR, ibid., p. 29. See also news reports by Odula (4 November 2019). Allocation of national funds shows that the budget for the judiciary has declined since the budget year 2014/15: See Mwiti (16 August 2018).

[178] Odula, ibid.

[179] Kameri-Mbote and Akech, p. 17f.

[180] DIHR, ibid., p. 29. Also implied in Kameri-Mbote and Akech, p. 17. On the concept of legal aid see Amondi (2014), p. 203f. According to Amondi the concept of legal aid is rooted in criminal law and refers to the system of providing free or inexpensive advice about law and if necessary legal representation, to those who cannot afford them. This concept has been extended to civil matters. In *Airey v Ireland* ((1979) European Human Rights Reports 305), the European Court of Human Rights ruled that the obligation of states to ensure access to the courts includes a right to free legal assistance in civil matters when the procedure involved is sufficiently complex to require legal assistance (also Amondi, ibid.). Mrs. Airey had sought a decree of judicial separation (divorce was illegal in Ireland) from her husband, but was not able to engage a solicitor for lack of funds. As legal aid was not available in Ireland for any civil matters, her request to be provided with free counsel was declined by the trial court, as was her appeal to the highest court. She moved to the European Court of Human Rights which held that the right to fair trial included access to affordable legal assistance and hence Mrs. Airey had the right to court-appointed counsel (Amondi, p. 204). Note: 'Wanjiku' is a female name in one of the Kenyan tribes (Kikuyu) which has been used to refer to the ordinary person since its use by former President Daniel Arap Moi in the 1990s when he dismissed attempts by civil society organizations to involve people in constitution-making by rhetorically asking: "What does Wanjiku know about the Constitution?" (See Kenya Human Rights Commission (November 2010).

[181] E.g. the National Poverty Eradication Plan 1999–2015, Poverty Reduction Strategy Paper (PRSP) of 2001–2004 and the Investment Program for Economic Recovery Strategy for Wealth and Employment Creation (ERS) 2003–2007.

[182] Amondi (2014), p. 206.

[183] E.g. under the International Covenant on Civil and Political Rights (1966), which enshrines the principles of equality before the law and the presumption of innocence, and includes guarantees of freedom from arbitrary arrest and detention and the right to a fair and public hearing by a competent, independent and impartial tribunal established by law, and ACHPR Principles and Guidelines on the Right to a Fair Trial and Legal Assistance in Africa, 2003, which obligate state parties to ensure that an accused person has a right to legal assistance assigned to him or her in any case where the interest of justice so require, and without payment by the accused if he or she does not have sufficient means to pay for it, respectively (Amondi, p. 209).

[184] Amondi (2014), p. 209.

many years would pass with formation of task forces and committees to realize it with little achieved.[185]

In 2010 a new Constitution was promulgated which addresses the question of dispensation of justice in a thorough and innovative way. It gives every person under the bill of rights the right to institute court proceedings whether by himself/herself or through representation (Art. 22 (2)) and to have the matter heard and determined by court (Art. 22 (4)) in a fair trial (Art. 25 (c)) as well as to receive redress/relief (Art. 23 (3)). Besides, it gives the High Court jurisdiction "... to hear and determine applications for redress of a denial, violation or infringement of, or threat to, a right or fundamental freedom in the Bill of Rights" (Art. 23 (1)). Further the State is charged with ensuring "... access to justice for all persons and, if any fee is required, it shall be reasonable and shall not impede access to justice" (Art. 48). Further, it also states that "Every person has the right to have any dispute that can be resolved by the application of law decided in a fair and public hearing before a court, or if appropriate, another independent and impartial tribunal or body" (Art. 50 (1)). In addition, the Constitution guarantees independence for the judiciary (Art. 160)) and spells out under article 159 (2) principles which shall guide it, namely "(a) justice shall be done to all, irrespective of status; (b) justice shall not be delayed; (c) alternative forms of dispute resolution including reconciliation, mediation, arbitration and traditional dispute resolution mechanisms shall be promoted; (d) justice shall be administered without undue regard to procedural technicalities; and (e) the purpose and principles of this Constitution shall be protected and promoted". Further, the Fifth Schedule read with article 261 provides for the timeframe for the enactment of various legislations aimed at implementing the Constitution among which is legislation meant to give effect to article 50 (i.e. the right to fair hearing) that had to be enacted within 4 years from the date of promulgation of the Constitution. Since traditional dispute resolution mechanisms are critical in the area of ABS in particular due to aTK it is important to note that in the Constitution of 2010 such mechanisms received *de jure* recognition and may be applied as long as they are not used in a way that "(a) contravenes the Bill of Rights; (b) is repugnant to justice and morality or results in outcomes that are repugnant to justice or morality; or (c) is inconsistent with this Constitution or any written law" (Art. 159 (3)).

That said, the Constitution lays down a fundamental and foundational framework for a robust judiciary. If the word and spirit of the new Constitution concerning the proposed judicial standards is fully implemented, Kenya will have a legal system capable of facilitating and guaranteeing access to justice, including recognition and enforcement of foreign judgements and arbitral awards. Despite the fact that the judiciary is still experiencing hitches in its reform agenda,[186] there are signs of enormous progress since the Constitution of 2010 was promulgated, in particular as

[185] Ibid, p. 206 ff.

[186] E.g. the judiciary still bemoans insufficient budgetary allocation (see Odula, Mwiti, fn 177 above) and reluctance by the executive to appoint judges (see Kakah (undated); Otieno (8 February 2020); Yusuf (8 June 2020).

far as enactment of legislations is concerned.[187] These improvements should be able to improve all landscapes of justice, including in the area of ABS, *albeit* related issues are new to the Kenyan legal system and hence there are inadequate legal personnel in matters relating to compliance and enforcement of ABS, in addition to more complex issues that are emerging, for instance the issues of DSI and synthetic biology.

9.2 General Issues

There is no NP complaint ABS law as the old law with all its weaknesses is still in existence. The scheme to revise it did not materialize, but as already mentioned there is a new effort in this regard, which should culminate in its revision or development of a stand-alone law. Hence, it goes without saying that although the main biodiversity Acts (WMCA and FCMA) and the TK Act are post-NP and, in addition, several *ad hoc* measures have been put in place to ameliorate the situation, there are still a number of challenges, of which many were described and suggestions made on how to deal with them back in 2009 in Kamau/Winter.[188]

First, legally speaking, the ABS measures are very fragmented thus disintegrating the mandates of the LAs with each establishing its conditions and procedural obligations. This in turn leads to multiple requirements[189] (e.g. PIC and/or permit and fees) and at times duplicated ones as well as procedures causing protracted waiting periods, high transactional costs, ambiguity, cumbersomeness and lack of transparency.[190] Although a multi-stakeholder ABS TC was formed in order to have a concerted approach of the (main) LAs and alleviate the stalemate caused by disunity in the ABS landscape the effort comes too late as the cooperation of the TC takes place at the end of the permitting process, during the application review for an access permit. If the LAs (which are the main stakeholders) are heard during this process, there is no point for each one of them to establish their own requirements and procedures.[191] On the other hand if applicants have to go through the usual procedure established by law and meet the requirements of every LA this committee (ABS TC) becomes redundant. All these result in legal uncertainty which has a negative impact not only for foreign but also local research.[192] As a solution Kamau/

[187] See e.g. Amondi (2014), p. 216ff; _____ (August 2015), p. 63ff.

[188] Kamau and Winter (2009), pp. 365–379.

[189] For example, the following can be integrated: a research licence from NACOSTI and a research permit from KWS; a research pass and a visa from the DoI.

[190] See Kamau and Winter (2009) in particular p. 370f.

[191] According to Gino Cocchiaro (Natural Justice, *pers. comms*, 6 April 2018) a lack of concerted approach also makes it hard for communities to be involved in the process, and research institutes and businesses etc. (which would otherwise be interested in aTK) are discouraged as there is so much for them to do (in order to get clearances) but not much clarity.

[192] Dr. Benson Mburu, *pers. comms*, 26 September 2017.

Winter (2009) suggests simplification of the access procedures either through coordination by a LA or concentration of licensing.[193] The ongoing work to improve the ABS landscape in Kenya foresees the creation of a one-stop/one-window shop[194] and therefore there is hope that this situation will change.

Second, the scope of activities subject to licensing as contemplated in the definition of the term 'access'[195] and the material scope as understood by the new draft legislations[196] and practice[197] are so broad.

Third, the creation of the county governments by the Constitution of 2010 which gave them a big control over biodiversity has also brought with it some challenges. It is now a requirement that the county government and the GR managers in-charge of management of game, forest, land and marine reserves where bioprospecting and collection of genetic resource shall take place, and the affected communities are involved in the PIC and MAT process. The Constitution further states that "Revenue raised nationally shall be shared equitably among the national and county governments" (Art. 202 (1)) and establishes the criteria to be taken into account in determining the equitable shares (Art. 203)). These requirements still have to be concretized in an ABS law and therefore there is no law at the moment to determine how ABS should take place in the new situation. This causes legal uncertainty.

Fourth, the use of terms is at times not uniform across the different laws. For example the use of 'access', 'prior informed consent' and 'benefit-sharing' in Regulations 2006 and the draft bioprospecting regulations; 'biological resources' in the WCMA and EMCA ("includes genetic resources") and draft bioprospecting regulations thereto ("means genetic resources"); and 'community' and 'genetic resources' in EMCA, PTKCEA and draft bioprospecting regulations. This can lead to confusion.

Fifth, there is no legal backing for most of the approaches of the new practice which can cause legal uncertainty. For example: (1) The circumvention of the requirement of r. 10 of Regulations 2006 that requires NEMA, upon receipt of an application for an access permit, to give notification at least in one newspaper with

---

[193] Ibid, p. 371ff.

[194] Mukonyi Kavaka Watai (Head of Bioprospecting, KWS), *pers. comms*, 20 September 2017 and Jane Nyandika (ABS Desk Officer, NEMA), *pers. comms*, 25 September 2017. According to *pers. comms* with Watai (22 December 2020), the permitting flow chat for the creation of an online system has been approved and is being reengineered to integrate all permitting institutions. The process will also be streamlined at the level of devolved governments, i.e. at the county governments. These are already establishing county technical committees for the purpose of streamlining ABS systems, e.g. through the creation of county registers for indigenous knowledge technologies, technology transfer/licensing of researchers etc.

[195] See Kamau and Winter (2009), p. 375f.

[196] The draft bioprospecting regulations include biotrade as a subject of ABS.

[197] The overview and flowchart of the ABS process show that a permit is needed for all of the following: biological resources, genetic resources, DNA/RNA extracts, bio chemical resources, derivatives, progeny, traditional knowledge, digital sequence information and associated information.

nationwide circulation and wait for a period of 21 days for representations from the public. (2) Requiring a permit for anything not included in law under material scope.

10 Conclusion

There is more that lies in a functioning regime than just a healthy look on the surface. As seen above, the aggregate of practical measures make the Kenyan ABS regime look complaint with most of the Nagoya Protocol obligations, and this can easily win it a clean bill of health. Legally though, the regime still yields numerous challenges e.g. due to fragmentation, a challenge which is historically grounded on competition of lead agencies. Likewise, the current ABS legislation is shallow and outdated because it was among the first generation ABS legislations that were enacted in reaction of unregulated bioprospecting and in anticipation of a great wealth of benefits from utilization of biological/genetic resources. Therefore, in spite of the effort practice makes to ease access, legal uncertainty among other weaknesses is maintained. As it takes two to tango the law needs to catch up with practice. A number of years have passed since the first attempt was made to give Kenya a comprehensive Nagoya Protocol compliant ABS law. Peradventure the current efforts will materialize but when this will happen is hard to predict. It only remains to hope that when the time for birth finally comes, the product will not be a still child.

Acknowledgements Sincere gratitude goes to DG NEMA Prof. Geoffrey Wahungu for facilitating my interviews and data collection at NEMA. The interviews and discussions conducted at different institutions in Kenya were very enriching to the study leading to this chapter. In particular, the author appreciates Mukonyi Kavaka Watai, Head of bioprospecting department at the KWS and Chair of the national ABS permitting committee not only for interviews conducted with him at different stages of this research but likewise for constant exchanges over emails. For different inputs the author would also like to thank Jane Nyandika and Joyce Imende of NEMA (ABS Desk Officers), Cicilia Wangari Githaiga (former senior Legal Officer, NEMA), Dr. Benson Kinyagia of NACOSTI and Stanley Atsali of KIPI. Besides, Gino Cocchiaro of Natural Justice was very helpful in giving enlightenment concerning the question on indigenous peoples and local communities over skype interview and several email exchanges. Your input Gino is appreciated. Finally, I would like to express my appreciation to Prof. Gerd Winter of the University of Bremen and Thomas Greiber of BfN for reviews and comments on this chapter. Of course any errors are solely the authors.

Addendum 1: Offences and Sanctions According to Different Legislations

Legislation	Contravention		Penalties & other remedies upon conviction	
	Offence	Provision contravened	Under the Act	May one be liable for penalty under any other law?
Science, Technology and Innovation Act, 2013	**s. 15** Accessing, handling, transacting, transferring or moving any specified . . . any material necessary for scientific research within, into or from Kenya without a license issued under the act	**s. 12** Requires any person undertaking scientific research in Kenya to obtain a license under the act for any of the acts mentioned under offence	**s. 15** • Fine not exceeding five million shillings • Or imprisonment for a term not exceeding 4 years • Or both • Confiscation of the materials related to the offence • Barring the person convicted from undertaking any of the activities prohibited without a license, including research	**s. 15** Yes. One may also in addition be liable to any other penalty which may be provided for in this act
Wildlife Conservation and Management Act, 2013	**s. 22 (9)** Contravention of any provisions of s. 22 and any rules made thereunder **s. 91** For the purpose of obtaining a license, permit knowingly or recklessly: (1) Makes a false statement or representation; Furnishes a document or information which is false;	**s. 22** Requires any person engaging in bioprospecting or export for the same involving any wildlife resource, or export for any other kind of research to obtain a permit and if the traditional use of the resource inspired the bioprospecting endeavor, or aTK will be used, to obtain PIC of the (relevant) local community	**s. 91 (d)** (for offences foreseen under ss. 22 & 91) • Fine of not less than two hundred thousand shillings • Or imprisonment of not less than 1 year • Or both	No indication. However, according to **s. 105**, in addition to any other remedy, the court before which a person has been charged of an offence under the act may order that a any license, permit or authorization given be cancelled, a thing connected to the offence be

(continued)

Legislation	Contravention		Penalties & other remedies upon conviction	
	Uses or furnishes a false, falsified or invalid license or permit or one is altered without authorization; Contravenes any condition or requirement of a license or permit			forfeited to the service and disposed and the costs for disposal be borne by the person convicted
	s. 92 Any offence in respect of an endangered or threatened species (ETS)	**s. 48** Prohibits any activity on listed species without a permit from the service	**s. 92** (for offence against ETS) • Fine of not less than 20 million shillings • Or imprisonment for life • Or both	
	s. 99 (1) (2) Trade in, import, export, re-export . . . of any specimen of a wildlife species into or from Kenya without a permit issued by the service under the act	**s. 85** Requires one to obtain a permit from the service for these activities	**s. 99 (3)** Depending on the category of wildlife: • Either a minimum of one or ten million shillings • Or imprisonment of a minimum two or 5 years • Or both • In addition, according to **s. 86** the service may revoke the permit	
	s. 102 (1)(a)(g) (h) Entering protected areas (national park or reserve) without a license or permit;	**Same** Prohibits such actions without a license or permit	**s. 102 (1)** • Fine of a minimum two hundred thousand shillings • Or imprisonment of a minimum 2 years • Or both	

(continued)

Legislation	Contravention		Penalties & other remedies upon conviction	
	undertaking any extractive activity in marine protected areas; undertaking any activity in wildlife protected areas contrary to the provisions of the act			
Forest Management and Conservation Act, 2016	**s. 64 (1) (a)** Removing any forest produce from a public or provisional forest without a license or permit	**s. 64 (1)** Requires any person removing any forest produce from a public or provisional forest to obtain a license or permit	**s. 64 (2)** • Fine not exceeding one hundred thousand shillings • Or imprisonment for a term not exceeding 6 months • Or both	No indication
Protection of Traditional Knowledge and Cultural Expressions Act, 2016	**s. 37 (2)** Making non-customary use of TK without authorization whether for commercial or non-commercial use	**s. 18 (1)** Prohibits, in any way, misappropriation, misuse, abuse, unfairly, inequitably or unlawfully accessing and exploiting TK **s. 18 (2)** Prohibits the using in a non-customary manner TK without the PIC of its owners	**s. 37 (2)** • Fine not exceeding one million shillings • Or imprisonment for a term not exceeding 5 years • Or both	**s. 41** Yes
	s. 37 (3) Failure to acknowledge the source of the TK		**s. 37 (3)** • Fine not exceeding one million shillings • Or imprisonment for a term not exceeding 5 years • Or both	

(continued)

Legislation	Contravention		Penalties & other remedies upon conviction	
	s. 37 (4) Distortion, mutilation or other modification or derogatory action in a way prejudicial to the cultural interests of the community concerned		**s. 37 (4)** • Fine not exceeding one million shillings • Or imprisonment for a term not exceeding 5 years • Or both	
	s. 37 (5) Making false, confusing or misleading indications or allegations which draw upon or evoke TK in a way that suggests an endorsement or linkage with the holders		**s. 37 (5)** • Fine not exceeding two million shillings • Or imprisonment for a term not exceeding 10 years • Or both	
	s. 37 (6) Acquiring and exercising IPRs over protected TK without authorization		**s. 37 (6)** • Fine not exceeding two million shillings • Or imprisonment for a term not exceeding 10 years • Or both	
	s. 37 (7) Disclosing, subsequently using or acquiring and exercising IPRs over secret TK without authorization		**s. 37 (7)** • Fine not exceeding two million shillings • Or imprisonment for a term not exceeding 10 years • Or both	
	s. 37 (8) Importing an article or other thing into Kenya that relates to TK of Kenya		**s. 37 (8)** • Fine not exceeding five hundred thousand shillings • Or imprisonment for a term	

(continued)

Legislation	Contravention		Penalties & other remedies upon conviction	
	knowingly, or reasonably having to have known, that the import would contravene this act had the thing or article been created in Kenya		not exceeding 3 years • Or both	
	s. 37 (9) Exporting an article or other thing out of Kenya that relates to TK for a non-traditional use whether for commercial or non-commercial use without authorization		**s. 37 (9)** • Fine not exceeding five hundred thousand shillings • Or imprisonment for a term not exceeding 3 years • Or both	
	s. 37 (10) Making or having in possession any contrivance used or intended to be used for the purpose of exploiting unauthorized traditional knowledge		**s. 37 (10)** • Fine not exceeding five hundred thousand shillings • Or imprisonment for a term not exceeding 3 years • Or both	
Regulations 2006	**r. 23** Contravention or failure to comply with any of the matters provided in the regulations	**r. 23 implies all provisions/entire legislation**	**r. 24** • Fine not exceeding three hundred and fifty thousand shillings • Or imprisonment for a term not exceeding 18 months • Or both	**r. 15 (2)(h)** Implied. It is stated: "The holder of an access permit shall abide by the laws of the country". This is a permit condition

Source: *Own Illustration*

References

Books, Book Chapters and Articles

Amondi C (2014) Legal aid in Kenya: building a fort for Wanjiku. In: Pal Ghai Y, Cottrell Ghai J (eds) The legal profession and the new constitutional order in Kenya. Strathmore University Press, pp 201–219

Angwenyi AN (2009) The law-making process of the access and benefit-sharing regulations – the case of Kenya. In: Kamau EC, Winter G (eds) Genetic resources, traditional knowledge and the law. Solutions for access and benefit sharing. Earthscan, London, pp 173–186

Fombad CM (2020) Corruption and the crisis of constitutionalism. In: Fombad CM, Steytler N (eds) Corruption and constitutionalism in Africa. OUP, pp 15–68

Kamau EC (2005) Environmental law and self-management by industries in Kenya. J Environ Law 17(2):229–244. Oxford University Press

Kamau EC, Winter G (eds) (2009) Genetic resources, traditional knowledge & the law. Solutions for access & benefit sharing. Earthscan, London, pp 365–379

Lettington R (2001) Access to genetic resources in Kenya. In: Nnadozie K (ed) African perspectives on genetic resources. A Handbook on laws, policies and institutions. Environmental Law Institute, Washington, DC

The Danish Institute for Human Rights (DIHR) (2011) Access to Justice and Legal Aid in East Africa. A comparison of the legal aid schemes used in the region and the level of cooperation and coordination between the various actors. DIHR

Waruhiu JW, Odhiambo Otieno JJ (2014) Access to justice: the paralegal approach. In: Ghai YP, Ghai JC (eds) The legal profession and the new constitutional order in Kenya. Strathmore University Press, pp 185–200

Watai MK, Kimutai V, Ndhine EO (2015) Research on genetic resources and indigenous knowledge in the framework of the Kenyan ABS law. In: Kamau EC, Winter G, Stoll P-T (eds) Research and development on genetic resources. Public domain approaches in implementing the Nagoya Protocol. Routledge, London and New York, pp 110–124

Zehle S (2001) Indigenous people in biodiversity process, background for the 6th international indigenous forum on biological diversity, October 15–19 2001. Institut für Ökologie und Aktions-Ethnologie

Online Articles and Papers

Kamau EC (February 2009) Sovereignty over genetic resources: Right to regulate access in balance. The case of Kenya, Revista Internacional de Direito e Cidadania, n. 3 pp 73–88, http://www.egov.ufsc.br/portal/sites/default/files/anexos/33276-42382-1-PB.pdf. Accessed 21 July 2020

Kameri-Mbote P, Akech M (March 2011) Kenya. Justice sector and the rule of law. Discussion paper. Open Society Foundations. https://www.opensocietyfoundations.org/uploads/c9762a24-341f-49ed-b4e7-f4657a1b7c3b/kenya-justice-law-discussion-20110315_0.pdf. Accessed 20 Nov 2020

Kenya Human Rights Commission (November 2010) Wanjiku's Journey, Tracing Kenya's Quest for a New Constitution and Report on the 2010 National Referendum. KHRC, Nairobi. https://www.khrc.or.ke/publications/38-wanjiku-s-journey-tracing-kenya-s-quest-for-a-new-constitution-reporting-on-the-2010-referendum/file.html. Accessed 20 Nov 2020

Kwayera J (13 December 1999) Gene Banking: UK Scientists Accused of 'biopiracy', All Africa News Agency (AANA) wire Nairobi. https://www.grain.org/en/article/2083-uk-scientists-accused-of-biopiracy-in-kenya. Accessed 21 July 2020

Manek M (April 2001) The Implementation of Biodiversity-Related Conventions – A Kenyan Case Study. http://www.vliz.be/imisdocs/publications/246420.pdf. Accessed 21 July 2020

Mbaria J (23 August 2004) KWS Seeks Millions from Procter & Gamble. The East African, https://www.grain.org/en/article/2149-kws-seeks-millions-from-procter-gamble. Accessed 21 July 2020

Legal Documents

County Governments Act No. 17 of 2012

Endorois Peoples' Biocultural Protocol., https://endorois.org/programmes/developing-endorois-peoples-biocultural-protocol/ (accessed 15 June 2020)

Environment Management and Coordination Act 1999 (EMCA)

Environmental Management and Co-ordination (Conservation of Biological Diversity and Resources, Access to Genetic Resources and Benefit Sharing) Regulations, 2006

Industrial Property Act, No. 3 of 2001

Nagoya Protocol on Access to Genetic Resources and the Fair and Equitable Sharing of Benefits Arising from their Utilization to the Convention on Biological Diversity, 2011, Canada: SCBD, https://www.cbd.int/abs/doc/protocol/nagoya-protocol-en.pdf (accessed 15 December 2017)

Protection of Traditional Knowledge and Cultural Expressions Act No. 33 of 2016., http://kenyalaw.org/kl/fileadmin/pdfdownloads/Acts/ProtectionofTraditionalKnowledgeandCulturalExpressionsAct_No33of2016.pdf (accessed 02 December 2017)

Science, Technology and Innovation Act No. 28 of 2013

The Convention on Biological Diversity, 1992, United Nations, https://www.cbd.int/doc/legal/cbd-en.pdf (accessed 15 December 2017)

The Forest Conservation and Management Act No. 34 of 2016

The National Museums and Heritage Act No. 6 of 2006

The Registered Land Act (Chapter 300 of the Laws of Kenya) of 1989 (Rev. 2010)

The Seeds and Plant Varieties Act of 1972

Wildlife Conservation and Management Act 2013

Court Cases

Abdikadir Sheikh Hassan and 4 Others v. Kenya Wildlife Service [Civil Case No. 2959 (High Court of Kenya, 1996)]

Airey v Ireland (1979) European Human Rights Reports 305

Policy Documents and Reports

_____ (August 2015) Implementing the total Constitution: Towards a normative approach. A Report on the status of Constitutional implementation to the Kenya Law Reform Commission, Nairobi, http://www.klrc.go.ke/images/images/downloads/implementing-the-total-constitution-towards-a-normative-approach.pdf (accessed 13 June 2020)

CBD (30 August 2007) Report on the legal status of genetic resources in national law, including property law, where applicable, in a selection of countries, UNEP/CBD/WG-ABS/5/5, https://www.cbd.int/doc/meetings/abs/abswg-06/official/abswg-06-abswg-05-05-en.pdf (accessed 13 June 2020)

Interim National Report., https://absch.cbd.int/countries/KE/CP (accessed 10 June 2020)

Ministry of Environment and Natural Resources (March 2000) The Kenya National Biodiversity Strategy and Action Plan, https://www.cbd.int/doc/world/ke/ke-nbsap-01-en.pdf (accessed 19 June 2020)

Republic of Kenya (2015) Fifth National Report to the Conference of Parties to the Convention on Biological Diversity, https://www.cbd.int/countries/?country=ke (accessed 19 June 2020)

UN General Assembly Resolution No. 2850 (XXVI), https://undocs.org/en/A/RES/2850(XXVI (accessed 14 July 2020)

Guides

Biological Authorization Procedures.: https://absch.cbd.int/database/MSR/ABSCH-MSR-KE-238497/1, or https://absch.cbd.int/api/v2013/documents/391E5288-4807-F215-A83C-1AAEE8CBC314/attachments/Biological%20Authorization%20Procedures.pdf (accessed 28 July 2020)

Bonn Guidelines, Appendix II. Available at https://www.cbd.int/doc/publications/cbd-bonn-gdls-en.pdf (accessed 9 June 2020)

NEMA (2014a) Access and Benefit Sharing from Utilization of Biological Resources & Associated Traditional Knowledge in Kenya. General Information Brochure, Nairobi

NEMA (2014b) Kenya's Access and Benefit Sharing Toolkit for Genetic resources and Associated Traditional Knowledge, Nairobi, https://absch.cbd.int/database/MSR/ABSCH-MSR-KE-238496/1, or https://absch.cbd.int/api/v2013/documents/F3AB1BBD-08C1-4E30-1BA7-6562A31098FE/attachments/ABS%20TOOL%20KIT%20FINAL.pdf (accessed 10 June 2020)

Newspaper Articles

Kakah M (undated) The High Court faults Uhuru over delay in appointing judges, Daily Nation, https://www.nation.co.ke/news/High-Court-faults-Uhuru-over-judges-appointments/1056-5446252-wicnct/index.html (accessed 13 June 2020)

Mwiti L (16 August 2018) Budget cuts hit Kenya's judiciary – but there's no 'global funding target', Africa Check, https://africacheck.org/reports/budget-cuts-hit-kenyas-judiciary-but-theres-no-global-funding-target/ (accessed 13 June 2020)

Odula T (4 November 2019) Kenya judiciary complain of budget cuts, Associated Press (AP) News, https://apnews.com/e604c1af743140dab9e8c8369f4fa955 (accessed 13 June 2020)

Otieno O (8 February 2020) Court finds Kenyatta at fault for not approving new judges, The East African. https://www.theeastafrican.co.ke/news/ea/Court-finds-Kenyatta-at-fault-for-not-approving-new-judges/4552908-5448336-ipo8op/index.html (accessed 13 June 2020)

Yusuf M (8 June 2020) Kenya's judiciary puts executive on the spot over appointment of judges, VOA News, https://www.voanews.com/africa/kenyas-judiciary-puts-executive-spot-over-appointment-judges (accessed 13 June 2020)

Evanson Chege Kamau Dr. iur., is Senior Research Fellow at the Research Centre for European Environmental Law (FEU), University of Bremen. He is also an Adjunct Lecturer at the CvO University of Oldenburg. His area of expertise is public international law, European law and intellectual property law. His research focuses on genetic resources, environment, development and traditional rights on biological resources and traditional knowledge. Kamau has over 17 years of working experience in the area of ABS in which he has published widely. For many years he has also worked as a freelance legal consultant for governments, African Union (hired by IUCN and Capacity Building Initiative resp.), private industries and research organizations. echege@uni-bremen.de.

The New Law and Practice on ABS in Viet Nam: Innovations and Compliance with the Nagoya Protocol

Trang Thi Huong Tran Guihou, Tu Ba Nguyen, and Cuc Dang Thu Nguyen

Abstract This chapter provides a comprehensive picture of ABS legislation and its practice in Viet Nam in compliance with the Nagoya Protocol. The chapter analyzes the legal framework of the system by discussing the key ABS laws and by-laws in existence, including Biodiversity Law 2008, Decree No. 59/2017/ND-CP, its guidance document for implementation and related laws. It covers the background and influence of the development of ABS legislation. It clarifies the geographical, material scope with list of activities within the scope of ABS law of Viet Nam. Clarification of personal scope shows the difference between domestic and foreign users in the registration and application process for access to GR. Clarification of temporal scope shows the timelines of effectiveness of ABS law's application. It also explains the special regime of ownership and providers of GR in Viet Nam, as well as, how traditional knowledge has been regulated. The chapter presents the institutional framework looking at the national focal point and competent national authorities on ABS and their competences. It discusses the regulatory tools, including PIC with access procedures, requirements and permits, model agreement and benefit-sharing. Also it explains in detail the minimum ratio of benefit-sharing and how to apply it in practice and notes the lack of a financial mechanism of receiving and distributing of monetary benefits that come to the State budget. In addition, the chapter indicates a range of compliance measures and shows the regime's strengths and weaknesses. After a short assessment, suggestions are provided for improvement of the ABS framework in Viet Nam.

Keywords ABS · Biodiversity law of 2008 · Decree 59 of 2017 · Permit/license

T. T. H. Tran Guihou (✉) · T. B. Nguyen · C. D. T. Nguyen
Ministry of Natural Resources and Environment, Ha Noi, Viet Nam
e-mail: tranthi.huongtrang@cidce.org

E. Chege Kamau (ed.), *Global Transformations in the Use of Biodiversity for Research and Development*, Ius Gentium: Comparative Perspectives on Law and Justice 95, https://doi.org/10.1007/978-3-030-88711-7_10

1 Historical Background of Development of Legislation

1.1 Background and Influence

Being a Southeast Asian country on the Indochina peninsula that is endowed with abundant and rich biodiversity, Viet Nam is classified as one of the world's biodiversity hotspots.[1] Viet Nam has 42,900 species, of which nearly 14,000 are plant species and 11,000 species of marine organisms a number of which are endemic.[2] This rich biodiversity is reflected by the presence of an abundance of endemic genetic resources (GR), and traditional knowledge associated with GR (TK). Biodiversity plays an important role for the economic development of the country, as the economy is still based on the utilization of natural resources, especially in the fields of agriculture, forestry and fisheries which account for a significant percentage of the national economy. Moreover, the rapid development of biotechnology, pharmaceuticals, cosmetics and trade is expected to bring great benefits for the country. Therefore, GR is one of the renewable resources which can be considered as key to sustainable development in Viet Nam, which meaningfully contribute to poverty reduction, promotion of national sustainable economic development and conservation of natural resources.

Being aware of the role, value of GR and unsustainable use and degradation of biodiversity, the government has committed to international treaties and promulgated policies on conservation and sustainable use of biodiversity resources. Viet Nam became a member of the Convention on Biological Diversity (CBD) in 1994.

The Biodiversity Law of Viet Nam was approved by the Vietnam's National Assembly on 13 November 2008. It addressed the issues of access to GR and benefit-sharing (ABS) in Viet Nam for the first time and therefore it is the first law that had provisions on ABS. Section 1 chapter V of this law (from article 55 to article 61) covers all fundamental elements of ABS. These regulations are consistent with basic principles of the CBD and the Nagoya Protocol. The provisions of ABS under the Biodiversity Law 2008 (BL) were guided by articles 18, 19 and 20 of Decree No. 65/2010/ND-CP (Decree 65) dated 11 June 2010 detailing and guiding the implementation of a number of articles of the Biodiversity Law.

By Resolution No. 17/NQ-CP dated 17 March 2014, Viet Nam adhered to the Nagoya Protocol (NP) and officially became a party on 12 October 2014. Following this development, the Prime Minister approved a project to strengthen management capacity of ABS in the period 2016–2025.[3]

Decree No. 59/2017/ND-CP (Decree 59) on the management of access to GR and the sharing of benefits arising from their utilization was issued on 12 May 2017 to implement the NP and clarify the conformity situation/question in line with the Protocol and the provisions of the BL on ABS. The ABS provisions of Decree

[1] IUCN (2016), p. 1.

[2] UNCTAD (2016a), p. 1.

[3] See Decision No. 1141/QD-TTg of the Prime Minister dated 27 June 2016.

59 replaced those of Decree 65. The current legal system on ABS has developed and resulted from contributions of a number of projects implemented and funded by international organizations and developed countries.

The following projects aimed to raise awareness, update and improve the regulatory system. They include:

- Project for building legislation on access to plant GR (2000–2001) implemented by the Viet Nam Association for Conservation of Nature and Environment (VACNE) with the Ministry of Science, Technology and Environment (MoST) and supported by the International Development Research Center of Canada;
- Capacity building for the development of a legislation on ABS (2002–2004) implemented by Viet Nam Environmental Administration, Ministry of Nature Resources and Environment (MONRE), MoST, VACNE, International Union for Conservation of Nature (IUCN) and German Technical Cooperation Agency (GIZ);
- Projects under the framework for the implementation of the National Action Plan for Biodiversity (1996–2005) implemented with various international partners.[4]

In addition, several projects implemented by MONRE sought to put in practice the Biodiversity Law, 2008 including a project of VACNE, IUCN, a project implemented by Institute of Strategy and Policy on Natural Resources and Environment and a background paper of theory, practice and recommendation for regime of management of ABS from 2009 to 2010.[5]

In cooperation with international organizations, several communities and companies, research institutions have begun to develop their ABS knowledge when related to access to GR for research projects on plant varieties, breeding, cosmetic and pharmaceutical products. However, the system for ABS is still relatively new in Viet Nam for both users and providers.

Since becoming a party to the NP, the Nature and Biodiversity Conservation Agency (BCA) under the Viet Nam Environment Administration has carried out a number of projects related to ABS in Viet Nam for capacity building, enhancing the legal system and raising awareness:

- Project with the Association of Southeast Asian Nations (ASEAN) Center for Biodiversity on "Building capacities of countries in support of the development and implementation of National ABS Frameworks". Viet Nam, Laos People's Democratic Republic, and Myanmar have been selected to participate in the activities of the project. The implementation will be carried out through collaboration between the United Nations Environment Programme (UNEP) Regional Office for Asia and the Pacific and the ASEAN Centre for Biodiversity. The project aims at providing technical support to countries for developing national

[4]ISPONRE (2010), p. 27.

[5]See http://isponre.gov.vn/home/du-an-de-tai-dang-thuc-hien/132-nghien-cuu-co-so-ly-luan-thuc-tien-va-de-xuat-co-che-quan-ly-hoat-dong-tiep-can-nguon-gen-va-chia-se-loi-ich-tu-nguon-gen-o-viet-nam.

ABS regulatory and institutional frameworks and piloting initiatives to implement the Nagoya Protocol on ABS for selected ASEAN member States. The project has a strong focus on promoting regional cooperation, knowledge sharing and learning on ABS within ASEAN member states and China.
- Project of UNEP and the Global Environmental Fund (GEF) on "Building Capacity for Regionally Harmonized National Processes for Implementing CBD Provisions on Access to Genetic Resources and Sharing of Benefits". The project is a strategic regional project that will build the capacities of government agencies and stakeholders in the many dimensions of access and benefit-sharing.
- Project with United Nations Development Programme (UNDP) via GEF[6] on "Capacity Building for the Ratification and Implementation of the Nagoya Protocol on Access and Benefit Sharing in Viet Nam". It is a 4 year's project of implementation in Hanoi and the province of Lao Cai. The project supported the establishment of a comprehensive legal, regulatory and administrative framework in full compliance with the NP in Viet Nam. Capacity at different levels of the government should be strengthened to improve understanding and implementation of the national ABS regime.
- The cooperating activities with BioTrade project of Helvetas to develop training materials for ABS.
- The activities and project with United Nations Conference on Trade and Development (UNCTAD) for BioTrade and ABS in 2016–2017 to improve capacity and revision of the law.[7]

1.2 Aims and Intentions of the Legislation

The ABS legislation affirms implementation of national sovereignty over all GR in the territory of Viet Nam. It creates a mechanism for all foreign individuals and organizations. At the same time, it encourages domestic organizations and individuals to carry out research and development activities on GR. It ensures that benefits arising from the utilization of GR must be shared in a fair and equitable manner among stakeholders, to contribute to biodiversity conservation and to promote scientific research on GR and the role of local communities in the conservation and sustainable use of biodiversity.

[6] See http://english.vov.vn/society/VietNam-promotes-competency-on-genetic-resources-access-305,149.vov.

[7] See http://unctad.org/en/Pages/MeetingDetails.aspx?meetingid=1156.

2 The ABS Framework

2.1 Key Statute and Its by-Laws

The Biodiversity Law is the key statute. It affirms national sovereignty over GR and national responsibility for conservation and sustainable use of GR in the territory of Viet Nam. The law requires prior informed consent (PIC) and mutually agreed terms (MAT) for access to GR under article 57. The basic steps of procedures for access to GR include: (1) Registering access to GR; (2) Entering into written contracts on access to GR and benefit-sharing with organizations, family households[8] or individuals assigned to manage GR; and (3) Application for permit/licenses for access to GR.[9]

Article 58 of the BL requires engagement of contracts on access to GR and benefit-sharing between GR users and providers and stipulates for main contents of the contract. Article 59 on licenses for access to GR provides for basic contents of MAT.

Article 61 of the BL provides for sharing of benefits from access to GR which states that benefits obtained from access to GR must be shared among the following parties: (a) The State; (b) Organizations, households and individuals assigned to manage GR; (c) Organizations and individuals permitted/licensed to access GR and related parties as prescribed in the permit/license. Benefits obtained from access to GR must be shared on the basis of contracts on access to GR and benefit-sharing and in accordance with relevant laws.

Because the BL only provides general principles that cannot be applied directly, it is necessary to have a sub-law to specify those general principles. Accordingly, Decree 65 was issued with some provisions on specific orders and procedures for access to GR and benefit-sharing. However, provisions of Decree 65 were still limited and unfeasibly concretized to be applied directly into practice. There was no applicant for permit/license to access GR during the validity of ABS provisions of the Decree 65 which lasted from 30 July 2010 until 1 July 2017. When Viet Nam adhered to the Nagoya Protocol in 2014, there was an urgent need to promulgate a specialized and detailed decree to concretize the ABS provisions of the BL and to implement the NP. To meet this need, Decree 59 was issued.

---

[8]Family households in which members have common property and jointly contribute their efforts and labor to their common economic activities in agricultural, forestry or fishery production or in a number of other production and/or business domains defined by law shall be subjects when participating in civil relations in such domains (Art. 106, Civil Code 2005).

[9]The difference between Step 1 and 3 is that: Step 1: registration is to register to access GR. The CNA recognizes the information of user supplies but does not require long procedures to evaluate the dossier, capacity of the users, possibility to grant license. Step 3: the user applies for the licenses with full dossier of application, the commission of appraisal should be organized to evaluate the dossier and apply (for commercial purposes) and consult the CNA to grant the license or not. Step 2: the contract on ABS will only come to full effect after the ABS license is issued.

Decree 59 has been in effect since 1 July 2017. The decree consists of 28 articles in 5 chapters, detailing orders and procedures of the ABS process from registration and negotiation of an ABS contract to application for access to GR and benefit-sharing, following up and monitoring compliance through information and reporting. Rights and obligations of the involved parties are also concretized, including the state management responsibility to designate a national focal point (NFP) and competent national authorities (CNA) to grant permits/licenses. To facilitate its implementation, the Decree provides 9 sample forms, which assist the application of the Decree in practice.

2.2 Scope

Vietnamese ABS legislation regulates activities of access to GR and associated traditional knowledge (aTK) and sharing of benefits arising from their utilization under the sovereignty of the Socialist Republic of Viet Nam.

Regarding the status of sovereignty over GR in Viet Nam, the BL provides that "The State uniformly manages all GR on Vietnamese territory" (Art. 55.1). This confirms State ownership of all GR in national territory on behalf of the people of Viet Nam. This ownership regime is consistent with article 53 of the Constitution of 2013: "Land, water resources, mineral resources, resources in the sea and airspace, other natural resources, and property managed or invested in by the State are public property, owned by all the people, and represented and uniformly managed by the State".

Exercising its property right, the State assigns 4 groups of organizations and individuals the management of GR, i.e.: Conservation area management units and organizations assigned to manage conservation areas; Heads of biodiversity conservation facilities, scientific research and technological development institutions; Organizations, households and individuals assigned to manage or use land, forests or water surface; Commune-level People's Committees (Art. 55.2). They have specific rights and obligations under article 56 BL. They are entitled to get benefits shared by users of GR as stipulated in articles 58 and 61 BL. They also become providers of GR following article 3.3 Decree 59.

However, it may be complicated in the case of access to GR if TK associated to it is included. Following the BL, GR is a tangible property under the ownership of the State, on behalf of all its citizens, TK is an intangible property that may belong to certain individual persons, communities or undetermined owners as the case is with folk knowledge or public TK.

ABS rules apply to all GR situated in Viet Nam whether native, wild, cultivated GR, with direct access or indirect access, for example access to plant directly or access to derivatives, or exotic species. This is because the ABS law of Viet Nam provides a definition of GR which is different from the one of the CBD. It states: "GR includes all species and genetic specimens in nature, conservation zones, biodiversity conservation facilities and scientific research and technological development institutions and in nature" (Art. 3.2 BL). For cultivated species, ABS rules apply to exotic species if these have been produced in Viet Nam for a long time.

ABS rules focus on research and development (R&D) activities which cover not only access to GR but also derivatives. However, they do not concern PIC and MAT deal with derivatives when they are accessed alone without GR. There is no provision on benefit-sharing agreement without PIC for access to be negotiated and entered into. The model ABS contract provided by the Decree 59 (Art. 15) always requires provider and PIC. For access to derivatives only or the case if access to GR was completed before the effect of the BL 2008 but the utilization of their derivatives to produce and commercialize products continues, if the user does not need to access GR directly but wish to share benefits, he/she can refer to article 10 of the NP on the Global Multilateral Benefit Sharing Mechanism.

To clarify this scope, the definitions of 'utilization' are stated by Decree 59. The Decree provides some definitions for 'utilization' which are verbatim to those of the NP. However, like the Nagoya Protocol, Decree 59, or any specific legal instrument does not specify R&D. It has many different interpretations. Being a supplying country of GR, current guidance document for implementation of Decree 59 determines: 'R&D' comprise creative and systematic work undertaken in order to increase the stock of knowledge—including knowledge of humankind, culture and society—and to devise new applications of available knowledge.

Decree 59 does not provide a list of activities considered as utilization of GR in scope of ABS rules. The existing guidance document for implementation of Decree 59 determines the following activities within the scope of ABS law of Viet Nam:

- Carrying out R&D on specific naturally separated compounds.
- Carrying out research on different extraction processes involved in plant extraction to different potential compounds.
- Plant and livestock breeding using biotechnology to create new plant and livestock varieties by recombinant DNA technology, genetic engineering, cell technology.
- Any biotechnology application that uses systems of biology, living organisms or their derivatives to create or modify certain products or processes for certain uses (including antibiotic production).
- Any biotechnology application that uses enzymes and protein technology to produce a natural conversion of vegetable oil components to another fatty acid and separate plant cells that allow the separation of hydrophilic and lipophilic fractions from kernels, leaves, seeds.
- Reproduction of genetically modified insects to control diseases such as malaria.
- Microorganism culture to produce potential chemical compounds.[10]

[10]See Ministry of Natural Resources and Environment (2019), p. 22.

There are some views that even if ‘digital sequence information’ (DSI)[11] is not within the definition of GR, it is within the scope of the NP insofar as it results from the utilization of the GR or subsequent applications and commercialization and therefore should be covered by benefit-sharing.[12]

In practice, according to a survey conducted by the authors, awareness of the value of genetic information is very limited in Viet Nam, or it is the first time for most of people to hear about the term DSI. However, there are two opinions which affirm the need of having management mechanism for DSI to protect rights and benefits of providers but still facilitate and promote research and information exchange of scientists.[13]

The question of DSI is not quite clear as the BL and its existing guiding documents are silent on this issue. Theoretically, the property right established includes tangible and intangible elements, so, the State’s property rights for GR cover both. The intangible property is the genetic information, which represents the real value of GR. But, the BL and its current mechanism of management pay more attention to the management of GR as tangible property and less to the intangible values. Moreover, management system is organized according to the geographic location and administrative jurisdiction, challenged to deal with intangible values.

Under the scope of the ABS framework falls likewise associated traditional knowledge (aTK). The BL defines aTK as knowledge, experience and initiatives of native people on the conservation and use of GR (Art.3.38) and grants it intellectual property protection. It also encourages and supports organizations and individuals to seek its protection.

Decree 59 does not regulate aTK but a guidance document on aTK is being developed[14] under which the main issues are being considered, including: Definition of scope of protection in a clear and effective manner (the new definition of the term

[11] In reading DSI terminology and other outcomes from the CBD/DSI/AHTEG 2018/1/4 dated 13–16 February 2018, DSI includes, among other things, the following: (a) The nucleic acid sequence reads and the associated data; (b) Information on the sequence assembly, its annotation and genetic mapping. This information may describe whole genomes, individual genes or fragments thereof, barcodes, organelle genomes or single nucleotide polymorphisms; (c) Information on gene expression; (d) Data on macromolecules and cellular metabolites; (e) Information on ecological relationships, and abiotic factors of the environment; (f) Function, such as behavioural data; (g) Structure, including morphological data and phenotype; (h) Information related to taxonomy; (i) Modalities of use“. For details see Lyal in this volume.

[12] The question on whether DSI is within the scope of the CBD and whether it falls under the sovereign rights of the provider country of the genetic resource is discussed in details by Lyal in this volume.

[13] The teamwork had interviewed targeted group on DSI, including: management agencies (officials of MONRE, MARD, DARD, DONRE), Providers (representatives of the National Park Management Board, Nature Reserve, research institutions, universities), Users (private companies, research institutes). However, DSI is a new concept. Only a few government officials (MONRE, MARD) and some researchers know about DSI through notification of CBD or by exchanging scientific information with foreign partners: 90% of interviewees answered that they have never heard about DSI before. The remaining other interviewees have no idea about DSI.

[14] As this chapter was being written the date of publication had not yet been announced.

'aTK', adding utilization of aTK as the trigger of access requirements, consider issues of different types of aTK (secret, widely disseminated...); Identification of aTK holders; Benefit-sharing mechanism; Role of State and local bodies; Consideration of supportive measures such as aTK registration.

For the personal scope, ABS legislation applies to individuals and organizations engaged in activities related to access to GR for the purpose of utilization for research or development of commercial products.

The term 'users' is translated in Vietnamese and understood as 'Accessor' by Decree 59. 'Accessor' refers to the individuals and/or organizations that carry out activities to access GR for utilization under sovereignty of the Socialist Republic of Viet Nam (Cl. 2, Art. 3). Users/Accessors can be institutes, research centers and universities; international research organizations; domestic companies; foreign companies; traders and intermediaries (in some cases); or individual researchers and scientists. Decree 59 differentiates between domestic and foreign users by procedures of registration and application for a permit/license to access GR.[15] This is a similar experience with ABS laws of other countries e.g. India.

To determine domestic and foreign users, the Investment Law, 2014 should be referred to. Following this law, foreign individuals are defined as individuals with foreign nationality while foreign organizations are organizations established under foreign laws (Art. 3.14) Domestic individuals are defined as individuals having Vietnamese nationality and domestic organizations are those established under Vietnamese law.

In order to get a permit/license to access GR, the users must first register for access to GR and then apply for a permit/license to access GR from the CNA. The user has responsibility to share benefits derived from GR with the provider.

It is necessary to clarify the role and position of traders and intermediaries who have access to GR (investigation, collection) to sell to others for use in R&D or biotechnology. These traders and intermediaries are not users/accessors under Decree 59 as they do not carry out R&D activities; they are not actors for registering and applying for a permit/license to access GR. However, most foreign users always access GR through traders and intermediaries. Within the value chain, the traders and intermediaries' presence is indispensable, for most of ABS cases. Users cannot go directly to work with the local providers. Because, most local providers are farmers and each farmer is only capable to provide a small quantity of raw GR samples without preprocessing and control of qualification. While the users are foreign companies conducting R&D need a certain quantity of GR samples which should be collected and cultivated under certain qualified technical standards. This makes most of the ABS cases with participation of traders and intermediaries complicated and difficult to clarify who are the providers, users/accessors for ABS contract.

---

[15]Conditions, when registration suffices—in comparison to the conditions when an application (which requires approval) is demanded are: Registration: it is just required to fill the form to inform the CNA about the intention/plan of access to GR and CNA recognizes the registration. Application for a license: CNA will consider the dossier of application whether it satisfies the conditions or not. The CNA should organize the appraisal committee to evaluate the dossier for grant of the license.

For example, one European company in agribusiness is seeking and selecting Vietnamese farmers to plant and harvest a number of plants under EU organic agricultural standards on the land of the farmers. In nature, the farmers are providers according to article 3.1 of Decree 59. This company pays the farmers to look after and collect plants. Then, the company sorts/classifies and preprocesses to sell plants to a biotechnology company, which carries out R&D to extract active ingredients for cosmetic production. The supply contract between this European company and the biotechnology company sets out conditions to be met and clearly defines that the sale of those plants is for R&D purpose. In nature, this European company is a trader/intermediary, not a provider. However, it is difficult and complicated for the biotechnology company to sign contracts directly with each farmer who is a provider of GR under Decree 59. The solution found was that the European company received an authorization contract with all the farmers to become an "umbrella" provider and signed ABS contract with the biotechnology company who registered and applied for the ABS permit/license.

The Biodiversity Law came in to effect in 1 July 2009, thus, registration before this date is not required. However, the law and its guiding documents are silent on the question of GR accessed before 2009 when its utilization is still being continued. There is no legal provision to receive such kind of benefit.

Organizations and individuals who have accessed GR from 1 July 2009 up to the date Decree 59 entered into force, and who wish to continue to utilize GR but have not yet been granted a permit/license, shall have to register and request/apply for a permit/license in accordance with the provisions of Decree 59.

In accordance with definitions of the Biodiversity Law and Decree 59, ABS regulation does not cover human GR. GR under Annex I of the FAO International Treaty on Plant Genetic Resources for Food and Agriculture are not exempted from regulation by the ABS regime of Viet Nam as it is not party to the treaty.

2.3 Institutional Framework and Functions

Decree 59 designates MONRE as the NFP on the Nagoya Protocol. The NFP has the following responsibilities:

- Implementing unified management and monitoring of granting, renewal and withdrawal of permits/licenses for access to GR;
- Acting as a focal point for liaising by providing information and coordinating information exchange with the Secretariat of the CBD via the ABS Clearing-House in accordance with the Nagoya Protocol; leading the development of a national report on the implementation of the Nagoya Protocol in Viet Nam; proposing and implementing decisions of the Conference of the Parties to the Nagoya Protocol; coordinating and organizing the implementation of national obligations to the Nagoya Protocol;

- Coordinating with other countries in implementing measures to comply with the Nagoya Protocol applied to the utilization of Vietnamese GR in foreign countries; and
- Organizing the implementation of bilateral and multilateral international cooperation for access to GR and benefit-sharing.

In addition according to article 6 of Decree 59, MONRE and Ministry of Agriculture and Rural Development (MARD) are mandated as National Competent Authorities (CNA) to grant, renew and withdraw permits/licenses for access to GR. Accordingly, MARD shall grant, renew and withdraw permits/licenses to access GR of agricultural crop varieties, livestock, aquatic species, and forest seedlings whereas MONRE shall grant, renew and withdraw permits/licenses to access remaining GR which are not under responsibility of MARD (Art. 6.1, Decree 59). It is not always easy for users to determine which one between MONRE and MARD is responsible for which GR. However, there are lists of plant varieties, animal/aquatic breeds and forest seedlings which can be used to determine what falls under which CNA's responsibility. The lists include:

1. List of plant varieties permitted for trade and production;
2. List of animal breeds permitted for trade and production;
3. List of aquatic breeds permitted for trade and production;
4. List of forest plant varieties permitted for trade and production;
5. List of precious and rare animal breeds banned from export;
6. List of precious and rare plants varieties banned from export;
7. List of fishery species for export under conditions, banned from export;
8. List of plant GR, restricted for international exchange; resources for international exchange in special cases;
9. List of precious and rare livestock gene sources to be preserved;
10. List of endangered, rare and precious forest animals and plants;
11. List of threatened, rare and precious species prioritized for conservation; and
12. List of wild plants and animals of the Annex of Convention on international trade in endangered species of wild fauna and flora.

MARD receives, appraises and grants ABS permits/licenses for GR under List 1–6. In addition, applications by organizations and individuals for GR under List 7–12 that are under utilization conditions and have been approved as animal breeds and/or plant varieties must be submitted to MARD. For all remaining cases applications shall be submitted to MONRE with "written consent" of other relevant competent agencies.[16]

Access to GR is subject to a number of procedures and requirements according to article 8 of Decree 59:

1. Registering access to GR to national competent authorities.
2. Making an agreement and signing the ABS contract with the provider.

[16]See Ministry of Natural Resources and Environment (2019), p. 26.

3. Requesting the commune-level People's Committee to certify the contract.
4. Submitting application dossiers for access to GR to national competent authorities.
5. Implementing permit/license on access to GR and benefit-sharing.

Depending on the type of GR, the purposes of utilization or intention of user, the procedures and requirements are different, namely, the difference between commercial and non-commercial purposes, between domestic users and foreign users, and special cases. The difference is in the timeframe of consideration, handling and approval of the dossier. For non-commercial purposes, the timeframe will be shortened to a maximum of 30 days, while for commercial purposes, the timeframe is 90 days.

For the domestic users, access to GR for utilization of a non-commercial purpose does not need a permit/license. However, transfer of GR abroad or utilization of a commercial purpose and commercial product development are subject to registration and application for an ABS permit/license.[17]

Foreign organizations and individuals wishing to access GR for any purpose must register and apply for an ABS permit/license. In addition, there is an obligatory requirement to enter into cooperation with a domestic scientific and technological organization. Without a written document of cooperation, the dossier of registration and application for a permit/license to access GR will be considered as insufficiently complete to be handled.

Decree 59 also accepts different ways of submitting application dossiers. They can be submitted either directly to the head office of the CNA, by post or online registration through online public service portals.[18] However, it should be considered that specific regulations apply for the collection of GR under the list of rare and precious plant varieties and livestock to be preserved and restricted to international exchange, and/or the list of species with priority of protection following Decree No. 160/2013/ND-CP), Decree No. 64/2019/ND-CP) and/or the list of wild plants and animals of the Annex of the Convention on International Trade in Endangered Species of Wild Fauna and Flora (CITES). These regulations include prohibition of breeding, rearing and artificial propagation, reproduction of endangered species of precious and rare wild fauna and flora, import or illegal release of genetic modified organisms (GMOs), and control risks from GMOs and alien species.

For permits/licenses for access to GR for commercial purposes, the Decree 59 requires a council of appraisal[19] for the dossier of application, but regulation of

---

[17] For more on commercial and non-commercial research purposes see below, Sect. 3.3.

[18] Online public service portal of MONRE at https://dichvucong.monre.gov.vn/Pages/Default.aspx; of MARD at https://dichvucong.mard.gov.vn/portaldvc/Home/default.aspx.

[19] A Council of Appraisal is set up by the national competent authority to appraise the application dossier for a license to access genetic resources for commercial research or; commercial product development purposes. The Council of Appraisal consists of representatives of the MONRE, MARD, other Ministries; Provincial People's Committee where the access to genetic resources is proposed to take place, and relevant Experts (Art.13.2, Decree 59).

Table 1 Handled dossiers by CNAs to date

By MONRE				
Year	Dossiers for non-commercial purposes		Dossiers for commercial purposes	Total permits/ licenses granted
	Decisions allowing taking of GR abroad	Licenses for access to GR	Licenses for access to GR	
2017	02	0	0	02
2018	22	0	0	22
2019	18	01	0	19
2020	13	0	02	15
By MARD				
Year	**Dossiers for non-commercial purposes**		**Dossiers for non-commercial purposes**	**Total permits/ licenses granted**
	Decisions allowing taking of GR abroad	**Licenses for access to GR**	**Licenses for access to GR**	
2017	0	0	0	0
2018	0	0	0	0
2019	0	03	0	03
2020	0	0	03	03

Source: *Own illustration*

the council's organization, operations, responsibilities and powers should be provided by MONRE and MARD. Until 11 September 2019, MONRE issued Circular No. 15/2019/TT-BTNMT on organization and operation of the council of appraisal for the application for permit/license to access GR for commercial purposes. The Circular came into effect on the 28 October 2019. MARD issued Circular No. 07/2020/TT-BNNPTNT dated 22 May 2020 for the same council of appraisal for application under its authority which came into force on the 6 July 2020. Council of appraisal is established by the Minister of MONRE or MARD for each dossier of application for a permit/license. Standing units for appraisal are assigned under MONRE and MARD. They are Viet Nam Environment Administration of MONRE following Circular No. 15/2019/TT-BTNMT; Viet Nam Administration for Fishery, Viet Nam Administration for Forestry, Department of Crop Development and Department of Livestock Production of MARD following Circular No. 07/2020/TT-BNNPTNT. From 2017 to date MONRE has dealt with 56 dossiers for non-commercial purposes and 02 for commercial while MARD has dealt with 03 dossiers for non-commercial purposes and 03 for commercial (Table 1). All have received permits/licenses and those handled by MONRE in addition permits/licenses for export.

For aTK, unclear demarcation of authority between MONRE and MOST can cause problems in case access to GR cannot be separated from TK associated with

GR. The BL says that TK copyrights[20] are protected by the State. The State encourages and supports organizations and individuals to register TK copyrights. And, authority of management of TK copyrights is assigned to MOST. Therefore, if there is no cooperation between MOST and MONRE, it will be difficult to perform competency of licensing access to GR and aTK.

This difficulty can be solved if we take copyrights (or similar IP tools) as one of the tools to protect aTK. In case, it cannot be considered as a proper tool to protect aTK, MOST can play a role of a checkpoint in the ABS regime. Granting permits to access aTK should be handled by MONRE at the same time with granting permit for access to GR.

3 Nature of Regulatory Tools for Access to In-Situ and Ex-Situ and aTK

3.1 What Does the BL Understand with 'Access'?

Before addressing the question of regulatory tools for access it is important to know what the Biodiversity Law (BL) understands with the term 'access'. According to clause 29 article 3 of the BL 'access' (to GR) "means activities of investigating and collecting genetic resources for research and development and production of commercial products". This definition of the BL is based on the translation of English word 'access'. However, with the original Vietnamese meaning,[21] it will not cover the meaning as the English equivalent. Therefore, the BL provides the definition of access to GR including two activities: investigating and collecting ('investigating' can be understood as bioprospecting). Like the Nagoya Protocol, Decree 59 focuses on access for 'utilization'.

3.2 Prior Informed Consent

In line with the ABS regime of Viet Nam, the contract serves both as PIC and MAT on benefit-sharing between parties. Those that may participate in the PIC process include public and private entities (including non-governmental entities), indigenous people, private individual and land users if they are defined as providers of

[20]The Biodiversity Law of 2008 uses TK copyrights as a protection right for aTK in contrast to international practice which associates copyrights to creations such as a book, a piece of music or a recipe.

[21]Cl. 29, Art. 3 of the BL: "*Tiếp cận nguồn gen* là hoạt động điều tra, thu thập nguồn gen để nghiên cứu phát triển, sản xuất sản phẩm thương mại" ('access' to genetic resources means activities of investigating and collecting genetic resources for research and development and production of commercial products).

GR. 'Providers' of GR participate by negotiating and signing ABS contracts with the user. The ABS contract signed and certified is one of the obligatory documents to obtain an ABS permit/license.

'Provider' refers to the individuals and/or organizations assigned to manage GR by the State, as specified in clause 2, article 55 of the BL and following the article 3.1 of Decree 59.

Clause 2, article 55 of the Biodiversity Law 2008 assigns organizations and individuals to manage GR as follows:

(i) Conservation area management units and organizations assigned to manage conservation areas shall manage GR in conservation areas;
(ii) Heads of biodiversity conservation facilities, scientific research and technological development institutions, and GR storage and preservation establishments shall manage their own GR;
(iii) Organizations, households and individuals assigned to manage or use land, forests or water surface shall manage GR assigned to them for management or use;
(iv) Commune-level People's Committees shall manage GR in their localities, except cases specified above (i–iii).

ABS law of Viet Nam does not differentiate providers as indigenous or other providers. All the above mentioned providers have the same rights and responsibilities following article 56 of the BL. The indigenous people who live in buffer zones of protected areas where GR is found cannot participate in the PIC unless they are mandated to manage and use land, forests and water surface. If the indigenous people are holders of aTK, they may participate in the PIC as providers of aTK, but there is still no legal provision dealing with this issue in detail. More on ABS contracts will be discussed below under model agreements in Sect. 3.4.

3.3 *Access Permit/License*

Article 3.7 of the Decree 59 defines a permit/license to access GR as a document issued by a competent national authority, which allows the user/accessor to carry out activities to access GR for the following purposes: research for non-commercial purposes; research for commercial purposes; and commercial product development.

The ABS regime of Viet Nam differentiates access to GR for commercial purposes or non-commercial purposes as described above. However, Decree 59 on ABS does not define what commercial or non-commercial purpose is.[22] Maybe reference should be made to the definition of 'commercial activity' under the Commerce Law of 2005. Accordingly, "commercial activity is an open economy

[22]On the contested debate on how to define (and who is to define, which law is to be applied (conflicts of laws)) this term in Western law see Godt and Burchardi (2018), p. 36.

activity for purpose of gaining a profit" (Cl. 1, Art. 3). Thus, "gaining a profit" is a key element of commerce. Access to GR for commercial purpose is the access for a certain benefit, by developing products and commercializing them to gain a profit. Therefore, the criterion for research to be considered as for non-commercial purposes is not gaining a profit.

Characteristics of access to GR for commercial utilization are to use GR to develop specialized bio-preparations, metabolic enzymes, boosting genes, or small molecules. Those preparations, enzymes or genes can be used in crop cultivation and plant protection, pharmaceutical development, specialized chemical production or in industrial processing. GR can also be transplanted into plants to obtain the desired characteristics to improve crop productivity, herbicide resistance and disease resistance.

Characteristics of access to GR for non-commercial utilization are often the utilization of GR to enhance knowledge or understanding of the natural world with various activities ranging from taxonomic research, crop/livestock science, genetic mapping to ecosystem analysis. However, the boundaries between commercial and non-commercial purposes as well as involved actors are often unclear. Companies can collaborate with Vietnamese science and technology organizations to conduct research for commercial purposes, and sometimes research for non-commercial purposes can have results commercialized.

For commercial purposes, determination of the starting point and endpoint of utilization becomes important. Even though the ABS law of Viet Nam does not mention the GR's value chain with starting point and ending point of utilization, the definitions of GR and biotechnology are critical factors to determine the 'utilization' of GR within ABS. They are also the basis for differentiating between ABS and other activities in the GR value chain.

Access to GR for R&D and biotechnology is different with that for collection of GR by traders to re-sell to a third party. Access to GR for collection to re-sell shall not be considered an ABS activity because there are no initial R&D and biotechnology elements included. Yet, the third party who re-buys GR for R&D or biotechnology is subject to ABS regulations. Therefore, the 'starting point' of 'utilization' of GR within ABS scope can be defined by commencement of activity of access to GR for R&D and biotechnology. By identifying an access to GR within the ABS scope, those who conduct such access shall be defined to be subject to the registration and application for a license to access.

It is not clear how to define the 'end(ing) point' of 'utilization' of GR in ABS. From a reading of Decree 59, it is implied that the 'end(ing) point' may be the moment at which the benefits to be shared to providers can be calculated (e.g. total annual revenue of a product generated from the utilization of GR), or the moment of transfer of accessed GR to a third party (Art. 22). Decree 59 states that the transfer to a third party is a transfer of all rights and obligations of users; and there is only one third party. For example, the user A signed an ABS contract with provider B. The parties have negotiated and agreed that the contract will be valid for 7 years and the duration of sharing benefits will be 7 years. It is understood that the 'end(ing) point' of 'utilization' will be at the end of the seventh year.

For non-commercial purposes, Decree 59 does not specifically regulate all cases defined in article 8 of the Nagoya Protocol. However, there are simplified procedures as already mentioned which offer a 30 days duration for appraisal of application dossiers for access to GR in order to facilitate and promote non-commercial research as defined in article 8a and b of the NP.

Decree 59 includes separated regulations for Vietnamese students, doctoral students and Vietnamese scientific and technological organizations who wish to transfer GR abroad (Art. 20). This is a simplified and shortened procedure to promote scientific research. Following this procedure, the dossier is handled within 15 days from the date of receiving a valid application. Up to now, there are more than 50 ABS cases which are handled and permitted/licensed under this procedure.[23]

The validity of the permit/license shall be decided by a CNA based on the proposed objective and the plan of access to GR in the application dossier for the license to access GR, but it shall be no longer than 3 years. The permit/license can be renewed without limit of times of renewal (Art. 17, Decree 59). The permit/license can be withdrawn if the users provide false information; the access and utilization of GR causes harm to people, environment, security, national defense or national interests of Viet Nam; and other violations in accordance with the law (Art. 18, Decree 59).

To be granted a permit/license to access GR, an organization or individual must meet requirements of Article 8, Decree 59 on registration with the CNA, signing ABS contract with provider, being certified by the Commune-level People's Committee and submission of application dossier to the CNA; and completion of the application dossier at the request of the CNA during the process of appraisal of application. In addition, access to GR must not fall under the following cases: the use of GR threatens to harm humans, the environment, security and defense or national interests. (Cl. 4, Art. 59, BL).

In the interest of the country and community, CNAs may grant licenses without having to seek the consent of organizations, households or individuals assigned to manage GR who are defined as providers of GR under article 3.3 of Decree 59. This effect only occurs if the "interest of the country and community" is substantiated but not always applied. However, BL and Decree 59 does not clarify and concretize which kind of interest of the country and community that gives a wide discretion to the State's authorities.

3.3.1 Contents of a Permit/License

The ABS law provides a form of permit/license, which includes minimum information: name of the competent national authority; information of the permit/license holder; guarantor organization; scope of access to and utilization of GR; name of

[23]Data of MONRE on the number of ABS cases which have been handled up to June 2020, *pers. comms*, anonymous.

accessed GR; quantity/amount, purpose, duration, place of access to GR; information of provider, information of the user/accessor of GR; place where GR activities will be implemented; information on the use of TK; information on the taking of GR out of the territory of Viet Nam; responsibility of the permit/license holder; information of implementing signed ABS contract; plan for access to the GR; compliance with reporting obligations and other relevant regulations; date of effect; and recipients of the of permit/license. These are quite relevant and consistent with article 17.4 of the Nagoya Protocol on the internationally recognized certificate of compliance.

3.3.2 Conditions of a Permit/License

There are specific requirements and conditions for different users and different utilizations. For example: any Vietnamese individual who wishes to access GR for commercial purposes must have a graduate academic degree (bachelor's degree or higher) in a field relevant to GR such as biology, biotechnology, pharmaceutical or agricultural sciences, be a member of a science and technology organization operating in accordance with the law of the country where the organization is established in the fields of biology, biotechnology, pharmaceutical, or agricultural sciences, and be guaranteed by such an organization. Foreign individuals should meet the same requirements mentioned above as Vietnamese individuals and as said above must, in addition, have cooperation agreements with a Vietnamese scientific and technological organization. Foreign organizations must meet the latter condition (Cl.2, Art. 9, Decree 59).

Thus, in the case of programs and projects involving both domestic and foreign individuals and organizations, foreign entities can register or authorize domestic entities that they cooperate with to register access to GR.

The level of cooperation between foreign individuals/organizations and Vietnamese scientific and technological organizations depends on demand of the foreign users and Vietnamese scientific and technological organizations. It can be made under a contract with strict binding terms or a memorandum of understanding to provide engagement of flexible cooperation.

Vietnamese scientific and technological organizations are determined by the Law on Science and Technology 2013, which refers to organizations that mainly function in scientific research, technological development research, and science and technology service activities, being established and operating according to laws (Cl. 11, Art. 3). Accordingly, one scientific and technological organization is defined based on its registration of operation with competent State management authority and its certificate of registration on scientific operation (Cl. 4, Art. 3).

In the case of joint programs or projects with the participation of many individuals and organizations, all (or several individuals and organizations) can apply jointly if they wish to be included in the application dossier and on the permit/license. Those who do not wish can make a letter of authorization for the wishing individual or organization to carry out the necessary procedures.

Decree 59 requires that a permit/licensee shall only utilize GR for the registered purposes; in case of a change of purpose from non-commercial to commercial utilization, the user must apply for a new permit/license (Art. 14).

In the same way, transfer of accessed GR or derivatives of accessed GR to a third party with changing of purposes of utilization specified in the granted license must involve the application of a new permit/license by the third party. This will include a new procedure of negotiation and signing of an ABS contract with the provider. The new license should be obtained prior to receiving GR or derivatives of GR. In case the transfer to the third party is without any change of purpose of utilization, the Decree only requires a written notification to the CNA which granted the permit/license.

The transfer of accessed GR to a third party shall include the transfer of the obligations of the user/accessor under the granted permit/license, and the signed contract between the user/accessor and provider, including the provisions of the sharing of benefits arising from the utilization of GR to the provider.

Concerning intellectual property rights (IPRs), their registration for innovative results from the utilization of GR and its derivatives must state clearly the source or origin of accessed GR, and comply with obligation of benefit-sharing under cl. 2, article 22 of Decree 59.

3.3.3 Other Permits/Licenses

Following article 16.3 of Decree 59, organizations and individuals who are granted permits/licenses to access GR have the right to transfer them abroad, except for cases where GR belong to the list of GR prohibited or limited for exporting. It may be understood that the permit/license to access GR would be replaced with the permit/license to export GR, except in case of limited or prohibited export. Therefore, Circular 43/2018/TT-BNNPTNT of the MARD dated 28 December 2018 on importing and exporting of plant varieties, breeds and crop plant GR, states that "the export and import of crop plant GR for the purpose of scientific and technical research and exchange shall comply with the Government's Decree 59/2017/ND-CP of 12 May 2017 on management of access to GR and benefit-sharing from GR" (Art. 6). But, the Circular only covers crop plant GR for scientific research and exchange. It is not clear if crop plant GR for commercial purposes still need an export permit/license. Anyway, plant varieties and crop plant GR are all under the mandate of the Department of Crop Development of MARD, which has power to handle the application at the same time.

For GR under the list of endangered, precious and rare forest flora and fauna contained in the annex of CITES, the export shall comply with the Government's regulations on CITES implementation that requires a permit/license for export. This permit/license may be understood as another permit/license different from that for access to GR.

3.4 Model Agreements

ABS law of Viet Nam provides a model agreement which is called ABS contract form under annex 3 of Decree 59 which is based on the basic requirements of the BL. An ABS contract is defined as a contract for access to GR and the sharing of benefit arising from their utilization. It means an agreement between the provider and the user/accessor on the terms and conditions for access to GR for the purposes of utilization and the sharing of benefits arising from their utilization (Cl. 8, Art. 3, Decree 59).

This is the basis for granting the permit/license—no permit/license is granted without an ABS contract. The contract serves both as the PIC of the provider and the MAT on benefit-sharing between parties. The user and the provider are obliged to fill the form of the ABS contract. Therefore, it can be understood that it is compulsory to use the model agreement/ABS contract. The contract leaves some places for parties to negotiate. The authority can control the agreement signed during the procedure for granting the permit/license to access GR. The CNA can require two parties to explain or revise before granting the permit/license. Basic requirements of ABS contracts under the BL include: the contract must be in writing and signed with the provider of GR after receiving a written confirmation on accessing GR from the CNA; and the contract must be certified by commune-level People's Committees of localities where GR are accessed.

The ABS contract must contain the following principal details: Purpose of access to GR; GR to be accessed and volume of GR to be collected; Place of access to GR; Plan of access to GR; The transfer of the results of survey and collection of GR to a third party; Activities; Participants and place for conducting R&D or production of commercial products using GR; Sharing of benefits with the State and involved parties, including the distribution of IPRs over invention results on the basis of access to GR and aTK copyrights on GR. Disputes over or complaints about access to GR and benefit-sharing shall be settled under Vietnamese law and treaties to which Viet Nam is a contracting party.

It is noted that the ABS contract shall be effective only after the CNA grants the permit/license to access GR. When the permit/license to access GR expires, the user will not be allowed to access GR, but benefit-sharing provisions of the contract will remain in force. When the permit/license to access GR is withdrawn, the user will not be allowed to continue with access or utilization of permitted/licensed GR, but must continue to fulfill the agreement on benefit-sharing of the accessed GR as stipulated under the ABS contract and must compensate for damage (if any) and restoration of the environment and biodiversity, according to the Vietnamese law.

By providing the sample ABS contract and providing conditions for entry into effect of the ABS contract, the CNA can control ABS contract. The discretion of agreement between the provider and user is limited.

4 Benefit-Sharing

The ABS law of Viet Nam defines both types of benefits from utilization of GR: monetary and non-monetary benefits (Art. 21, Decree 59):

- Monetary benefits may include: Access fees/fee per sample collected; Payment of royalties; Permit/license fees in case of commercialization; Lump sum or milestone payments; Other monetary benefits arising during the utilization of GR.
- Non-monetary benefits may include: Sharing of research results; Rights to be involved in collaboration on the R&D and production of commercial products; Rights to have access to scientific and technical information related to the GR; Technology transfer to providers of GR; Training and capacity building for providers in R&D of GR; Joint IPRs corresponding to the percentage of contributions to innovative results based on access to GR; Other non-monetary benefits.

For monetary benefits, Decree 59 stipulates that the share of monetary benefits of the product generated from the utilization of GR shall be not less than 1% of the total annual revenue of such product. The share of monetary benefits for the provider when obtained through the transfer of GR or derivatives thereof, or the utilization of IPRs based on the GR, shall be not less than 2% of such total transfer value or total revenue from the use of intellectual property rights (Art. 22).

Decree 59 only sets up the minimum ratio to be shared by the users and gives leeway to parties of an ABS contract to negotiate where, how and how much benefit is to be shared. The Decree mentions both monetary and non-monetary benefits and does not oblige sharing of monetary benefits. Therefore, it is possible that parties agree on non-monetary benefits only, but these kinds of non-monetary benefits should be convertible into a certain monetary amount to satisfy the requirement of the minimum ratio of not less than 1% or 2%. It is noted that "the total annual revenue of such product" means the revenue of the user from selling the products utilizing permitted/licensed GR. The total monetary benefit is shared as follows:

- When the provider is Commune-level People's Committees, or Protected Area's Management Board, or state-managed facilities for storing or preserving GR, or biodiversity conservation facilities, or institutes for research and technology development: 30% of the shared money shall be paid to the GR provider as prescribed in cl. 1 and 2 of article 22, Decree 59. The remaining 70% of the shared money shall be paid into the state budget to be used for conservation and sustainable use of biodiversity.
- When the provider is an individual or a household or an organization assigned to manage GR by the State: 50% of the shared money shall be paid to the GR provider as prescribed in cl. 1 and 2 of article 22, Decree 59 and 50% into the state budget to be used for conservation and sustainable use of biodiversity.

Beneficiaries of non-monetary benefits include: domestic providers, domestic partners of the foreign user/accessor, and other involved organizations and individuals. The origin of the accessed GR should be clearly stated when publishing any results of the scientific research or applying for IPRs for any innovative results based on access and utilization of GR.

ABS law of Viet Nam is still lacking a financial mechanism of receiving and distributing monetary benefits hence all monies go to the State budget. There is no separate fund for GR or biodiversity conservation and sustainable use. Likewise, there is no provision to guide on how the benefits shared from utilization of GR will be used for biodiversity conservation and sustainable use purposes. Currently, there are three related funds which can be considered for receiving and distributing such benefits: the Forest Protection and Development Fund, the Viet Nam Environmental Protection Fund, and the Aquatic Resource Protection and Development Fund.

5 Provider and User Compliance Measures

The NP has no specific and strict regulations on compliance. Therefore, the national law of each country needs to include more specific and clear regulations to support the maximum level of enforcement possible. Article 15 NP requires parties to designate a NFP and one or more CNAs on ABS, and article 17 further requires measures for monitoring the utilization of GR which include checkpoints, issuance of permit/license. Being mainly a provider country Viet Nam seems to focus on compliance measures for providers but not for users. A number of them (provider compliance measures) will be considered in the following.

5.1 Reporting Measures

Decree 59 regulates the responsibility of MONRE to set up and operate electronic portal, registration and reporting on GR and ABS via electronic information network. Moreover, being a focal point, MONRE is responsible for liaising, providing information and coordinating information exchange with the Secretariat of the CBD via the ABS Clearing-House in accordance with the Nagoya Protocol; leading the development of a national report on the implementation of the Protocol in Viet Nam; proposing and implementing decisions of the Conference of the Parties of the CBD; coordinating and organizing the implementation of national obligations.

Within 15 working days from the date of granting, renewal or withdrawal of a permit/license to access GR, MONRE shall be responsible to upload information for publishing or cancelling the internationally recognized certificate of compliance for access to GR and benefit-sharing and other issues related to the ABS Clearing House of the CBD in accordance with the NP. Within 5 working days from the date of granting, renewal or withdrawal of a permit/license to access GR, MARD shall have the responsibility to send 1 original copy of such decision to MONRE to update the national database. MONRE is developing a Vietnamese clearing house on ABS which is expected to function as a platform for information on ABS and addressing cases of non-compliance (Art. 24, Decree 59).

Responsibilities of the permit/license holder include reporting in writing to the CNA about activities of access to GR stipulated in granted permits/licenses and the status and result of utilization of GR every 2 years (Art. 25, Decree 59). They shall also submit irregularly reports upon request by the CNAs. Vietnamese students, doctoral students, and Vietnamese scientific and technological organizations wishing to transfer GR abroad shall submit reports on their study or research results to the CNAs which allowed them to transfer GR abroad within 3 months of completion of the program or project.

MONRE issued Circular No. 10/2020/TT-BTNMT on will issue a circular to stipulate the regime and forms of report on access to GR and the sharing of benefits arising from their utilization. The Circular came into effect on the 16 November 2020.

5.2 *Contractual Clauses*

The ABS model contract provides for the responsibilities of users/accessors on informing and reporting; they must report following the provisions of the Decree. The responsibility of reporting and accounting is specified in article 6 of the model contract. The user/accessor must maintain and accurately update the full accounting records and reports related to the contract, including: (1) Transactions that have been made; (2) A separate report on receipts and invoices; (3) Accounting records must be available and accessible for checking, and made under generally accepted accounting standards; (4) Reports on and records of revenues derived from the exploitation and utilization of GR/aTK must be accessible to ensure correct payments; (5) The user must archive accounting records and reports related to specimens for a certain number of years[24] after expiration or termination of the contract.

5.3 *Permit Conditions Used*

For renewal of a permit/license to access GR, the users/accessor must submit a report on access to and utilization of GR at the time of requesting for permit/license renewal and a signed contract with provider (Art. 17, Decree 59). Therefore, if the users/accessors fail to report, they will be at risk of refusal for renewal of the permit/license. In addition, Decree 59 provides measures for withdrawal of permits/licenses to enforce compliance. The cases considered to withdrawal the permit/license

[24]The contract does not specify the number of years as this would be determined by agreement between the parties to the contract.

include: supplying false information, ABS activities harm to people, environment, security, national defense or national interests of Viet Nam; access and utilization of GR out of the scope of the permit/license; and other violations in accordance with the law (Art. 18). However, it also gives excessive discretion to the CNA by giving it power to withdraw a license "in case of other violations", which would include violation in reporting and accounting and other violations of the ABS contract.

The consequence of withdrawal of a permit/license is that, from the date of issuing the decision of withdrawal, the permit/license holder shall not be allowed to continue accessing or utilizing permitted/licensed GR, sharing benefits arising from accessed GR, and compensating and recovering any damages to the biodiversity and environment (if any).

5.4 Patents/Intellectual Property Rights Requirements

ABS rules require that registration for IPRs for innovative results from the utilization of GR and its derivatives must state clearly the source or origin of accessed GR (Cl .3, Art. 14, Decree 59). This is understood as an administrative duty of both the State competent agency in charge of management of intellectual property—which is expected to become one checkpoint, and inventors who claim patents. However, there is no practice of implementation of disclosure of the origin of GR.

5.5 Administrative Sanctions

Administrative sanctions for violation of ABS law are not provided directly by Decree 59 but can be meted based on Decree155/2016/ND-CP of 18 November 2016 on sanctioning administrative violations in the field of environmental protection. This Decree provides specific sanctions applicable to ABS violations. Its article 44 establishes monetary penalties ranging from VND 5 million to 50 million. In this regard, parties may be fined as illustrated in Table 2 depending on the circumstances.

Additional sanctions to the ones specified above (Table 2) include: taking away the right of the use of a permit/license to access GR from 6 to 12 months for violations specified by cl. 2 of article 44[25] and confiscating exhibits and means of

[25]Violations are: no compliance with the provisions on a contract of access to GR and benefit-sharing; no compliance with the provisions on the control for the investigation and collection of GR; no engagement of a written contract with organizations, households and individuals assigned to manage such GR; no implementation of procedures for having ABS contract certified by the competent authorities; no implementation of sharing of benefits arising from utilization with the involved parties following the permit/license and ABS contract; failure to report, as prescribed in the law, to the competent agencies on the results of research, development and commercial

Table 2 Summary of sanctions for ABS administrative violation

Fine to be paid	Wrongful action committed
Warning	No notification to the state competent agencies on exchange, transfer and provision of GR to other organizations and individuals to use for purposes of research, development and commercial production.
Warning	No notification about the process and results of research, development and commercial production, benefits arising from the development and commercial production as required
VND5,000,000 to VND10,000,000	No compliance with the provisions on a contract of access to GR and benefit sharing
	No compliance with the provisions on the control for the investigation and collection of GR
	No engagement of a written contract with organizations, households and individuals assigned to manage such GR
	No implementation of procedures for having ABS contract certified by the competent authorities
	No implementation of sharing of benefits arising from utilization with the involved parties following the permit/license and ABS contract
	Failure to report, as prescribed in the law, to the competent agencies on the results of research, development and commercial following permit/license for access to GR
	Use of a permit/license of access GR for the wrong purpose
VND10,000,000 to VND30,000,000	Exchange, transfer and provision of GR not in accordance with the law
VND30,000,000 to VND50,000,000	Access to GR without permit/license of the state competent authorities

Source: *Own illustration*

administrative violations for cases specified by cl. 3 and 4 of article 44[26] Decree No. 155/2016/ND-CP.

Remedial measures are to withdraw results arising from illegal activities of access to GR within the time set by the competent agencies. In fact, until now, there is no report or assessment on implementation of Decree No. 155/2016/ND-CP.

Although the Vietnamese Civil Code includes enforcement procedures for cases of non-compliance with national laws, such provisions would not be effective for ABS, especially cases where permit/licensed GR have been exported for utilization abroad. The Vietnamese Civil Code would not be enforceable in other countries' territories.

following permit/license for access to GR; use of a permit/license of access GR for the wrong purpose.

[26]Violations are: Exchange, transfer and provision of GR not in accordance with the law; access to GR without permit/license of the state competent authorities.

5.6 Involved Actors of Compliance

In addition to compliance with compulsory regulations, users/accessors (big companies) are now voluntarily promoting ethical biotrade. Ethical biotrade comes not only from the sense of self-respect and ethics of scientists, researchers and traders, who want to be responsible for biodiversity conservation, but also from requirements of consumers for big brands. Consumers in developed countries not only pay attention to the quality, design and price of products, but also require manufacturers and suppliers to ensure that measures for environmental protection and biodiversity conservation have been taken. Consumers only accept 'green', 'clean' products that are environmentally friendly and natural, but will boycott products from 'unethical biotrade'. According to market rules, in order to consume products, improve the brand image of the company in the eyes of consumers, big companies want to comply with the laws of the country which provided GR and have evidence of compliance with access to GR which is validated with an internationally recognized certificate of compliance.

The idea of ethical biotrade was first introduced in Viet Nam in 2003 by UNCTAD and further developed through the BioTrade project implemented by Helvetas.[27] In Viet Nam, this term refers to the collection or cultivation and subsequent processing and commercialization of products derived from biodiversity, conducted in line with environmental, social and economic criteria.[28] The relationship between biotrade and ABS has already been identified as an important factor.[29] ABS principles under the NP and the implementation of certain of its aspects will impact biotrade business and activities.[30]

Activities with certification and other recognition in line with Union of Ethical BioTrade's standard and UNCTAD's principles and criteria are very effective voluntary measures for compliance. In fact, biotrade companies are regularly audited to verify their compliance with biotrade principles and standards that already include certain ABS regulations. Therefore, a shortened and simplified procedure for permit/license for access to GR for biotrade would promote interest, activities and investment in the biotrade in Viet Nam. The simplified procedure would also reduce the administrative burden for CNAs. The benefits shared voluntarily during the biotrade value chain can supplement the benefits shared under ABS regulations.[31]

---

[27] See UNCTAD (2016b), p. 22.

[28] See Union for Ethical BioTrade (UEBT) (2018), p. 5.

[29] See http://unctad.org/en/PublicationsLibrary/UNCTAD_DITC_TED_2012_6.pdf.

[30] UNCTAD (2016b), pp. 30–33.

[31] Notes from the minutes of the workshop on "Access and benefit-sharing and BioTrade", 27–29 June 2016, Hanoi, Vietnam.

6 Regime's Strengths and Weaknesses

The ABS regime of Viet Nam has a number of strengths. First, it is quite a comprehensive legal framework. The BL and the new Decree 59 provide feasible principles, procedures, and regulations on basic elements of ABS to make ABS activities work in practice. The related ABS definitions are clear and concrete to support implementation. PIC is provided by the regulations on registration for access to GR and application for the permit/license to access GR with applicable and feasible procedures. MAT is provided by regulation of the ABS contract. A model agreement is provided which has applicable terms and conditions. A detailed mechanism of benefit-sharing between stakeholders is established. With the establishment of MONRE as the NFP and MONRE and MARD as the CNAs institutional arrangements also comply with the basic requirements of the NP.

Second, Decree 59 provides a range of forms in detail for implementation of facilitation provisions in practice. The number of applications and permits/licenses granted to access GR is increasing gradually, especially for non-commercial purposes.

Third, there is government support for development and implementation of ABS legislation and policy. By Resolution No.17/NQ-CP of the Government of 17 March 2014, the Prime Minister approved the project to strengthen the management capacity of ABS in the period 2016–2025. The implementation of this project is continuing and has received support from international organizations.

Fourth, there is a force of government officials and experts who are actively and positively in the development and improvement of the ABS regime.

However, in spite of its strengths, the ABS regime proves some weaknesses. For example, it still lacks some concrete regulations on aTK, DSI, mechanism of receiving benefits shared to the State budget and their distribution directly to biodiversity conservation and sustainable use. Besides, some provisions of the BL became outdated with the requirements of Decree 59. Because, Decree 59 is a guiding law instrument which must be consistent with the provisions of the BL, this causes potential conflict and struggle in the implementation of the ABS rules. This is proved by the procedure for obtaining permits/licenses for access of GR. The step for certification of ABS contract by the Commune People's Committee makes the licensing procedure burdensome and complicated. Because, Commune People's Committees are the lowest level of administrative authority, they have very limited capacity. Therefore, the original objective of the BL of involving local communities during ABS procedures through these Committees has not been achieved. In addition, the bureaucratic administrative system is still a major restraint to the ABS procedures, which discourage applicants and weaken compliance with the ABS laws in Viet Nam.

In addition, punishment for non-compliance is not clear enough and the capacity of the government's authority is weak to enforce. This causes most of the domestic users of GR for commercial purposes not to comply with ABS law. Some domestic users wishing to comply with the ABS law have concerns of unfair competition between users complying and not complying.

There is still limited understanding of ABS by government officials who develop the legal texts related to ABS implementation. For example, unclearness of the Circular 43/2018/TT-BNNPTNT of the MARD dated 28 December 2018 on importing and exporting of plant varieties, breeds and crop plant GR is whether crop plant GR for commercial purposes still needs an export permit/license or not.

There is still limited awareness of ABS issues by government officials, providers and Vietnamese users.[32] That causes challenges for effective implementation of ABS law.

7 Assessment and Suggestions

It can be said that the ABS law of Viet Nam is "nice" on the 'surface' but still has problems when put in practice. Decree 59 was issued with the efforts of policy makers in order to give it maximum concretization and make it directly applicable without need for the Minister's circular. However, the implementation of the Decree is being obstructed by the administrative system which requires more specific guidance in detail, including forms for reporting, finance for appraisal councils, fees of application for permit/license, and receiving monetary benefits to the State budget.

There are opinions from the Vietnamese companies that the CNA should have tools for carrying out statistics and monitoring domestic companies and enterprises which have GR utilization activities. It should have a mechanism to supervise compliance with ABS law. Sanctions for violation of ABS law should be specific and effective for both ABS State management agencies and the ABS users.[33]

The sustainable solution is to raise awareness of all involved stakeholders, especially decision makers and government officials on ABS. Once they are aware of the importance of the development of the ABS regime and following demands from practice, this force will make the ABS regime in Viet Nam work effectively. Awareness-raising via communication and dissemination of information can be

[32] The teamwork had taken a small survey on DSI and ABS in scope of the ABS project, in which total 130 copies of questionnaires were sent to 130 individuals and organizations; total feedbacks were 48, in which, 27 were from research institutions and universities, 7 from management State agencies, 5 from management board of protected areas, 4 from NGOs and associations and 5 from enterprises/companies. Therefore, the feedbacks did not meet 50% of the total sent question which showed that ABS was not paid attention to by the targeted groups. In addition, the received answers show that there is limited awareness on ABS issues because the questionnaires were about basic understanding of ABS e.g. regulations on ABS, genetic resources, providers, users, access to genetic resources, commercial and non-commercial purposes, scope of ABS regulation, *ex-situ* genetic resources collection. However, the received answers were varied, including many not answered or incorrectly answered.

[33] The teamwork had interviewed targeted group on ABS and had this result.

conducted through publication of ABS factsheet, guidance document for implementation, FAQ (frequently asked questions), ABS handbook and webpage (https://vietnamabs.gov.vn/). Concurrently, capacity building should be strengthened by training of trainers, and serial trainings for different target groups. In addition, policy and legal framework should be possibly updated and revised for better effectiveness, ABS rules for aTK should be developed, and national and international cooperation should be enhanced for effective implementation of ABS regulations.

8 Conclusion

It has not been long since Viet Nam adhered to and ratified the Nagoya Protocol. Viet Nam has made progress in developing legislation and policy to implement the Nagoya Protocol. A new Decree on ABS, which implements the requirements of the Nagoya Protocol, was issued and came into force in 2017. This has increased the demand for the country's GR. Decree 59 can be recognized as a landmark ABS legislation development of Viet Nam which put a basis of legal certainty, transparency and clarity, a prerequisite for ABS implementation of the Nagoya Protocol.

It has been only 2 years since the Decree 59 came into effect. There is no official assessment of the implementation of the Decree in practice. However, the number of applications for permits/licenses to access GR and the workshops and seminars that have been held shows an improvement in capacity building, awareness-raising and that the Decree is coming alive.

The analysis of laws on ABS also raises a number of issues that need to be improved for effective implementation of the national ABS law and the Nagoya Protocol. Recently, the government of Viet Nam planned to revise the BL. This will be a good opportunity to amend the provisions on ABS to meet more effectively the requirements of the Nagoya Protocol and fill the gaps of Decree 59 as well as overcome existing constraints which cannot be solved by the existing Decree 59. This is because the Decree being a by-law document issued by the government has a lower legal validity than the Biodiversity Law, especially in regard to the demarcation of assignments among the authorities, responsibilities between relevant ministries, their cooperation in inter-ministerial ABS issues and enhancing compliance and enforcement through checkpoints.

A sound and effective ABS national legal framework will facilitate the implementation of the Nagoya Protocol and also promote GR and biodiversity conservation and sustainable use in the country.

References

Publications

Godt C, Burchardi M (2018) Expert opinion on the liability of registered collections, CvO University Oldenburg/BfN, https://uol.de/f/2/Dokumente/Godt_Burchardi_2018_Haftung_Registrierter_Sammlungen_nach_VO_EU_Nr._511_2014_.pdf. Accessed 4 Sept 2020

IUCN (2016) Protecting important biodiversity hotspots, hotspots of international biodiversity conservation, Vietnam Ecology & Nature Protection Handbook, International Business, USA, Washington DC, USA-Viet Nam, https://www.iucn.org/sites/dev/files/content/documents/cepf_factsheet_layout_revised_feb_2016.pdf. Accessed 12 Nov 2020

UNCTAD (2016a) Facilitating biotrade in a challenging access and benefit sharing environment. https://unctad.org/system/files/official-document/webditcted2016d4_en.pdf. Accessed 19 Nov 2020

UNCTAD (2016b) The interface between access and benefit-sharing rules and biotrade in Viet Nam, UNCTAD/WEB/DITC/TED/2016/9, https://unctad.org/system/files/official-document/webditcted2016d9_en.pdf. Accessed 5 Nov 2020

Union for Ethical BioTrade, (UEBT) (2018) ABS in Vietnam 14/5/2018, documents handed over at the conference "Branding and Biodiversity" -"Beauty of Sourcing with Respect", Paris, 19–20 June

Legal Documents

Biodiversity Law of 2008, https://absch.cbd.int/database/record/ABSCH-MSR-VN-202094. Accessed 5 Nov 2020

Circular 15/2019/TT-BTNMT of the Minister of Natural Resources and Environment dated 11 September 2019 on organization and operation of the council of appraisal for the dossier of application for access to GR for commercial purposes

Circular 43/2018/TT-BNNPTNT of the Minister of Agriculture and Rural Development dated 28 December 2018 on importing and exporting of plant varieties, breeds and plant's GR

Decree No. 64/2019/ND-CP

Law on Commerce of 2005

Law on Investment of 2014

Law on Science and Technology of 2013

Ministry of Natural Resources and Environment (2019) Guidance document for the implementation of Decree 59/2017/ND-CP, Hanoi, https://vietnamabs.gov.vn/wp-content/uploads/2019/06/GUIDANCE-DOCUMENT_ABS_English.pdf. Accessed 12 Nov 2020

Decree No. 155/2016/ND-CP of the Government dated 18 November 2016 on sanctioning administrative violations in the field of environmental protection

Decree No. 65/2010/ND-CP of the Government dated 11 June 2010 on detailing and guiding the implementation of a number of articles of the Biodiversity Law

Decree 160/2013/ND-CP of the Government dated 12 November 2013, on criteria to define species and management of species under list of species with priority of protection

Decision No. 1141/QD-TTg of the Prime Minister dated 27 June 2016 on approving the project to strengthen management capacity of ABS in the period 2016-2025
Decree No. 59/2017/ND-CP of the Government dated 12 May 2017 on the management of access to GR and the sharing of benefits arising from their utilization., https://vietnamabs.gov.vn/en/van-ban-phap-luat/decree-no-59-2017-np-cp/. Accessed 5 Nov 2020
Circular 07/2020/ TT-BNNPTNT dated 22 May 2020 on organization and operation of the council of appraisal for the dossier of application for access to GR for commercial purposes
Circular No. 10/2020/TT-BTNMT of MONRE dated 16 November 2020 on report on access to GR and the sharing of benefits arising from their utilization

Other Documents

CBD/DSI/AHTEG 2018/1/4 (13–16 February 2018) Report of the Ad Hoc Technical Expert Group on Digital Sequence Information on genetic resources, https://www.cbd.int/doc/c/4f53/a660/20273cadac313787b058a7b6/dsi-ahteg-2018-01-04-en.pdf. Accessed 27 June 2018
ISPONRE (2010) Internal report of project on research on theory and practice for proposing a mechanism of ABS management in Viet Nam
Notes from the minutes of the workshop on "Access and benefit sharing and BioTrade" 27-29 June 2016, Hanoi, Vietnam, UNCTAD/BCA/BIG/SECO

Websites

http://english.vov.vn/society/VietNam-promotes-competency-on-genetic-resources-access-305149.vov. Accessed 12 Nov 2020
http://isponre.gov.vn/home/du-an-de-tai-dang-thuc-hien/132-nghien-cuu-co-so-ly-luan-thuc-tien-va-de-xuat-co-che-quan-ly-hoat-dong-tiep-can-nguon-gen-va-chia-se-loi-ich-tu-nguon-gen-o-viet-nam. Accessed 12 Nov 2020
http://unctad.org/en/Pages/MeetingDetails.aspx?meetingid=1156. Accessed 12 Nov 2020
http://unctad.org/en/PublicationsLibrary/ditcted2014d6_en.pdf. Accessed 12 Nov 2020
http://unctad.org/en/PublicationsLibrary/UNCTAD_DITC_TED_2012_6.pdf. Accessed 12 Nov 2020
https://dichvucong.mard.gov.vn/portaldvc/Home/default.aspx. Accessed 19 Nov 2020
https://dichvucong.monre.gov.vn/Pages/Default.aspx. Accessed 19 Nov 2020

Tran Thi Huong Trang Guihou PhD in Law from University of Limoges, France (2012). She has 19 years of working experience in environmental law, specializing in biodiversity and ABS Law; winner IUCN Alexandre Kiss award (2006); member of Hanoi Bar Association since 2007. She prepared research documents for adhering to Nagoya Protocol (2012-2014) by Viet Nam and developed and revised Decree 59/2017/ND-CP on ABS. Currently she works as an independent legal consultant on ABS, biodiversity, environmental protection and sustainable development. tranglpsd@gmail.com.

Nguyen Ba Tu master's degree in Biology, Viet Nam Academy of Science and Technology (2009). He is senior official of Viet Nam's Ministry of Nature Resources and Environment, acting national focal point to ABS-CH and Cartagena Protocol emergency contact point, Biosafety-CH NFP. He worked as a Researcher at Institute of Biotechnology for 5 years (2005–2010), Viet Nam Environment Administration under the Ministry of Nature Resources and Environment since 2011; he was involved in the development of legal documents, policies and guidelines on ABS and

Biosafety management of Genetically Modified Organisms in Viet Nam. He was also a member of the ABS Appraisal Committee and Biosafety Management Committee. tunguyen.bca@gmail.com.

Nguyen Dang Thu Cuc MA, obtained her undergraduate in life sciences and graduate in development studies at the HAN University of the Netherlands and the Graduate Institute of Geneva, Switzerland respectively. Started working at the Biodiversity Conservation Agency under Vietnam Environment Administration, Ministry of Natural Resources and Environment since 2009; was involved in development of legal documents and policies on biodiversity conservation and in particular ABS. She acted as the national competent authority to the Cartagena Protocol on Biosafety (2014–2019) and the Nagoya Protocol on ABS (2017–2019). nguyendt.cuc@gmail.com.

The Fastest Animals Are Not the Fastest Over Time: Malaysia Adopts a Comprehensive ABS Legislation After a Long Steady Effort

Evanson Chege Kamau

Abstract Malaysia did not have a law to regulate access to genetic resources, associated traditional knowledge (aTK) and benefit-sharing at a Federal level prior to the Nagoya Protocol. Only two States, Sabah and Sarawak, had such a law. Following the assent of the bill of 2017 regulating access and benefit-sharing (ABS) within the entire territory the country now has a relatively comprehensive law in this area. For instance, besides access measures, it also implements user measures for genetic resources and aTK from other countries utilized in Malaysia, which is at the moment uncommon in legislations of traditional provider countries, except the few EU member States also considered as such. Besides, its implementation of article 4 of the Protocol by exempting all specialized ABS instruments from its scope is exemplary. None of the other legislations interrogated in the current study has so clearly and expressly exempted all specialized international ABS instruments. In addition, it takes a stance concerning genetic resources accessed without prior informed consent (PIC) and mutually agreed terms (MAT) prior to the Protocol requiring them to be regularized if put into new research and development (R&D) uses or used for a new product. This chapter examines this law in a view to find out, *inter alia*: (1) how the different elements or aspects of the Protocol are implemented domestically, (2) how the bill copes with issues that were unresolved in the Protocol or that have emerged after it e.g. the regulation of Digital Sequence Information, (3) how the different understanding of terminology e.g. 'commercial' and 'non-commercial research', 'biological resources' and 'genetic resources' is dealt with and the impact this has on ABS, and (4) whether and how the extension of Malaysian rights over genetic resources is construed. The chapter also examines the general compliance of the regime with the Nagoya Protocol and makes suggestions for improving its weak points.

Keywords Sarawak · Sabah · Malaysian Act 795 of 2017 · Access and benefit-sharing · Genetic resources

---

E. C. Kamau (✉)
Faculty of Law, University of Bremen, Bremen, Germany
e-mail: echege@uni-bremen.de

E. Chege Kamau (ed.), *Global Transformations in the Use of Biodiversity for Research and Development*, Ius Gentium: Comparative Perspectives on Law and Justice 95, https://doi.org/10.1007/978-3-030-88711-7_11

1 Development of Legislation

1.1 Status of Biodiversity

Malaysia is a Federation comprising Peninsular Malaysia and the States of Sabah and Sarawak. It is one of the 12 megadiverse countries of the world with exceedingly rich biodiversity comprising various ecosystems and species.[1] Its terrestrial ecosystems consist of an immense variety of wild plants and animals including dipterocarp forests and montane forests, while the coastal and marine areas house important ecosystems such as mangrove forests and coral reefs.[2] These natural ecosystems contain a diverse array of floral and faunal communities.[3] There are an estimated 15,000 species of vascular plants in Malaysia, with about 8300 species in Peninsular Malaysia and about 12,000 in Sabah and Sarawak. Fauna diversity includes 307 known species of mammals, 30 of which are endemic to Malaysia, 785 species of birds, 242 species of amphibians and 567 species of reptiles, as well as 2068 species of freshwater and marine fishes.[4] This diversity has suffered different pressures as a result of population growth, habitat fragmentation, invasive alien species, pollution, poaching, increasing competition for land as well as climate change, a general lack of awareness on the importance of biodiversity as well as significant knowledge gaps leaving many species vulnerable with some facing threats of extinction.[5] The rich biological resources have also given rise to a rich cultural heritage of sustainable use amongst the indigenous people of Malaysia, especially those dependent on the forest for their livelihood.[6] The indigenous people of Sarawak, for example, have for generations used the sago of a palm (*Eugeissona utilis*) found in the forest.[7] However, Malaysia did not have a single comprehensive legislation relating to conservation and management of biological diversity as a whole, which was done through sectoral laws, and none at all at the Federal level to regulate access to genetic resources (GR), associated traditional knowledge (aTK) and benefit-sharing.

---

[1] Ministry of Natural Resources and Environment (2014), p. 6.

[2] Ministry of Natural Resources and Environment (2016), pp. 10, 21–24. See also Malaysia National Biodiversity Policy (1998), p. 5ff.

[3] Ibid.

[4] Ibid.

[5] Ibid.

[6] Malaysia National Biodiversity Policy (1998), p. 7.

[7] Ibid.

1.2 Background

Malaysia ratified the Convention on Biological Diversity (CBD) in 1994 after which a number of efforts were initiated at the national level to implement the Convention.[8] These initiatives were led by the Ministry of Science, Technology, and the Environment (MOSTE) under which a National Committee on Biological Diversity (NCBD) was established in 1994 to coordinate these efforts.[9] Under this committee was, besides, a National Technical Committee on Biological Diversity and its three task forces working on specific issues: the country study on biological diversity, the national policy on biological diversity, and access to genetic resources.[10] In addition, a national Genetic Modification Advisory Committee (GMAC) was established.[11] In 1998 the National Policy on Biological Diversity was launched which provided important guidelines and strategies to States and institutions in Malaysia to take affirmative actions to safeguard the country's biodiversity heritage and ensure that it is among others utilized in a sustainable manner for the continued progress and socio-economic development of the country.[12] Two out of the 15 strategies related directly to access and benefit-sharing (ABS).[13] The State of Sarawak passed a State law (Sarawak Biodiversity Center Ordinance of 1997 as amended in 2003) and established the Sarawak Biodiversity Council as the competent authority to regulate access to genetic resources (GR) and implement legislation for protection of biodiversity, and the Sarawak Biodiversity Center to assist in the implementation of the legislation.[14] In 1998 it enacted the Sarawak Biodiversity (Access, Collection and Research) Regulations and replaced them with Sarawak Biodiversity Regulations in 2004 which introduced a permit system to regulate access to GR.[15] In the same vein the State of Sabah enacted in 2000 the Sabah Biodiversity Ordinance and established the Sabah Biodiversity Center in 2001.[16] Also the government undertook a number of steps at the national level to finalize major commitments including the access to genetic resources Bill.[17] Throughout 1999 nationwide consultations took place between States on ABS issues which resulted to the first national (federal) draft framework.[18] The draft was submitted to MOSTE for further deliberations and actions which led to among others consultations with *inter alia* ministries and agencies responsible for the management of biodiversity, industries and

[8]Carrizosa et al. (2004), p. 246.

[9]Ibid.

[10]Ibid.

[11]Ibid.

[12]Ibid.

[13]SaBC & GDF (2010), p. 4.

[14]Nordin et al. (2012), p. 185.

[15]Ibid, p. 185 f.

[16]Carrizosa et al. (2004), pp. 246, 255.

[17]Ibid, p. 246.

[18]Ibid, p. 255.

non-governmental organizations.[19] There is an information gap in literature concerning what transpired for a long duration of time but there is an indication of a possible stalemate caused by a number of controversial issues as well as considerable work that had to be accomplished before the Bill could be adopted in all states and passed into law.[20] One of the difficulties involved the establishment of institutional structure for the implementation of the law[21] in a federal setup. In 2010 the government and the United Nations Development Programme (UNDP) Malaysia country office approved a project on capacity development for the formulation of a policy and regulatory framework for ABS of biological resources in Malaysia with an overall objective to support the establishment of a regulatory framework for ABS and the relevant capacity building in this area.[22] Among others the project carried out some background studies on institutional arrangements for implementing ABS in Australia and in India in order to tap experiences from countries which had a similar federal government system as Malaysia.[23] Once ready the draft had to go through stakeholder consultation and revisions before it could be tabled in parliament.[24]

It was only in 2017 when the Bill was passed into law. Therefore, ABS is now regulated by the so-called Access to Biological Resources and Benefit Sharing Act 795 of 2017.[25] In the states of Sarawak and Sabah it will enter into force "on such date as the Minister may, after consultation with the State Authorities, appoint by notification in the Gazette" (s. 1 (3)). This is because these two states have a different constitutional status *vis-à-vis* the Federal Government according to an agreement signed with the Federation prior to their becoming members which accorded them special rights.[26] The other States also have rights over their biological, biochemical, and genetic resources found in public lands based on the Constitution which accords ownership of land to the individual state.[27]

[19] Ibid, pp. 252, 255.

[20] See Carrizosa et al. (2004), p 253f; Schriver (2012), p. 2ff.

[21] Carrizosa et al. (2004), p. 253.

[22] For details of the project see Schriver (2012).

[23] Ibid, p. 2.

[24] Ibid, p. 11.

[25] The Malaysian Act has a broad material scope covering 'biological resources' with 'genetic resources' subsumed in the former according to its definition in s. 4.

[26] Carrizosa et al. (2004), p. 245.

[27] Ibid.

2 Scope of ABS Legislation

The material scope of the Act applies to biological resources, traditional knowledge associated to such biological resources and human genetic resources[28] when they are accessed for use in research whether with a commercial or commercial potential purpose or non-commercial purpose. According to s. 4 on definitions the concept of 'biological resource(s)' (BR) is broad and includes:

(a) the genetic resources, organisms, microorganisms, derivatives and parts of the genetic resources, organisms, microorganisms or derivatives;
(b) the populations and any other biotic component of an ecosystem with actual or potential use or value for humanity; and.
(c) any information relating to paragraphs (a) and (b).

In terms of geographical scope the Act extends to all the genetic and biological resources including wild, domesticated and cultivated species of flora and fauna, both *in-situ* and *ex-situ* conditions found within the territory of Malaysia, which should be understood as the country's terrestrial and marine jurisdictions.

The Act will not apply to BR used as commodity for the purpose of direct use or consumption as determined by the Competent Authority (CA).[29] However, the transitional provisions of s. 63 (3)–(4) suggest that if such a commodity is later used in research and development (R&D) the person engaging in such activities must enter into a benefit-sharing agreement (BSA) with the resource provider and receive a permit from the CA. In other words such an action will be caught by the ABS legislation.

The Act establishes a temporal scope which is the date of its coming into force. As a consequence, "any person who, on the date of the coming into operation of this Act, is accessing a biological resource or traditional knowledge associated with a biological resource to which this Act applies with the consent of the resource provider shall, within twelve months from the date of the coming into operation of this Act, make an application for a permit under section 12 or 15, as the case may be". Besides, it applies retroactively for BR and aTK possessed by anyone before its entry into force if no BSA exists.

Excluded from its scope is specific biological or genetic resource covered by and for the purpose of any specialized instrument on access to biological resources and benefit-sharing to which Malaysia is a party (s. 2) and to the use and exchange of BR

---

[28] The scope of material application of the Act to 'human genetic resources' is derived from s. 12 (2) (f) and 15 (3) (g) on conditions for a permit application for commercial, commercial potential and non-commercial purposes respectively, which state that the CA may approve the application if it is satisfied that "in the case of access to **human genetic resource**, its use is not for the purposes contrary to ethical values or the public interest" (*stress in bold added*). The inclusion of human genetic resources in the scope of an ABS regime is unusual as the CBD expressly excludes them from its scope (see Decision II/11. 2 http://www.cbd.int/decision/cop/default.shtml?id=7084) and hence from the main text itself.

[29] Syafira (18 December 2012).

by the indigenous and local community (ILC) among themselves in the exercise of their traditional and customary practices (s. 5 (2) (g)).

3 Institutional Framework

The Act has established a national competent authority (NCA), a competent authority (CA) and an advisory committee (AC). The NCA is established under s. 7 (1) of the Act and consists of the Secretary General of the Ministry responsible for natural resources and environment, who shall be the Chairman, and such number of persons to be appointed by the Minister (s. 2). According to s. 8 it has the following functions:

(a) to coordinate the implementation and enforcement of the provisions of this Act by the Competent Authorities;
(b) to determine the fees payable upon consultation with the Competent Authorities;
(c) to communicate with other countries and the secretariat established under any treaty, agreement, convention or protocol relating to access and benefit sharing in relation to a biological resource, as appropriate, on matters under this Act;
(d) to implement and to fulfil the requirements under any treaty, agreement, convention or protocol relating to access and benefit sharing in relation to a biological resource to which Malaysia is a party where such treaty, agreement, convention or protocol relates to the purposes of this Act;
(e) to create awareness and to provide training, education and information relating to access and benefit sharing in relation to a biological resource;
(f) to keep and maintain a register of permits issued by the Competent Authorities and information relating to the permit;
(g) to establish measures under subsection 30 (1) with the aim of monitoring and tracking of a biological resource or traditional knowledge associated with a biological resource accessed;
(h) to support customary laws and practices of indigenous and local communities, and the development of community protocols and procedures by the indigenous and local communities, as the case may be;
(i) to establish and maintain a clearing house mechanism under section 32;
(j) to act as Competent Authority for all *ex-situ* collections where the origin of the biological resource cannot be ascertained with due diligence and which do not come within the jurisdiction of any other Competent Authority; and.
(k) to do such other things as it deems fit to enable it to perform its functions effectively or which are incidental to the performance of its functions.

The CA means "... the Competent Authority as specified in the First Schedule" (s. 4). There are 14 CAs (see Table 1) which are responsible for issuing access permits and performing other functions (see below): one responsible for three regions and the other 13 for one region each (see Table 1). Each one of them is responsible for all biological resources covered under the Act in the region specified

Table 1 Competent authorities for permit

Competent authority	Biological resource	Region
Ministry of Federal Territories	All covered under this act	Federal Territories of Kuala Lumpur, Labuan and Putrajaya
Johor state economic planning unit	All covered under this act	Johore
Kedah state economic planning unit	All covered under this act	Kedah
Kelantan state economic planning unit	All covered under this act	Kelantan
Melaka state economic planning unit	All covered under this act	Malacca
Negeri Sembilan state forestry department	All covered under this act	Negeri Sembilan
Pahang state economic planning division	All covered under this act	Pahang
Penang state economic planning unit	All covered under this act	Penang
Perak state economic planning unit	All covered under this act	Perak
Perlis state economic planning unit	All covered under this act	Perlis
Selangor state economic planning unit	All covered under this act	Selangor
Terengganu state economic planning unit	All covered under this act	Terengganu
Sabah biodiversity council	All covered under this act	Sabah
Ministry of urban development and natural resources	All covered under this act	Sarawak

Source: *Act 795, First Schedule*

in the first schedule of the Act (s. 9 (1)). A CA has powers to establish an advisory body consisting of indigenous and local communities (ILCs) to deal with issues of BR and aTK relating to such communities whose advice shall be sought and taken into account (s. 9 (2)). The CA may establish such committee to facilitate the carrying out of its functions under the Act if it deems it necessary (s. 9 (3)). It shall consult and seek the advice of the NCA in the exercise of its powers and performance of its functions under the Act (s. 9 (5)).

According to s. 10 competent authorities have the following functions:

(a) to deal with all the applications for access to a biological resource or traditional knowledge associated with a biological resource[30] within its jurisdiction and the

[30] See pre-conditions for access to BR and/or aTK of ILCs under Sect. 4.2 below.

sharing of benefits arising from the utilization of the biological resource or traditional knowledge associated with a biological resource;
(b) to maintain a record of all access applications and decisions relating to such applications, including the permits issued;
(c) to prepare an annual report concerning such access applications and decisions relating to such applications including permits issued and to submit the report and copies of the permits to the National Competent Authority on or before such date as the National Competent Authority may determine; and.
(d) to do such other things as it deems fit to enable it to perform its functions effectively or which are incidental to the performance of its functions.

In addition, they have powers to do all things necessary or expedient for or in connection with the performance of their functions under the Act.

An advisory committee (AC) shall be established by the NCA according to s. 11 (1) of the Act from persons with experience, knowledge and expertise on matters relating to the scientific, legal, technical, ethical and other relevant disciplines. The AC shall provide the NCA with advice related to their expertise upon request (s. 11 (4). A member of the committee shall hold office for a term not exceeding 2 years unless such a member resigns or vacates his office or his appointment is revoked (s. 11 (3)).

4 Requirements, Procedures and Conditions for Access to *In-Situ* and *Ex-Situ* Genetic Resources

4.1 Permit

Access to BR and/or aTK, including by an authorized intermediary,[31] is subject to an access permit. A permit may be issued either for commercial or potential commercial purpose, or for non-commercial purpose.

A permit for commercial or potential commercial purpose is issued by the CA with particulars as may be determined by the NCA in consultation with the CA (s. 13 (1)). An application must be sent to the CA in the form and manner as may be prescribed and the prescribed fee paid (s. 12 (1)). The CA may approve the application if all matters listed under s. 12 (2) are satisfied which include the establishment of a BSA, prior informed consent (PIC) and numerous environmental precautionary requirements. Upon approval of the application the CA shall grant an access permit which shall be evidence of PIC of the CA and ILC (see Sect. 4.2 below), the origin of the BR, the person to whom the permit has been granted, the conclusion of a BSA with the resource provider, and the use for which the permit is

[31] 'Authorized intermediary' "means any person named by the applicant for a permit under sections 12 or 15 to take the biological resource or traditional knowledge associated with biological resource on its behalf" (s. 4).

issued (s. 24). If any of the grounds is not satisfied, the CA shall refuse the application except in regard to s. 12 (c) and (d)[32] if it is satisfied that the application for permit does not undermine the conservation and sustainable use of biodiversity (s. 12 (4)). The CA may also deny the application if, after consultation with the NCA, it is evident that the applicant is from a jurisdiction that does not have adequate and effective user compliance measures (s. 12 (3) (b)).

The CA must state the grounds for refusal and if the application qualifies approve it with or without conditions and issue a permit (s. 12 (5)). Additional conditions on the permit may be imposed, or amendment or revocation of any conditions any time after the permit has been issued (s. 13 (3)). In such circumstances the CA shall give the permit holder a written notice of its intention within which time the permit holder will have an opportunity to make representations and after which the CA shall decide whether to impose, vary or revoke any of those conditions (s. 13 (4), (5)). The CA shall inform the permit holder of its final decision as soon as practicable and the decision shall take effect on a date specified in the written notice (s. 13 (6)).

A permit application for non-commercial purpose must be sent to the CA in the form and manner as may be prescribed and shall be accompanied by a copy of a statutory declaration duly completed as specified in the second schedule and the prescribed fee (s. 15 (1)). According to the second schedule the permit holder must swear that s/he shall:

(a) not use the accessed BR and/or aTK for commercial purposes;
(b) give a written report of the results of the research to the CA;
(c) undertake to offer a taxonomic duplicate of the BR collected to the CA;
(d) undertake not to give the BR to any person other than the CA without its prior written permission; and.
(e) undertake not to carry out, or allow others to carry out, R&D for commercial or potential commercial purposes on any BR, or derivative, or in relation to aTK, unless s/he obtains a permit for commercial or potential commercial purposes in accordance with section 12 and enters into a BSA in accordance with subsection 22 (1) of the Act.

An undertaking for non-commercial purpose must be carried out in collaboration with a public institution of higher education, public research institution or government agency (s. 15 (2)). However, the CA may exempt this condition if it is satisfied that:

(a) the applicant is a non-profit organization based or registered in Malaysia;
(b) local researchers are involved in the activity; and.
(c) a program for capacity building is included in the activity.

The rest of the procedure from review, grounds for approval/refusal of the application for permit and penalties (s. 15 (3)–(7)), issuing of the permit (s. 16)

[32](c) the application is not for any threatened taxa; and (d) the application is not for any endemic species, rare species or any species protected under any federal or state law.

and transmitting of the decision is the same as that for commercial or potential commercial purpose, except the grounds under s. 12 (2) (a)[33] and 15 (3) (a)[34] differ due to differing purposes. This also shows that the exemption of non-commercial research from the conclusion of a BSA as a condition for a permit is the only procedural difference between the two types of researches. The conditions differ but the procedures are the same. However, an access permit requirement is exempted for employees, students and researchers of institutions of higher learning and research and government agency or anyone under their authority within Malaysia subject to PIC of ILC as well as for the exchange of BR among themselves for non-commercial purpose, and for any person in or outside Malaysia who accesses a biological resource from a permit holder or employees and students of institutions of higher learning and research and government agency within Malaysia at their request for the purpose of carrying out or continuing any research for non-commercial purpose (s. 18).

At any time the CA may revoke any permit, whether for commercial or potential commercial or non-commercial purpose (s. 28). The grounds for revocation which can be relied on accumulatively or not are listed under subparagraph (a)-(e).[35] If the CA revokes the permit, it shall notify the permit holder and the NCA of its decision and grounds. Upon the notification the permit holder must surrender the permit, any research results and related documents, and the accessed biological resource to the CA that issued the permit. However, the person aggrieved may in accordance with s. 29 appeal against the decision of the CA in court within 30 days from the time of receiving a written notification of revocation.[36] The court shall confirm or set the decision of the CA aside.

Circumstances where a permit is not required are listed under s. 18 of the Act. Accordingly, no permit is required—.

(a) by any person employed or studying and carrying out research for non-commercial purpose in or under the authority of, a public higher education institution, public research institution or government agency within Malaysia, subject to such conditions as may be prescribed and subject to the prior informed consent of the relevant indigenous and local community for any access to a biological resource and traditional knowledge associated with a biological resource referred to in subsection 23 (1);

[33]"... a benefit sharing agreement has been established in accordance with section 22".

[34]"the application is not for commercial or potential commercial purpose".

[35](a) the permit holder has contravened any provisions of this Act; (b) the permit holder has contravened any of the conditions of the permit; (c) the permit was issued as a result of false, misleading or inaccurate information; (d) the permit was obtained improperly or illegally; or (e) the permit holder has been convicted of an offence under this Act.

[36]For the purposes of s. 29, "'Court' means the High Court of Malaya and the High Court in Sabah and Sarawak or either of the High Court in Malaya and the High Court in Sabah and Sarawak, as the case may require" (s. 29 (3)).

(b) for the exchange of biological resources between persons within a public higher education institution, public research institution or government agency within Malaysia or between such institutions or agencies within Malaysia for non-commercial purpose, unless otherwise required by a competent authority and subject to such conditions as may be prescribed; or.
(c) by any person or institution in or outside Malaysia who accesses a biological resource from a permit holder under subsection 15 (1) or the person or institution under paragraph (a), at the request of such permit holder or the person or institution, for the purpose of carrying out or continuing any research for non-commercial purpose.

4.2 Prior Informed Consent of and Benefit-Sharing Agreement with ILCs

For access to a BR on land to which an ILC have a right as established by law, or to the aTK of such a community, prior informed consent (PIC) must be obtained from the relevant community (s. 23). Such PIC shall be obtained in accordance with customary laws and practices, protocols and procedures of ILC, as the case may be. If access is for commercial or potential commercial purposes the applicant must enter into a BSA with the relevant ILC. This process is executed through the representative, organization or body identified in accordance with the customary laws and practices, protocols and procedures of the said indigenous and local communities. However, if such representation cannot be identified in regard to aTK, the process shall be executed with the holders of the aTK within the ILC and, if the holders of the aTK cannot be identified, with the Federal Government or State Authority.

If the same aTK is shared by more than one ILC, the applicant must obtain PIC and enter into a BSA with the duly identified representative or organization of all the holders of the aTK. However, if it is not practicable in all the circumstances of the case to ascertain all such holders, and this is proven to the satisfaction of the CA, the applicant shall obtain the PIC of, and enter into BSA with the duly identified representatives or organization of such holders as the applicant may ascertain. Should another ILC also claim ownership, the CA shall determine such claim in consultation with the ILC whose PIC has been obtained and BSA entered into, and if it proves the claim, declare such an ILC entitlement to the benefits due. Further, the CA shall determine the quantum or nature of benefits such an ILC is entitled to in consultation with all the ILCs concerned.

5 Permit Conditions

5.1 Third Party Transfer

A permit for commercial or potential commercial purpose cannot be transferred or any right, duty, liability or obligation under the permit assigned to a third party (s. 14 (1)).

For a permit for non-commercial purpose no person or institution referred to in s. 18 (a) and (b) shall transfer any BR and/or aTK, or results of research in relation to the BR and/or aTK to a person or institution other than that referred to in s. 18 (a) and (b) without the prior approval of the CA (s. 19 (1)). If the CA approves the transfer it shall prescribe the conditions under which the transfer may take place and, in addition, the transferee shall be required to apply for a permit under the Act unless the CA decides otherwise (s. 19 (2)). Besides, the permit holder or the person or institution referred to in s. 18 (a) and (b) shall notify the CA of the access (s. 19 (3)).

5.2 Change of Intent

For commercial or potential commercial purpose no change of use is allowed in relation to the BR as specified in the permit issued under s. 13 (1) except upon a fresh application being made and a permit issued under s. 13 (§ 2). A holder for a permit for non-commercial purpose, whether the initial holder or a transferee, is not permitted to carry out, or allow others to carry out, R&D for commercial purposes on any BR and/or aTK, unless the person obtains a permit for commercial or potential commercial purpose under s. 13 and concludes a BSA in accordance with s. 22 (s. 20).

5.3 Records of BR and aTK

A permit holder is obliged to keep a record of the BR and/or aTK accessed for 20 years after the end of the period of use as long as the BR and/or aTK is in his possession (s. 26). The records shall indicate the description of the BR and/or aTK, including available unique identifiers; the date or dates of access; the place of access; the quantity or size of the biological resource (such as weight or physical dimension); the common and scientific name of, or given to, the biological resource; the location where the biological resource is kept; and the particulars about any subsequent physical disposition of the biological resource, including the names and addresses of others having possession of the BR or a part thereof. The permit holder shall furnish the NCA and CA with a copy of the records within 30 days after the BR is taken or within a period determined by the NCA and CA.

5.4 *Disposal of BR*

According to s. 27, the permit holder may offer the BR for which s/he has a record to the CA if s/he does not intend to keep it. However if the CA does not agree to take it, the permit holder shall dispose the BR in a manner as may be determined by the CA and the permit holder shall forthwith furnish a report of the disposal of the BR to the CA. There is no information whether the 20 years' obligation to keep records ceases after disposal.

6 Benefit-Sharing

As already said and according to s. 22, an applicant for a permit for access to BR and/aTK for commercial or potential commercial purpose must enter into a BSA with the resource provider. The BSA shall be based on mutually agreed terms and must provide for fair and equitable benefit-sharing. Where the Federal Government or State Authority is not the resource provider, it may require the applicant to pay a percentage of any monetary benefits derived under the BSA as it may determine. Such monetary benefits shall be deposited into a fund as may be established by the Federal Government or State Government. The CA shall use any payment or any part of the payment received towards the conservation of biodiversity and the sustainable use of its components and for such other incidental expenses. Benefits shared with the ILC for utilization of their aTK shall be applied for the interest of ILC taking into account the advice of the advisory body established under s. 9 (2) (s. 23 (5)). Benefits shared can be of a monetary or non-monetary character.

7 Sanctions

The Act establishes sanctions for offences committed which are illustrated in Table 2 below.

8 Provider and User Compliance Measures

Section 30 of the Act gives the NCA in cooperation with the CA power to establish measures for the purpose of monitoring and tracking BR and aTK. Such measures shall include designation of checkpoints and the requirement to produce the permit at the checkpoint established.[37] The checkpoints in turn are required to inform the

[37] At the time of writing this chapter checkpoints had not been established.

Table 2 Offences and sanctions

Section of Act	Offence	Sanctions	
		Individual	Body corporate
s. 12 (6)	Accessing BR or aTK for commercial or potential commercial purpose without a permit or contravention of any condition imposed on the permit	Fine not exceeding 500,000 ringgit[a] or imprisonment for a term not exceeding 10 years or both	Fine not exceeding 5,000,000 ringgit
s. 14 (1), (2)	Transferring a permit or assigning any right, duty, liability or obligation under the permit to any other person	Fine not exceeding 200,000 ringgit or imprisonment for a term not exceeding 10 years or both	Fine not exceeding 500,000 ringgit
s. 15 (7)	Accessing BR or aTK for non-commercial purpose without a permit or contravention of any condition imposed on the permit	Fine not exceeding 100,000 ringgit or imprisonment for a term not exceeding 7 years or both	Fine not exceeding 1,000,000 ringgit
s. 16 (7)	Change of use/intent in regard to BR issued under a permit for non-commercial purpose without making a fresh application or contravening additional conditions or amendments or revocations made on the permit	Fine not exceeding 100,000 ringgit or imprisonment for a term not exceeding 7 years or both	Fine not exceeding 500,000 ringgit
s. 21 (1)	Supplying, selling, offering or advertising for sale, offering for profit, gain or benefit BR or aTK by a person who is not an authorized intermediary to a person who requires a permit under the Act	Fine not exceeding 500,000 ringgit or imprisonment for a term not exceeding 10 years or both	Fine not exceeding 5000,000 ringgit
s. 28 (5)	Failure to comply with the obligation to surrender the permit, any research results and related documents, and the accessed BR after revocation of a permit	Fine not exceeding 200,000 ringgit or imprisonment for a term not exceeding 10 years or both, and for continuing offence a further fine not exceeding 5000 ringgit for each day the offence continues after conviction	Fine not exceeding 500,000 ringgit, and for continuing offence a further fine not exceeding 10,000 ringgit for each day the offence continues after conviction
s. 31 (2)	Failure to notify the NCA within 30 days about an application for a patent based on BR or aTK	Fine not exceeding 100,000 ringgit or imprisonment for a term not exceeding 7 years or both	Fine not exceeding 500,000 ringgit

(continued)

Table 2 (continued)

Section of Act	Offence	Sanctions	
		Individual	Body corporate
s. 41 (2), (3)	Refusal/failure of the person in control/charge of a conveyance to stop it and open all parts for examination by an enforcement officer	Fine not exceeding 100,000 ringgit or imprisonment for a term not exceeding 7 years or both	
s. 44 (9), (11)	Failure to provide or be responsible for free passage for repatriation of BR imported illegally to Malaysia and, where required, proper maintenance and storage during its transportation	Fine not exceeding 100,000 ringgit or imprisonment for a term not exceeding 7 years or both	Fine not exceeding 150,000 ringgit
s. 53 (1)	False, inaccurate of misleading declaration	Fine not exceeding 50,000 ringgit or imprisonment for a term not exceeding 3 years or both	Fine not exceeding 500,000 ringgit
s. 53 (2)	Altering, forging, mutilating or defacing a permit, certificate or registration or knowingly making use of such a permit	Fine not exceeding 100,000 ringgit or imprisonment for a term not exceeding 7 years or both	Fine not exceeding 500,000 ringgit
s. 62 (3)	The regulations to be adopted may prescribe any act in contravention of the regulations to be an offence	Fine not exceeding 200,000 ringgit or imprisonment for a term not exceeding 10 years or both	Fine not exceeding 500,000 ringgit

Source: *Own illustration*
[a]1 ringgit is equivalent to approximately 0,24 USD (0,20 EUR) as at 13 October 2020

NCA and the relevant CA in writing of the production of the permit. Checkpoints shall include offices or authorities dealing with applications for patent, product registration or approval and public research funding grants.

Besides, s. 31 requires any person applying, locally or abroad, for a patent based on BR or aTK to notify the CA in writing 30 days from the date of application. Further, s. 34 obliges any person, before using or commercializing BR or aTK accessed from a party to the Nagoya Protocol (NP), to provide a permit or its equivalent to the relevant authority as evidence that he/she has complied with the law or other regulatory requirements of that party. However, this is only imposed if such a party subjects access to a permit. A notification by a CA of another NP party to the CAN that it has not issued such a permit or its equivalent will be *prima facie* evidence of the notification. To impose compliance the NCA has been empowered to establish relevant measures.

In addition, s. 35 gives power to the Minister to appoint enforcement officers with powers to investigate any case relating to the commission of an offence under the Act. The enforcement officer may at any time enter any premises for the purpose of inspecting,[38] verifying[39] or collecting any BR (s. 38) and may seize and detain a thing suspected to have been used in the commission of an offence (s. 40–41). These powers do not limit similar powers conferred on any person under any other written law (s. 36). The Act establishes the procedure to be followed by the enforcing officers, CA, affected persons and the magistrate adjudicating the case (s. 42–44).

9 Regime's Strengths, Weaknesses and Assessment

9.1 Concerning Compliance with the Nagoya Protocol

The Act has tried to comply with the NP. It establishes national competent authorities and defines their functions (ss. 7–11) in line with article 13 NP, including which authority is responsible for which BR and in which region (First Schedule). Information on how to lodge applications for permits, the requirements and grounds for denial (ss. 12–16) or revocation (s. 28) in line with article 6.3 NP is provided as well as the obligation of the CA to issue a permit at the time of access as an (written) evidence of its decision to grant it (ss. 17, 24).

It takes measures to ensure that aTK of the ILCs is accessed with PIC and MAT and according to their customary laws, community protocols and procedures and BSA concluded with them according to mutually agreed terms (s. 23) in line with articles 7 and 12.

Measures to comply with article 4 NP, which exempts specialized instruments for ABS that are consistent with, supportive of and do not run counter to the objectives of the CBD from regulation under the NP, have been taken under s. 2. As said, the Act exempts from its scope not only the multilateral system of the International Treaty on Plant Genetic Resources for Food and Agriculture but any specialized ABS instrument to which Malaysia is a party.

The Act differentiates research for commercial, potential commercial and non-commercial purpose through definition (of non-commercial, s. 4) but through the elaboration of the term 'access' and 'research and development' the two types of researches seem to have been thrown into the same pot; the difference fades out. Besides, the differentiation is of no gain as the procedure for access is not varied. However, some exemptions have been created for use of BR by and exchange between employees, students and researchers of institutions of higher learning and government agency or anyone under their authority for non-commercial purposes

[38]To be inspected is any BR, conveyance, machinery, contrivance, equipment, book, record, document or other article as considered necessary.

[39]To be verified is the accuracy of records or statements or any information given to an enforcement officer.

and a possibility of exchange between these and persons in Malaysia and abroad. In addition, a list of access to BR that may be exempted by the Minister from the scope of the Act is also provided under s. 60.These measures are supportive of article 8 (a) NP.

Measures to implement articles 15 and 16 NP on compliance with measures of another party (s. 34) as well as article 17 on monitoring, tracking (s. 30–31) and enforcing compliance (ss. 35–49) have been put in place.

In spite of the said, the Act also has shortcomings. For example, it subjects human genetic resources to the ABS regime contrary to the CBD and its NP. The scope is generally very broad covering 'biological resources' instead of 'genetic resources'. It seems like the definition of the former tries to capture the broadened scope based on the NP e.g. to include derivatives and to already regulate information.

There is nothing said concerning opportunities to seek justice in case of disputes from mutually agreed terms or mutual recognition and enforcement of foreign judgements, nor is jurisdiction to which disputes shall be subject defined, or options for dispute resolution or applicable law as required by article 18 NP. Perhaps general provisions of the law dealing with these issues apply and thus the Act chooses to be silent but it needs to be clear on this. There is an element of access to justice under s. 29 which creates an opportunity for anyone to seek recourse in court of law if aggrieved by the decision of the CA to revoke a permit. This is also an indication that courts can be approached for other ABS related matters.

The Act states that benefits deposited in the fund shall be invested into conservation and sustainable use of biodiversity (s. 22 (5)), which is in line with article 9 NP. It is still to be seen how this provision will be applied in practice as the percentage to be used for this purpose is not defined. The provision just says "any payment or any part of the payment".

9.2 *Other Issues*

Besides terms defined in the CBD and NP e.g. 'biological resources', 'derivative' (both are broader than the CBD definitions of the same and include information) and 'genetic resource' (which is verbatim to the CBD definition except it adds 'fungi' to the material it includes), the Act provides critical definitions under s. 4 such as 'authorized intermediary', 'local community' 'non-commercial research' which implies any research for academic purposes or non-profit oriented, 'research and development' etc. The term 'access' is not defined under definitions (s. 4) but is explained under s. 5 (1). Accordingly, a person is said to have access to BR when he/she takes[40] it for purposes of R&D and there is reasonable prospect as determined by the CA that the BR will be subject to R&D.

[40]The term 'take' is defined under s. 4 and means collection, harvesting, picking, catching, obtaining etc. of BR. A list of activities involving taking of BR for uses that do not comprise access is provided under s. 5 (2).

Also the Act protects the rights of small farmers to save, use, exchange and sell farm-saved seed or propagating material and to undertake conventional breeding or traditional practices in use in agriculture, horticulture, poultry farming, dairy farming, animal husbandry or bee keeping (s. 6).

Although the law generally looks legally certain, there are elements which might cause some uncertainty. For example, the right of the CA to impose additional conditions or amend any condition after the permit has been issued any time (s. 13 (3)). Also the Act does not give fixed timeframe for the CA to inform the permit holder of its final decision after the permit holder has made representations against imposition of new conditions, amendment or revocation of a permit but just says as soon as practicable (s. 13 (6)).

10 Conclusion

For the first time Malaysia has a stand-alone ABS legislation covering the whole federation. Before 2017 issues concerning access to biological resources and benefit-sharing were regulated through sectoral laws except in the states of Sabah and Sarawak, which had enacted stand-alone laws.[41] The Act, as seen above, defines the scope, establishes national authorities to regulate ABS, states the requirements for grant of access permits and lays down procedures for applications and their review, differentiates requirements, conditions and procedures for commercial/ potential commercial and non-commercial research etc. Besides it has also defined violations and sanctions thereto, a clearing house mechanism and, unlike most traditional provider countries' laws, has user (compliance) measures as well as monitoring and tracking measures to support compliance. The law is clear and to a great extent gives a sense of legal certainty.

The law has the potential to ease access to BR and aTK and ensure fair and equitable sharing of benefits. Nonetheless, some issues are still incomplete and need elaboration. For example, it is not clear which particulars the applicant must submit with the application, the duration, terms, conditions and restrictions to be imposed upon issuance of a permit, the minimum terms for a BSA, or the fees to be charged for a permit. Some organs have also not yet been created. As already said the Act empowers the Minister to make regulations to clarify and specify such issues. As this chapter is been written the regulations are not existent. Likewise, there is currently no information concerning the legislation or country profile in relevant issues including on the website of the Secretariat of the CBD (except the national focal point). Likewise, there are no guidelines or booklet to help a user to navigate through the system. Probably this is normal for a country that did not have a law in this area prior to the Act of 2017. However, these issues will need to be addressed swiftly to operationalize the Act and to test its strengths as well as unveil the weaknesses. The

[41] Nordin et al. (2012), p. 185.

situation is likely to change once the implementation of the legislation picks speed and regulations are made as foreseen in s. 62 (1) of the Act.

It is true that Malaysia has taken a long time to deliver a legislation to govern the use of biodiversity for research and development and the sharing of benefits arising therefrom, but as the fable about the hare and the tortoise teaches, over the long-run the race will go to the slower, steadier animal. Malaysian long-time and steady efforts have delivered one of the best existing ABS laws.

References

Bibliographical Material

Carrizosa S, Brush SB, Wright BD, McGuire PE (eds) (2004) Accessing biodiversity and sharing the benefits: lessons from implementation of the convention on biological diversity. IUCN, Gland, Switzerland and Cambridge, UK

Nordin R, Hassan KH, Zainol ZA, Shapiee R, Jalil F, Witbrodt MA, Hassan MS (2012) Regulating access to genetic resources and the sharing of benefit through the Sarawak access benefit-sharing legislation. Res J Appl Sci 7(3):185–193

SaBC (Sabah Biodiversity Centre) & GDF (Global Diversity Foundation) (2010) ABS and Traditional Knowledge in Sabah. In: Sham MA (ed) Biocultural diversity and conservation learning platform briefing notes, Vol 2. Compiled from a seminar on ABS and Traditional Knowledge in Sabah, 16 March 2009, Kota Kinabalu, Sabah

Online Sources

Schriver C (2012) Institutional arrangements for implementing ABS law and its subsidiary regulations. http://www.ketsa.gov.my/ms-my/Documents/Pengumuman/Final%20Report%20IA_ABS_Draft.pdf. Accessed 14 Oct 2020

Syafira S (2012 December 18) I am drafter: Access and benefit sharing law in Malaysia. http://syafi-ct.blogspot.com/2012/12/i-am-drafter-access-and-benefit-sharing.html. Accessed 26 Nov 2020

Policy Documents and Legislation

Access to Biological Resources and Benefit Sharing Act 795 of 2017. Available at http://extwprlegs1.fao.org/docs/pdf/mal176890.pdf. Accessed 12 Oct 2020

Malaysia National Biodiversity Policy (1998) https://www.cbd.int/doc/meetings/nbsap/nbsapcbw-seasi-01/other/nbsapcbw-seasi-01-my-en.pdf. Accessed 10 Oct 2020

Ministry of Natural Resources and Environment (2014) National Biodiversity Planning to Support the Implementation of the CBD 2011–2020 Strategic Plan in Malaysia (NBSAP). https://www.informea.org/en/action-plan/malaysia-nbsap-v2-2016. Accessed 10 Oct 2020

Ministry of Natural Resources and Environment (2016) National policy on biological diversity 2016–2025. https://www.cbd.int/doc/world/my/my-nbsap-v2-en.pdf. Accessed 10 Oct 2020

Evanson Chege Kamau Dr. iur., is Senior Research Fellow at the Research Centre for European Environmental Law (FEU), University of Bremen. He is also an Adjunct Lecturer at the CvO University of Oldenburg. His area of expertise is public international law, European law and intellectual property law. His research focuses on genetic resources, environment, development and traditional rights on biological resources and traditional knowledge. Kamau has over 17 years of working experience in the area of ABS in which he has published widely. For many years he has also worked as a freelance legal consultant for governments, African Union (hired by IUCN and Capacity Building Initiative resp.), private industries and research organizations. echege@uni-bremen.de.

Access and Benefit-Sharing Law and Policy in South Korea

Jae-Hyup Lee and Ah Young Cho

Abstract Although South Korea, which is regarded as a user country, put forth a great deal of effort to establish the access and benefit-sharing (ABS) regime at the national level, stakeholders' opinions as to how to incorporate the Nagoya Protocol into domestic law were so different in the process of making law that enacting the ABS Act was delayed until January 2017. Besides, some ministries involved in genetic resources related work had enacted their laws before the enactment of the ABS Act. As a result, provisions of the ABS Act reflect these problematic circumstances. This chapter provides an analysis of the provisions of the ABS Act of Korea and discusses the limitations and legal suggestions to overcome the current situation. Weaknesses of the regime could be summarized as (1) dispersed authorities and laws, (2) non-legally binding reporting system, (3) non-legally binding mutually agreed terms and (4) an obscure concept of 'provider'. For a more effective regime, it is necessary to build an integrated platform that enables comprehensive management and support of domestic and international genetic resources. In addition, the Korean government can play a facilitating role by establishing guidelines for companies and researchers.

Keywords Korean ABS Act · Domestic implementation · Access and benefit-sharing · User country · Compliance

1 Introduction

By July 2020, 126 countries have ratified the Nagoya Protocol since it entered into force on 12 October 2014. In the early stages, the number of parties increased, especially in developing and least developed countries, which have (or are believed to have) many genetic resources. Recently, the number of ratifying developed

J.-H. Lee (✉) · A. Y. Cho
Seoul National University School of Law, Seoul, South Korea
e-mail: jhyup@snu.ac.kr

E. Chege Kamau (ed.), *Global Transformations in the Use of Biodiversity for Research and Development*, Ius Gentium: Comparative Perspectives on Law and Justice 95, https://doi.org/10.1007/978-3-030-88711-7_12

countries has also increased. These international trends have influenced South Korea's policy on genetic resources, since South Korea has a stake in genetic resources (GR) and has undergone many changes in the surrounding environment related to GR since it became a party to the Convention on Biological Diversity (CBD) in 1995. As a hosting nation of the CBD Conference of the Parties (COP) 12 and the Conference of the Parties serving as the Meeting of the Parties to the Nagoya Protocol (COP-MOP) 1 in Pyeongchang in 2014, South Korea put forth a great deal of effort to establish the ABS regime at the national level in line with international trends. Korea signed the Nagoya Protocol on 20 September 2011, and the Korean National Assembly proposed legislation on genetic resources access and benefit-sharing (ABS), "Act on Access and Utilization of Genetic Resources and Sharing of Benefits" (ABS Act), in 2013. However, this bill did not become law until 2017. Despite Korea's continued efforts to embrace the international regime associated with the Nagoya Protocol, it took a considerable amount of time to become a party to the Nagoya Protocol and even longer to enact domestic ABS legislation to implement the Nagoya Protocol. This delay resulted from the complexities of reflecting the interests of the state and impact on industries from the perspective of South Korea as a 'user country'.

South Korea imports a large amount of genetic resources for academic study and commercial use from resource-rich countries. In addition, the biotechnology industry has grown into a major industry in Korea. In the cosmetics industry, where exports have increased rapidly, many products are made using overseas GR. For these reasons, South Korea took steps to incorporate the interests of domestic bio-industry. According to the results of a survey conducted on bio-industry-related companies, 41.5% of companies responded that they had used genes or their derivatives in the process of research and development (R&D) and commercialization, of which 59.1% (24.5% of all bio-industry-related companies) had used foreign genes or their derivatives.[1] The companies cited the main reasons for using overseas GR as "absence of domestic alternative resources", "expensive domestic genetic resources costs", and "insufficient domestic supply of genetic resources".[2] When companies or researchers in Korea import a large amount of genetic materials from abroad, they are more affected by the legislative, administrative or policy measures of the country that provides the genetic materials than by Korean domestic ABS law. In addition, provider countries may be more reluctant to export GR to Korea if Korea does not have any measure of sharing benefits arising from the utilization of GR. Therefore, delayed ratification of the Nagoya Protocol did not help protect the interests of Korea's bio-industry. Eventually, the Korean government responded to these concerns by becoming a party to the Nagoya Protocol and enacting the ABS Act in January 2017.

[1] Korea Institute of Intellectual Property (2013), p. 74.

[2] Ibid, p. 82.

2 Development of Legislation

As said, Korea is generally regarded as a user country of genetic resources, and the government of Korea has primarily focused on its role as a user country due to its industrial structure and limited land space and mountainous terrain. Others suggest that Korea could be classified as a rich provider country since Korea benefits from diverse biotic resources due to a unique geography featuring the coexistence of forest and coastal ecosystem even with its relatively small size and mountainous terrain.[3] It is evident that new endemic species of plants and animals are constantly being discovered in Korea, and it has significant biological diversity. However, it is difficult to say Korea is a provider country because genetic resources in Korea are insufficient and expensive to provide in large quantities. Also, from a political standpoint, bio-industry is an ascendant industry in Korea, and a tremendous amount of GR is needed for its development. According to a survey of bio-industry-related companies and universities, the size of the bio-industry in Korea is 8.9 trillion Korean Won (7475 billion USD) in 2016.[4] Sixty-seven per cent of the biological resources used in Korea are imported from abroad, and 78% of those used for cosmetics are imported.[5] This exemplifies how Korean industry depends on overseas GR. As a consequence, Korean companies and researchers who work with GR have a strong interest in the development of the ABS system.

Korea enacted laws and regulations for protecting the environment and species even before Korea ratified the CBD. However, these laws and regulations only protected biological diversity indirectly and lack comprehensive management of biodiversity. They focus on the protection of specific species, and biodiversity policies had been dispersed throughout government departments.[6]

In order to implement the CBD and integrate biodiversity policies systemically, Korea enacted the "Act on Conservation and Use of Biodiversity" in February 2012. The primary purposes of the Act are to (i) contribute to the enhancement of biodiversity by creating a national management system, (ii) promote sustainable use of biological resources, and (iii) cooperate with the international mechanisms including the Convention.[7] As one of the three pillars of the CBD is the fair and equitable sharing of the benefits arising out of the utilization of genetic resources, the Act also contains some provisions about ABS-related issues. For example, its article 19 provides for the obligation of sharing profits from biological resources. Moreover, some ministries involving GR related work enacted their laws. However, the provisions of these laws are declarative, fragmented, and not enough to implement the Nagoya Protocol in Korea. Although several laws establish requirements of ABS related issues or refer to the Nagoya Protocol, they lack unity because different

[3]Union for Ethical BioTrade (2016), p. 1.

[4]Ministry of Trade, industry and Energy (2017).

[5]Korea Biotechnology Industry Organization (2017), p. 1.

[6]Republic of Korea (2014), p. 28.

[7]Act on Conservation and Use of Biodiversity, Art. 1.

ministries enacted relevant provisions according to their jurisdictions. Besides, the Nagoya Protocol requires State Parties to take legislative, administrative, or policy measures to implement a fair and equitable sharing of benefits resulting from the use of genetic resources. Therefore, the necessity of enacting a new law to implement the Nagoya Protocol had increased. As a result, the legislative process to implement the Nagoya Protocol began with the draft in 2013, but the process required much time due to many concerns such as the impact on domestic industry. The National Assembly was finally able to enact the ABS Act in 2017 after an extensive policy debate about the ABS system to balance the interests of South Korea as primarily a user country with provider country equities.

The purposes of the ABS Act are "to provide for necessary matters for implementing the Nagoya Protocol on Access to and the Fair and Equitable Sharing of Benefits Arising from their Utilization to the Convention on Biological Diversity; and for those for promoting the fair and equitable sharing of the benefits arising from accessing and using genetic resources and traditional knowledge associated with genetic resources thereby contributing to the conservation and sustainable use of biological diversity, improving the lives of the people, and promoting international cooperation".[8] When we compare it to the Act on Conservation and Use of Biodiversity, the ABS Act reaffirms one of CBD's three pillars as elaborated in the Nagoya Protocol and has a more specific objective than that of the said Act.

3 The ABS Framework

3.1 Scope

According to article 3, the ABS Act shall not apply to genetic resources existing in areas beyond national jurisdiction (ABNJ), such as Antarctica. In other words, the ABS Act sets its geographical scope of application within national jurisdictions.

The definition of GR follows 'genetic resources' as defined in article 2, s. 4 of the Act on Conservation and Use of Biodiversity. The Act defines genetic resources as "materials which have practical or potential value, among plants, animals, and microorganisms or other genetic material which becomes genetic origins including a genetic functional unit". This definition of genetic resources is not very different from the one of the CBD and the Nagoya Protocol. However, neither 'derivatives' nor 'digital sequence information' (DSI) are found in any provisions of the ABS Act. The DSI of GR has been registered in the open database so far and is freely available, but whether the Nagoya Protocol applies to DSI is under discussion. In this respect, it seems that Korea aligns itself with the user country perspective. However, in stipulating the types of GR to which the ABS Act does not apply, 'derivatives' are not included explicitly in the same manner, as no conclusion was made regarding

[8] Act on Conservation and Use of Biodiversity, Art. 1.

derivatives, which was the biggest issue in the Nagoya Protocol's subsequent discussion. This, in turn, reflects Korea's ambiguous attitude toward derivatives as a user country, and commentators expect that the matter will be clarified in the future.[9] In addition, the ABS Act defines traditional knowledge (TK) without mentioning 'indigenous and local communities (ILC)', which arguably do not exist in Korea. This approach seems to be an attempt to protect TK that can actually be found in Korea.

In terms of the temporal scope of the ABS Act, article 1 of Addenda of the ABS Act stipulates that "this Act shall enter into force on the date the Protocol becomes effective in the Republic of Korea: provided, that articles 9 through 11, 15, 16, 22 and 28 shall enter into force one year after the date the Protocol becomes effective in the Republic of Korea". In other words, the provisions of the ABS Act about prohibition of access to and utilization of domestic genetic resource(s) and procedural compliance regarding access to and utilization of foreign genetic resource (s) shall begin to apply for a person who accesses genetic resource(s) for their utilization after the ABS Act enters into force. The Presidential Decree of the ABS Act also has a similar addendum, which states that duty-related matters are deferred for 1 year so that the applicable provisions of the law come into force 1 year after enactment. In consequence, the ABS Act and the Presidential Decree entered into force entirely on 17 August 2018. This might be in order to minimize confusion among stakeholders, including workers and researchers of bio-industry.

Article 4 of the ABS Act enumerates exemptions of applicability, as follows: (1) human genetic resource(s); (2) genetic resource(s) existing in an area beyond national jurisdiction, such as Antarctica; (3) genetic resource(s) accessed for any purpose other than utilization;[10] (4) genetic resource(s) subject to other international treaties related to access to genetic resource(s) and benefit-sharing. Also, the Act does not apply to genetic resource(s) for which patent rights have been granted. This is according to a unique provision found in article 3 which says that "this Act shall not apply to genetic resource(s) granted patent rights pursuant to Article 87 (1) of the Patent Act". This is the result of reflecting the opinion of the Korean Intellectual Property Office that it may not be appropriate for the government to regulate privately held genetic resources.

---

[9] Kang (2018), p. 598.

[10] "Utilization" means conducting research and development on the genetic and biochemical composition of genetic resource(s), including through the application of biotechnology (Art. 2 Act on Conservation and Use of Biodiversity). There is no catalogue of activities listed in the Act, but a guideline published by NIBR in Dec 2019 provides the list of "Guidance document on the scope of application and core obligations of Regulation (EU) No 511/2014" as an example.

Table 1 Competent national authorities and their fields of jurisdiction

Ministry	Fields of jurisdiction
Ministry of Science and ICT	Biological resources pursuant to the act on the securement, management, and utilization of biological research resources
Ministry of Agriculture, Food and Rural Affairs	Agricultural biological resources pursuant to the act on the protection, management and use of agricultural biological resources
Ministry of Health and Welfare	Pathogen resources pursuant to the act on the promotion of collection, management and utilization of pathogen resource
Ministry of Environment	Biological resources in the field of wildlife in accordance with the wildlife protection and management act and biological resources in accordance with the act on the conservation and use of biological diversity
Ministry of Oceans and Fisheries	Marine biological resources pursuant to the act on the securement, management, and utilization of marine biological resources

Source: *Own illustration*

3.2 Institutional Framework

Korea has two national focal points, the Ministry of Environment and the Ministry of Foreign Affairs. The functions of each ministry are distinct. The Ministry of Foreign Affairs liaises with the Secretariat of the Convention on Biological Diversity. On the other hand, the Ministry of Environment disseminates information on access to genetic resource(s) and benefit-sharing. However, among all parties to the Nagoya Protocol, Korea is the only country that has two national focal points, while the other parties have one national focal point.[11] According to article 13 of the Nagoya Protocol, each Party shall designate "a" national focal point on access and benefit-sharing, so an in-depth discussion is needed to decide whether having two focal points is appropriate or not.

There are five competent national authorities based on each ministry's field of jurisdiction. The competent national authorities (competent national authority or CNA) and their fields of jurisdiction are delineated as in Table 1.

The duties of the head of the CNA are to process a report on access or report on changes in accordance with article 9 (reporting access to domestic genetic resources), to prohibit or restrict access and utilization of domestic genetic resources in accordance with article 12 (prohibition of access to and utilization of domestic genetic resources), and to support the fair and equitable sharing of benefits from domestic genetic resources. Other necessary matters relating to access and utilization of domestic genetic resources shall be prescribed by Presidential Decree. Article 3 of the Presidential Decree stipulates further matters, which are to investigate and manage the current status of utilizing domestic GR, and to investigate and manage

[11] The Access and Benefit-Sharing Clearing-House, https://absch.cbd.int/.

the establishment of the agreement on sharing benefits from domestic GR and the implementation status thereof.

With regard to checkpoints, there are six checkpoints in Korea, adding the Ministry of Trade, Industry, and Energy to the CNA list. Its field of jurisdiction is biological resources according to the Act on the Securement, Management, and Utilization of Biological Research Resources. This is because the Ministry of Trade, Industry, and Energy has some affiliated research institutions. The duties of the checkpoints provided by the ABS Act are processing procedural compliance reporting, monitoring and recommending procedural compliance, supporting persons who utilize foreign genetic resources domestically, and other matters prescribed by Presidential Decree.

4 Nature of Regulatory Tools for Access to Genetic Resources and Associated Traditional Knowledge

According to article 2, the definition of ‘access’ is “acquiring specimen or substance of genetic resources, or collecting genetic resources and information regarding traditional knowledge associated with genetic resources”. Moreover, this provision also states that genetic resources and traditional knowledge associated with genetic resources shall be referred to as “genetic resources”.

4.1 Reporting Access to Korean Genetic Resources and Prior Informed Consent

Instead of requiring a benefit-sharing agreement prior to access, Korea only requires a report to the competent authority regarding access. Article 9 states that “where foreigners, overseas Koreans, foreign institutions, international organizations, and those other persons designated by Ordinance of the Ministry of Environment seek access to domestic genetic resource(s), they shall report such access to the head of the Competent National Authority”. Article 4 of the Presidential Decree of the ABS Act requires that a person who seeks to report access to domestic genetic resource (s) pursuant to article 9 (1) of the ABS Act shall submit a report on access to domestic genetic resource(s) stating as follows: (1) the reporting person’s name and affiliation (in the case of a corporation, its corporate name and the name of its representative), address (in the case of a corporation, the location of its place of business), date of birth or corporate registration number, and contact number; (2) name of the genetic resource(s) to be accessed (common name and scientific name of the relevant species), quantity, and concentration; (3) methods for accessing the relevant genetic resource(s) (including information on third parties, if genetic resource(s) are provided through third parties contracts, etc.) and period of use;

(4) the name and affiliation of the person providing the relevant genetic resource (s) (in the case of a corporation, its corporate name and the name of its representative), address (in the case of a corporation, the location of its place of business), date of birth or corporate registration number, and contact number; (5) the purpose(s) for accessing the relevant genetic resource(s); (6) method for utilizing genetic resource (s), including biotechnology that will be applied to the relevant genetic resource(s); (7) country(ies) seeking to utilize the relevant genetic resource(s); and (8) whether mutually agreed terms are established regarding the relevant genetic resource(s), and the details thereof.

Meanwhile, reporting in accordance with the ABS Act is deemed completed if (1) an approval is obtained in accordance with article 11 (2) of the Act on the Conservation and Use of Biological Diversity or where reporting is completed in accordance with article 13 (1) of the same Act; (2) an approval is obtained in accordance with article 18 (1) of the Act on the Conservation, Management, and Use of Agricultural Bio-Resources; (3) a permission is granted under article 11 (1) of the Act on Securing, Management, and Use, etc. of Marine and Fisheries Bio-Resources or where approval is obtained in accordance with Article 22 (1) of the same Act; or (4) an approval or permission is obtained under article 16 (1) or 18 (1) of the Act on the Promotion of Collection, Management, and Utilization of Pathogen Resources or where reporting is completed in accordance with Article 16 (2) of the same Act (article 9 (2)).

However, the text of article 9 leaves some ambiguity as to the meaning of 'report'. When a certain matter is notified to the administrative agency, in principle, the reporting obligation is fulfilled if the notification meets the procedural requirements. This is understood as a mitigated regulatory measure that distinguishes it from a permission. However, even when a report is purportedly submitted, it only has legal effect as notice when the administration has accepted the report. Some believe that 'report' under article 9 should be interpreted as 'a report requiring acceptance'. This means that the head of the State responsible authority who has been notified in accordance with article 4 (3) of the Presidential Decree, which embodies article 9 (1) of the ABS Act, states "[U]pon receipt of a report under paragraph (1), the head of the Competent National Authority shall notify the person who has filed the report as to whether such report is accepted, within 30 days from receipt thereof. In such cases, he/she shall issue a declaration certificate for access to domestic genetic resource(s), as prescribed by Ordinance of the Ministry of Environment". Others hold the contrary opinion that a report, rather than a regulatory approach such as PIC, which is generally applied to genetic resource access, should be interpreted as a "self-satisfying report" since the original intent of the choice of the word 'report' means a "self-satisfying report" when we look through the history of the enactment of the ABS Act.[12]

Article 10 of the ABS Act stipulates exemptions from the requirement to report access to domestic genetic resource(s). It states that "notwithstanding article 9 (1),

[12]Park (2018), p. 243.

the head of the Competent National Authority may, as prescribed by Presidential Decree, seek to simplify procedures for reporting access to genetic resource(s) or waive such reporting following consultation with the head of the relevant central administrative agency". These are the cases (1) where it is deemed that expedited access to or utilization of the genetic resource(s) is required for the development of therapeutic drugs and food security in the event there are threats to the life and health of humans, animals, and plants, and (2) where genetic resource(s) are accessed for non-commercial purposes such as pure research; provided, that if the intended purpose has changed, the access shall be reported without delay in accordance with article 9 (1).

However, one of the problems in the interpretation of article 10 is the distinction between the research of commercial and non-commercial nature, and article 8 (a) of the Nagoya Protocol has the same problem. The commerciality distinction is difficult for the following reasons. First, both types of research usually require access to the same genetic material resources. Second, both types of research can benefit the conservation and sustainable use of biodiversity. Third, regardless of whether a primary agent of the research is a public institution or a private company, it is possible to participate in commercial research as well as non-commercial research. Lastly, research which started with non-commercial purposes can become commercial research. As a result, article 8 (a) of the Nagoya Protocol requires the provision of "'simplified access procedures" for purely scientific research and other research for non-commercial purposes, and it is necessary to make additional comments regarding cases of "change of intent (purpose)" that differ from the time of application. However, the Nagoya Protocol is silent on the criteria to distinguish between the two. The ABS Act also has no criteria.

On the other hand, article 12 specifies cases which access and utilization of domestic GR can be prohibited or restricted. The head of the CNA may seek to prohibit or restrict access to and utilization of domestic genetic resource(s), following consultation with the head of the relevant central administrative agency, if such access or utilization (1) actually harmed or is deemed to risk harming the conservation and sustainable use of biodiversity, and (2) has or is deemed to risk having a negative socio-economic impact on the value of biodiversity.

4.2 *Procedural Compliance Reporting on Foreign Genetic Resources*

A person who intends to access foreign genetic resources for utilization in Korea has a duty to report to the head of the national checkpoint, as prescribed by Presidential Decree, that he/she has complied with the procedures established by the country providing such resource(s) (Article 15 (1)).

The compliance report procedure is prescribed by article 6 of the Presidential Decree. The person shall submit a procedural compliance report to the head of the relevant national checkpoint within 90 days from receipt of consent to access foreign genetic resource(s) from the providing country of the relevant foreign GR. And this report shall include information as follows: the reporting person's name and affiliation (in the case of a corporation, its corporate name and the name of its representative), address (in the case of a corporation, the location of its place of business), date of birth or corporate registration number, and contact number; name of the providing country of the foreign genetic resource(s); name and address of the person providing the foreign genetic resource(s); name of the providing country's institution granting the prior informed consent, date of grant, and grant number; name of the foreign genetic resource(s) granted by the prior informed consent (common name and scientific name of the relevant species), quantity, or concentration; purpose for utilizing the foreign genetic resource(s) granted by the prior informed consent; and whether mutually agreed terms are established regarding the relevant genetic resource(s), and the details of such terms if concluded.

According to article 16 of the ABS Act, the head of the national checkpoint may monitor whether a person utilizing foreign genetic resource(s) domestically has complied with the procedures prescribed in article14 (1). Where necessary as a result of the monitoring conducted, the head of the national checkpoint may recommend a person utilizing foreign GR domestically to comply with the procedures prescribed in article 14 (1). A person utilizing foreign GR domestically shall cooperate in monitoring. However, there is a criticism that the national checkpoints will have little effect on countries that do not have such laws or regulations related to ABS, since the procedural compliance report obligation applies only to a person who uses GR from overseas provider countries that established ABS procedures.[13]

4.3 *Application for Verifying the Establishment of Mutually Agreed Terms*

The Nagoya Protocol relies heavily on a private contract based on the so-called mutually agreed terms (MAT) for the regulatory function of benefit-sharing between providers and users. Parties to the Nagoya Protocol are obliged to adopt appropriate measures to provide that MAT have been established as required by relevant national laws or regulatory requirements (article 15 (1) of the Nagoya Protocol). However, under the ABS Act, there is room for MAT to be considered established as long as an agreement between the parties is recognized to exist, regardless of the content, form, and timing of the MAT since the ABS Act does not have any explicit provisions related to MAT. For instance, according to "Form 1. Enforcement Rule of the Act on Access to and Utilization of Genetic Resources and Benefit-Sharing", applicants should answer to the question whether MATs are established or not. In addition, the Presidential Decree states that where a person in receipt of a declaration certificate

[13] Kim (2017), p. 168.

establishes mutually agreed terms after the certificate is issued, he/she may request the head of the Competent National Authority to verify the conclusion of mutually agreed terms.[14] This means that MAT could be established even after issuing PIC.[15] Also, there is no provision about the content. However, this may contravene the purpose of the Nagoya Protocol and the ABS Act because article 6(3)(g) of the Nagoya Protocol stipulates that each Party requiring prior informed consent shall establish clear rules and procedures for requiring and establishing MAT. Therefore, there is a great need for an appropriate provision. The principles of fair and equitable benefit-sharing based on the idea of equity and private autonomy are in a strained or conflicting relationship with each other; the intervention of governments that are parties to the Nagoya Protocol to MAT cannot be avoided although MAT comes under the principle of private autonomy.[16]

5 Sanctions and Compliance with the Nagoya Protocol

According to article 26 of the ABS act, any person who accesses or utilizes genetic resource(s) prohibited or restricted from access or utilization, in violation of article 12 (1), shall be punished by imprisonment for not more than 3 years or by a fine not exceeding 30 million Korean Won. Anyone who violates the reporting obligations prescribed by law shall be subject to administrative fines (not exceeding ten million Korean Won shall be imposed on any person who fails to file a report under articles 9 (1) and 15 (1), and not exceeding five million Korean Won shall be imposed on any person who fails to file report on changes under article 9 (3)), in accordance with article 28. Also, used genetic resources shall be confiscated under article 27. There is no punishment provision for the illegal use of foreign resources. Even if a domestic user is prohibited from accessing or using a GR or accesses and uses a restricted GR, it cannot be punished under the ABS Act. In order for the Nagoya Protocol to work effectively, efficient enforcement must be secured, and such enforcement is ultimately determined by whether punishment provisions and reimbursement laws actually work. However, it is difficult to expect fair compensation decisions and enforcement when each national judicial system will have its own laws, enforcement practices, and jurisdiction over a limited subset of ABS transactions.

[14] Article 4(4) of the Presidential Decree.

[15] However, if establishing MAT is a prerequisite for access, it can have a negative impact, such as procedural delays in the use of genetic resources. See Oh (2018), p. 303.

[16] See Oh (2018), p. 304 and Park (2015), p. 101.

6 Benefit-Sharing

According to article 2 of the ABS Act, ‘benefits’ means both monetary and non-monetary benefits. The range of monetary benefits includes, but is not limited to, royalties and income from utilizing genetic resource(s), and non-monetary benefits include, but are not limited to, sharing of research results and transfer of technology.

In terms of benefit-sharing, article 11 of the ABS Act stipulates “providers and users of genetic resource(s) shall agree to share the benefits of domestic genetic resource(s) fairly and equitably”. Originally, the draft bill of Ministry of Environment required that providers and users “shall try to agree to share the benefits of domestic genetic resource(s)”, and “the head of the competent national authorities may assist in ensuring that the benefit-sharing is concluded fairly and equitably”. Some held the opinion that article 5 of the Nagoya Protocol only obliges Parties to take legislative, administrative, and policy measures in order to share profits, so the purpose of article 11 of the draft bill stipulated that the efforts of users and providers and the support of the government be reasonable.[17] However, the provision was finally amended to “shall agree to share”. Using this stronger language theoretically increased the importance of agreement on benefit-sharing. Nonetheless, there is no separate sanction for breach of the obligation.

7 Regime’s Strengths and Weaknesses and Suggestions

The new Korean ABS regime shows several positive features. First, an ‘integrated reporting system’ established and operated by the Minister of Environment in accordance with article 7 of the Presidential Decree following consultation with the heads of the CNA and the national checkpoint provide both foreign users of Korean GR and Korean users of foreign GR with simple and easy online reporting services.[18] With this system reporting on access to domestic genetic resource (s) under article 9 and procedural compliance reporting on access to foreign GR under article 15 can be managed electronically and efficiently. This lessens the burden of and the confusion for a person who seeks to access and utilize Korean domestic GR.

Moreover, it is not the Korean government but the users who decide where the reports are submitted among five competent national authorities. This is a very unique system where huge discretion is given to the users for convenience in determining the competent national authority.[19]

[17] The National Assembly of the Republic of Korea, Environment and Labor Committee (2016), pp. 28–30.

[18] Ministry of Environment and Ministry of Foreign Affairs (2018), p. 7.

[19] Park (2019), p. 277.

Second, article 18 (1) of the ABS Act prescribes a Council to deliberate on determining any unclear scopes of authority between CNA or national checkpoints (Presidential Decree 11 (1)). If the work of ABS, which is dispersed across each ministry, can be integrated into a single governance structure, experts and administrative departments that have in-depth knowledge and understanding of GR can provide specialized expertise in completing the work.

Third, a specialized institution provides administrative and technical support to main government agencies. The Genetic Resources Information Center (GRIC) and databases related to GR are vital to serve that purpose.[20] The GRIC was established in March 2018 to expertly perform duties related to ABS system in accordance with article 17 (1) of the ABS Act, and its duties are (1) collecting, managing, investigating, and providing information related to access to and utilization of domestic and foreign genetic resources and benefit-sharing, (2) providing information on domestic genetic resources in regard to the Access and Benefit-Sharing Clearing-House in accordance with article 14 of the Nagoya Protocol, and (3) other necessary matters relating to the duties of the National Focal Point, Competent National Authorities, and National Checkpoints as prescribed by Presidential Decree which are matters concerning administrative and technical support, international cooperation (including information exchanges with foreign genetic resources information management organizations for investigation of the utilization status of the domestic genetic resource(s) abroad), and public relations, in order for providers and users of genetic resource(s) to raise awareness of access to and utilization of genetic resource(s) and benefit-sharing.

Fourth, a systematic and streamlined information system has been developed to collect and integrate the scattered biological data in Korea. Foreign users should use the retrieval system when they try to find an appropriate competent national authority that they shall report to access Korean genetic resources and check if the competent national authority has the genetic resources. The government tries to integrate the database of each ministry into one system called KOBIS (the Korea Bio-Resource Information System) for efficient acquisition and systematic management of biological data. Each database is detailed in the table below (Table 2).

However, the regime also suffers from several weaknesses. First, the dispersed authority and laws are the most severe problem. Multiple national authorities compete over the jurisdiction of ABS-related issues, and the relevant laws are fragmented. There are five competent national authorities and six national checkpoints. This situation may present an obstacle to an organized and efficient establishment of the ABS system based on the Nagoya Protocol.

Korea's biodiversity laws, especially ABS-related legislation, were scattered according to each ministry's jurisdictions and laws, and thus the policy on ABS system lacked integrated and consistent management before the ABS Act was enacted. Although it was expected that integration would be possible according to

[20] Park (2018), pp. 245–249.

Table 2 The composition of Korea Bio-Resource Information System and its databases

	Governmental Branch	DATABASE	Progress
KOBIS ("KOREA BIO-RESOURCE INFORMATION SYSTEM") Integrates databases into one network system	Ministry of Environment	CBD-CHM KOREA	– Founded/ Managing National CHM – Connecting different data base – Manages biological diversity based on function (Agricultural, Dry or Wet land/ Forest/ Interior water)
	Ministry of Science and CT	ARIS	– Focusing on research resources – Networking small unit data banks – Database for researchers (Differentiating public/research database)
	Ministry of Agriculture, Food and Rural Affairs	BRIS	– Provides detailed and focused data(in Agricultural field) – Provides integrated information based on the actions made (Collecting/Managing/Distributing/Transfer of information/ Utilization)
	Ministry of Health and Welfare	1.KBN 2.BIMS 3.0 3.Vector Bank 4.CODA	1. KBN provides unified data base of 17 individual unit collections of human resources 2. BIMS (Biobank Information Management System) managing R&D research on human derives/ epidemic or clinical information 3. Provides data on carrier insects of diseases 4. Clinical & Omics data archive
	Ministry of Oceans and Fisheries	MBRIS	Comprehensive information system for efficiency on marine biological resources Combines fragmented information of marine resources, building standard data system for marine resources

Source: *Own illustration*

the enactment of the ABS Act, the result is not yet perfect. Many scholars have criticized this situation in Korea.[21]

In Korea, there are six more laws related to ABS besides the ABS Act. A common distinctive feature of these laws is that ministries try to maximize the definition of GR under their jurisdictions in order to expand their leverage on GR. Foreign users

[21] See Park (2018, 2019).

who want to access domestic GR might get confused about the institution to which their accesses must be reported depending on the location of the genetic resources (e.g., inland, forest, ocean) or the purpose of their use. For example, marine pathogens could belong to pathogen resources as well as marine biological resources at the same time. This may be a factor in avoiding access and use of domestic GR.[22] In addition, each major legislation related to the conservation of biodiversity has different timelines for plan implementation; even their starting points are all different.

Second, Korea only requires reporting to the CNA and does not require a benefit-sharing agreement prior to access, when foreigners try to access Korean GR. This can be regarded as one of the Korean government's attempts to increase the number of instances of access and utilization of Korean domestic genetic resources. This decision seems to be based on the judgment that Korea does not have abundant GR. However, commentators criticize that the current access reporting system gives up sovereign rights to domestic biological gene sources and should be reexamined.[23] Also, more endemic species and GR may be discovered in Korea, and new materials could be developed from the resources.

Third, MAT is not legally binding. Although the 'Benefit-Sharing of Domestic Genetic Resources' provision of the initial bill was finally revised to "shall agree to share the benefits of domestic genetic resource(s)", no sanctions can be imposed even if the obligation is violated because there are no relevant provisions in the ABS Act. In addition, only the presence or absence of the MAT is required in filling out the access report to domestic resources in accordance with article 9 of the ABS Act and article 4 of the Presidential Decree.[24]

Fourth, the ABS Act and Presidential Decree do not have a provision on "provider" for benefit-sharing entities. This causes a serious lack of clarity on the procedural order and method of access reporting.[25] In Korea, there has been no single case where an internationally recognized certificate of compliance has been registered for the issue of PIC and MAT for Korea's access to genetic resources.[26] To avoid uncertainty, it may be advisable for Korea to amend ABS-related legislation to set out explicit provisions concerning delegation or entrustment of authorization of providers.[27]

[22] Park (2018), pp. 245–249.

[23] Ibid, p. 243.

[24] Ibid, pp. 244–245.

[25] Park (2019), p. 274.

[26] Ryu (2018), p. 125.

[27] Park (2019), p. 275.

8 Conclusion

The Nagoya Protocol still has a need for international consensus, and this lack of consensus may be complemented through follow-up measures such as legislation on genetic resources. Therefore, the analysis and review of the ABS legislation of the Parties to the Nagoya Protocol is important information for practitioners and users, and provides insight into understanding the role of national interests in the international community.

In the previous sections, we analyzed the provisions of the ABS Act of Korea, discussed its problems, and suggested how to overcome these problems. In response to the international trend to strengthen sovereignty over biological resources, Korea enacted the ABS Act and attempted to establish an institutional basis for reporting access to domestic GR and sharing profits from their utilization. Responding to the ABS system can be considered in two ways: (1) to protect domestic biological resources as a provider country, and (2) to promote access to foreign biological resources as a user country. There are some deficiencies from both perspectives.

One of the most problematic issues is multiple authorities and decentralized laws. This not only prevents foreign users from understanding how they could access Korea's GR, but can also increase administrative tasks, additional costs, and risk management burdens due to the possibility of overlapping and potentially conflicting work done by different ministries. However, the development of life science and technology has expanded its scope in newly developed industry, and there are many stakeholders and governmental organizations that manage and assist these industries in Korea. Therefore, it is urgent to build an amicable system of cooperation and coordination across these stakeholders and institutions. To do so, it is necessary to build an integrated platform that enables comprehensive management and support of domestic and international biological genetic resources. As mentioned above, the multi-layered structure of a number of competent national authorities and national checkpoints clearly imposes limitations on efficient ABS and causes work duplication among the ministries. But it seems impossible to integrate them completely into one agency. Rather, it will be realistic to create a control tower for them as a long-term goal.

The one-year grace period in the ABS Act was not sufficient, and many actors in impacted industries were not well-prepared. An actor importing, transferring, utilizing, or selling a biological, genetic source depends upon having a clear interpretation and best practice of which governmental authority is implementing the Nagoya Protocol and when it will be applied. Governments must play a facilitating role by establishing guidelines for companies and researchers that provide a clear interpretation of representative cases. Companies and researchers can use them as a basis for consultation and negotiation with overseas providers without doubting the credibility of the legal regime providing the framework for the ABS exchange.

References

Articles

Kang MK (2018) A comparative study on the legislation relating to Nagoya protocol in Korea and China under the Korea-China FTA. Chin Law Rev 36:587–608

Kim DS (2017) Analysis of the access and benefit-sharing framework in the Republic of Korea. Korean J Int Econ Law 15(3):157–185

Oh SY (2018) Establishing clear rules and procedures for requiring mutually agreed terms in Korea. Environ Law Rev 40(2):289–310

Park JW (2015) Korean experience in implementing Nagoya protocol and the related legal issues. Environ Law Rev 37(1):67–112

Park JW (2018) The significance and limitations of the "act on access to utilization and benefit-sharing of genetic resources". Environ Law Rev 40(2):225–261

Park WS (2019) A study for enhancing Korea's access to genetic resources system. Joongang Law Rev 21(1):245–286

Ryu AR (2018) A study on process and cases of ABS to genetic resources and associated traditional knowledge in India. Environ Law Rev 40(3):123–151

Reports

Korea Biotechnology Industry Organization (2017) Report on the promotion of domestic user perceptions on responding to the Nagoya protocol. National Institute of Biological Resources, Incheon

Korea Institute of Intellectual Property (2013) A study on the trend analysis and legal protection of new intellectual properties - a study on current condition of foreign country's system for genetic resources and effective. Korean Intellectual Property Office, Daejeon

Ministry of Environment and Ministry of Foreign Affairs (2018) A guide to act on access to and utilization of genetic resources and benefit-sharing in Republic of Korea. Ministry of Environment and Ministry of Foreign Affairs, Seoul

Republic of Korea (2014) The Fifth National Report to the Convention on Biological Diversity. https://www.cbd.int/doc/world/kr/kr-nr-05-en.pdf . Accessed 3 Aug 2020

The National Assembly of the Republic of Korea Environment & Labor Committee (2016) Review report on a bill on ABS. The National Assembly of the Republic of Korea, Seoul

Union for Ethical BioTrade (2016) ABS in South Korea. https://static1.squarespace.com/static/58bfcaf22994ca36885f063e/t/5b1699540e2e72611be71b5b/1528207701797/UEBT-ABS+South+Korea-2016.pdf. Accessed 4 Dec 2019

Webpages

Ministry of Trade, industry and Energy (2017) Domestic Bio Industry Survey as of 2016. http://www.motie.go.kr/motie/ne/presse/press2/bbs/bbsView.do?bbs_seq_n=159959&bbs_cd_n=81&currentPage=1&search_key_n=&cate_n=1&dept_v=&search_val_v=. Accessed 3 Aug 2020

The Access and Benefit-Sharing Clearing-House. ABS National Focal points. https://absch.cbd.int/search/nationalRecords?schema=focalPoint. Accessed 3 Aug 2020

Jae-Hyup Lee is Professor of Law at Seoul National University. He has widely published in both Korean and English in the areas of international and comparative environmental law and energy law. As a legal counsel to the Ministry of Foreign Affairs and Trade of Korea, he has participated in a number of international conferences and negotiations related to biodiversity, biosafety, climate change, and trade. He is currently Editor-in-Chief of the *Journal of Korean Law*, and the Director of the Center for Energy & Environmental Law and Policy at Seoul National University. jhyup@snu.ac.kr.

Ah Young Cho is currently a senior researcher at the satellite geophysics laboratory at Seoul National University (SNU) as well as a PhD candidate in Environmental Law at SNU Law School. She graduated from Columbia Law School (2018–2019, LLM) and was a visiting scholar at the same school (2019–2020). She received a BS in Environmental Earth Science and a J.D. from SNU. Prior to starting an LLM program at Columbia Law School, she worked as a research scientist and lawyer at the Ocean Policy Institute/Korea Institute of Ocean Science and Technology, a governmental research institute. ahyoung0707@gmail.com.

ABS in Australia: A Story of Early Success and Faltering Progress

Geoffrey C. Burton

Abstract Australia was the first Organization for Economic Co-operation and Development (OECD) country to introduce Access and Benefit-sharing (ABS) laws to implement the Convention on Biological Diversity (CBD) article 15—Access to Genetic Resources. To do so, it followed a path of legal and operational innovation, contributed to the further international evolution of ABS through the development of world's best ABS practice—the CBD Bonn Guidelines on Access to Genetic Resources and the Fair and Equitable Sharing of Benefits Arising out of their Utilization (Bonn Guidelines) and later contributed to the development of the Nagoya Protocol on Access to Genetic Resources and the Fair and Equitable Sharing of Benefits Arising from their Utilization to the Convention on Biological Diversity (Nagoya Protocol). Yet eight years after signing the Protocol Australia has yet to ratify the treaty. This chapter outlines Australia's development and progress on ABS, describes its innovations and contribution to the global ABS system, identifies the remaining steps needed to fulfil its obligations prior to ratification and suggests why progress has stalled. It will, moreover, identify increasing signs that Australia's State Governments and its research community are losing patience with this delay and have begun taking concrete steps to achieve compliance with the Nagoya Protocol.

Keywords Genetic resources · Nagoya Protocol · Bonn Guidelines · Ratification · Voumard Inquiry

G. C. Burton (✉)
Jean Shannon & Associates, Sydney, NSW, Australia
e-mail: geoff@jeanshannon.com

E. Chege Kamau (ed.), *Global Transformations in the Use of Biodiversity for Research and Development*, Ius Gentium: Comparative Perspectives on Law and Justice 95, https://doi.org/10.1007/978-3-030-88711-7_13

1 Background: Geography, Biodiversity, Indigenous Peoples and Industry

Australia's continental area of 7.7 million square kilometers is similar in size to that of the United States of America. Australia's marine jurisdiction, however, at 8.2 million square kilometers is larger than its terrestrial extent and ranges from tropical waters near the equator to Antarctica.[1]

Australia is one of the 17 megadiverse countries.[2] Geographical isolation has produced abundant biodiversity, often unique and sometimes ancient or rare. Moreover, Australia's indigenous peoples' 50,000 years of human settlement[3] has given them an intimate knowledge of biodiversity; its properties and management. Accordingly, indigenous communities hold an extensive body of traditional knowledge (TK) associated with genetic resources (GR). Moreover, indigenous communities own, manage or have recognized land rights over 27% of Australia.[4] These rights vary from full ownership of lands and waters to rights limited to simply using the land for traditional cultural practices. This requires any Australian ABS law to take this into account.

Australia is a wealthy country with a developed economy.[5] It has growing biotechnology and life-science industries. In 2016 the sector was valued at AU $100 billion with 1000 companies, including 300 biotechnology companies and 400 medical technology companies. The sector employs 48,000 people.[6] Biodiscovery is an important element as reflected by its inclusion in Australia's National Biotechnology Strategy with its stated objective of enhancing access to Australia's GR.[7]

2 Evolution of ABS in Australia

The evolution of access and benefit-sharing (ABS) in Australia at national level has a long history. Australia became a party to the Convention on Biological Diversity (CBD) on 29 December 1993. Its first legislative step to implement its commitments

[1] See https://www.ga.gov.au/scientific-topics/national-location-information/dimensions/oceans-and-seas.

[2] See https://www.australianwildlife.org/wildlife.

[3] See http://www.australasianscience.com.au/article/science-and-technology/when-did-aboriginal-people-first-arrive-australia.html.

[4] See https://aiatsis.gov.au/explore/articles/land-rights.

[5] In 2019, Australia was the world's 14th largest economy at US$1.4 trillion and a per capita income of US$54,000.00. See https://www.austrade.gov.au/International/Invest/Resources/Benchmark-Report.

[6] See https://www.ausbiotech.org/biotechnology-industry/fast-facts.

[7] See https://www.cbd.int/doc/measures/abs/msr-abs-au4-en.pdf, p. 26.

under the CBD was to introduce an overarching national environmental law: The *Environment Protection Biodiversity Conservation Act* 1999 (EPBC Act). Provision for an ABS scheme was included at section 301.

2.1 Section 301 of the EPBC Act

As the EPBC Act was being prepared for implementation, the question arose of how to implement Australia's particular obligations under articles 15 and 8 (j) of the CBD. Article 15 enables parties to the CBD to provide facilitated access to genetic resources and requires their prior informed consent and to be undertaken on mutually agreed terms for the fair and equitable sharing of benefits. Article 8 (j) encourages equitable benefit-sharing with the holders of traditional knowledge associated with biological diversity. Section 301 of the incoming Act, however, made provision for the control of biological resources in Commonwealth jurisdictions to be made by subordinate legislation (regulations). Nor did it detail how this might be undertaken. Nevertheless, section 301 did provide that the regulations include:

- the equitable sharing of the benefits arising from the use of biological resources:
- the facilitation of access to such resources in Commonwealth (Federal) areas;
- the right to deny access to such resources; and
- the granting of access to such resources and the terms and conditions of such access.[8]

2.2 Policy Investigation

To develop such a governance scheme the Government decided to initiate a national inquiry into ABS (the Voumard Inquiry 2000[9]). This would be a top-down process but involving comprehensive consultation. The Inquiry would be headed by an independent chairperson and supported by an advisory committee, including representatives from the scientific, research, biodiversity-conservation and indigenous peoples' communities. The opportunity was also taken to include a member of the CBD Secretariat in their private capacity.

At the outset, and under Ministerial direction, officials of the Department of Environment and Heritage supported the Inquiry as the primary means of securing evidence from stakeholders and distilling a consensus on the best means of meeting the Commonwealth and State and Territory Government's agreed goal set out in

[8] Section 301 of the Act, see https://www.legislation.gov.au/Details/C2018C00440/Download.

[9] See the Voumard Report.

Australia's 1996 *National Strategy for the Conservation of Australia's Biological Diversity* at Objective 2.8. This was to:

> ensure that the social and economic benefits of the use of genetic material and products derived from Australia's biological diversity accrue to Australia.[10]

In approaching the task of recommending a domestic legal structure to secure this objective, the Inquiry recognized that Australia's geographical, biological, political and ethnic circumstances created an unusual situation whereby its own internal dialogues paralleled the developing dialogue between developed and developing countries. These parallels were subsequently reflected in elements of submissions made to the Inquiry. Accordingly, the Inquiry was conscious that the functionality of any future scheme it recommended would depend, in part, on its compatibility with future requirements for international research and investment partnerships, respect for traditional indigenous knowledge (TK) and on emerging international agreements on ABS.[11]

So, Australia was in a position to mirror many of the features and concerns of developed-economy 'user' countries while at the same time being; a biodiversity-rich, resource providing country and one with indigenous communities holding traditional knowledge associated with genetic resources. Accordingly, Australia was able to view issues and concerns from the perspective of both developed and developing countries and was therefore well placed to reconcile those concerns domestically and to address them internationally in negotiations underway within the CBD.

3 Ownership of Genetic Resources in Commonwealth Areas[12]

3.1 Commonwealth Areas

The National Government of Australia is often referred to as the 'Commonwealth Government' or informally as the 'Federal Government'. Accordingly, land or waters owned or managed by the Commonwealth is referred to in the regulations as 'Commonwealth areas'.

Australia's States and territories manage the bulk of land while the national (Commonwealth) Government is responsible for the greater part of its marine jurisdiction. On land, the Commonwealth's terrestrial jurisdiction is limited. It

[10] See National Strategy for the Conservation of Australia's Biological Diversity, p. 35.

[11] Executive staff of the Inquiry headed the Australian delegations to CBD discussions and negotiations on international cooperation on ABS.

[12] Under the EPBC Act section 525 (pp. 503–504) a "Commonwealth Area" includes land owned or leased by the Commonwealth, its external territories, and offshore waters but excludes the Australian Capital Territory and some waters given to the States to administer.

largely consists of national parks leased from indigenous peoples, various defense lands and Australia's offshore islands. Otherwise, lands, enclosed waters and waters between the low-water mark and the three-mile limit are the jurisdiction of State and Territory Governments.[13]

Australia is a common law country. It has a strong tradition of respect for private property. Accordingly, under Australia's Constitution the Commonwealth Government is constrained from taking any genetic resources (GR) it does not own without paying the owner compensation on just terms—but it is free to regulate its use.[14] Thus, any Commonwealth ABS law regulating the taking of resources must guarantee shared benefits flow to private landholders where they exist. In this case, with some caveats relating to the operation of indigenous peoples' Native Title rights in Australia,[15] privately owned land within Commonwealth administered lands and waters is largely confined to indigenously owned or managed lands or waters and leased back to the Commonwealth.

With this in mind, regulations introducing ABS would have to set out two pathways for benefit-sharing; one for Commonwealth owned or managed lands and waters and one for indigenously owned and managed lands and waters within the Commonwealth's jurisdiction. In the latter case, and with Article 8 (j) taken into account, the role of the Commonwealth would be to protect the interests of indigenous owners by ensuring fair dealing by users.

3.2 Ownership of State and Territory Genetic Resources

As a federated nation Australia is made up of 6 sovereign States and 2 territories. Under the operation of the Australian Constitution, each State is sovereign except for those powers reserved for the national Government. Australia's 2 territories, the Northern Territory (NT) and the Australian Capital Territory (ACT) have been accorded self-Government and may pass such laws as they will, provided those laws do not conflict with Commonwealth law.

Only two Australian jurisdictions, Queensland and the NT, have enacted specific ABS laws to date. In both cases they respect common law principles in relation to private property. In Queensland the Government chose to apply ABS law to public lands and waters and does not extend it to private property. Thus, Queensland does

---

[13] Commonwealth of Australia Constitution Act 1900.

[14] This principle is enacted at sub-section 51 (31) of the Commonwealth of Australia Constitution Act 1900. See https://www.aph.gov.au/About_Parliament/Senate/Powers_practice_n_procedures/Constitution.aspx.

[15] Native Title: put simply, native title is the application of the principle in British common law that if land is possessed and possession has not been defeated, then the possessor owns the land. Thus, if indigenous Australians maintain possession of the land then they have native title to it. When native title is recognized, any acquisition by the Commonwealth requires compensation on just terms. See section 18 of the Native Title Act 1993, https://www.legislation.gov.au/Details/C2017C00178.

not seek to regulate or obtain any benefit from access to GR found on private property. The NT took a different approach in its ABS law. Its ABS law applies to all land and waters within its jurisdiction but requires that any shared benefits derived from GR provided by private property owners must flow to the owners under a benefit-sharing agreement (BSA).

4 Origins of Australia's ABS Laws: Commonwealth Public Inquiry on Access to Biological Resources in Commonwealth Areas (the Voumard Inquiry)

Under the active guidance of the Inquiry Chairman, Mr. John Voumard, and with the support of the Inquiry's Expert Reference Group providing him with expertise on; indigenous, legal, scientific, research, and environmental issues, the Inquiry recommended a practical, legally-robust access and benefit-sharing framework to be introduced by regulation under the Environment Protection Biodiversity Conservation Act 1999.[16]

The Inquiry's consultation process was extensive. It involved advertising in national media for written submissions. It held public hearings around Australia, met with State Governments, and with Federal agencies including those representing agriculture, education, science and intellectual property. This process also included meeting with representatives of individual companies and with members of the scientific research and environment communities and as well as representatives of the broader biotechnology industry. The Inquiry particularly sought input from indigenous representative organizations and in order to meet with the elders of indigenous communities directly, it flew to regions with indigenously owned land located in Commonwealth areas (federal areas). These were located in Australia's NT and on the southern coast of Australia.

The result was a information rich process and led to the development of principles and objectives derived from stakeholder concerns. These concerns may be summarized as follows:

Scientific Research Community The scientific research community involved in 'pure' or non-commercial research was concerned that:

- their freedom to conduct research and publish should not be constrained;
- access to biological resources should not be limited;
- any approval process introduced should not result in undue delay or costs; and
- the contribution of science to the development of public policy, at all levels of Government, was being inhibited by the continued lack of public investment in the study and identification of Australia's extensive biodiversity.

[16] The Voumard Report.

On this last point, various stakeholders made the simple aphorism that "you cannot manage what you do not know you have" and its corollary, "or how it works". They pointed out continued underfunding of taxonomy had led to significant unintended economic consequences. They cited the example of the miss-identification of imported Red Fire Ants leading to a costly, multimillion dollar eradication campaign.[17] Consequently, they argued that any scheme for the successful utilization of the nation's GR should be supported by increased public investment in morphological and molecular taxonomy.

Biotechnology Industry The Australian biotechnology industry and representatives of the applied sciences shared some of the concerns of the broader scientific community but were particularly sensitive to the:

- lack of public and private capital investment into the application of biotechnology into the genetic and biochemical makeup of species;
- burden of any imposed regulatory costs and delays to research;
- importance of the continued conservation of areas accessed to enable re-collection as required;
- lack of development capital to advance discoveries to commercialization;
- under appreciation of the contribution of molecular taxonomy to classical or morphological taxonomy;
- importance of legal certainty for the acquisition, possession and utilization of samples of biological resources;
- the cost of patent protection and its defense; and
- the value of a transparent and robust regulatory process to prevent perceptions of 'bio-piracy'.

However, their strongest concern was that any benefit-sharing requirement introduced must be realistic about the value and likelihood of the successful commercialization of genetic resources.

They made the point that the probability of success in bio-discovery was low, the time scales involved were long, the process involved significant additional investment along a lengthy, complex, commercialization pathway, and usually required establishing partnerships with overseas companies with expertise in final stage manufacturing, product approval and marketing. In short, and in particular, the development of a new pharmaceutical derived from nature was a process best described as "a low probability - high reward equation". It was pointed out that this level of inherent uncertainty is well illustrated by the fact that even after extensive investment, and successful Phase 3 clinical trials, 50% of drugs put

[17] The failure to promptly identify Fire Ants has incurred substantial eradication costs, with AU$28 million spent in 2001 and, by 2016, this figure had risen to AU$328 million. See Consultancy Report for the Steering Committee for the National Red Imported Fire Ant Eradication Program, p. 19.

forward to the United States Food and Drug Administration (USFDA) for evaluation and subsequent drug registration nevertheless fail.[18]

Conservation Groups Environment and conservation groups were concerned that the taking of biological samples must not damage the environment and that the abundance of species should not be put at risk. They asserted that the principles of ecologically sustainable development, including the precautionary principle, and, in certain situations, the need for environmental assessment, should guide the Inquiry's recommendations. Moreover, they asserted that where samples were to be accessed from indigenously owned or managed land, it should only be obtained with the owners' consent and with all rights of indigenous peoples being respected.

Aboriginal Land Owners Within Commonwealth Jurisdiction Aboriginal elders and their representatives were clear: Any system put in place must respect the rights and culture of indigenous Australians. Accordingly, indigenous land owners must have the right to grant or deny access to their land and waters. Any economic benefits derived from the use of their biological resources should be shared in accordance with a benefit sharing agreement (BSA) negotiated between the users and the providing local community. The Northern Land Council, as a statutory Aboriginal representative body, reinforced these views. It strongly cautioned that care should be taken to ensure that unrealistic expectations were not created and thereby create community discord. It had been its earlier experience that unrealistic expectations over mining royalties caused great social conflict among and between local indigenous communities.

External Territory of Norfolk Island The Australian external territory of Norfolk Island is located in the Pacific Ocean, 1600 kilometers from the Australian mainland. It has an unusual history. The Island was given to the descendants of Tahitians and British naval mutineers by Queen Victoria in 1856 and has been administered by the Commonwealth of Australia as an external territory since 1914. For some years it was a self-governing community until 1 July 2016, when the Commonwealth assumed full responsibility for direct governance.

Following consultation with the then Norfolk Island Government, the Inquiry accepted the principle that benefits derived from any utilization of the Island's land or marine genetic resources should go to the benefit of the island community. This was not withstanding that half of the island was a Commonwealth national park and that the Commonwealth would likely undertake any necessary permitting and associated benefit-sharing contract negotiations. Such permitting and benefit-sharing agreements would therefore, effectively, be conducted on behalf of the island.

[18] Over the intervening period, this approval rate does not seem to have changed appreciably. See https://blogs.sciencemag.org/pipeline/archives/2019/05/09/the-latest-on-drug-failure-and-approval-rates.

Agricultural Representative Agencies The then Department of Agriculture Fisheries and Forestry expressed the view that nothing should be done that would inhibit Australia's access to overseas plant genetic resources for food and agriculture or conflict with the forthcoming International Treaty on Genetic Resources for Food and Agriculture (ITPGRFA)—should Australia sign the Treaty in its final form.[19] The department also made the point that any system introduced by regulations under the EPBC Act must not conflict with the provisions of Australia's *sui generis* intellectual property laws concerning the registration and protection of plant varieties. These laws implement the provisions of the International Union for the Protection of New Varieties of Plants Treaty (UPOV). Australia became a member of UPOV on 1 March 1989 and introduced a Plant Breeder's Rights Act in 1994 to comply with its provisions.[20]

5 Development of Underlying Principles: Addressing Stakeholder Concerns

In response to stakeholder input and after exhaustive analysis, the Inquiry report made 73 recommendations to the Government to create a practical legal framework for facilitated access to GR and to establish an equitable benefit-sharing scheme.[21] These recommendations and the legal scheme it recommended for introduction were based on principles[22] it developed in response to stakeholder concerns and were consistent with Australian common law and the supervening provisions of the EPBC Act.

These principles and baseline facts were that:

- the Commonwealth of Australia has jurisdiction over biological resources in Commonwealth areas;
- Australia is both a provider and user of biological resources and has responsibilities to both;
- commercial and non-commercial scientific research should be treated differently;
- Australia is a mega-biodiverse country with much of its biodiversity undescribed or not understood and this therefore creates an obstacle to efficient research and utilization;
- any regulatory regime should only apply to organisms found to be native or endemic to Australia and not those imported or bred;

[19] The Treaty was adopted on 3 November 2001 after 7 years of negotiations. See http://www.fao.org/3/Y2650e/Y2650e01.htm#3.

[20] See https://www.upov.int/edocs/pubdocs/en/upov_pub_423.pdf.

[21] See the Voumard Report, pp. 6–12.

[22] The principles outlined were the product of extensive discussions by the author and other Inquiry staff with members of the Expert Reference Group and with officials of the Commonwealth Attorney Generals Department. They underpin the Voumard Report recommendations.

- the scheme to be introduced should not conflict with existing treaties to which Australia is a signatory or has ratified;
- benefit-sharing contracts should recognize the differential contributions made by the provider and the user in the negotiated apportionment of benefits;
- benefit-sharing contracts should allow for the fact that industrial sectors have different research, commercialization and production timescales and levels of profitability;
- non-commercial or 'pure' research outcomes are public goods and should be encouraged by the removal of regulatory obstacles as much as possible;
- the Commonwealth's fiduciary responsibility would be best met by:
  - use of negotiated benefit-sharing agreements instead of a cost-recovery approach;
  - introducing minimal fees to encourage research; and
  - waiving of fees for non-commercial research.
- creating a competitive research advantage for Australia by:
  - introduction of a transparent regulatory system;
  - minimizing delay by prompt processing of applications;
  - use of low fees;
  - providing guidance to parties by publication of model contracts to minimize transaction costs; and
  - obviating biopiracy accusations by use of electronic publication of permits to objectively demonstrate and verify lawful possession and use- and thus demonstrate legal certainty while minimizing due diligence costs and delay.
- biological resources found on indigenous peoples' owned land and waters in Commonwealth areas are the property of the owners and accordingly any benefits obtained under a negotiated benefit-sharing agreement belong to the owners;
- the principal role of Government in relation to aboriginal owned land is to ensure fair dealing in recognition of the often-occurring disparity between negotiation resources available to owners and access proponents;
- any use of traditional knowledge (TK) must be disclosed to indigenous owners by access proponents to enable an informed decision on access and benefit-sharing;
- any refusal by owners to provide access to biological resources is to be final; and
- the scheme to be introduced would not apply to human remains.

Government officials took onboard those principles and included them in the development of the drafting instructions for the ABS regulations.[23]

---

[23] The author was responsible for the development of the draft regulations from the drafting instructions.

6 Acceptance of the Bonn Guidelines

Running in parallel with the inquiry process and subsequent development of draft EPBC Act regulations, the Australian Government was actively participating in the development and adoption of the 2002 Bonn Guidelines. The guidelines' final negotiated terms were consistent with Australian Government policy development on ABS and with its emerging draft EPBC Act regulations. Moreover, the terms of the Bonn Guidelines were also consistent with the objectives of the Commonwealth, State and Territory Government's National Strategy for the Conservation of Australia's Biological Diversity and thus opened the way to further cooperation between the Commonwealth Government and its State and Territory Governments.

Accordingly, on 11 October 2002, the national Government concluded an agreement with all member States and territories for a common implementation of the Bonn Guidelines. This was set out in the Council of Australian Governments' *Nationally consistent approach for access to and the utilization of Australia's native genetic and biochemical resources* signed by the Australian Prime Minister and heads of all State and Territory Governments.[24] The stated goal of the Nationally Consistent Approach is:

> To position Australia to obtain the maximum economic, and social and economic benefits from the ecologically sustainable use of its genetic and biochemical resources whilst protecting our biodiversity and natural capital.

The Nationally Consistent Approach included 14 principles and 11 common elements to be reflected in each jurisdiction's subsequent ABS law and administration.[25]

7 The Commonwealth ABS System

The 2002 draft law was finalized and introduced in 2005 as a new part to the EPBC Act regulations 2000. This was Part 8A, Access to biological resources in Commonwealth areas with the stated aim of giving effect to the Bonn Guidelines and the Commonwealth Government's commitments to the nationally consistent approach.[26]

The system established by the regulations implements the Bonn Guidelines in detail and serves as a model ABS law.[27] It sets out the purpose of the ABS

---

[24] Council of Australian Governments' nationally consistent approach for access to and the utilization of Australia's native genetic and biochemical resources.

[25] See ibid.

[26] See http://www.environment.gov.au/node/14439, and https://www.legislation.gov.au/Details/F2014C00950.

[27] Burton (2009). This sets out a line-by-line analysis of the congruence between the regulations and the provisions of the Bonn Guidelines.

regulations as being: the conservation of biological resources in Commonwealth areas and their sustainable use, the provision of equitable benefit-sharing, respecting indigenous peoples' traditional knowledge, providing legal certainty, minimizing cost, and contributing to Australia's Nationally Consistent Approach.[28]

The system can be described in the following terms:

(1) Access permit

A permit issued by Australia's National Competent Authority (NCA) as the Minister's delegate is needed for access to native biological resources for the purpose of research and development (R&D) on its genetic or biochemical makeup. The permit applies to Commonwealth areas only. The permit is free for non-commercial purposes, e.g. taxonomic science, and only a nominal amount of AU$50 is charged for access for commercial purposes. This minimizes cost to researchers.

(2) Rapid approval

Rapid approval of a commercial application is granted by the NCA if there is no likely environmental damage and a BSA has been made. Approval can be made within 10 working days or as little as 2 days in an emergency. Prompt decision-making reduces costs.

(3) Simplified arrangements

Special, simplified arrangements are in place for non-commercial access. These do not require a complex procedure but evidence of the permission of the resource provider and signing a short legal document as described in Sect. 10.5 below. This approach eliminates the need for time-consuming, negotiated benefit-sharing contracts.

(4) Exemption by the Minister

Additional flexibility is provided whereby the Minister responsible for the legislation may exempt a Government research body from the operation of the regulations if he or she is satisfied that the research body already meets the purposes of the ABS regulations or has entered into a long-term BSA. This reduces future 'red tape'.

8 Indigenous Peoples' Land Leased to the Commonwealth

The bulk of private land leased to the Commonwealth Government is owned by Australia's indigenous peoples: its Aboriginal and Torres Strait islander communities. The land is managed as jointly-managed national parks. For a permit to be granted for access to such lands, it is conditional on evidence that a BSA has been reached between the traditional owners and the prospective user. Such an agreement must demonstrate that it has been reached with the provider's prior informed consent and on mutually agreed terms. Benefits derived from such agreements must go to the

---

[28] Section 8A.01 "Purpose of Part 8", p. 55, https://www.legislation.gov.au/Details/F2014C00950.

owners.[29] As already said, the Government must play its role in these circumstances to ensure that the indigenous peoples' rights are respected through fair dealing and that this has taken place subject to their prior informed consent (PIC) to a negotiated BSA. To this end, regulations 8A.10 and 8A.15 require the Minister to give consideration to whether:

- the indigenous provider (owner) has adequate knowledge of the regulations and was able to enter into reasonable negotiations;
- the indigenous provider was given adequate time to consider the application for a permit and time to consult with the indigenous community;
- if the traditional owners are represented by a Land Council, the views of the Council;
- the provider was given adequate time to consider the views of any representative Aboriginal or Torres Strait Islander body if the area is one where native title exists; and
- the indigenous provider received independent legal advice.

To assist both resource providers and resource users, two model benefit-sharing agreements have been published. One is for public lands and waters in Commonwealth areas and one is for indigenous peoples' lands in Commonwealth areas.[30]

The system introduced by these ABS regulations has now been operating successfully for 16 years with permits being granted at the rate of 2 to 3 each month. In 2019, 38 access permits were issued.[31]

9 Drivers for the Commonwealth ABS System

In summary the drivers for the Commonwealth ABS system were:

- Australia's constitutional arrangements;
- its geographical distribution of land and waters;
- its overarching responsibilities under the Convention on Biological Diversity (CBD);
- compatibility with the Bonn Guidelines and Australia's national strategies;
- the conservation and sustainable use of biodiversity;
- facilitating access and use;
- protecting private and indigenous peoples' land rights;
- respecting indigenous peoples' traditional knowledge;

---

[29] Sections 8A.07 to 8A.10 refers to legal protections for indigenous providers. See https://www.legislation.gov.au/Details/F2014C00950.

[30] Example of model agreement, http://www.environment.gov.au/system/files/pages/e3584028-d083-4aec-acdd-c0aa635a529f/files/commonwealth-and-access-party-model-benefit-sharing-agreement-2012.pdf.

[31] Department of Environment and Energy 2018/19 Annual Report.

- creating legal certainty for access and subsequent use;
- minimizing costs to all parties;
- creating a national competitive advantage for the commercialization of genetic resources; and
- incentivizing investment in R&D and creating new biodiversity knowledge.

10 Challenges and Innovations

Behind the development of the ABS regulations lay a series of conceptual and practical challenges. They arose out of the Inquiry's consultations with stakeholders and required the development of innovative solutions.

10.1 Definition of Terms: 'Access' and the Concept of 'Utilization'

The first of these challenges flowed from the uncertainty created by international and domestic debates on the meaning of CBD article 15. This centered on the difficulty of determining the scope of genetic resources and thus determining the limits of article 15 in its application to genetic resources' derivatives or their components. The definitional stumbling block lay in the CBD definition of 'genetic resources' which itself turned on the definition of 'genetic material': 'genetic material' being defined in article 2 as: "*...any material of plant, animal, microbial or other origin containing units of heredity*". The leading question was then, what about coverage of biological material not containing units of heredity such as ribonucleic acid (RNA) derived from deoxyribonucleic acid (DNA)? This was especially important as much of the potential economic value of genetic resources lies in RNA proteins produced by DNA, with antibacterial or anticancer properties.

Australia did not seek to resolve this dilemma, but took an alternative approach to defining the scope of GR. It did not attempt to determine whether derivatives were "in-scope" or not. Rather, the Australian approach attempted to show there was another way to think about the issue and, at the same time, to provide a pragmatic and realistic way forward.

Instead, Australia focused on the purpose for which GR were accessed or taken. This new consideration led to the examination of the intent behind the taking of any biological resources containing GR for their R&D use.

The consideration of 'use', in turn allowed 'intended use' to be described by real-world scientific conduct and practice: i.e., on GR's utilization. This approach then allows for a description to be made in law that might end confusion and ambiguity about what is being regulated under article CBD 15.

Accordingly, Australia took as an initial working definition, the following definition of 'access to biological resources':

> **access to biological resources** means the taking of biological resources of native species for research and development on any genetic resources, or biochemical compounds, comprising or contained in the biological resources. . .[32]

This was subsequently included in its ABS regulation at 8A.03, "Meaning of access to biological resources".

In this way, the taking and specified use of native species' biological material would be regulated. The reference to 'biological resources' as the supervening element of 'access' is consistent with article 2 of the CBD. Article 2 defines biological resources as meaning it:

> . . .includes genetic resources, organisms or parts thereof, populations, or any other biotic component of ecosystems with actual or potential use or value for humanity.

Thus, the adoption of this definition of 'access' in the regulations sets the scope of the activity to encompass R&D on genetic resources and includes any biochemical compounds found within a biological resource.[33] This successful thematic approach was later reflected in the definition of 'utilization' in the Nagoya Protocol (NP).[34]

10.2 *Electronic Verification of Permits*

To address the difficulty and cost of verifying the legal provenance of collected GR, it was decided to develop and pilot a low-cost system of public electronic verification of access permits. This would have several advantages for: users of genetic resources, for providers of genetic resources, for domestic and foreign regulators, and for those investors and communities concerned with the possibility of biopiracy. Concerns about biopiracy or the unlawful taking and use of resources could then be pre-empted and legal certainty demonstrated.

Such public verification was achieved by building an internet-accessible verification website. Australia's NCA would put requisite access permit details in the

[32] See https://www.legislation.gov.au/Details/F2014C00950.

[33] Australia decided to confine ABS to its native species to avoid deriving benefits from foreign species and thus encouraging countries to obtain benefits from utilization of Australian native species found in their own jurisdictions. Author was director of Australian Access Task Force from 1999 to 2006.

[34] Article 2 of the Nagoya Protocol defines 'utilization of genetic resources' as: "means to conduct research and development on the genetic and/or biochemical composition of genetic resources, including through the application of biotechnology as defined in Article 2 of the Convention".

form of signed copies of issued permits on the website. It would show to whom, for what, and under what conditions a permit to collect resources was issued. It would show if it was for commercial or non-commercial purposes and whether a BSA was in place.[35] Where reporting obligations were required, details of progress could be added.

In this way, any prospective partner in the development of new products derived from genetic resources, possible future research partners, university technology transfer departments or anyone with obligations to undertake due diligence could do so at any time, from anywhere, and at no cost. Moreover, any question about provenance or lawful possession, or any suggestion of biopiracy would be resolved quickly and at no cost to those involved.

An additional benefit was that parties interested in the subject matter of the research being undertaken would also have the information necessary to contact the permit holder directly should they so wish.

The risk of biopiracy accusations, whether the unlawful taking of resources or where no law existed or the taking without permission of owners or indigenous peoples, was recognized during the Inquiry as an obstacle to R&D and one which was difficult for researchers to satisfactorily resolve.[36] It was considered that by using this level of transparency, Australia might gain a comparative advantage in research and benefit-sharing over jurisdictions unable to address the issue quickly and satisfactorily.

This innovation was begun as a pilot program to be continued until the establishment of the foreshadowed international ABS Clearing-House Mechanism that would serve a similar function and turn permits into International Certificates of Compliance. It should be noted that the continued publication of signed Commonwealth ABS permits contains additional details beyond the minimum requirements of article 17(4) of the Protocol, thus providing even more confidence to users or their prospective partners in material collected under Commonwealth permits.

Moreover, Australia also provides foreign regulators discharging their compliance obligations under the Protocol with instant access to current and historic permit details.[37] This is even though, as a Protocol-signed but unratified country, Australia is unable to create International Certificates of Compliance via the Protocol's Clearing-House Mechanism.[38]

---

[35] Copies of all permits issued since 2006, http://www.environment.gov.au/resource/list-permits-issued.

[36] The author had discussions at the time with representatives of other Governments and foreign companies about their experiences with accusations of biopiracy.

[37] All such permits issued year by year, https://environment.gov.au/resource/science-and-research-permits-2019.

[38] The International Certificate of Compliance can only be created when a party to the NP sends its permit to the ABS Clearing-House Mechanism.

10.3 Model Contracts

To facilitate access and reduce transaction costs for both providers and users, Australia developed and published two model benefit-sharing agreements, one for indigenously owned land and waters and one for Commonwealth-only owned lands and managed waters.[39] These contracts were themselves modelled on current private international law contacts used in common law countries. Part of their intended utility lay in their inclusion of typical 'boiler plate' clauses.[40] The availability of such clauses was intended to assist in reducing the time parties might spend in contract negotiations over such provisions.

Both model contracts also set out how equitable benefit-sharing capacity may differ by industry sector. Illustrative payment thresholds and sliding royalty scales for different industry sectors were included in the model. The model contracts also allow for non-financial benefits to be valued and taken into consideration. This is significant as negotiated non-financial benefits are often direct and certain whereas the possibility of economic returns to share is uncertain and difficult to quantify in advance.

10.4 Consultation and Informed Consent

Significant to any contract for access to genetic resources found on indigenous peoples land or waters is a requirement that the Commonwealth Minister be satisfied the owners have given their PIC and that any intended use of traditional knowledge associated with genetic resources has been properly disclosed to the owners. To this end the regulations set out objective criteria for the Minister's satisfaction. Moreover, the regulations require that any BSA reached with indigenous owners must provide for reasonable benefit-sharing arrangements, including protection for, recognition of and valuing of any indigenous people's knowledge to be used.[41]

10.5 Permits for Non-commercial Use

Access permits are free for non-commercial use. To minimize cost and administration burdens placed on researchers such permits do not require complex benefit-sharing contractual agreements. Instead, to protect the interests of resource owners,

---

[39] A copy can be found on the World Intellectual Property Organization Contractual Data Base.

[40] The term "boiler plate clauses" refers to commonly used and accepted terms in contracts dealing with common issues and requirements.

[41] See Regulations 8A.08 "Benefit-sharing agreements" and 8A.10 "Informed consent", https://www.legislation.gov.au/Details/F2020C00778.

the regulations introduced a further innovation. This was the introduction of a simplified benefit-sharing arrangement in the form of a legal document commonly known in common law countries as a Statutory Declaration made under the Statutory Declarations Act 1959.

Under this arrangement, they must show in a signed Statutory Declaration[42] that:

- they have the written permission of the provider before any transfer of samples to third parties;
- it is for non-commercial purposes;
- they will publish or share their research results with the provider or publish them;
- they will offer a taxonomic duplicate of any new species found to a public collection; and,
- negotiate a commercial benefit sharing agreement with the provider if they later wish to commercialize, i.e. if the use intent changes.

The Act provides penalties for false statements. A person who intentionally makes a false statement in a Statutory Declaration is guilty of an offence, punishable by up to 4 years in prison.[43]

10.6 Progress to a Digital Atlas of Living Australia

A number of stakeholders, including the Australian Government's Biological Resources Study (ABRS), drew the Inquiry's attention to the fact that much of Australia's biodiversity is undescribed and this was an obstacle to biodiscovery.

The ABRS estimates that 75% of Australia's biodiversity remains to be discovered and described.[44] Without information about the location, distribution and identity of species, the task of biodiversity conservation is handicapped and the risk of the misallocation of conservation efforts is significantly increased. This situation also impacts on biodiscovery as targeted sampling, re-collecting and the establishment of intellectual property is made more difficult, uncertain and costly.[45] To address this issue the following was done:

---

[42] See http://www.environment.gov.au/topics/science-and-research/australias-biological-resources/permits.

[43] The Act and penalty provision, https://www.legislation.gov.au/Details/C2004C00189.

[44] See https://www.environment.gov.au/science/abrs/about.

[45] It is a condition of patent approval that an invention must be able to be made from the description given in the patent application. For an invention whose starting point is a living organism, it is essential that the organism is properly identified. Otherwise replication is not possible. Misidentification can lead to patent invalidity and substantial economic losses.

Australasian Virtual Herbarium In 2001 the National Biotechnology Strategy provided pilot funding to establish an Australian Virtual Herbarium (AVH)[46] through the Department of the Environment. This would digitize data on the holdings of Australia's various herbaria via the Council of the Heads of Herbaria (CHAR). This initiative was successful and has since been expanded to include New Zealand. This created the Australasian Virtual Herbarium. In 2020 the now, Australasian Virtual Herbarium[47] (AVH), holds accessible data on over eight million plant, algae and fungi specimens from both countries.

Prime Ministers Science and Engineering Council (PMSEIC) Funding Support In 2005 the growing awareness of the importance of biodiversity across Australia led the Government to fund the establishment of the Atlas of Living Australia.[48] The Atlas, to be managed by the country's Commonwealth Scientific Industrial and Research Organization (CSIRO), would aggregate and foster the digitization of all information on Australia's biodiversity. The Atlas currently provides free, online access to more than 91 million records about Australia's biodiversity. It aids biodiscovery, biotechnology research, supports basic research, environmental monitoring, conservation planning, education, and biosecurity activities.

The initial step of AU$35 million funding was separate to the National Inquiry. It was taken on the recommendation of the Prime Ministers Science and Engineering Council (PMSEIC).[49] The PMSEIC report highlighted the importance of Australia's 80% rate of species endemism and its more than half a million species to future biodiscovery and the future of biotechnology. The decision demonstrates the link between biodiversity and its utilization for biodiscovery and hence its importance as a driver for research and innovation.[50]

11 Delay and Ratification

The Australian Government signed the Nagoya Protocol (NP) in 2012 but has not yet ratified. NP compliance options for ratification and published by the Australian Government include: requiring Government research funding be conditional on use of lawfully obtained genetic resources, the restriction on the importation of

[46] The Australia Virtual Herbarium changed its name to the Australasian Virtual Herbarium with the addition of New Zealand. The term 'Australasian' means Australia and New Zealand and sometimes other islands in the region as well.

[47] See https://avh.ala.org.au/#tab_simpleSearch.

[48] See https://www.ala.org.au/.

[49] A copy of the PMSEIC Report has been removed and can no longer be found at the PMSEIC website but can be found at http://absbonn.pbworks.com/f/Australia+biodiscovery+report+2005.pdf.

[50] The author was policy advisor to the PMSEIC Working Group.

unlawfully obtained GR for R&D, and the creation of International Certificates of Compliance and increased penalties for breaches of Australian ABS regulations. Initial progress towards ratification was made by the 2014 release of a model system for implementing the NP for comment.[51] However, since then no further public development of the model has occurred.

The Australian Government's published position has been that its domestic ABS measures are consistent with the Protocol and it is developing its approach to implementation and ratification.[52] This does not, however, explain the 8-year delay since Australia signed the Protocol. Interviews in 2016 with officials not authorized to speak on behalf of the Government offered a more prosaic explanation. This is that the Government was reluctant to legislate compliance measures as opening the door to amendment of the EPBC Act may lead to pressure for other undesirable amendments.[53]

Since then, the situation has changed. In 2020 events have occurred which demonstrate the changing priorities of the national Government on environmental management. The first of these was the 2019 independent review of the EPBC Act recommending sweeping changes to the Act.

Following the publication of the Review in June 2020[54] the Government introduced the *Environment Protection and Biodiversity Conservation Amendment (Streamlining Environmental Approvals) Bill 2020* on 27 August 2020. This Bill would amend the EPBC Act to devolve Commonwealth environmental approvals down to States and territories.[55] The interim review also foreshadows wholesale streamlining of the EPBC Act of which devolution of responsibilities was a key step, but an examination of the review report finds it is silent on the issue of the NP ratification.

A further indication of a shift in priorities is evidenced in the latest Department of Agriculture, Water and the Environment Annual Report 2019–20 on the operations of the EPBC Act. The section dealing with ABS regulations under Part 8A of the EPBC Act regulations is limited to a single sentence—showing that the department simply issued 25 permits for access to biological resources in Commonwealth areas in 2019–20.[56]

Ahead of the 2020 United Nations Summit on Biodiversity Australia joined the United States, Brazil, China, Russia, Indonesia and India to decline to sign the Leaders Pledge for Nature. From press reports, the Prime Minister's spokesman

[51] https://www.allens.com.au/insights-news/insights/2014/05/australia-releases-model-for-implementing-the-nagoya/.

[52] See http://www.environment.gov.au/system/files/pages/9fc06ac0-f5af-4b47-a80f-d9378088d743/files/nagoya-factsheet.pdf.

[53] Interviews conducted in July 2019 by the author with officials—not named for privacy reasons.

[54] Copy of the Independent Review of the EPBC Act Interim Report https://epbcactreview.environment.gov.au/resources/interim-report/executive-summary.

[55] See Environment Protection and Biodiversity Conservation Amendment (Streamlining Environmental Approvals) Bill 2020.

[56] Department of Agriculture, Water and the Environment Annual Report 2019–20.

indicated the principal objection was the commitment in the pledge to be carbon neutral by 2050.[57]

Taken together these examples suggest the Government does not currently apply any significant priority to ratification of the NP. Nevertheless, there is growing pressure for ratification.

12 Pressure for Change

During the May 2019 annual conference of the Australia New Zealand Marine Biotechnology Society, research scientists from the Commonwealth Scientific and Industrial Research Organization (CSIRO) stated to the author that Chinese researchers advised them that Australia's non-ratification of the Protocol was preventing them from entering into research partnerships with CSIRO. Researchers from Tasmania, and South Australia separately also stated that the absence of Protocol implementation was creating difficulties for establishing international collaborative research more broadly.

12.1 State Responses: Western Australia

The States of Western Australia, South Australia, Tasmania and New South Wales had indicated in earlier consultations with Commonwealth officials that State ABS legislation would be contingent on the national Government implementing its domestic compliance obligations and ratification of the Protocol.[58] This caution by State Governments may also be wearing thin.

For example, the Government of Western Australia now relies on the management of ABS or bioprospecting on public lands through the grant of licenses under section 72 of its Biodiversity Conservation 2018 regulations which provides:

> Bioprospecting license
>
> (1) A bioprospecting license authorizes the holder of the license to take fauna or flora for a bioprospecting activity.
> (2) A bioprospecting license does not authorize a person to take fauna or flora that belongs to a threatened species unless an authorization relating to the threatened species has been given to the person by the Minister under section 40 of the Act.
> (3) A bioprospecting license does not authorize a person to take flora on private land.
> (4) Without limiting any other provision of these regulations, a bioprospecting license may include a condition that requires the holder of the license to enter into an arrangement

[57] Refusal to sign, https://www.sbs.com.au/news/australia-is-refusing-to-sign-a-global-pledge-to-protect-earth-s-biodiversity.

[58] Informal discussions with the author.

> with the CEO for the sharing of profits or for the provision of other benefits to the State.[59]

The terms of the section above give the Government the flexibility to meet the various conditions set out in both the Bonn Guidelines and the Protocol by administrative action.

12.2 State Responses: Queensland

The experience of the Queensland State Government provides additional direct evidence of pressure for change and of a State Government's response. Queensland undertook a comprehensive review of its Biodiscovery Act in 2016. The Queensland Government responded to that review in 2018 with legislative and policy changes. The changes introduced reduced complexity and cost to users of its genetic resources. They were made with the stated aim of positioning Queensland as the preferred location to undertake biodiscovery in Australia so that biodiscovery entities could successfully collaborate internationally, and could access markets and do so with world leading research and commercialization.

In 2019 Queensland went further. It is the first Australian State to undertake concrete action to more closely align its ABS laws with the NP. On 20 August 2020, the Queensland Parliament passed the *Biodiscovery and Other Legislation Amendment Act* 2020, with its amendments coming into force on 30 September 2020.[60] Notable in the amendments to the principal Act are the inclusion of provisions protecting the rights of Aboriginal and Torres Strait Islanders as custodians of traditional knowledge (TK) to be used in biodiscovery. The Act at Part 2A requires that users must enter into a BSA with the custodians for any use of TK. The Act further enables the Minister to establish a code of traditional knowledge practice to guide parties involved. In addition, the Minister cannot approve any biodiscovery application, unless he is satisfied that the proponent does not intend to use traditional knowledge or, if they do, they comply with their traditional knowledge obligations.[61]

---

[59] See https://www.legislation.wa.gov.au/legislation/statutes.nsf/law_s50938.html.

[60] See https://environment.des.qld.gov.au/licences-permits/plants-animals/biodiscovery/biodiscovery-act-reform#:~:text=A%20suite%20of%20priority%20amendments%20to%20the%20Biodiscovery,for%20the%20use%20of%20traditional%20knowledge%20in%20biodiscovery.

[61] Ibid.

13 Conclusion

Australia's future ratification of the NP and implementation depends on continued pressure from its biotechnology research communities and from its State and Territory Governments.

To date, Australia's early successes, innovations and its contributions to the development of a compatible international ABS system has not provided sufficient momentum to ensure Australia's timely ratification of the Protocol and its domestic implementation. Unless it does so, Australia will lose the competitive advantages established through its implementation of the Bonn Guidelines and forego the benefits of the Protocol's international framework to better secure compliance with Australia's own ABS laws and to foster future investment in R&D.

References

Book Chapter

Burton G (2009) In: Kamau, Winter (eds) Genetic resources, traditional knowledge and the law. Solutions for access and benefit sharing. Earthscan, London, pp 271–310

Legislations

Commonwealth of Australia

Commonwealth of Australia Constitution Act 1900., https://australianpolitics.com/downloads/constitution/constitution_full-text.pdf, p 17 (accessed 26 November 2020)
Environment Protection and Biodiversity Conservation Amendment (Streamlining Environmental Approvals) Bill 2020., http://www5.austlii.edu.au/au/legis/cth/bill/epabcaeab20201067/ (accessed 7 December 2020)
Environment Protection Biodiversity Conservation Act 1999., https://www.legislation.gov.au/Series/C2004A00485 (accessed 26 November 2020)
Environment Protection Biodiversity Conservation regulations 2000., https://www.legislation.gov.au/Details/F2016C00914 (accessed 7 December 2020)
Native Title Act 1993., https://www.legislation.gov.au/Details/C2017C00178 (accessed 28 October 2020)
Plant Breeder's Rights Act 1994., https://www.legislation.gov.au/Details/C2018C00361 (accessed 7 October 2020)
Statutory Declarations Act 1959., https://www.legislation.gov.au/Details/C2004C00189 (accessed 7 December 2020)

State

Western Australia

Biodiversity Conservation Regulations 2018., https://www.legislation.wa.gov.au/legislation/statutes.nsf/law_s50938.html (accessed 7 October 2020)

Queensland

Biodiscovery Act 2004., https://www.legislation.qld.gov.au/view/whole/html/inforce/current/act-2004-019 (accessed 7 December 2020)
Biodiscovery and other legislation Amendment Act 2020., https://www.legislation.qld.gov.au/view/html/asmade/act-2020-027/lh (accessed 7 December 2020)

Reports/Inquiries

Consultancy Report for the Steering Committee for the National Red Imported Fire Ant Eradication Program., https://www.daf.qld.gov.au/__data/assets/pdf_file/0010/1479790/National Red Imported-Fire-Ant-Eradication-Program-Efficiency-and-Effectiveness-Review-.pdf (accessed 12 October 2020)
Council of Australian governments' nationally consistent approach for access to and the utilization of Australia's native genetic and biochemical resources, http://www.environment.gov.au/system/files/resources/bbfbde06-d13a-4061-b2f9-c115d994de2d/files/nca.pdf (accessed 31 August 2019)
Department of Agriculture, Water and the Environment Annual Report 2019-20 on the operations of the EPBC Act, https://www.transparency.gov.au/annual-reports/department-agriculture-water-and-environment/reporting-year/2019-20-24 (accessed 31 October 2020)
Department of Environment and Energy 2018/19 Annual Report, https://www.awe.gov.au/sites/default/files/2020-01/annual-report-2018-19-environment-and-energy.docx (accessed 07 December 2020)
Independent Review of the EPBC Act Interim Report., https://epbcactreview.environment.gov.au/resources/interim-report/executive-summary (accessed 30 October 2020)
National Strategy for the Conservation of Australia's Biological Diversity., http://www.environment.gov.au/archive/biodiversity/publications/strategy/pubs/national-strategy-96.pdf (accessed 26 November 2020)
PMSEIC Report., http://absbonn.pbworks.com/f/Australia+biodiscovery+report+2005.pdf (accessed 1 November 2020)
The Voumard Report., https://www.environment.gov.au/system/files/resources/d0f84da6-eb69-4053-8d96-ec294da649bc/files/abrca.pdf (accessed 16 September 2020)
World Intellectual Property Organization Contractual Data Base., https://www.wipo.int/tk/en/databases/contracts/texts/australiamodel.html#_Toc170201080 (accessed 31 August 2020)

Website Links

https://www.ga.gov.au/scientific-topics/national-location-information/dimensions/oceans-and-seas (accessed 15 October 2020)
http://www.australasianscience.com.au/article/science-and-technology/when-did-aboriginal-people-first-arrive-australia.html (accessed 31 August 2019)
https://aiatsis.gov.au/explore/articles/land-rights (accessed 31 August 2019)
https://www.austrade.gov.au/International/Invest/Resources/Benchmark-Report (accessed 28 October 2020)
https://www.ausbiotech.org/biotechnology-industry/fast-facts (accessed 31 August 2019)
https://www.cbd.int/doc/measures/abs/msr-abs-au4-en.pdf (accessed 31 August 2019)
https://www.legislation.gov.au/Details/C2018C00440/Download (accessed 28 September 2020)
https://www.legislation.gov.au/Details/C2017C00178 (accessed 28 October 2020)
https://www.regional.gov.au/territories/norfolk_island/ (accessed 5 October 2020)
https://www.britannica.com/place/Norfolk-Island#ref5310 (accessed 10 October 2020)
https://www.upov.int/edocs/pubdocs/en/upov_pub_423.pdf (accessed 17 October 2020)
http://www.environment.gov.au/system/files/resources/bbfbde06-d13a-4061-b2f9-c115d994de2d/files/nca.pdf (accessed 31 August 2019)
https://www.legislation.gov.au/Details/F2014C00950 (accessed 31 August 2019)
http://www.environment.gov.au/resource/list-permits-issued (accessed 30 November 2020)
https://environment.gov.au/resource/science-and-research-permits-2019 (accessed 15 October 2020)
https://www.legislation.gov.au/Details/F2020C00778 (accessed 30 October 2020)
http://www.environment.gov.au/topics/science-and-research/australias-biological-resources/permits (accessed 31 August 2020)
https://www.environment.gov.au/science/abrs/about (accessed 30 November 2020)
http://www.environment.gov.au/system/files/pages/9fc06ac0-f5af-4b47-a80f-d9378088d743/files/nagoya-factsheet.pdf (accessed 31 August 2019)
https://www.sbs.com.au/news/australia-is-refusing-to-sign-a-global-pledge-to-protect-earth-s-biodiversity (accessed 4 December 2020)
https://www.legislation.wa.gov.au/legislation/statutes.nsf/law_s50938.html (accessed 31 October 2020)
https://environment.des.qld.gov.au/licences-permits/plants-animals/biodiscovery/biodiscovery-act-reform#:~:text=A%20suite%20of%20priority%20amendments%20to%20the%20Biodiscovery,for%20the%20use%20of%20traditional%20knowledge%20in%20biodiscovery (accessed 31 October 2020)
https://www.aph.gov.au/About_Parliament/Senate/Powers_practice_n_procedures/Constitution.aspx (accessed 31 October 2020)
http://www.environment.gov.au/system/files/pages/e3584028-d083-4aec-acdd-c0aa635a529f/files/commonwealth-and-access-party-model-benefit-sharing-agreement-2012.pdf (accessed 28 November 2020)
https://www.australianwildlife.org/wildlife (accessed 31 August 2019)
http://www.fao.org/3/Y2650e/Y2650e01.htm#3 (accessed 15 October 2020)
https://www.allens.com.au/insights-news/insights/2014/05/australia-releases-model-for-implementing-the-nagoya/ (accessed 4 December 2020
https://blogs.sciencemag.org/pipeline/archives/2019/05/09/the-latest-on-drug-failure-and-approval-rates (accessed 4 December 2020)
http://www.environment.gov.au/node/14439 (accessed 31 August 2019)
https://avh.ala.org.au/#tab_simpleSearch (accessed 30 November 2020)
https://www.ala.org.au/ (accessed 30 November 2020)

Geoffrey C. Burton provides expert advice on ABS systems to countries of the Asia-Pacific. He was Australia's ABS National Competent Authority responsible for ABS law and policy from 1999 to 2006. He led Australia's negotiations on ABS at the CBD and co-chaired the initial international regime negotiation. While a UN University Adjunct Senior Fellow, Mr Burton contributed to several books on ABS and authored over a hundred articles and presentations on ABS implementation. Mr Burton addressed the UN in New York (2006) and the Brazilian Congress (2012). In 2020 Mr Burton assisted Samoa with its forthcoming ABS law. geoff@jeanshannon.com.

The ABS Compliance Regime of the European Union

Gerd Winter

Abstract According to article 15 of the Convention on Biological Diversity (CBD) States have sovereign rights over their genetic resources. This entails that States may regulate access to their genetic resources and demand the sharing of arising benefits. These rights and corresponding obligations have been specified in the Nagoya Protocol of 2010. States providing genetic resources have only limited legal powers and actual means to monitor and control the utilization of accessed resources if they are moved to and utilized in other States. The Nagoya Protocol therefore obliges user States to take measures to ensure compliance with provider State regulations. The European Union has pioneered in establishing a quite ambitious compliance regime. At present it is experiencing a phase of adjusting practices to the new rules, but also of learning about difficulties that may lead to revisions of the legal bases. The following contribution summarizes the EU compliance regime as it stands, focusing on its objectives, the basic rights and obligations of users concerning access, utilization and benefit sharing, as well as the powers and obligations of administrative oversight. It also addresses problems experienced in practice and discusses possible practical solutions, ending with a short reflection on ideas about a more fundamental transition from the current bilateral to a more multilateral model of ABS regulation.

Keywords Access · Utilization · Benefit-sharing · Due diligence · European Union

1 Overview of Legislation of Provider and User States

In most legal systems of the EU Member States the genetic potential of organisms living *in-situ* within their borders are considered in principle to be free goods (*res nullius*). This means that foreign and domestic researchers are free to use, breed,

G. Winter (✉)
Research Center for European Environmental Law, University of Bremen, Bremen, Germany
e-mail: gwinter@uni-bremen.de

E. Chege Kamau (ed.), *Global Transformations in the Use of Biodiversity for Research and Development*, Ius Gentium: Comparative Perspectives on Law and Justice 95, https://doi.org/10.1007/978-3-030-88711-7_14

re-combine and commercialize the genetic makeup at will. The benefits arising from such research and development (R&D) need not be shared with the State of origin. They are essentially reaped by the researchers and developers are entitled to claim exclusive rights on the genetic resource or product.

A few EU Member States have established or are about to establish legislation requiring prior consent for access to their genetic resources and prior consent or mutual agreement on the sharing of benefits derived from R&D on the genetic resources. These States—acting in a role of so-called provider States—include Bulgaria, Croatia, France, Malta, and Spain, as well as—focusing on traditional knowledge—Finland and Sweden.[1] Within the European Economic Area (EEA) and European Free Trade Association (EFTA) Norway and Switzerland have provider regimes *albeit* of a more lenient version where the executive is empowered to introduce it occasionally.[2]

While the EU did not harmonize the regulation by Member States on ABS concerning their GR it did so with regard to the utilization of GR within Member States. In other words, harmonization has targeted Member States in their role of user States, not of provider States. The EU was moved to focus on the user side by the Nagoya Protocol (NP) to the Convention on Biological Diversity (CBD) of 2010 of which the EU is contracting party. According to the NP the user States must take measures to ensure that the genetic resources utilized in their territory have been accessed and utilized in compliance with the regulations of the provider States and mutually agreed terms (MATs) have been concluded on the sharing of benefits.[3]

The EU enacted a legislative regulation that lays out the user State obligations for all Member States.[4] An executive regulation of the Commission further specifies how certain obligations shall be implemented.[5] In addition, the Commission published a guidance paper that concentrates on delimiting the scope of application and the core obligations of the EU compliance regime.[6] Issues remaining for the Member States to regulate include the allocation of competences to public

[1]On France and Spain see Mahop and Silvestri in this volume.

[2]For Norway see Act of 19 June 2009 No. 100 Relating to the Management of Biological, Geological and Landscape Diversity, Art. 57 f., 61a An executive order of 2016 has been introduced related to the protection of traditional knowledge associated with genetic resources. For Switzerland see Federal Act on the Protection of Nature and Cultural Heritage of 1 July 1966, Art. 23 f. An executive order, the Nagoya Ordinance 2015 provides in article 8 that a user must record information about his/her access and pass it to subsequent users. He/she must notify the competent authority of the information before commercializing a product developed on the basis of utilized GR.

[3]Art. 15 Nagoya Protocol.

[4]Regulation (EU) No 511/2014 of the European Parliament and of the Council on the compliance measures for users from the Nagoya Protocol on Access to Genetic Resources and the Fair and Equitable Sharing of Benefits Arising from their Utilisation in the Union, OJ 2014 L 150/ 59.

[5]Commission Implementing Regulation (EU) 2015/1866 laying down detailed rules for the implementation of regulation (EU) No 511/2014, OJ 2015 L 275/4.

[6]Commission Guidance Document on the scope of application and core obligations of Regulation (EU) No 511/2014 . . ., OJ C 2021 13/1. In the following cited as Commission Guidance.

authorities, their provision with enforcement powers and introduction of sanctions to address cases of non-compliance.[7]

2 Objectives

The objectives of Regulation (EU) 511/2014, as stated in article 1, are to ensure compliance with the ABS provisions of the Nagoya Protocol and, thereby, to contribute to the conservation of biological resources and the sustainable use of its components. While these objectives reflect the three basic objectives of CBD article 1, as reinforced and specified by article 1 NP, it is noteworthy that the Regulation does not venture into the transfer of relevant technologies, capacity building and appropriate funding which is also part of the objectives and provisions both of the CBD and NP. This gap could be an obstacle to trust-building with provider States, because like in other instances the related provisions which sound so promising are set aside when it comes to implementation. Legally the lack of harmonization by EU law means that it is for the Member States to make the promises a reality. It appears, however, that hardly any Member State has addressed the issue in its legislation other than in the context of general research collaboration and development aid policy.

In philosophical terms the focus on ABS and disregard for technology, capacity and finance transfer indicate a shift from distributional to commutative justice. Commutative justice means a fair quid pro quo between two equal parties: the countries holding genetic resources (GR) are conceived as owners of a commodity that can be sold in exchange for non-monetary and monetary benefits. In contrast, distributive justice starts with the observation that the two parties are unequal because they are at different stages of development, and engages in assisting the weaker side.[8] The two different contexts may also influence the basic expectations and attitudes of interactions at both law-making and law-applying levels: In a commutative framing the parties will negotiate about the costs of conservation, the value of the genetic resources, the costs of involving provider State personnel, the value of the knowledge shared, etc., finally translating qualitative into monetarized assessments. In contrast, in a distributive framing the unequal starting points of the parties are addressed apart from or in addition to the fair pricing of GR as commodities. Joint ventures of R&D and the sharing of R&D results will rather be seen as based on normative principles and flanked by more general technical, scientific and financial support. Even the often cited need to build trust can be seen differently in the two contexts: Is trust the outcome of deals, or of reason? It is open to discussion if

[7]See e.g. for Germany, Gesetz zur Umsetzung der Verpflichtungen nach dem Nagoya-Protokoll v. 25.11.2015.

[8]For the meaning and Aristotelian origin of the two concepts see the instructive article of Chroust and Osborne (1942); on their application in the ABS context see Stoll (2009).

commutative or distributive justice will better serve the final goal of the CBD, the conservation and sustainable use of GR.

3 Overview of Instruments

The instruments Regulation (EU) 511/2014 employs in view of its objectives consist of basic duties placed on users of genetic resources together with administrative oversight of their fulfilment.[9] There are two categories of basic duties: to ensure that the GR or traditional knowledge associated with GR were lawfully acquired, and that they are lawfully utilized. There are also two instruments of oversight: declarations by users and checks by competent authorities. Additional provisions such as on registered collections and recognized best practices are aimed at simplifying the obligations of both users and administrative bodies. The system can best be explained if the different stages of dealing with genetic resources from access to benefit-sharing are distinguished and the different basic duties and means of administrative oversight are related to them (see Table 1). On the basis of this doctrinal structure of the compliance regime the following questions can be posed: What activities are subject to what basic duties? (Sect. 4) What activities are subject to what means of administrative oversight? (Sect. 5) What basic duties and administrative oversight apply to activities subsequent to research and development? (Sect. 6) What specifics apply to traditional knowledge? (Sect. 7).

4 What Activities Are Subject to What Basic Duties?

Basic duties are imposed on users of GR. It must therefore be determined, how the term genetic resources is defined, what genetic resources are in or out of scope of the compliance regime, who is a user, and what content basic duties have.

[9]It is noteworthy in doctrinal terms that the distinction between basic obligations of addressees and administrative instruments of oversight is a modern regulatory technique. See for instance Directive 2010/75/EU on industrial emissions, articles 11 and 14 which impose basic obligations on operators and relates to them the administrative instrument which is a prior authorization in that case. Older regulations used to only address administrative bodies and provide them with powers to intervene. In such a system the Regulation (EU) 511/2014 might have required that a user needs, besides the provider State's PIC, another permit from EU MS authorities which would be granted if the provider State permit was obtained or other conditions fulfilled.

Table 1 Basic duties and administrative oversight at various stages of access, utilization and commercialization of genetic resources

Administrative oversight		Checks		Checks		Checks[a]	Checks[a]
Declarations		Due diligence declaration[b] about		Due diligence declaration[b] about			
		Access	Utilization	Access	Utilization		
Basic duties of actors	Compliance with access conditions	Compliance with		Compliance with utilization conditions		Compliance with conditions for applications and commercialization	Compliance with benefit sharing conditions
		Access conditions	Utilization conditions				
Activities	**Access**	**R&D**		**Final stage of product development**		**Subsequent applications and commercialization**	**Sharing of non-monetary and monetary benefits**
Actors	Researcher, intermediary (collections, brokers)	Researcher and developer		Researcher and developer		Researcher, developer, marketer	Researcher, developer, marketer

Source: *Own illustration*

[a] Contested, see further Sects. 6.1 and 6.2

[b] Declaration simplified if GR are acquired from registered collections and, under conditions, other intermediaries

4.1 *What Are Genetic Resources?*

According to article 2 CBD and article 2 (1) and (2) Regulation (EU) 511/2014 'genetic resources' means any material of plant, animal, microbial or other origin containing functional units of heredity and being of actual or potential value. ABS is thus related to the genetic potential of organisms.

It is debated if GR are only material or also information. In the latter case the sovereign rights of countries of origin could be construed to establish some kind of intellectual property in non-invented or 'wild' information. However, the fact that the definition of GR points to material speaks against such a proposition. A country of origin may nevertheless introduce a new kind of intellectual property but this would only be enforceable within its own territory, not in other countries. A more reasonable way for provider countries to ensure control of their GR as information is by establishing appropriate conditions within access permits and mutually agreed terms (MAT).[10] Such conditions could be conceived as constituting quasi-property rights.

4.2 *What Kinds of Genetic Resources Are Included in or Excluded from Scope?*

Delineations of scope of the compliance regime are drawn in a material, temporal and geographic dimension.

4.2.1 Material Scope

Excluded from scope are GR for which ABS is governed by specialized international instruments (Art. 2 (2)). These comprise the plant varieties that are included in the special ABS system of the International Treaty on Plant Genetic Resources for Food and Agriculture (ITPGRFA). For human pathogens the compliance regime is somewhat attenuated corresponding to the restrictions of provider States' rights to regulate ABS in emergency situations.[11]

Following a resolution of the Parties to the CBD human genetic resources are out of scope of the NP as well as the EU regime.[12] However, human microbiota are not genetic resources of a human being as such and therefore considered not to be covered by that exemption. The Commission Guidance suggests that a difference should be made between R&D on the microbiota aiming at general knowledge—

[10]Commission Guidance, para. 2.3.5. See further below.

[11]See Art. 4 (8) and the WHO Resolution establishing the Pandemic Influenza Preparedness (PIP) Framework, www.who.int/influenza/resources/pip_framework/en/.

[12]See consideration (19) of Regulation (EU) 511/2014.

which would then not be covered by the exemption—and R&D aiming at knowledge about the unique composition and functions of the microbiota with respect to the individual person.[13] The difference is plausible but will be difficult to draw in practice.

4.2.2 Temporal Scope

The EU compliance regime covers only those GR or associated traditional knowledge that were accessed as of 12 October 2014,[14] this being the date of entering into force of the Nagoya Protocol (Art. 17 (2)).[15] This time limit is problematic because compliance with ABS requirements was obligatory already under article 15 CBD which entered into force as early as in December 1993.

Provider States may however have regulations in force that extend ABS requirements to GR accessed earlier. Such incongruence where an obligation under the law of one State is not mirrored by an obligation of another State is quite frequent in many regulatory regimes. In view of the principle of territoriality of jurisdictions and the principles of the conflict of administrative law[16] the regulation established by the provider State cannot be enforced outside the jurisdiction of that State, except if the other State's law is ready to assist. Without that researchers who acquired GR earlier will not be monitored by authorities based in the EU. However, in such a case it is advisable that they rectify the situation by applying for retrospective approval and MAT negotiations.[17]

4.2.3 Geographical Scope

The geographical scope is determined by the locations of utilization and access.

The EU regime is applicable to any utilization of GR within the EU. It may however be that an EU based researcher is obligated by MAT to ensure that cooperating researchers outside the EU comply with the MAT conditions for utilization.

The regime only applies to GR accessed from provider States that are at the time of access contracting parties to the Nagoya Protocol. Once again researchers may be confronted with the incongruent situation that a provider State operates an access

---

[13] Commission Guidance, para. 2.3.1.7.

[14] Art. 17 (2) Regulation (EU) 511/2014.

[15] Art. 2 (1) refers to access in a State of origin. This means that GR which were accessed before 12 October 2014 and taken from a collection after that date would not be in scope of the EU Regulation.

[16] Cf. Godt et al. (2020), p. 54.

[17] Such rectification is provided in some Latin-American States as a genuine category of administrative law. See e.g. the *regularidade de acesso* according to Art. 42 Decreto N° 8.772, de 11 de Maio de 2016 of the President of Brazil.

regime without being contracting State of the NP and hence lacking support by the EU compliance regime.

The compliance regime relates to GR that fall under the sovereign rights of a provider country (Art. 2 (1)). Such GR are those which occur under *ex-situ* or *in-situ* conditions or were acquired in accordance with the CBD (CBD Art. 15 (3)). *In-situ* means "conditions where genetic resources exist within ecosystems and natural habitats, and, in the case of domesticated or cultivated species, in the surroundings where they have developed their distinctive properties".[18] This excludes from the scope alien species unless they are self-sustaining in the wild or subject to breeding activities.[19] "Acquired in accordance of the CBD" is broad language and not further defined but certainly refers to the authority of States under Art. 15 CBD to determine access and in consequence to their ABS legislation.

4.3 Is Access Subject to EU Based Duties?

Article 4 (1) imposes basic duties on users. It appears as if this implies that the access as such is not in the scope of EU law because it requires due diligence only at the stage of utilization. This duty has nevertheless an advance effect on the stage of access because due diligence to ensure compliance is formulated retrospectively. This means that researchers who utilize accessed GR later on are under obligation to respect the provider State requirements not only by command of provider State law but also by command of EU law. The only contrast is that other than provider State law EU law does not provide checks of due diligence at the access stage.[20]

4.4 What Is a User of Genetic Resources? What Is Utilization?

The EU compliance regime primarily addresses 'users', i.e. those "natural and legal persons who utilize genetic resources or traditional knowledge associated with genetic resources" (Art. 2 (4)), utilization meaning "to conduct research and development on the genetic and/or biochemical composition of genetic resources, including through the application of biotechnology" (Art. 2 (5)).

---

[18] Art. 15 (3) together with Art. 2 12th indent CBD.

[19] Similar Commission Guidance, para. 2.1.4., although somewhat overcomplicating the issue.

[20] Cf. the empty upper left box in Table 1.

4.4.1 Is Basic Research out of Scope?

Access must be with the intention to 'utilize' the GR, which, as said, means to conduct research and development (R&D). It has been discussed if the "and" in the formula "research and development" means that both R and D must come together so that "pure" R (i.e. R without D) would be excluded from the scope of the regulation. But there is good reason to also include basic research as such. Grammatically the "and" can mean both cumulation and alternativity. But even if the "and" was understood cumulatively basic research must to a large extent be regarded as being part of development because research and development often overlap. Most significantly, basic research nowadays routinely sequences the genome and normally stores the sequence data in publicly accessible data bases from which they can be accessed for development purposes.[21] Therefore, also such basic research should be made subject to the EU access regime. Only if the sequencing of a genome is exclusively aimed at the taxonomic determination of an already known species, such as when a biotope is mapped for an environmental impact assessment, this should not be counted as research in the ABS sense.[22]

Of course, however, a provider State is free to exclude basic research from prior consent, or, in application of article 8 (a) NP, provide for simplified access procedures.[23] This would also have to be taken into account by the EU based supervisory authorities.

4.4.2 Are GR Used as Research Tools Out of Scope?

As said, according to article 2 (5) 'utilization' means "to conduct research and development on the genetic and/or biochemical composition of genetic resources". Research "on" GR has been distinguished from research "with" GR which excludes from scope those GR that are used as a research tool, such as, for instance, if the function of a gene is searched by comparison with others or by a catalytic effect they may possess.[24] However, this should only be accepted if the properties of the GR used as a tool are common knowledge. If research is carried out to discover the relevant properties such as in the development of a new testing or reference tool this must be regarded as research "on" the GR.

---

[21] Contrastingly, the Commission Guidance in para. 2.3.5. excludes sequenced genomic data from the scope.

[22] Similar Commission Guidance para. 2.3.3.1.

[23] Simplified procedures are for instance instituted in Brazil, see Massini Mozini in this volume.

[24] Commission Guidance, para. 2.3.3.2.

4.4.3 How Is R&D Distinguished from Uses of GR "as Such"?

The fact that access is meant to be "for" utilization implies that there can be more purposes that are not subject to ABS conditions. They are not easy to categorize. They are sometimes termed bulk use or traded commodity. What is meant is, for instance, the consumption (such as of plants), the propagation without breeding, the use as material (such as wood in construction), and the use as cultural object (such as pet animals). Maybe, alluding to the doctrinal classification of *utere* and *fruere* one might term such uses "use and enjoyment of GR as such". While *utere* and *fruere* of biological resources have since long adequately been framed by the traditional private law of property and contract the genetic potential of organisms has remained out of scope of those legal forms and were found to require new forms of disposition rights—such as the ABS regime.

Access normally refers to the taking of a sample from some location. It has been disputed that the term also covers a situation where a biological resource was taken for other purposes, such as for consumption, and only later on becomes an object of biological R&D. The word defining "access" used by Regulation (EU) 511/2014 is acquisition, which seems to point to a physical taking. In contrast, the Commission's understanding is that subsequent R&D is a change of utilization intent.[25] However, a change of intent (such as from non-commercial to commercial) presupposes according to article 8 (a) NP that the GR was first accessed for utilization. The case of subsequent R&D on a commodity should therefore rather be categorized as access to GR in a non-physical intellectual sense.[26]

4.4.4 Is a Researcher a User if Acquiring GR from Intermediaries?

A user is also a researcher who acquires GR from intermediaries such as collections or brokers. His/her basic obligations will be examined below (Sect. 4.5.2).

4.4.5 Is a Researcher a User If Working on Information Obtained from Public Databases?

It is an open question how to classify research on genetic information that was obtained from public databases or other publication media and describe GR that were acquired from a provider state. It has been argued that such research on "pure" information is not research on GR and therefore not utilization because utilization

[25] Commission Guidance, para. 2.3.1.3.

[26] Similar Greiber (2012), p. 90. See also the definition of "accesso" as "pesquisa ou desenvolvimento tecnológico realizado sobre amostra de patrimônio genético" (research or technical development on a sample of the genetic patrimonium) in the Brazilian Lei 13.123, Art. 1 VIII.

must be directed to GR as material.[27] However, this perception ignores the fact that the information represents the material, and that the information can through synthesis be transformed back into material. If, for instance, the genome of a plant is sequenced and a gene found to code for a specific function this information can be used to insert the gene into another plant. This is certainly R&D on the GR although a pure informational phase intervened. That the sequence and its function were published does not change the fact that they were utilized. It only has an effect on the way if and how the utilization is directed by access conditions.[28]

4.5 What Basic Obligations Does a User Have Concerning Access Conditions?

Regulation (EU) 511/2014 lays out two kinds of obligations on users. One shall make sure that the GR and traditional knowledge was acquired in accordance with the provider State requirements, while the other shall ensure that the utilization itself complies with possible conditions. The first will now be discussed, the second in Sect. 4.6.

4.5.1 Ascertaining the Lawfulness of Access

Users of GR bear a basic obligation which is laid out as follows (Art. 4 (1))[29]:

> Users shall exercise due diligence to ascertain that genetic resources and traditional knowledge associated with genetic resources which they utilize have been accessed in accordance with applicable access and benefit-sharing legislation or regulatory requirements, [. . .].

By demanding "due diligence to ascertain" this provision does not set out an obligation in effect but one in conduct. This is of consequence in cases where the access legislation and practice of a provider State is unclear or difficult to find out. The user must undertake all reasonably expectable steps such as to search for the competent authority and the relevant regulations of the provider State (which may be available from the ABS Clearing-House (ABSCH) website), and inquire about interpretations and practices. If in spite of such efforts it remains unclear whether PIC and MAT are required he/she may go ahead and take or otherwise acquire the envisaged sample.[30]

[27] This understanding seems to be shared by the European Commission, see Guidance para. 2.3.5.

[28] For these question see further Sect. 4.5.3 below.

[29] Art. 4 (1) Regulation (EU) 511/2014.

[30] There may of course be other conditions to be respected outside the ABS regime, such as concerning environmental protection, access to protected areas, general oversight of research activities, international trade, etc.

However, the obligation can become one of effect after access has been completed. Article 4 (5) states:

> When the information in their possession is insufficient or uncertainties about the legality of access and utilization persist, users shall obtain an access permit or its equivalent and establish mutually agreed terms, or discontinue utilization.

Hence, if uncertainty about the legality of access persists the user must still obtain an access permit and agreement. This appears to be incompatible with the due diligence concept because with that the parties should acquiesce with the result of good efforts. A compromise solution could be that the user may wait for new evidence to appear, and if that is the case (such as by a clarifying sublegal norm), rectification is still to be tried and may also be ordered by the competent authority (Art. 9 (6)), but (absent *mens rea*) no fine can be imposed.[31]

4.5.2 Acquisition from Intermediaries

If a user acquires GR from intermediaries he/she is still subject to the basic duty to ascertain if the GR were lawfully accessed. In doing this he/she can rely on the documents that must be transferred to him/her according to article 4 (3). However, if the sample is acquired from a registered collection the burden is alleviated. The researcher is then considered to have exercised due diligence as regards seeking information (Art. 4 (7)).[32] In case the intermediary (such as a trader of raw materials) acquired the GR not for utilization and hence does not possess access documents the researcher must go back to the provider state and ask for PIC and MAT. Or the intermediary does this for him/her, maybe even prophylactically if considering to sell the GR for R&D.

Collections can be registered if they meet certain requirements. First, if accessing GR by its own personnel the collection must like anybody else abide by articles 4 (1) and (3), i.e. it must check and document whether its samples have been accessed in compliance with the provider state regulations. In addition, if transferring to third persons samples that were collected by other persons (which is the normal case) it may do so only with documentation evidencing compliance of access with provider state requirements and with mutually agreed terms (article 5 (3) (b)).[33]

The first collection to be registered was the German Collection of Microorganisms and Cell Cultures (DSMZ).[34] It saves the legal documents such as the prior

[31] For the enforcement practices of the German competent authority see Greiber and Frederichs in this volume.

[32] The situation is similar if the researcher is presented documents that have been recognized as best practices (Art. 8 with 9 (1)), although the presumption of due diligence is weaker because the documents only "may reduce the risk of non-compliance" (Art. 8 with 9 (1)).

[33] See the quite ambitious provisions on requests for inclusion in the list of collections as well as on verification and checks in articles 2–4 Commission Regulation (EU) 2015/1866.

[34] Cf. https://www.dsmz.de/collection/nagoya-protocol/registered-collection. Provider States sometimes insist on conditions concerning the handling of information from R&D, such as by prohibiting

consent, mutually agreed terms and material transfer agreements (MTA) pertinent to the access, transfer and utilization of the GR. The documents are made accessible at the collection's website. By signing a standard material transfer agreement (SMTA) called Terms and Conditions the user acquiring a GR (or progeny) from the collection is obliged to read, copy, store and transfer further the information and adhere to the conditions contained in it.[35] This last clause is a viral clause that transfers the rights and duties contained in the PIC, MAT and MTA to the user.[36] At the stages of due diligence declarations the user may submit the information to the competent authority. If he/she does so this will be regarded as fulfilling due diligence concerning the lawful access to the utilized GR (Art. 4 (7)).

One may have doubts what the advantage for collections is, given the costs of fulfilling the registration requirements. It is at least a service for their customers which could be remunerated by corresponding fees, and may also build trust with provider States when the collection itself or other users wish to acquire GR from provider States. From the provider States' perspective it may be considered as a shortcoming that the collections are freed from any duty to monitor whether the users of collections actually abide by the conditions transferred to them. Such supervisory responsibility would however seriously hinder the functioning of the collections.

4.5.3 Acquisition from Databases and Other Publication Media

If a researcher performs research with genetic information downloaded from a database he/she is, as said above (Sect. 4.4.5), a user and thus bound to ascertain the lawfulness of access of the GR. However, this burden is alleviated if the database does not indicate any conditions for the utilization. In that situation due diligence does not require any further search. If the data base however does associate information with the genetic information such as, for instance, by referring to the state of origin of the pertinent GR the user must practice due diligence to ascertain compliance of access with provider state requirements. It is difficult to advise what has to be done in such situation. I suggest that the user needs to check whether the provider State has an ABS regime in force. If that is the case he/she should identify possibilities of general restrictions of access, but for lack of more precise indications would not be required to investigate if individualized PIC and MAT exist.

the publication and uploading of sequence data to public databases, commanding the earmarking of published information to require consent for commercialization, obliging users to share new R&D results, etc.

[35] Art. 10 of the Terms and Conditions of DMSZ. They are accessible at https://www.dsmz.de/terms.

[36] It could be argued that the clause only binds the researcher in relation to the collection and not also in relation to the provider State. However, if the collection agreed in MAT to practice a viral clause this implies that the MAT extend to the researcher.

4.5.4 Documentation Duties

Once prior consent and the conclusion of MAT have been completed users are obliged to keep certain information. The most important items include either an internationally recognized certificate of compliance[37] or, in its absence, the date and place of access, the description of the genetic resources, the source from which they were directly obtained, access permits, and mutually agreed terms. The user must keep this information for 20 years.[38]

The information must be transferred to subsequent users. This shall ensure that the entire process from access via utilization to commercialization can be followed. It (at least ideally) allows to determine if a product was generated from accessed GR and hence benefits must be shared.

4.5.5 Transfer of Rights and Obligations to Third Parties

It is common practice that users cooperate with other researchers and developers, and for that purpose exchange the accessed GR with them. This entails the risk that third users escape the reach of the access conditions, such as that no commercialization is allowed or benefits must be shared. Provider States react to that risk by various means. They sometimes exclude transfer to third parties altogether, make it dependent on subsequent consent, or—more proactively—establish a so-called viral clause meaning that the user must pass the rights and obligations he/she contracted over to the third party, and inform the provider State about the names and coordinates of the same.[39]

Regulation (EU) 511/2014 does not take position in this regard. It only ensures compliance with such conditions by requiring that the relevant information, including about utilization conditions, must be kept and transferred to third parties.[40] The formulation that the documents must be "transferred" establishes an obligation of form not of content. This means that it neither requires that subsequent users shall be bound by PIC and/or MAT nor that the previous user is responsible to supervise compliance by the subsequent user. These are matters left to the MAT between provider State and user.[41]

[37] Further in this chapter see Sect. 5.3 below.

[38] Art. 4 (3) Regulation (EU) 511/2014.

[39] See for instance the viral clause in DFG (2020) Clause 7.

[40] Art. 4 (3) Regulation (EU) 511/2014.

[41] See the variants of possible formulations in DFG (2020) Clause 7.

4.6 *What Basic Duties Has a User in Respect of the Utilization Itself?*

4.6.1 Are Utilization Conditions in the Scope of Basic Duties at all?

Prior consent and MAT requirements are often used by provider States as a leverage of directing the utilization of the accessed GR. The final objective of such conditioning can be to ensure the sharing of monetary benefits, but also to participate in non-monetary benefits such as joint R&D, capacity building and technology transfer. Conditions to this effect may include that the user shall conduct certain R&D activities in collaboration with provider State R&D institutions, report about R&D results, concentrate on basic (taxonomic etc.) research, be allowed to also conduct applied research and development, publish or not R&D results, etc.

The question arises to what extent those conditions shall come under the compliance regime of Regulation (EU) 511/2014. In contrast to article 4 (1) which demands to ascertain that the utilized GR have been lawfully accessed, article 4 (2) requires that the utilization itself must comply with MAT.[42] This is even an obligation in effect, not one attenuated by a due diligence clause, which is understandable because the user would in the normal case know if MAT exist or not.

A qualification must however be made. As indicated, article 4 (2) only speaks about accordance of utilization with mutually agreed terms. Does this mean that conditions attached to PIC but not included in MAT are out of the basic obligations? Such situation may well arise where provider States ease access by only requiring PIC and not MAT for basic research but, for instance, attach a condition to the PIC decision excluding R&D that aims at commercializable products. I believe that would be a very artificial reading if a related basic obligation was denied. MAT in the sense of article 4 (2) should be read to also include PIC conditions. After all the user also agrees with one-sided conditions by applying and accepting PIC, thus in a way constituting "mutually agreed terms".

4.6.2 How Are Conditions Concerning Commercial Utilization to Be Understood?

Provider States sometimes set a condition that only non-commercial R&D shall be allowed while a change of intent to commercial R&D is made dependent on prior consent.[43] Such directions are hard to follow if it is not defined what commercial R&D shall mean. One could understand "commercial" to characterize the content of the R&D activities having in mind that commercial R&D generates marketable results and non-commercial basic knowledge. This may work for taxonomic research, but once research starts with sequencing genomes and identifying

[42] It appears logical that the same is true if the conditions are established as part of the PIC.

[43] Related clauses are, for instance, set by Costa Rica. See Cabrera Medaglia in this volume.

functions of genes—a process that could also be termed basic research—the results are already marketable because they can be sold and may even be patented.[44] A researcher is therefore well advised not to accept restrictions concerning the commercial/non-commercial content of his/her research. Instead he/she may agree with a clause that requires prior consent for any commercialization of results while the R&D itself shall be unrestricted. Such terms will in the normal case also be in the interest of the provider State because the same could draw considerable shares in benefits from R&D that generates marketable products. A researcher who is confronted with a restriction excluding commercial research may refer back to the provider State asking if the clause could be understood to mean commercialization as an intention rather than a characteristic of the utilization itself.

4.6.3 Are Utilization Conditions Binding if Flagged in Public Databases?

As explained above R&D with information downloaded from public data bases is utilization of GR so that the user is bound to ascertain the lawfulness of the same. In the normal case of using information from databases, however, the database will not flag any conditions attached to the relevant information. It is then impossible for the user to identify any conditions so that he/she is freed from the due diligence obligation. It is beyond the scope of this article whether the databases and other information media should be regulated to enable such tracing in order to pay tribute to corresponding clauses of provider State regulations.[45]

5 What Activities Are Subject to What Means of Administrative Oversight?

The observance of the basic duties by users is supervised by responsible administrative agencies. Two instruments stand out: declarations of users at and checks by competent authorities.

5.1 What Declarations Have to Be Made at What Stages of R&D?

Users must declare to the competent authority that they comply with the provider State requirements. This must be done at two stages of the R&D process: the receival

[44] See further v. Kries and Winter (2015).

[45] See further Lyal in this volume.

of research funding and the final product development (Art. 7 (1) and (2)). Through the declarations the user confirms that the requirements of article 4 are fulfilled informing about the existence of a certificate of compliance or the information listed in article 4 (3) about the kinds and origin of the GR. If MAT were concluded this fact must also be declared (Art. 7 (2)). The competent authority may be satisfied with such declarations, or ask the user to submit the relevant documents for further checking.

Annexes II and III of Commission Regulation (EU) 2015/1866 provide templates for the two declarations. In addition, the EU Commission offers an IT tool called DECLARE that users may employ for their declarations. It is not mandatory, leaving Member States discretion to operate their own formats. The declaration is addressed to the competent Member State authority but forwarded to the Commission and the ABS Clearing-House (Art. 7(3)).[46]

The declaration at the stage of final product development could cause a problem if the R&D process involved many stages. It will then be difficult to trace the product back to a duly accessed GR, especially if the product developer is not identical with the original accessor of the GR. This is particularly evident in animal breeding where multiple stages of reproduction may diminish the influence of an original contribution. Article 7 (2) Regulation (EU) 511/2014 which refers to "a product developed via the utilization ..."[47] just postulates causality without giving more concrete guidance. There is a need to develop cut-off criteria in such situations. For the moment the exclusion of minimal contributions from benefit-sharing claims could be inferred applying the general legal principle "*de minimis non curat praetor*" and referring to the stipulation in article 5 Nagoya Protocol that benefit-sharing shall be fair and equitable.

5.2 *What Additional Checks May the Competent Authorities Undertake?*

In addition to monitoring through receiving declarations the competent authorities are mandated to perform checks of users. Such checks shall be made either according to risk-based plans or in cases of suspicion of non-compliance, such suspicion possibly emerging from information the competent authority collected itself or was provided by third persons or provider States (Art. 9).[48]

---

[46] See further on the practices of monitoring the contribution by Greiber in this volume.

[47] See other language versions of Art. 7 (2): "un produit élaboré par le biais de l'utilisation ..."; Produkt "das durch Nutzung von genetischen Ressourcen ... entwickelt wurde"; "un producto elaborado mediante la utilización ...".

[48] Concerning the execution of checks in Germany (by the BfN) see Greiber and Frederichs in this volume.

The checks of users also address the lawfulness of the utilization itself, because according to article 9 (3) (b) the authority shall check "a user's compliance with this Regulation", "this Regulation" covering also article 4 (2) and thus utilization practices.

Such checking of utilizations themselves is in line with the NP requirements. Article 17 (1) (a) NP provides that checkpoints would collect information including on the utilization, and have functions relevant to the utilization at any stage from research to commercialization.

Ambitious as this compliance system appears an empirical survey of 2019 of the EU Commission shows that there are still many shortcomings in actual practice. This is due to the novelty of the ABS regime. It is still widely unknown to researchers, and informed supervisory personnel is often lacking.[49] It should however be noted that the DSMZ has created a task force which actively disseminates relevant information and offers roundtables for scientists and science managers.[50]

5.3 *What Authorities Are Competent for What?*

The administrative implementation of the EU ABS regime is organized as a multilevel system. Basically, the Member States are in charge of administrative oversight. They must establish, empower and mandate competent authorities to receive due diligence declarations and conduct the required compliance checks. In Germany the Bundesamt für Naturschutz (Federal Agency for Nature Conservation—BfN) is the competent authority. The national focal point is the Bundesministerium für Umwelt, Naturschutz und nukleare Sicherheit (Federal Ministry for Environment, Nature Conservation and Nuclear Safety—BMU).[51]

The EU Commission is the competent authority for receiving due diligence declarations concerning projects directly funded by the EU (Art. 7 (1)).[52] In addition, and more importantly, the Commission has some higher level functions facilitating or supervising the Member States authorities, such as the establishment and maintenance of the register of collections (Art. 5 (1)) and of a list of competent authorities of the Member States (Art. 6 (1), (2)), granting recognition of best ABS practices of associations of users or other interested parties, and maintaining an internet based register of best practices (Art. 8).

At the international level the ABS Clearing-House (ABSCH) that was established on the basis of article 14 NP must be kept informed about EU practices, such as by transmission of prior consents and MAT which through such notification are given the status of an internationally recognized certificate of compliance (Art. 17 (2) with

[49] Report from the Commission COM/2019/13 final.

[50] See German Nagoya Protocol HUB, accessible at www.dsmz.de.

[51] Gesetz zur Umsetzung der Verpflichtungen nach dem Nagoya-Protokoll v. 25.11.2015.

[52] The compliance checks however remain in the competence of the Member States.

Art. 6 (3) (e) NP), as well as by transmission of the information submitted together with the due diligence declarations at the stages of research funding and premarketing (Art. 7 (3)). The ABS Clearing-House does not have powers to act in case of non-fulfilment of NP requirements. In such case it nevertheless has implied competences not to grant the status of an internationally recognized certificate and to communicate with the national and EU focal point in order to rectify any insufficiency.

5.4 Transparency and Public Participation of ABS Proceedings

Article 15 establishes a consultation forum with participation of interested parties as a means to further the implementation of the compliance regime. Regulation (EU) 511/2014 does not provide for participation of the public in administrative procedures. It is for the contracting parties of the NP to introduce related provisions, such as, for instance, when the competent authorities intend to take correcting measures. Article 9 (3) (b) at least acknowledges that compliance checks may react to public information about cases of non-compliance. Concerning transparency Article 10 (2) refers to the general regime on access to information as established by Directive 2003/4/EC.

6 What Basic Duties and Administrative Oversight Apply to Activities Subsequent to Utilization?

At the stages after research and development the generated knowledge and developed products may be further used, multiplied, patented, and brought to markets. Regulation (EU) 511/2014 recaps these activities by the terms 'subsequent applications and commercialization' and 'benefit sharing'. It shall be asked once more, what basic duties actors have, and how they are supervised by public authorities.

6.1 What Basic Duties and Administrative Oversight Apply Concerning Subsequent Applications and Commercialization?

What 'subsequent applications and commercialization' means is nowhere defined in the Regulation. But the term can be imagined to comprise, for instance, the propagation, publication, patenting, registration for marketing, marketing and selling of research results and developed products.

Subsequent applications and commercialization are activities that are frequently juxtaposed to the notion of utilization. This can be deduced from the definition of utilization in article 3 (5) which only refers to research and development, and the definition of MAT in article 3 (6) which refers to "utilization as well as subsequent applications and commercialization". The distinction of the two terms can also be found in article 5 (1) NP which refers to "benefits arising from the utilization of genetic resources as well as subsequent applications and commercialization".

The distinction has consequences for the EU compliance regime. Article 4 (2) which commands the basic duties only refers to utilization. Does this mean that subsequent applications and commercialization are out of scope of the basic duties? This would also exclude those activities from the declaration duties under article 7 (1) and (2) as well as from any verification checks under article 9. For instance, a MAT condition may allow the development of a product but disallow any commercialization. If the actor wanted to bring the product on the market he/she would not be bound (by EU law) to respect such condition. Neither would the supervisory authority have powers to intervene. If the user duly submitted the declaration under article 7 (2) this would anyway only relate to the final development of a product, not however to the subsequent commercialization.[53]

This consequence is hardly reasonable. It would deprive provider States of one of their main tools of ensuring benefit-sharing, namely to (co-)direct the commercialization of R&D results. This consequence could be avoided by a more open interpretation of the term utilization. 'Utilization' could be understood to be the generic term for all activities from research and development through to the application for intellectual property rights (IPRs), the registration (if any) of a product and the offering of a product on the market.[54] The activities towards the end of the valorization chain could all be included in the term of development. 'Development' is an indeterminate notion that is often used in law covering a broad range of preparing natural assets for human use. Think, for instance, of the term in the context of development policies which are also invoked in article 22 (2) and 23 NP and obviously include commercialization activities. The fact that subsequent applications and commercialization are mentioned in juxtaposition to utilization at some points of both Regulation (EU) 511/2014 and the NP can be understood as a reminder that they should be taken into account in contexts where they become important, such as when MAT conditions must be observed and benefits shared. Such understanding would also help to cope with the odd fact that article 3 (6) speaks of "benefits arising from the utilization of genetic resources", omitting subsequent applications and commercialization, although the most interesting benefits for many provider States – monetary ones – arise from such subsequent activities.

---

[53] If that yields monetary benefits the duty to share will still be under oversight of the competent authority (see below), but the bringing on the market itself could not be forbidden.

[54] See similar but with a different construction Godt et al. (2020), pp. 44–48 who start from a distinction between 'utilization' and 'subsequent applications and commercialization' but extend 'utilization' to reach into the premarketing phases e.g. of product registration.

Article 4 (2) should therefore be read to refer to benefits arising from utilization *including* subsequent applications and commercialization. This would imply that it is part of the basic duty of users to ascertain the lawfulness of access and to respect any conditions of the provider State concerning subsequent applications and commercialization. While the due diligence declaration would not cover those activities the competent authorities would still have powers to check their lawfulness.

Concerning geographical scope it should be noted that subsequent applications and commercialization are subject to the EU compliance regime only if occurring within the EU territory. This is also the case if the sales occur on foreign markets but the marketing activities (such as the registration or exportation of a product) start from the EU, or if the revenue flows back to EU based actors. However, if products are transferred to foreign actors and by them brought to foreign markets this must be supervised by the responsible foreign country. There is an obvious gap of control, however, if that country is not a contracting State of the NP and does not have its own compliance regime.

6.2 What Basic Duties and Administrative Oversight Apply Concerning Generated Benefits?

Benefits can be non-monetary and monetary. They may emerge during the R&D activities as joint research projects or lump-sum payments in anticipation of the market value of R&D results, or at the end of R&D activities in the form of sharing knowledge resulting from R&D, joint publications, intellectual property rights or as revenue from product sales. If such benefits are addressed by PIC and/or MAT conditions the question arises if they are an object for EU law based basic duties and administrative oversight.

The basic duty under article 4 (1) demands users to exercise due diligence not only concerning access but also concerning benefit-sharing, asking to ascertain that "benefits are fairly and equitably shared upon mutually agreed terms, in accordance with any applicable legislation or regulatory requirements". In doctrinal terms this is a duty under EU law assisting the parallel duty based on provider State law.

The wording "benefits are [...] shared" indicates that the basic duty aims at checking the reality of benefit-sharing rather than a simple look at whether PIC and MAT texts contain conditions on benefit-sharing. Concerning the duty to keep and transfer information this comprises information on "mutually agreed terms, including benefit-sharing arrangements, where applicable" (Art. 3 (3) (b) (vi)), but not on actual practices of benefit-sharing.

Concerning administrative oversight the due diligence declarations at research and premarketing stages can be helpful concerning benefits arising during the R&D processes, but not concerning the sharing of benefits arising from knowledge results and final products because these accrue after the stage of their final development.

There is nevertheless oversight also at the stage of generated benefits based on the general checks on user compliance under article 9 which refers to article 4 and thus to the duty of users to diligently ascertain lawful benefit-sharing.

7 What Specifics Apply for ABS Concerning Traditional Knowledge?

Both the basic duties and instruments of administrative oversight are also applicable to access to and utilization of traditional knowledge associated with genetic resources. This corresponds to the related provisions of the Nagoya Protocol which extends the compliance obligations also to associated traditional knowledge.[55] While the Nagoya Protocol does not give a definition, other than that the knowledge must belong to indigenous or local communities and be related to genetic resources, the EU Regulation goes a bit further by specifying that the knowledge must be "relevant for the utilization of genetic resources" and "as such described in the mutually agreed terms applying to the utilization of genetic resources". The latter clause can cause confusion if the knowledge is objectively relevant for the utilization of genetic resources but not described "as such". It is submitted that according to Nagoya Protocol standards benefit-sharing would also apply in this case of *de facto* utilization.

Article 7 NP prescribes that for access to associated traditional knowledge prior informed consent of the indigenous or local community must be obtained. This is mandatory and often comes in addition to the normal PIC and MAT of/with the provider State. The procedure can be demanding because the unwritten customary deliberation and decision practices may first have to be studied and then set in motion. The standard of due diligence under article 4 (1) and the corresponding administrative oversight are then difficult to determine.

8 Conclusion

Regulation (EU) 511/2014 is an exemplary step forward of user States which strive to ensure compliance with provider State ABS regimes. The instruments foreseen are best to be ordered if related to the overall process of valorization of GR and aTK from basic research and development down to commercialization and benefit-sharing. Two instruments can be distinguished: First, basic duties are imposed on users to ensure that the utilized GR and aTK were accessed respecting the provider State requirements, and that the utilization itself including subsequent applications, commercialization and benefit-sharing comply with PIC and MAT conditions. In

[55] Art. 7 NP.

doctrinal terms these duties are EU law flanking the corresponding duties under provider State law. As such they serve as points of reference for the second instrument, administrative oversight. Administrative oversight consists of checking powers and obligations which are facilitated by declarations the users have to submit at the stages of research funding and of final product development concerning due diligence in fulfilling their basic duties. The overall supervisory structure is shown in Table 1 above.

Some controversies of interpretation and practice have been addressed and solutions proposed, including that the basic duties and administrative checks should extend to utilization, subsequent applications, commercialization and benefit-sharing, that GR in form of information while not protected as such by sovereign rights can be conditioned by provider State's PIC and MAT, that such conditions loose traceability in public data-bases (which poses the question of reforming data-base practices), and that in cases of multiple contributions to the development of products the causation of accessed GR or aTK must at least be acknowledged if its characteristic trait is still present in the final product.

If these problems and others which were not addressed in the present article are not solved in pragmatic ways more fundamental alternatives may more and more be propagated. They start with a critique of the quid pro quo character of the present ABS concept where GR and aTK are offered in exchange for the promise that benefits are shared. This implies that the provider States strive to control the R&D process in order to ensure benefit-sharing. However, such control becomes the more difficult the more the R&D process is "socialized" because more researchers and developers come in and more GR and aTK are built into final products. User States step in to ensure compliance with provider State regimes but only with limited effect in case of "socialized" utilization. While the present quid pro quo system could then be reserved for cases where products are brought to the market only after very few steps of R&D, for the other cases pool approaches may be considered. They reach from regional or species-specific pools of genetic resources and related information[56] to more radical systems that allow for free utilization of any genetic resource both in form of material and information, giving up direct links between provider State and final products. In such systems products arising from the R&D on genetic resources would be subject to a charge which flows into a global fund or regional or species-related funds from which nature protection can be financed and provider States are supported that are engaged in the conservation of genetic resources.[57] However, it goes beyond the scope of this contribution to discuss such approaches in more detail.[58]

[56]See further the various contributions in Kamau and Winter (2013) (eds), pp. 221–222, and Dedeurwaerdere et al. (2013).

[57]For a prominent variant see Ruiz Muller (2015).

[58]See further Winter (2021).

Acknowledgements I am grateful to Thomas Greiber and Evanson Chege Kamau for their in-depth reading and commenting of earlier drafts. Of course, any remaining mistakes and contestable interpretations are mine.

References

Books/Book Chapters and Articles

Chroust A-H, Osborne DL (1942) Aristotle's conception of justice. Notre Dame Law Rev 17:129–143, http://scholarship.law.nd.edu/ndlr/vol17/iss2/2. Accessed 10 Dec 2020

Dedeurwaerdere T, Broggiato A, Louafi S, Welch EW, Batur F (2013) Governing global scientific research commons under the Nagoya protocol. In: Morgera E, Buck M, Tsioumani M (eds) The 2010 Nagoya protocol on access and benefit sharing. Nijhoff, Leiden, pp 389–421

Godt C, Šušnjar D, Wolff F (2020) Umsetzung des Nagoya Protokolls in EU- und nationales Recht. Transposition of the Nagoya Protocol into EU- and National Law. Ein Alternativvorschlag zur Umsetzung der EU VO 511/2014. An alternative proposal for the transposition of EU Reg. 511/2014. Nomos, Baden Baden

Greiber T et al. (2012) An explanatory guide to the Nagoya Protocol on Access and Benefit Sharing, IUNC Environmental Policy Paper No. 83

Kamau EC, Winter G (2013) Common pools of genetic resources. Equity and innovation in international biodiversity law. Routledge, London, pp 193–223

Kries Cv, Winter G (2015) Defining commercial and non-commercial research and development under the Nagoya protocol and in other contexts. In: Kamau EC, Winter G, Stoll P-T (eds) Research and development on genetic resources, Public domain approaches in implementing the Nagoya Protocol. Routledge, London, pp 60–74

Ruiz Muller M (2015) Genetic resources as natural information. Implications for the convention on biological diversity and Nagoya protocol. Routledge, London

Stoll P-T (2009) Access to genetic resources and benefit sharing – underlying concepts and the idea of justice. In: Kamau EC, Winter G (eds) Genetic resources, traditional knowledge and the law. Earthscan, London, pp 3–18

Winter G (2021) Problems and solutions of access to genetic resources and benefit sharing: a theoretical perspective - Part I. law. Environ Dev J (LEAD) 17(1):72, available at http://www.lead-journal.org/content/a1705.pdf. Part II is available as advance article in 17/0 LEAD (2021), p 3, accessible at http://www.lead-journal.org/current_issue.htm

Legal Documents

Regulation (EU) No 511/2014 of the European Parliament and of the Council of 16 April 2014 on compliance measures for users from the Nagoya Protocol on Access to Genetic Resources and the Fair and Equitable Sharing of Benefits Arising from their Utilization in the Union. OJ 2014 L 150/59

Commission Implementing Regulation (EU) 2015/1866 of 13 October 2015 laying down detailed rules for the implementation of Regulation (EU) No 511/2014 of the European Parliament and of the Council as regards the register of collections, monitoring user compliance and best practices. OJ 2015 L 275/4

Commission Guidance document on the scope of application and core obligations of Regulation (EU) No 511/2014 of the European Parliament and of the Council on the compliance measures for users from the Nagoya Protocol on Access to Genetic Resources and the Fair and Equitable Sharing of Benefits Arising from their Utilisation in the Union, OJ 2021/C 13/01

Directive 2010/75/EU of the European Parliament and of the Council of 24 November 2010 on industrial emissions (integrated pollution prevention and control) (Recast). OJ L 334

(Brazilian) Decreto N° 8.772, de 11 de Maio de 2016 of the President of Brazil, https://www2.camara.leg.br/legin/fed/decret/2016/decreto-8772-11-maio-2016-783077-publicacaooriginal-150340-pe.html (accessed 6 December 2020)

(Brazilian) Lei 13.123, de 20 de MAIO, http://www.planalto.gov.br/ccivil_03/_ato2015-2018/2015/lei/l13123.htm (accessed 6 December 2020)

(German) Gesetz zur Umsetzung der Verpflichtungen nach dem Nagoya-Protokoll zur Durchführung der Verordnung (EU) Nr. 511/2014 und zur Änderung des Patentgesetzes sowie zur Änderung des Umweltauditgesetzes v. 25.11.2015, BGBl. I Nr. 47, S. 2092

Nagoya Protocol on Access to Genetic Resources and the Fair and Equitable Sharing of Benefits Arising from their Utilization to the Convention on Biological Diversity., https://www.cbd.int/abs/doc/protocol/nagoya-protocol-en.pdf (accessed 9 December 2020)

(Norwegian) Act of 19 June 2009 No. 100 Relating to the Management of Biological, Geological and Landscape Diversity., https://absch.cbd.int/database/record/ABSCH-MSR-NO-203981 (accessed 6 December 2020)

(Swiss) Federal Act on the Protection of Nature and Cultural Heritage of 1 July 1966., https://www.admin.ch/opc/en/classified-compilation/19660144/index.html (accessed 6 December 2020)

(Swiss) Nagoya Ordinance 2016, Ordinance on Access to Genetic Resources and the Fair and Equitable Sharing of Benefits Arising from their Utilisation of 11 December 2015 (Status as of 1 January 2017), https://www.admin.ch/opc/en/classified-compilation/20150120/index.html (accessed 10 December 2020)

WHO Resolution establishing the Pandemic Influenza Preparedness (PIP) Framework., www.who.int/influenza/resources/pip_framework/en/ (accessed 9 December 2020)

Reports

Report from the Commission to the European Parliament and the Council - Regulation (EU) No 511/2014 of the European Parliament and of the Council of 16 April 2014 on compliance measures for users from the Nagoya Protocol on Access to Genetic Resources and the Fair and Equitable Sharing of Benefits Arising from their Utilization in the Union, COM/2019/13 final, https://ec.europa.eu/transparency/regdoc/rep/1/2019/EN/COM-2019-13-F1-EN-MAIN-PART-1.PDF (accessed 9 December 2020)

Model Clauses

DFG (2020) Model clauses for mutually agreed terms on access to genetic resources and benefit sharing, Deutsche Forschungsgemeinschaft, https://www.dfg.de/download/pdf/dfg_im_profil/gremien/senat/biologische_vielfalt/191212_model_clauses.pdf (accessed 10 December 2020)

Website Links

https://www.dsmz.de/collection/nagoya-protocol/registered-collection (accessed 9 December 2020)
https://www.dsmz.de/terms (accessed 9 December 2020)
www.dsmz.de (accessed 9 December 2020)

Gerd Winter was born in Diepholz, Germany, in 1943. He is Research Professor of Public Law at Research Center for European Environmental Law, University of Bremen. He has taught and published extensively on German, comparative, European and international environmental law, including on property, distributional and risk related problems of biotechnology (see https://www.gerd-winter.jura.uni-bremen.de/veroeffchronol.html). He also served as consultant on public and environmental law reform in transition countries as well as counsel in numerous administrative and court proceedings concerning environmental issues. gwinter@uni-bremen.de.

Access and Benefit-Sharing Regime of Spain: Striking the Right Balance Between Its Interests as a Provider and a User of Genetic Resources

Luciana C. Silvestri

Abstract Spain anticipated the possibility of regulating the access to and utilization of its genetic resources through Law No. 42 of 2007, a power that it had not exercised until recently. Newly adopted Royal Decree No. 124/2017 related to the access to genetic resources deriving from wild taxons and to the control of their utilization, soundly fulfills obligations set under the Nagoya Protocol and the European rules on the matter. In addition, the Decree perfectly strikes the balance between Spain's interests as provider and a user of genetic resources. After an illustration of the uniqueness of Spanish biodiversity and the country's commitment to the access and benefit-sharing negotiations, the chapter analyses how access to Spanish genetic resources is regulated. It also elaborates on the national measures taken to fulfill the obligations set under Regulation (EU) No. 511/2014 of the European Parliament and of the Council of 16 April 2014 on compliance measures. The case of Spain may well serve as an example to other countries which, like Spain, have a dual role, being both provider and user of genetic resources.

Keywords Access and benefit-sharing legislation · Genetic resources · Spain Royal Decree No. 124/2017

1 Background

Spain is an important provider of *in-situ* genetic resources. Its unique geographical situation, in conjunction with its great climate variability and distinctive archipelagoes such as Macaronesia and the Balearic Islands turn Spain into the most biodiverse country in Europe. Its biological richness is truly amazing: it features the highest number of vascular plants (8000–9000 species) in Europe and the

L. C. Silvestri (✉)
Institute of Human, Social and Environmental Sciences, National Scientific and Technical Research Council (CONICET), Mendoza, Argentina
e-mail: lsilvestri@mendoza-conicet.gob.ar

E. Chege Kamau (ed.), *Global Transformations in the Use of Biodiversity for Research and Development*, Ius Gentium: Comparative Perspectives on Law and Justice 95, https://doi.org/10.1007/978-3-030-88711-7_15

Mediterranean region, and a rate of endemism of 20 to 25%.[1] If fauna species are considered, some 70,000 are known to exist, a number that accounts for half of those present in the entire continent.[2]

Spain is also one of the countries with the greatest cultural diversity in Europe. It boasts a rich cultural heritage shaped by the influence of the various peoples inhabiting the Iberian Peninsula over the centuries: Iberians, Celts, Phoenicians, Greeks, Romans and Arabs. The country is also the holder of a significant level of biodiversity, which has resulted in vast traditional knowledge (TK) associated with the use of genetic resources, including plants, animals and fungi.[3] Presently, for example, some 1200 plant species are still used for medicinal purposes[4] and around 500 wild plant species are estimated to be edible.[5]

Given its significant biological richness and the vast TK associated with its utilization, Spain, as any other typical provider country, wishes to benefit from the utilization of its genetic resources and ensure effective compliance with its access and benefit-sharing regime (ABS).

This role as a 'provider' of genetic resources and its resulting positioning on the issue manifested itself in at least three instances. First, when Spain volunteered to host and finance the fourth meeting of the Convention on Biological Diversity (CBD) Working Group which took place in Granada in 2006. This initiative contributed to solving the conflict of interests existing between developed countries—which wanted to prioritize work on protected areas—and developing countries, usually important providers of biodiversity—which were mostly interested in negotiating and adopting a legally-binding regime on ABS which would allow them to benefit from the utilization of their resources. The Spanish contribution constituted a furtherance of the interests of the latter group of countries, without neglecting progress in the activities related to protected areas required by the more developed countries.

Second, during the negotiations that culminated in the adoption of the Nagoya Protocol (NP), Spain, as most biodiversity-rich countries, strongly supported the view that the outcome of the negotiations should necessarily be a legally-binding regime on ABS. This collided with the vision of the European Commission and some European Union (EU) Member States, who did not accept until the very last minute of the negotiations that their aftermath, if it was to be successful, involved the implementation of a strong regime capable of promoting and safeguarding the fair and equitable sharing of benefits arising out of the utilization of genetic resources.[6]

Third, once the NP was adopted and discussions on possible legal alternatives to comply with it were initiated at the European level, Spain strongly opposed the

---

[1] Biodiversity and Natural Heritage Strategic Plan 2011–2017 (2011), p. 103095.

[2] Organismo Autónomo Parques Nacionales (2002), p. 18.

[3] Pardo de Santayana Gómez de Olea et al. (2012), p. 9.

[4] Fernández López and Amezcúa Ogayar (2007).

[5] Morales et al. (2011), p. 175.

[6] Lago Candeira and Silvestri (2012), pp. 270–271.

European Commission's proposal on the issue and argued for the implementation of stricter compliance measures. In particular, Spain argued that the famous 'due diligence approach'; which was later included in Regulation (EU) No. 511/2014[7] softened NP provisions on users' compliance and entailed a system that was totally different from the one internationally agreed.[8]

In addition to being a provider of genetic resources and identifying itself with negotiating positions that are in line with such reality, Spain is also an important user of genetic resources. In this sense, statistics show that, roughly, 1130 companies in Spain were exclusively focused on biotechnology research and development in 2016, employing, altogether, 10,700 highly qualified researchers.[9] As a consequence of this, Spain supported an international ABS regime that entails transparent, predictable and efficient legal ABS national frameworks and that does not unnecessarily impede the development of its fast-growing biotechnology industries.

2 Legal Framework

2.1 Legislation

Spain has been a party to the CBD and to the NP since December, 1993[10] and October, 2014,[11] respectively.

The Spanish legislation applicable to ABS may be classified into regulations of community origin, aimed at promoting a coherent application of the NP in the EU, and regulations of a strictly domestic nature. Community-origin regulations comprise Regulation (EU) No. 511/2014 of the European Parliament and of the Council of 16 April 2014 on compliance measures for users of the Nagoya Protocol on Access to Genetic Resources and the Fair and Equitable Sharing of Benefits Arising from their Utilization in the Union and Commission Implementing Regulation (EU) 2015/1866 of 13 October 2015 laying down detailed rules for the implementation of Regulation (EU) No. 511/2014 of the European Parliament and of the Council as regards the register of collections, monitoring user compliance and best practices. These two are legally-binding measures for all EU Member States, including Spain. In addition to these, the EU Commission developed a

[7] Regulation (EU) No. 511/2014 on compliance measures for users from the Nagoya Protocol on Access to Genetic Resources and the Fair and Equitable Sharing of Benefits Arising from their Utilization in the Union.

[8] Silvestri and Lago Candeira (2015), p. 223.

[9] Asociación Española de Bioempresas (2018), p. 29.

[10] Ratification Instrument to the Convention on Biological Diversity, held in Rio de Janeiro on June 5, 1992.

[11] Ratification Instrument of the Nagoya Protocol on access to genetic resources and equitable sharing of the benefits deriving from their utilization to the Convention on Biological Diversity, held in Nagoya on October 29, 2010.

non-legally-binding Guidance document on the scope of application and core obligations of Regulation (EU) No. 511/2014 of the European Parliament and of the Council on compliance measures for users of the Nagoya Protocol on Access to Genetic Resources and the Fair and Equitable Sharing of Benefits Arising from their Utilization in the Union.[12]

In turn, regulations of a distinctly Spanish origin are Law No. 42/2007 of 13 December, on Natural Heritage and Biodiversity,[13] modified by Law No. 33/2015[14] of 21 September and Royal Decree No. 124/2017[15] of 24 February, related to the access to genetic resources deriving from wild taxons and to the control of their utilization. They were adopted, at least as far as the access to genetic resources is concerned, by virtue of the power the parties hold to regulate access based on their sovereign rights over their natural resources.[16] Both pieces represent cases of basic legislation adopted by the State, who has competence to adopt this kind of regulations for environmental protection pursuant to the Spanish Constitution.[17]

The former regulates access to genetic resources in Spain and sets forth the compliance measures and sanctions provided for in Regulation (EU) No. 511/2014. The Decree, in turn, expounds on articles 71, 72 and 74 of the Law by listing access procedures to Spanish genetic resources. It also designates the competent authorities for the application of Regulation (EU) No. 511/2014 in Spain and establishes the specific procedures for the processing of due diligence declarations, among other matters.

Apart from the state regulations, the Autonomous Communities of Spain also have competence to issue environmental protection legislation.[18] Their legislation must conform to basic state laws and may establish more restrictive environmental requirements.[19] Precisely this is contemplated by Law No. 42/2007, which disposes that the Autonomous Communities may set conditions on the access to *in-situ* genetic resources when their collection requires special protection in order to ensure their conservation and sustainable use, besides the requirements of basic state legislation.[20]

Prior to the modification of Law No. 42/2007 and to the adoption of Royal Decree No. 124/2017, certain Communities had partially regulated access to the genetic resources found in their territories. So did, for example, Andalusia, Catalonia and the

[12]For details on the EU regime on access and benefit-sharing see Winter in this volume.

[13]https://www.boe.es/buscar/doc.php?lang=es&id=BOE-A-2007-21490.

[14]https://www.boe.es/buscar/pdf/2015/BOE-A-2015-10142-consolidado.pdf.

[15]https://www.boe.es/buscar/doc.php?id=BOE-A-2017-2743.

[16]NP, Article 6.1.

[17]Spanish Constitution, Article 149.1.23.

[18]Spanish Constitution, Article 148.1.9.

[19]Spanish Constitution, Article 149.1.23.

[20]Law No. 42/2007, Article 71.10.

Canary Islands.[21] However, in the present, and considering that the new state legislation is in force, the Autonomous Communities must adjust their respective regulations thereto.

2.2 *Scope of the Legislation*

Spanish legislation on access to genetic resources is applicable to those resources deriving from wild taxons found either in *in-situ* or *ex-situ*[22] conditions. The regime will be applicable to those resources which are subjected to 'utilization'; that is to say, only when research or development on the genetic and/or biochemical composition of the genetic resources is intended, including the application of biotechnology.[23]

Because there may exist gray areas in which it is difficult to determine whether there effectively is utilization of genetic resources or not, Spain uses the standard laid down by the Guidance document on the scope of application and core obligations of Regulation (EU) No. 511/2014 for such purpose. According to this standard, it must be defined whether the activity to be conducted on the genetic resources will enable a new understanding of the characteristics of the resource or not, and whether this understanding may, even if only potentially, enable future developments.[24] If the answer is affirmative, such activity will be considered as 'utilization' and will be then covered under the Spanish access regime. Otherwise, it will not.

Thus, by way of example, activities such as the supply and processing of raw materials for the extraction of essential oils for use in cosmetic products would not be covered under Spanish legislation, as it is understood that, in such a case, the processes used do not give origin to new knowledge on the characteristics of the resource. Neither would be the application of biotechnological processes in the context of industrial production when the resources used do not constitute, in themselves, the subject of research or development. An example of this would be, for instance, the incorporation of yeasts in the winemaking process, provided that the yeast is used as it is found and it is not the subject of research and/or development.

Exceptions to the application of the regime are:[25] (a) access to genetic resources exclusively for taxonomic purposes;[26] (b) plant genetic resources for food and

[21] Lago Candeira and Silvestri (2012), p. 287.

[22] Law No. 42/2007, Article 71 and Royal Decree No. 124/2017, Article 3.1.

[23] NP, Article 2.c and Royal Decree No. 124/2017, Article 2.

[24] Guidance document on the scope of application and core obligations of Regulation (EU) No. 511/2014 section 2.3.3.

[25] Law No. 42/2007, Article 71. 4 and Royal Decree No. 124/2017, Article 3.3.

[26] Exclusively taxonomic purposes means "*the application of principles and methods for the identification, delimitation and classification of living beings, which requires the study of their phylogenetic relationships as well as of the evolutionary and ecological processes that biodiversity*

agriculture covered under Law No. 30/2006, of 26 July, on seeds and nursery plants and on plant genetic resources; (c) fish genetic resources regulated by Law No. 3/2001, of 26 March, on State Marine Fishing; (d) animal genetic resources for food and agriculture which will be regulated by future specific legislation; (e) collection of materials/samples and their maintenance in germplasm banks or in *ex-situ* collections with the exclusive purpose of their conservation and f) the production and commercialization of seeds and forest materials/plants covered under Royal Decree No. 289/2003, of 7 March, on commercialization of reproduction forest materials, as long as there is no utilization of the genetic resources and no transfer to third parties for a different use.[27]

In practice, the exceptions set forth for plant and animal genetic resources[28] present serious difficulties to the effective application of the ABS regime.[29] In the case of the former, the legal reference to the regime established in Law No. 30/2006, in which the access to plant genetic resources is regulated, among other things, proves to be somewhat futile. This is due to the fact that the said law excludes from its scope of application a certain type of plant genetic resource: the crops contained in Appendix 1 of the International Treaty on Plant Genetic Resources for Food and Agriculture (ITPGRFA).[30] In this case, there is no specific state regulation on the subject and, therefore, there is a legal gap for access to this kind of resources.[31] Likewise, although Law No. 30/2006 is applicable to the access to the plant genetic resources not included in Appendix 1 of the Treaty, and, thus, the legal reference would be to a certain extent effective in relation to them, the conditions for access to this kind of resources are not regulated[32] and this also represents an insurmountable obstacle.

In the case of the exception envisaged for animal genetic resources for food and agriculture, there exists no specific regulation governing this aspect, which represents another loophole in the legislation.

Finally, it is worth pointing out that, in relation to the temporary scope of application of Royal Decree No. 124/2017 and the obligation to request

---

has generated, by using morphological, physiological, genetic, behavioral and environmental data." Royal Decree No. 124/2017, Article 2.3.

[27] If transfer is made to a third party for the purpose of utilization, it must be indicated that any use of the genetic resource will require the prior authorization of access in accordance with Law No. 42/2007, of 13 December, and with the Royal Decree No. 124/2017 (Article 3.3).

[28] Law No. 42/2007, Articles 71.4 a and c.

[29] Silvestri and Lago Candeira (2015), p. 219.

[30] Law No. 30/2006, Article 45.3.

[31] In other words, Law 42/2007 excludes from its scope plant genetic resources for food and agriculture and states they are regulated by law 30/2006. Law 30/2006 itself excludes from its scope plant genetic resources for food and agriculture included in the ITPGRFA. This means that access to *in-situ* plant GR covered by the ITPGRFA would then again be covered by law 42/2002. In practical terms this is a buried cross reference and have the undesirable effect of the inexistence of an specific regulation for *in-situ* access to plant GR covered by the International Treaty.

[32] MAGRAMA (2015), p. 208.

authorization to access Spanish genetic resources, the following provision applies:[33] those users accessing a genetic resource found in a Spanish *ex-situ* collection or which is under the power of a company, institution, or research center, or of any physical or legal person after the Decree's effective date, that is to say, after 15 March 2017, must request an access permit from the corresponding authority. However, if the user of the resources is the very company, institution or research center, and it had them under its control prior to that date, the user may utilize them or continue to utilize them without the need to request an administrative authorization. For *in-situ* access to Spanish genetic resources, the provisions of the Royal Decree logically apply as of the effective date thereof.

2.3 Institutional Framework and Access Procedures

The territorial administrative organization of Spain comprises 17 Autonomous Communities with constitutional competence to manage their own environment.[34]

Accordingly, they are responsible for providing the prior informed consent (PIC) and establishing mutually agreed terms (MAT) for the access to genetic resources located within their respective jurisdictions.[35] They also have competence to authorize access to the genetic resources provided they are endemic to the Community.[36] In order to fulfill such responsibility, the Communities must appoint a competent regional authority to manage access to genetic resources[37] and communicate its contact information to the State administration, a procedure which all Communities have already carried out.

If genetic resources are geographically scattered on the territory of more than one Community, the provision of the consent and the establishment of the conditions will be under the responsibility of the agency determined by the corresponding Community or Communities from where genetic resources are accessed. In this case, the access permit will be granted by the General Directorate of Environmental and Natural Environment Quality and Assessment pertaining to the Ministry of Agriculture and Fisheries, Food and Environment;[38] that is to say, by the national competent authority.

When resources fall within the following categories, the responsibility for granting PIC and negotiating MAT pertains to: (a) Spanish maritime jurisdiction (General Directorate for the Sustainability of the Coast and the Sea) (b) state public assets (agency of the Public National Administration to which the assets are

[33]Royal Decree No. 124/2017, third final provision.

[34]Spanish Constitution, Article 148.1.9.

[35]Law No. 42/2007, Article 68.2.

[36]Royal Decree No. 124/2017, Article 5.2.

[37]Law No. 42/2007, Article 71.5.

[38]Royal Decree No. 124/2017, Article 5.1.d.

assigned) and (c) state-owned or state-related institutions for *ex-situ* conservation (the institution's managing agency).[39] In such cases, the General Directorate of Environmental and Natural Environment Quality and Assessment is the competent authority providing access; that is to say, it is the authority which authorizes access once consent has been provided and the conditions negotiated, as may be the case.[40]

There are two types of procedures for access to genetic resources, depending on whether they are intended to be used for non-commercial (or purely scientific) purposes or for commercial purposes. In the first case, the access application, together with a sworn statement, must be addressed to the competent national or regional authority,[41] which, once the application and the statement are reviewed, will contact the agency in charge of providing the PIC and establishing MAT.[42] This agency must issue a preliminary report on the application and, if deemed necessary, establish additional specific conditions on the distribution of benefits.[43] If all the requirements are met, the competent authority will grant the access permit within a maximum period of two months.[44] Once the user has obtained the authorization, they will be able to effectively access the resources.[45]

In such a case when the resources are to be used for commercial purposes, the user must, first of all, obtain the PIC and negotiate MAT with the corresponding agency in each case. Having obtained the documentation accrediting PIC and MAT exist, the user will request authorization for access from the competent authority.[46] Once the documentation has been revised, the authority will issue the corresponding permit within a maximum period of six months.[47] Only then will the user be able to effectively access the resources.[48]

2.4 Traditional Knowledge

As far as access to TK in possession of indigenous communities is concerned, there are no such communities in Spain, for which reason the international legislation[49] on the issue is not applicable in the Iberian country.

[39] Law No. 42/2007, Article 71.3 and Royal Decree No. 124/2017, Article 5.1.a.

[40] Royal Decree No. 124/2017, Article 5.1.

[41] Royal Decree No. 124/2017, Article 6.2.

[42] Royal Decree No. 124/2017, Article 5.1.

[43] Royal Decree No. 124/2017, Article 6.4.

[44] Royal Decree No. 124/2017, Article 6.5.

[45] Royal Decree No. 124/2017, Article 6.1.

[46] Royal Decree No. 124/2017, Article 7.2.

[47] Royal Decree No. 124/2017, Article 7.6.

[48] Royal Decree No. 124/2017, Article 7.1.

[49] NP, Articles 7 and 12.

As said, however, it does have a rich cultural heritage associated with the utilization of biodiversity and local communities that preserve it, augment it and pass it on to future generations. The problem lies in establishing how that body of knowledge and those communities should and can be protected in the framework of the NP. This is so because there exist no legal definitions, either at the international or at the national level, as to what a local community consists of. Consequently, as long as the issue has not been resolved in both instances, Spain will not effectively know which situations fall within the international regime and, therefore, which ones it should contemplate in its legislation.[50]

2.5 *Access Permits*

There are two types of access permits in Spanish legislation: for non-commercial and commercial purposes. Purposes are considered non-commercial when the research results is not directed to the protection of a product or process by means of intellectual property rights, especially of patents, or the commercialization of a product or process.[51] On the other hand, it is estimated that the purposes are commercial when access to the genetic resources is sought to obtain a development or a product for its commercialization or sale, or to obtain a patent or a product to which access restrictions will be applied by means of intellectual or industrial property rights.[52]

The minimal contents that access permits for non-commercial and commercial purposes must include are set forth in Appendices 2 and 4 of Royal Decree No. 124/2017, respectively. Both permits must include information on the following elements:[53] (a) the competent authority furnishing access, (b) the agency which provided the PIC and established MAT, (c) the permit's reference number, (d) date of granting, (e) permit's date of expiration, if applicable, (f) data of the individual or institution in favor of which the permit is granted, (g) a statement that the permit is issued as proof of the fact that PIC has been given and that MAT have been established for the utilization of the specified genetic resources, (h) a description of the genetic resources covered under the permit and (i) the purposes for which the permit is granted (non-commercial/commercial).

If authorization is requested for non-commercial purposes, the following requirements must be complied with:[54] (a) if, during the course of the research, a possibility arises for use of the resources for commercial purposes, a new application must be made setting forth such purposes, (b) origin of the resource as Spanish must always

[50] Lago Candeira and Silvestri (2012), p. 292.

[51] Royal Decree No. 124/2017, Article 2.1.

[52] Royal Decree No. 124/2017, Article 2.2.

[53] Royal Decree No. 124/2017, Appendices 2 and 4.

[54] Royal Decree No. 124/2017, Appendix 2 and Article 6.2.

be indicated, (c) the genetic resource will not be provided to any non-authorized person and, in any case, the assignment of the resource to a third party will be carried out under the same terms as the original authorization and according to the declaration made by the user, and (d) the competent authority will be informed of the final research results.

If authorization is requested for access to genetic resources for commercial purposes, apart from the requirements listed above, common to both types of permits, the following requirements must be met:[55] (a) the indication that the resource used is Spanish, (b) the obligation to inform the competent authority that some commercial activity deriving from the utilization of the resource has taken place, for example a patent application, the registration and/or commercialization of the product, etc., (c) the obligation to inform the net benefits obtained by the product for the calculation and annual liquidation of the annual benefits while the patent is in force or the product is available in the market, and (d) the requirement that any assignment of the resource to a third party will be carried out under the same conditions as those imposed in the authorization process and in the MAT. As these are minimal contents, the competent authority granting access may add any further specific requirements deemed convenient.

The permits granted will be notified by the competent authorities providing access to the national focal point,[56] the General Directorate of Biodiversity and Environmental Quality of the Ministry for Ecological Transition, which, in turn, notify them to the Access and Benefit-Sharing Clearing House designated by the Protocol. Up to June of 2019, Spain had notified 40 access permits on genetic resources to the ABS Clearing House, all of them requested for non-commercial purposes.[57]

2.6 Benefit-Sharing and Model Agreements

Spanish legislation stipulates the obligation to establish MAT for the distribution of benefits deriving from the utilization of genetic resources.[58] It is explained above in Sect. 2.3 at which point of the access procedure this must take place for both commercial and non-commercial purposes, and with which authority the agreement should be negotiated.

To this it only remains to be added that, when access is sought for non-commercial purposes, the agency in charge of negotiating MAT may demand the following two additional access conditions: (a) to deposit duplicates of the material being accessed in an *ex-situ* collection located in Spain and (b) to

[55] Royal Decree No. 124/2017, Appendix 4.

[56] Law No. 42/2007, Article 71.6 and Royal Decree No. 124/2, Article 10, subsections 2 and 3.

[57] ABS Clearing House. Profile: Spain.

[58] Law No. 42/2007, Article 71 and Royal Decree No. 124/2017, Articles 6 and 7.

collaborate with a Spanish scientific institution when the access to the resources is sought by a foreign person.[59] In any case, the list of possible benefits is illustrative and does not exclude the possibility for other type of conditions to be negotiated upon granting access to genetic resources for exclusively scientific purposes. In like manner, in the case access to genetic resources is requested for commercial purposes, the discretion of the parties to reach an agreement on those benefits they deem more advantageous will prevail.

In any event, besides the freedom to negotiate and establish different types of opportunities, the legislation contemplates that the benefits deriving from the utilization of genetic resources be mainly destined to the conservation of biodiversity and the sustainable use of its components,[60] thus connecting the access and benefit distribution mechanism to these two objectives, precisely as stipulated by the NP.[61]

Finally, it should be highlighted that the legislation contemplates the development and approval of guidelines for the establishment of MAT that include the distribution of benefits when access to genetic resources is sought for commercial purposes.[62] Such guidelines, however, have not yet been developed.

2.7 *Compliance Measures*

The follow-up of the utilization of genetic resources in Spain[63] is carried out in accordance with EU legislation -Regulation (EU) No. 511/2014 of the European Parliament and of the Council and Commission Implementing Regulation (EU) 2015/1866- and with the current Spanish legislation, particularly Law No. 42/2007 and Royal Decree No. 124/2017.

The regulations of EU origin involve two monitoring mechanisms which are interconnected. The first one compels any user of genetic resources to proceed in conformity with the 'due diligence' principle.[64] This obligation, in turn, involves two additional obligations. Firstly, the user must "seek, maintain and transfer" a set of documentary evidence that ensures that the access and utilization of genetic resources and associated traditional knowledge (aTK) have been conducted according to the current regulatory measures of the provider country.[65] The documentation to be sought, maintained and transferred includes the internationally recognized certificate of compliance, the information and documents accrediting

[59] Royal Decree No. 124/2017, Article 6.4.

[60] Law No. 42/2007, Article 71.7.

[61] NP, Articles 1 and 9.

[62] Royal Decree No. 124/2017, Article 7.4.

[63] NP, Articles 15, 16 and 17.

[64] Regulation (EU) No 511/2014, Article 4.1.

[65] Regulation (EU) No 511/2014, Article 4, items 1 and 3.

date, place, description, source, etc. of the accessed resources and aTK, the access permit, etc.[66]

Secondly, the user of genetic resources and aTK must "declare" that they will act or have acted with due diligence when they are beneficiaries of research funds involving the utilization thereof or when they manufacture a product based thereon.[67] To these two instances of 'declaration of due diligence', originated in EU law, Spanish legislation adds a third one: the obligation to declare that the due diligence obligation has been met when applying for a patent developed out of the utilization of genetic resources or aTK.[68]

This last instance of control is critical, as, during the negotiations leading to the adoption of the NP, resource provider countries demanded that the point designated for checking legal access and utilization was precisely the patent office of each user country. They understood that it was during the stage of application and granting of a patent based on the development of genetic resources that the true economic potential of the resources was activated; that is to say, the possibility to benefit out of its utilization and the actual opportunity to conduct effective monitoring.[69]

Spanish legislation provides in detail how due diligence declaration must be conducted in the three situations entailing the obligation to declare: the granting of research funds, the manufacture of a product and the application for a patent, whenever they involve the utilization of genetic resources or aTK. Such declaration must be submitted through the electronic office of the Ministry of Agriculture and Fisheries, Food and Environment,[70] presently called Ministry for Ecological Transition (MITECO by its Spanish acronym). Meanwhile, the institutions financing research projects, the competent public authority in charge of authorizing the commercialization of the product and the Spanish Registry of Trademarks and Patents will verify, as applicable, the effective submission of the due diligence declaration at MITECO's electronic office.[71]

The legal effects of failing to prove the submission of the due diligence declaration at MITECO's electronic office vary depending on the situation. If this occurs during the granting of research funds, the financing institution will not make the final payment for the project,[72] and it will even be entitled to demand the total or partial return of the funds received.[73] In turn, if the case is the approval of a product for commercialization, failure to submit the due diligence declaration will entail the halting of the procedure.[74] Moreover, if the internationally recognized certificate of

[66] Regulation (EU) No 511/2014, Article 4.3.

[67] Regulation (EU) No 511/2014, Article 7, items 1 and 2.

[68] Royal Decree No. 124/2017, Article 14.3.

[69] Buck and Hamilton (2011), p. 53.

[70] Regulation (EU) No. 124/2017, Article 14, items 1.a, 2.a and 3.a.

[71] Royal Decree No. 124/2017, Article 14, items 1.c, 2.b and 3.b.

[72] Royal Decree No. 124/2017, Article 14.1.c.

[73] Royal Decree No. 124/2017, Article 14.1.d.

[74] Royal Decree No. 124/2017, Article 14.2.b.

compliance indicates that the authorized utilization of the resources or knowledge is exclusively non-commercial, the products will be prevented from authorization or release into the market.[75] Lastly, failure to submit accreditation of due diligence during application for a patent will not put the validity of the process into question nor will it cause the process to be halted, but the Registry of Trademarks and Patents will have to cooperate with MITECO and inform them of such irregularity so that they may take the appropriate measures.[76]

Regarding the second monitoring mechanism on the utilization of genetic resources and aTK, that is to say, the controls implemented by the administration so as to verify that the users of genetic resources effectively comply with the due diligence obligation,[77] the legislation of Spanish origin establishes that they will be conducted by the competent autonomous agency according to the user's corporate domicile.[78] Exceptionally, the General Directorate of Environmental and Natural Environment Quality and Assessment pertaining to the Ministry of Agriculture and Fisheries, Food and Environment will control users when they are public national institutions.[79]

Controls will be carried out when the authority has information related to a probable failure on the part of the user, for example in the case they have received substantiated concerns from a third party or on the basis of a monitoring plan which is to be devised, applying risk criteria.[80] This plan will be drawn up by the General Directorate of Environmental and Natural Environment Quality and Assessment in coordination with the rest of the public administrations involved. The goal is to reduce the risk of utilization, in Spain, of genetic resources and aTK obtained illegally.[81]

Finally, it is worth mentioning two important issues regarding compliance measures. First, the span of control extends to those activities conducted in Spain and involving the utilization of genetic resources -whether Spanish or foreign- or aTK associated with the use of genetic resources when aTK comes from third countries that are party to the NP.[82] Second, the legislation provides for a regime of sanctions for those users who contravene the obligations stipulated for access to Spanish genetic resources and for the utilization of genetic resources and aTK, pursuant to

[75]Royal Decree No. 124/2017, Article 14.2.b.

[76]Royal Decree No. 124/2017, Article 14.3.b and 14.4.

[77]Regulation (EU) No 511/2014, Article 9.

[78]Royal Decree No. 124/2017, Articles 15.1 and 13.2.a.

[79]Royal Decree No. 124/2017, Articles 15.1 and 13.1.e.

[80]Regulation (EU) No. 511/2014, Article 9.3, items a and b and Royal Decree No. 124/207, Article 15, items 2 and 3.

[81]Royal Decree No. 124/2017, Article 16. 1.

[82]Royal Decree No. 124/2017, Article 12.

the provisions of Regulation (EU) No. 511/2014.[83] The sanctions include fines ranging from 100 to 2,000,000 Euros, depending on their seriousness.[84]

3 Assessment

Spain anticipated the possibility of regulating the access to and utilization of its genetic resources through Law No. 42 of 2007,[85] a power it had not exercised until then. After the passing of Decree No. 124/2017, following the implementation of the NP and the European legislation on the issue,[86] Spain has been able to effectively regulate access to the genetic resources under its sovereignty.

The Spanish legal framework has proven to comply with the obligations imposed by the NP, mainly those concerning the characteristics that must be present in national legislative, administrative and political measures regulating access to genetic resources.[87] In fact, the administrative procedures stipulated in Spanish legislation are clear, transparent and efficient, they are not arbitrary and they provide legal certainty. The country has also developed a website containing all the necessary information, presented in Spanish and in a clear and comprehensive manner, to access its genetic resources.[88]

In addition, the regime also complies with the obligations imposed by the NP in relation to the need to designate a national focal point as well as competent authorities for granting access to genetic resources.[89] In fulfilling its commitments on the issue, Spain has taken into account its own internal regime of environmental competencies set forth by the Spanish Constitution[90] and the national legislation.[91] Consequently, it has assigned the Autonomous Communities the responsibility for granting PIC for access to the resources and the establishment of MAT when the resources are located within their territories as well as the competence to issue the access permit when the resources are endemic to a single Community.[92]

Furthermore, it is worth highlighting that a simplified procedure for access to genetic resources for non-commercial purposes has been established, as provided for in the NP.[93] Considering that more than 90% of all access permits issued until the

[83] Law No. 42/2007, Articles 80, items 1.u and 1.v and Royal Decree No. 124/2017, Article 19.

[84] Law No. 42/2007, Article 81.

[85] Law No. 42/2007, Article 68.2.

[86] Mainly, Regulation (EU) No. 511/2014.

[87] NP, Article 6.3.

[88] MITECO Website.

[89] NP, Article 13.

[90] Spanish Constitution, Article 148.1.9.

[91] Law No. 42/2007, Article 68.2.

[92] Law No. 42/2007, Article 68.2 and Royal Decree No. 124/2017, Article 5.2.

[93] NP, Article 8 (a).

present day seek non-commercial purposes,[94] the fact that the legislation has, in effect, distinguished different types of procedures and of access permits according to the different purposes sought has proven to be strategic as well. Otherwise, this type of access applications for research purposes would have been unnecessarily complicated.

Along these lines, the fact that the Spanish regime directly and categorically excluded research on genetic resources from its scope of application when the research undertaking involves purely taxonomic purposes also proves to be a judicious choice.[95] Precisely this type of basic investigative pursuits, often carried out by doctoral and post-doctoral candidates in their quest for an academic university degree, and which generate useful knowledge for the conservation and sustainable use of biodiversity, should not be hindered by absurd and bureaucratic administrative requirements.[96] For the same reasons, the fact that the collection of materials and samples and their conservation in germplasm banks or in *ex-situ* collections has been excluded from the regime's material scope of application has proven to be strategic and logical, as such activities are conducted for conservation purposes.[97]

Similarly, the fact that the regime establishes that the benefits derived from the utilization of genetic resources be mainly destined to the conservation of biodiversity and the sustainable use of its components when access to them is granted by the Public National Administration also constitutes a satisfactory measure.[98] In fact, this is in line with the spirit of the NP which expressly links the sharing of the benefits derived from the utilization of genetic resources with those two objectives.[99]

Finally, it should be pointed out that Spanish legislation satisfactorily complies with the obligations regarding control of the utilization of genetic resources and aTK taking place in Spain. In this sense, the legislation faithfully captures the requirements of the NP regarding compliance measures for genetic resources[100] and aTK,[101] and the designation of a checkpoint.[102] It also satisfies the community regulations related to the issue, mainly articles 4, 7 and 9 of Regulation (EU) No. 511/2014.

It should be acknowledged, however, that Spain has been more ambitious than the Regulation in relation to the compliance measures stipulated therein, as strictly Spanish regulations *also* require the submission of the due diligence declaration in the case of an application for a patent, when it is based on the research or

[94] ABS Clearing House database: Profile Spain. http://www.absch.cbd.int/countries/ES/IRCC.

[95] Royal Decree No. 124/2017, Article 3.2.

[96] Prathapan et al. (2018), p. 3.

[97] Royal Decree No. 124/2017, Article 3.3.

[98] Law No. 42/2007, Article 71.7.

[99] NP, Articles 1 and 9.

[100] NP, Article 15.

[101] NP, Article 16.

[102] NP, Article 17.

development of genetic resources or aTK.[103] Thus, Spain aspires not to leave the requirements of the NP on the scrapheap, especially the one concerning the designation of an effective checkpoint for monitoring of the utilization of genetic resources.[104]

4 Conclusions

Spain was able to take advantage of the opportunity arising from the adoption of the NP and the European regulations subsequently issued to effectively and efficiently regulate the access to genetic resources and the control of the utilization of—both domestic and foreign—resources and aTK that takes place in the country.

The balance created between the measures regulating access to genetic resources and those controlling their utilization, as well as the use of aTK in Spanish territory, may well serve as an example to other countries which, like Spain, have a dual role, being both provider and user of genetic resources.

To conclude, Spanish legislation has been adopted very recently indeed and, therefore, it will take a number of years to gauge its effects. The soundness and coherence it boasts are clear signs that ABS will be efficiently and successfully implemented in Spain. This will, in turn, contribute to strengthening the value and actual implications of the NP worldwide.

References

Bibliographical References

ABS Clearing House database: Profile Spain. http://www.absch.cbd.int/countries/ES/IRCC (accessed 9 November 2020)

Asociación Española de Bioempresas (ASEBIO) (2018) Situación y tendencias del sector de la biotecnología en España 2017. http://www.saludcastillayleon.es/investigacion/fr/documentacion/informes-interes.fichiers/1361532-INFORMEASEBIO2017_Vf.pdf. Accessed 30 Oct 2019

Buck M, Hamilton C (2011) The Nagoya Protocol on access to genetic resources and the fair and equitable sharing of benefits arising from their utilization to the convention on biological diversity. RECIEL 20:47–61

Fernández López C, Amezcúa Ogayar C (2007) Plantas medicinales y útiles en la Península Ibérica 2.400 especies y 37.500 aplicaciones. Herbario Jaén, España

Lago Candeira A, Silvestri L (2012) Challenges in the implementation of the Nagoya Protocol from the perspective of a member state of the European Union: the case of Spain. In: Morgera E,

[103] Royal Decree No. 124/2017, Article 14.3.

[104] NP, Article 17.1.a.

Buck M, Tsioumani E (eds) The 2010 Nagoya Protocol on access and benefit-sharing in perspective. Leiden, Brill/Nijhoff, pp 269–228

MAGRAMA (2015) Informe Nacional sobre el Estado de la Biodiversidad para la Alimentación y la Agricultura. Ministerio de Agricultura, Alimentación y Medio Ambiente, Madrid. http://www.fao.org/3/CA3496ES/ca3496es.pdf. Accessed 30 Oct 2019

MITECO (Ministry for Ecological Transition) (2019) website, http://www.miteco.gob.es/es/biodiversidad/temas/recursos-geneticos/. Accessed 30 Oct 2019

Morales R, Tardío J, Aceituno L, Molina M, Pardo de Santayana M (2011) Biodiversidad y etnobotánica en España. In: Viejo-Montesinos JL (ed) Biodiversidad: Aproximación a la diversidad botánica y zoológica de España. Real Sociedad Española de Historia Natural, Madrid, pp 157–207

Organismo Autónomo Parques Nacionales (2002) La Naturaleza de España. Ministerio de Medio Ambiente, Madrid

Pardo de Santayana Gómez de Olea M, Morales Valverde R, Aceituno Mata L, Molina Simón M, Tardío Pato FJ (2012) Etnobiología y biodiversidad: El Inventario Español de los Conocimientos Tradicionales. Revista Ambienta 99:6–24

Prathapan D, Pethiyagoda R, Bawa K, Raven P, Priyadarsanan D, and 172 co-signatories (2018) When the cure kills - CBD limits biodiversity research. Science 360:1405–1406

Silvestri L, Lago Candeira A (2015) Implementing the Nagoya Protocol in Spain: challenges perspectives. In: Coolsaet B, Batur F, Broggiatto A, Pitseys J, Dedeurwaerdere T (eds) Implementing the Nagoya Protocol. Comparing access and benefit-sharing regimes in Europe. Brill/Nijhoff, Leiden, pp 210–226

Legal References

European Union. Regulation (EU) No. 511/2014 on compliance measures for users from the Nagoya Protocol on Access to Genetic Resources and the Fair and Equitable Sharing of Benefits Arising from their Utilization in the Union. Official Journal of the European Union. L 150, 20 May 2014

European Union. Commission Implementing Regulation (EU) 2015/1866 of 13 October 2015 laying down detailed rules for the implementation of Regulation (EU) No. 511/2014 of the European Parliament and of the Council as regards the register of collections, monitoring user compliance and best practices. Official Journal of the European Union. L 275, 20 May 2015

European Union. Guidance document on the scope of application and core obligations of Regulation (EU) No. 511/2014 of the European Parliament and of the Council on compliance measures for users of the Nagoya Protocol on Access to Genetic Resources and the Fair and Equitable Sharing of Benefits Arising from their Utilization in the Union. Official Journal of the European Union, C 313/ 01, 27 August 2016

Spain. Spanish Constitution of 1978. Official Journal of Spain No. 311, 29 December 1978, pp 29313–29424

Spain. Ratification Instrument to the Convention on Biological Diversit. Official Journal of Spain No. 27, 1 February 1994, pp 3113-3125

Spain. Law No. 3/2001, of 26 March, on State Marine Fishing. Official Journal of Spain No. 75, 28 March 2001, pp 11509–11532

Spain. Royal Decree No. 289/2003, of 7 March, on commercialization of reproduction forest materials, as long as there is no utilization of the genetic resources and no transfer to third parties for a different use. Official Journal of Spain No. 58, 8 March 2003, pp 9262–9299

Spain. Law No. 30/2006, of 26 July, on seeds and nursery plants and on plant genetic resources. Official Journal of Spain No. 178, 27 July 2006, pp 28165–28178

Spain. Royal Decree No. 1274/2011, of 16 September, approving the Spanish Biodiversity and Natural Heritage Strategic Plan (2011 – 2017). Official Journal of Spain No. 236, 30 September 2011, pp 103071–103280

Spain. Ratification Instrument of the Nagoya Protocol on access to genetic resources and equitable sharing of the benefits deriving from their utilization to the Convention on Biological Diversity. Official Journal of Spain No. 202, 20 August 2014, pp 65825–65844

Spain. Law No. 42/2007, of 13 December, on Natural Heritage and Biodiversity, modified by Law No. 33/2015, of 21 September. Official Journal of Spain No. 227, 22 September 2015, pp 83588–83632

Spain. Royal Decree No. 124/2017, of 24 February, related to the access to genetic resources deriving from wild taxons and to the control of their utilization. Official Journal of Spain No. 62, 14 March 2017, pp 18478–18499

United Nations. Convention on Biological Diversity. 5 June 1992. UNTS, vol. 1760, I-30619

United Nations. Nagoya Protocol on Access to Genetic Resources and the Fair and Equitable Sharing of Benefits Arising from their Utilization to the Convention on Biological Diversity. 29 October 2010. UNTS, vol. N/D, I-30619

Luciana C. Silvestri is an environmental law researcher at the National Scientific and Technical Research Council (CONICET) in Argentina. She served as a legal and technical adviser to the Spanish Ministry of Environment for access and benefit-sharing (ABS) international negotiations. She was one of the experts contributing to the GEF UNEP IUCN Regional Project on ABS for Latin American and the Caribbean, where she focused on ensuring effective legal ABS frameworks and adequate institutional capacities in eight countries of the region. She also served as an expert to the Informal Advisory Committee to the ABS Clearing-House of the Nagoya Protocol. She has extensive expertise in international cooperation, international negotiations, biodiversity protection and ABS. lsilvestri@mendoza-conicet.gob.ar.

The Post Nagoya Protocol ABS Regime in France: Exploring the Extent to Which It Upholds the Obligations of the Protocol

Marcelin Tonye Mahop

Abstract The Republic of France is fully committed to the implementation of the 1992 Convention on Biological Diversity (CBD) and its Protocol on access and benefit sharing (ABS) both as a sovereign country and as a member of the European Union (EU). In connection with France's status as an EU member State, the country is legally bound by Regulation (EU) 511/2014 on ABS which implements user measures of NP in the EU. In 1994 and 2016 respectively, France became party to the CBD and the Nagoya Protocol on Access to Genetic Resources and the Fair and Equitable Sharing of Benefits arising from their Utilization to the Convention on Biological Diversity. France considers itself as a megadiverse country, based on its unique position in Europe and the rich biodiversity found in its overseas territories. However, as a technologically advanced country, France is equally a major user of genetic resources. This chapter explores how France domesticates the NP on ABS from the perspectives of both a supplier and a user of genetic resources and associated traditional knowledge. It examines the extent to which the ABS regulatory infrastructure of France that has been constructed after the entry into force of the NP in 2014, complies with the NP. The chapter demonstrates that the post NP ABS regulatory framework of France is *de jure* compliant with the letter of the NP. However, questions may be asked about the *de facto* compliance, which is, whether the spirit of the NP is properly embedded in the ABS regulatory framework of France. In their current versions, the ABS regulatory instruments of France do not appear to be highly reassuring to other supplier countries of GR that they are capable to ensure that all the users of illegally and/or inappropriately accessed resources are seriously scrutinized.

Keywords Nagoya protocol · ABS regulations · Genetic resources · Megadiverse countries · Compliance

M. T. Mahop (✉)
Sustainability Research Institute (SRI), School of Earth and Environment (SEE), Faculty of Environment, University of Leeds, Leeds, UK
e-mail: m.tonyemahop@leeds.ac.uk

E. Chege Kamau (ed.), *Global Transformations in the Use of Biodiversity for Research and Development*, Ius Gentium: Comparative Perspectives on Law and Justice 95, https://doi.org/10.1007/978-3-030-88711-7_16

1 Introduction

As a sovereign country and as a member of the European Union (EU), the Republic of France is fully committed to the implementation of the 1992 Convention on Biological Diversity (CBD) and its protocols on biosafety and on access and benefit-sharing (ABS). France became party to the CBD by ratification on 29 September 1994, party to the Cartagena Protocol on Biosafety by approval on 11 September 2003 and party to the Nagoya Protocol on Access to Genetic Resources and the Fair and Equitable Sharing of Benefits arising from their Utilization to the Convention on Biological Diversity by ratification on 29 November 2016. Because of its rich and diverse biological, natural and cultural heritage, France is committed to the objectives of these Multilateral Environmental Agreements (MEAs). The wealth of biological and cultural diversity has led the country to proclaim itself as a megadiverse country despite not being recognized as such among the megadiverse countries.[1] According to the fifth National report to the Convention on Biological Diversity, the unique geographical position of France in Europe and overseas provides France with the amount of biodiversity wealth worth the status of a megadiverse country.[2] France's position in continental Europe lies at the confluence of four of the 11 biogeographical regions of Europe (Atlantic, Alpine, Continental and Mediterranean) making it one of the most diverse countries in Europe.[3] Moreover, the French overseas departments, territories and collectivities are situated at different latitudinal ranges and belong to a wide range of biogeographical regions including the Islands of Mascareignes, Comoros, the Caribbean, the South Pacific, the Austral and Antarctic Islands, and sub-boreal North America biome.[4] French territory is thus located in five of the world's 34 terrestrial biodiversity hotspots identified by the WWF and the IUCN.[5]

The rich biodiversity and natural heritage of France is however subject to various threats. Some of the prominent threats are demographic changes as well as changes in production and consumption patterns that are linked to the degradation and fragmentation of natural habitats, soil degradation and water pollution, increase of exotic invasive species, over exploitation of natural resources and climate change.[6] Arguably, the combined effects of these threats have impacted on several species with e.g. the IUCN red list reporting in 2014 that out of 160 species of orchids

[1] Megadiverse countries are referred to as countries, which harbor the majority of the earth's species with a large number of endemic species in them. Currently, there are 17 scientifically recognized megadiverse countries, and France (metropolitan or any of its overseas territories) is not in the list. https://www.worldatlas.com/articles/ecologically-megadiverse-countries-of-the-world.html (accessed 28 May 2019).

[2] Ministère de l'Ecologie, du Développement durable et de l'Energie (2014).

[3] Ibid.

[4] Ibid.

[5] Ibid.

[6] Ibid.

studied in France metropolitan, 27 were under threat of extinction (about 17%) and 4 were classified as endangered.[7] If France does not avert the loss of its biodiversity, domestic needs of genetic resources for research and development in many sectors may grow increasingly reliant on plant collection missions and access to genetic resources from other countries in search for breakthrough and blockbuster innovative products. To address the above mentioned threats, France has taken domestic measures with e.g. the development of the 2011–2020 National Biodiversity Strategy (NBS)[8] whose objectives are firmly aligned with the Aichi Biodiversity Targets of the 2011–2020 CBD Strategic Plan for Biodiversity.[9] Adopted in 2011, the aims of the NBS of France are to protect, restore and increase the diversity of life in mainland France and the overseas territories.[10] The objective of the NBS is to keep ecosystems functioning, adapting and evolving in the long term. The NBS intends to promote reduction of direct and indirect impacts on biodiversity, sustainable use of living and equitable distribution of the benefits that they deliver. The NBS comprises 18 objectives among which, objective 13 is dedicated to ensuring the fair and equitable sharing of the benefits arising from the utilization of biodiversity at all levels. This objective is pursued in fulfilment of the Aichi Biodiversity Target 16 which stipulates that: 'By 2015, the Nagoya Protocol on Access to Genetic Resources and the Fair and Equitable Sharing of the Benefits Arising from their Utilization is in force and operational, consistent with national legislation'. As a logical step in the operationalization of the Aichi Biodiversity Target 16 after ratification of the Nagoya Protocol, France has formulated its domestic ABS regulatory framework, which is designed to be compliant with the NP.

This chapter explores how France domesticates the Nagoya Protocol on ABS from both supplier and user perspectives of genetic resources and associated traditional knowledge. It examines the ABS regulations adopted by France after the adoption of the Protocol in 2010 and its entry into force in 2014. In particular, the chapter explores the approach of France as a member of the European Union, in defining and delineating the scope of access-oriented ABS regulatory framework, the institutional framework including the designation and roles of different national competent authorities; the national approach in addressing the prior informed consent (PIC) and the negotiations of mutually agreed terms (MAT) (the ABS contracts) and the compliance measures enshrined in the ABS regulatory framework. Discussions on these issues will attempt to highlight the strengths and weaknesses of the current ABS regulatory framework of France and deliver a view on the extent of overall compliance of the domestic ABS regulations and practice with the letter and spirit of the NP.

---

[7] UICN France and MNHN (2014).

[8] Ministère de l'Ecologie, du Développement durable des Transports et du Logement (2011).

[9] Convention on Biological Diversity (2011).

[10] Ibid, above at Sect. 8.

2 The Access and Benefit-Sharing Regulatory Framework of France

After the entry into force of the Nagoya Protocol on 12 October 2014, France became full party to the Protocol by ratification on 29 November 2016. However, an early signal of the level of consideration given by France to this MEA and therefore of the country's intent to join it and fulfil its obligations through domestic regulations was the signing of the NP on 20 September 2011. Following the signing and prior to its ratification, France as a sovereign country and as a member of the European Union (EU), embarked in what may be viewed as a dual process of domestication. This dual process resulted in what is effectively a post Nagoya Protocol ABS regulatory framework that rests on two principal pillars each with its constitutive instruments:

- The first pillar of the ABS regulatory framework of France is the access regulations. Through its access regulations, France theoretically constructs domestic ABS regulatory instruments from the standpoint of a supplier country that is rich in biological and cultural diversity. This pillar comprises three key instruments:
  - Loi No2016-1087 du 08 Aout 2016 pour la Reconquête de la Biodiversité, de la Nature et des Paysages, Title V: Accès aux ressources Génétiques et Partage Juste et Equitable des Avantages,[11] codifié aux articles L412–3 à L412–20 du Code de l'environnement
  - Décret No 2017–848 du 09 Mai 2017 Relatif à l'Accès aux Ressources Génétiques et aux connaissances traditionnelles associées au partage des avantages découlant de leur utilisation.[12] This decree provides for the designation of the relevant Competent National Authorities (CNA) responsible for the implementation of the monitoring of utilization and compliance provision of the 2014 EU ABS Regulation and for the registration of collections. It also includes a model contract for benefit sharing from the utilization of traditional knowledge associated with genetic resources
  - **Arrêté du 13 septembre 2017 fixant le contrat type de partage des avantages découlant de l'utilisation de ressources génétiques prélevées sur le territoire national,[13] mentionné à l'article R. 412–20 du code de l'environnement**
- The second pillar of the ABS regulatory framework of France is the EU ABS regulations, which are essentially measures that users of genetic resources and associated traditional knowledge must comply with when utilizing[14] genetic

[11] In the remainder of the chapter, this will be referred to as the 2016 ABS Law.

[12] In the remainder of the chapter, this will be referred to as the 2017 ABS Decree.

[13] In the remainder of the chapter, this will be referred to as the 2017 ABS regulations.

[14] Based on the definition of Utilization of article 2 of the NP.

resources and associated traditional knowledge in France. The principal instruments attached to this pillar are:

- Regulation (EU) 511/2014 of the European Parliament and of the Council of 16 April 2014 on compliance measures for users from the Nagoya Protocol on Access to Genetic Resources and the Fair and Equitable Sharing of Benefits Arising from their Utilization in the Union
- COMMISSION IMPLEMENTING REGULATION (EU) 2015/1866 of 13 October 2015 laying down detailed rules for the implementation of Regulation (EU) No 511/2014 of the European Parliament and of the Council about the register of collections, monitoring user compliance and best practices
- Guidance document on the scope of application and core obligations of Regulation (EU) No 511/2014 of the European Parliament and of the Council on the compliance measures for users from the Nagoya Protocol on Access to Genetic Resources and the Fair and Equitable Sharing of Benefits Arising from their Utilization in the Union.

Discussions in the subsequent sections will focus on these instruments as the principal constituents of the post Nagoya Protocol ABS regulatory framework in France.

2.1 *Scope of the ABS Regulatory Framework*

The Code de l' Environnement of France describes genetic resources (GR)—which are the components of biodiversity alongside natural terrestrial and marine spaces and resources and air quality—as components of the national heritage, which provides both ecosystemic and use value.[15] Based on this definition, the 2016 ABS law sets out the conditions of access to France GR for their utilization and for the fair and equitable sharing of the benefit derived from such utilization.[16] Additionally and when relevant, the 2016 ABS law applies to the utilization of the traditional knowledge associated with GR, in accordance with the Convention on Biological Diversity.[17] Access to GR for their utilization on the one hand and the utilization of the associated traditional knowledge (aTK) are therefore the principal elements that constitute the material scope of the 2016 ABS law of France.[18] The 2016 ABS law does not appear to cover collection of bulk commodities of biological resources and the possibility for their utilization as genetic resources as defined by the law.[19]

[15] Principes généraux, Art. L110-1 Code de l'Environnement.

[16] Art. L 412-3 of the 2016 France ABS Law.

[17] Ibid.

[18] Art. L 412-5 –I of the 2016 France ABS Law.

[19] Art. L. 412-4 of the Law on 'definitions', provides an elaborate definition of utilization as: research and development activities on the genetic or biochemical composition of all or part of

However, it must be stressed that the law has provisions on change of intent especially in regard to new uses of previously accessed resources, which requires a fresh access application or declaration.[20] This provision could perhaps be used to enforce the ABS law in the event that bulk biological materials collected for other purposes, find their way in utilization schemes as defined by the NP. In the relation to GR accessed from collections, the law defines change of intent as any research and development with a direct commercial application objective which is dissimilar to the utilization disclosed by the applicant and covered in the initial access application.[21] Furthermore, in relation to the transfer of materials to third party, such action must be accompanied by the transfer of the obligations outlined in the original access authorization, as any change of intent in relation to utilization requires a fresh application.[22] This raises the question as to what happens to resources not accessed from collections or to biological materials collected in bulk, and subsequently used as GR. In addition, the term user is not defined, and the law fails to distinguish between the access applicant and user of GR and aTK under France sovereignty raising the question of who is going to be checked when it comes to checking compliance with the ABS rules of France for materials that are utilized in France. It can however be inferred from the wordings of article 412-13 and article R.412-18[23] of the 2017 implementing decree that any person (French national or foreign individual or legal entity) with the ability to utilize the GR as defined by the law is eligible to access the resources required as long as they comply with the regulations in force.

There are several exemptions[24] from the scope of the ABS law in France, which are:

- human genetic resources.
- genetic resources in areas beyond national jurisdiction.
- genetic resources covered by specialized ABS instruments that are consistent with, and do not run counter to the objectives of the Convention on Biological Diversity.
- genetic resources of cultivated or domesticated species that are used as models in R&D activities (i.e. model species).
- traditional knowledge associated with genetic resources that may not be attributed to one or more communities of inhabitants.

---

animals, plants, micro-organisms or other biological material containing units of heredity, in particular through the application of biotechnology, as well as the valorization of these genetic resources, their applications and commercialization.

[20] Art. L 412-17-III of the 2016 France ABS law.

[21] Art. L. 412-6-2 of the 2016 France ABS Law.

[22] Art. L412-17-III of the 2016 France ABS law.

[23] Arts. R.412-13 and R.412-18 of the 2017 ABS Decree stipulate that ‘any’ person wishing to access the genetic resources.

[24] Art. L 412-5-II of the 2016 ABS Law.

- traditional knowledge associated with genetic resources, whose properties are well known, and that has been used repeatedly and for a long period of time outside the communities of inhabitants that have initially developed such knowledge; and
- traditional knowledge covered by the value enhancement measures that are defined by article 640-2 of the Code rural et de la pêche maritime.

In addition, access to GR for their utilization and utilization of aTK in the interest of national defense and national security are exempted from the scope of the law.[25] Furthermore, some categories of genetic resources that are regulated by specific ABS regulations are excluded from the scope of the law.[26] These categories of resources are:

- genetic resources of domesticated or cultivated species.
- genetic resources of wild relatives of cultivated crop and domesticated animal species.
- genetic resources used in forestry.
- genetic resources collected by laboratories for the prevention, surveillance and eradication of health risks that threaten plant and animal health as well as the food safety of animals; and
- genetic resources collected by laboratories for the prevention, surveillance and eradication of serious health risks that threaten human health.

The access procedures laid down in the 2016 ABS law of France apply when access is sought for GR and/or the utilization of aTK in France, both in the mainland France and in the overseas territories (subject to their competences), after the entry into force of the ABS law on 9 August 2016. Any materials accessed before that cut-off date is not subject to the prescribed ABS rules. However, concerning the resources held into collections that were constituted before the 9 August 2016, any subsequent access to such collection and any new uses of those resources are regulated by the law.[27] The fact that France comprises the mainland and the overseas departments and regions adds a layer of complexity regarding the geographical scope of its ABS regulatory framework.[28] Based on the principle of legislative identity that is enshrined in article 73 of the French Constitution and unless there is a special derogation, any regulatory framework applicable in mainland France is applicable to the overseas regions and territories.[29] In relation to the access pillar of the ABS regulatory framework, the 2016 ABS law empowers the overseas territories, if they so wish, to exercise the functions of competent national authorities (CNA), meaning that they can process applications for access to GR and utilization

[25] Art. L 412-5-II-3 of the 2016 ABS Law.

[26] Art. L 412-5-III of the 2016 ABS Law.

[27] Art. L 412-6 of 2016 France ABS law.

[28] Art. L 412-7 to L 412-9 of the 2016 ABS Law.

[29] Chiarolla (2015), pp. 77–114.

of aTK aimed at their territories.[30] An interpretation of the way the 2016 ABS law applies to the mainland and the overseas territories was attempted by a 2017 study[31] according to which, the 2016 ABS law applies to the five overseas regions and departments (Martinique, Guadeloupe, French Guyana, Réunion and Mayotte) and to Saint-Martin and Saint-Pierre-et-Miquelon. Furthermore, the law applies partially in New Caledonia and French Polynesia only the definitions of the terms *'communities of inhabitants'* and the provisions pertaining to the utilization of aTK.[32] With regards to the application of the EU regulations in France, they are in force in mainland and in the five overseas regions and department.[33] Additionally, they apply to the French Southern and Antarctic Lands, Wallis and Futuna, Saint-Pierre-et-Miquelon, Clipperton and Saint-Martin. However, EU ABS regulations do not apply to New Caledonia, French Polynesia and Saint-Barthélemy.[34]

2.2 The Institutional Framework for ABS in France

The ABS regulatory framework of France has clearly identified the institutions that are responsible for handling applications for access to GR and utilization of aTK, the issuance of the various forms of access permits and the roles and functions of the relevant CNAs. The agency bearing the ultimate responsibility for issuing the access permits is the ministry in charge of the environment.[35] This agency is responsible for 'granting access', which is a role assigned to the CNA in accordance with the NP.[36] However, although the ministry in charge of the environment makes the final decision to grant or refuse access, it is expected to consult with other entities as part of the decision-making process. When access is sought for GR in a national park,[37] the ministry in charge of the environment is obliged to share the application with the management of the park and seek their opinion on the permit application. The same applies if access is sought for GR in the territories of the communities of inhabitants.[38] In this case, the ministry must seek the view of the moral authority identified in article L.412-10 as the representative of the communities. With regards

[30] Art. L 412-15 of the 2016 ABS law of France. The territories concerned are the regional councils of Guadeloupe and Reunion; the assemblies of French Guyana and Martinique and the overseas departments of Mayotte.

[31] Aubertie (2017).

[32] Ibid, p. 105.

[33] Ibid.

[34] Ibid.

[35] Arts. R.412-13, R.412-18, R.412-28 of the 2017 ABS Implementing decree of France. It must be stressed that in France, the ministry in charge of the Environment is Le Ministere de la Transition Ecologique et Solidaire.

[36] Art. 13.2 of the Nagoya Protocol.

[37] Art. L.412-8 I of the 2016 ABS law.

[38] Ibid.

to access to the associated traditional knowledge[39] of the communities of inhabitants, the ministry in charge of environment is obliged to seek the opinion of the moral authority that represents the communities of inhabitants.[40] It is important to stress that the opinion of this moral authority will derive from its consultations with the communities of inhabitants. As indicated above, the overseas territories can decide if they so wish, to assume the responsibilities of the ministry of environment pertaining to the grant of the access permits within their territories.[41]

The ABS regulatory framework has also designated the relevant competent national authorities in fulfilment of France obligations under article 6 of the 2014 ABS user regulations of the European Union. With regards to monitoring compliance with access regulations in relation to the utilization of GR or aTK at the stage of research funding application, the ministry in charge of research is the CNA to receive the declaration of due diligence (DDD) with all the information identified in article 4 of the EU ABS Regulation.[42] The ministry in charge of the environment is the designated CNA to receive the DDD at the stage of final development of a product based on the GR or aTK or when seeking market approval for the commercialization of such products.[43] With regards to the DDD at the final stage of product development, users provide the information required to the National Industrial Intellectual Property Institute (for example when patents are sought over the product developed on the basis of the GR or TK) or to the national market approval agency (if market approval is sought for the commercialization of the products).[44] These agencies cannot use this information in their normal examination processes and should simply transfer this information to the relevant ministry in charge of the environment. Finally, the ministry in charge of research is the designated CNA for the registration of collections,[45] therefore for the implementation at the national level of article 5 of the 2014 EU ABS Regulation on registered collections.

The agencies mentioned above have responsibility over two broad types of access approaches, each of which results in the grant of some form of access permit by the ministry in charge of the environment:

---

[39] Art. L.412-4-5 of the 2016 ABS law defines aTK as: knowledge, innovations and practices relating to the genetic or biochemical properties of the resources, its use or its characteristic, which are held in an ancient and continuous manner by one or more communities of inhabitants, as well as the evolutions of this knowledge and practices when they emanate from these communities of inhabitants. The term 'moral authority' has not been defined in the ABS law or the decree. The only thing said is that for such identification, the business name, legal status, address etc. of the moral authority should be provided.

[40] Art. L.412-9 and L. 412-10.

[41] Art. L. 412-15 of the 2016 ABS Law.

[42] Art. D. 412-39-I of the 2017 ABS Decree.

[43] Art. D. 412-39-II of the 2017 ABS Decree.

[44] Art. L. 412-18-I-2 of the 2016 ABS law.

[45] Art. D. 412-41 of the 2017 ABS Decree.

- the declarative approach to access
- the authorization approach to access

The declarative approach covers access applications to GR for utilization for the purposes of enhancing the knowledge on and understanding of biodiversity, conservation and valorization of GR with no specific or direct objective of commercialization of the outcomes of such utilization.[46] Furthermore, access to GR in situations of emergency pertaining to human, animal or plant health other than the public health issues enshrined in article L.1413-8 of the Public Health Code, are also eligible for the declarative approach.[47] This declarative approach does not require negotiations of a benefit-sharing contract. Provided the ministry in charge of the environment is satisfied with the application, it issues a receipt to the applicant as a certification of completion of his declarative application and approval of the activities. In the event of an incomplete declaration, the ministry calls the applicant to complete the information required. The approach of the ABS regime of France in framing the declarative scheme to access to GR can be viewed as the national domestication of article 8 of the Nagoya Protocol on special considerations. Indeed, in what is the simplified approach to encourage research which contributes to the conservation and sustainable use of biological diversity, the wording of article L 412-7-I of the 2016 law is closely aligned to the wording of article 8 (a) of the NP.

The authorization approach covers access applications to GR for any type of utilization other than those considered under the declarative approach.[48] The 2016 ABS law and 2017 ABS implementing decree do not typically mention commercialization or commercial utilization of genetic resources as one of the key features of the authorization approach. One can however assume that, the mention of '*no specific and direct objective of commercialization*'[49] of genetic resources, which excludes commercialization from the declarative approach, is a signal that commercial use of genetic resources is the principal trigger of the authorization approach. Concerning the application process under the authorization approach, the access permit can only be granted after the applicant has successfully negotiated a benefit-sharing contract, based on the model contract.[50] Permits' applicants can only commence negotiations of the benefit-sharing contract when instructed to by the minister in charge of the environment, and this will happen when the ministry is satisfied with the completeness of the application.[51]

[46] Arts R.412-12 of the 2017 ABS decree. The criteria the applicant must fulfil in the submission / application are detailed in Article R.412-13 of the 2017 ABS Decree.

[47] Art. L. 412-7 III of the 2016 ABS Law.

[48] Art. R. 412-18 of the 2017 ABS Decree.

[49] Art. R. 412-12-3 of the 2017 Decree.

[50] Art. R. 412-20 of the 2017 ABS Decree. The standard contract was adopted by the 2017 ABS regulations establishing the standard contract for the sharing of benefits arising from the use of genetic resources.

[51] Art. R 412-19 of the 2017 ABS Decree.

Lastly, there is a special application process for the utilization of aTK of the communities of inhabitants of Guyana and Wallis and Futuna Islands.[52] After evaluation by the ministry in charge of the environment of the application, consultations between the moral authority established under article L.412-10 and the communities and successful negotiation of a benefit-sharing contract between the moral authority and the user, the ministry of environment will be able to grant the permit to utilize aTK.[53] It must be stressed that negotiations and signing of the benefit-sharing contract between the moral authority and the user are based on the PIC of the communities of inhabitants given to the moral authority through consultations.

After granting any of the permit mentioned above (receipt under the declarative process, the access permits to GR or to aTK under the authorization approach), the ministry of environment can transmit these documents to the ABS Clearing-House,[54] an act that transforms them into the internationally recognized certificate of compliance pursuant to article 17.4 of the Nagoya Protocol.

3 The Nature of the Regulatory Tools for Access to France's Genetic Resources and Associated Traditional Knowledge

3.1 PIC in the ABS Regulatory Framework of France

Access to GR for utilization and utilization of aTK of the communities of inhabitants via the permitting schemes discussed above is subject to the prior informed consent (PIC) of the ministry in charge of environment, which is the CNA responsible for the administration of the access schemes. Concerning the declarative scheme, there is no typical mention of the requirement for the PIC of the CNA for access to GR. It is only after conclusion of negotiations and signing of the ABS contract setting out the specific conditions of use that were consensually agreed and the benefits-sharing terms, that the CNA can approve the authorization permit. The ABS regulatory instruments of France do not link specific GR to the various types of authorization. The only distinction is on the two schemes for access: the declarative or the authorization scheme. The key feature of demarcation between these two schemes is on the ultimate aim of the utilization of genetic resources, in particular whether or not there is any form of commercial development in sight.[55] Concerning the

[52] Art. R. 412-28 of the 2017 ABS Decree.

[53] Art. R. 412-33 of the 2017 ABS Decree.

[54] Arts R. 412-15-I-1; R. 412-24-I-1 and R. 412-37 of the 2017 ABS Decree.

[55] Art. 412-12-3 of the 2017 ABS Decree- and possible interpretation in the Sect. 2.2 above on how the decree might be viewed to have approached the distinction between commercial and non-commercial use.

utilization of aTK of the communities of inhabitants, the PIC of the communities,[56] is a key pre-requisite of the authorization/permit decision to be made by the CNA. PIC of the communities is obtained by the legal authority through extensive consultations with the communities concerned, in order to gain their views on the project or activities that are linked to their traditional knowledge.[57] Throughout the consultative process, all aspects of the application for the utilization of aTK are presented and detailed to the communities of inhabitants in the best possible manner by the moral authority, including through the use of local languages. Once the communities have given their PIC to the moral authority, the later negotiates and signs an ABS contract with the applicant.[58] Thereafter, the applicant sends the signed copy of the contract to the CNA with any additional minutes from the consultation process.[59] The CNA will decide by approval or refusal of the permit based on the full application and PIC of the communities materialized through the benefit-sharing agreement signed between the moral authority and the user. As mentioned earlier, in relation to aTK, this process is seemingly only applicable in the overseas territories (the islands) of Guyana and Wallis et Futuna.

3.2 Decision on Declaration and Authorization in the ABS Regulatory Framework of France

Two main types of access permits are available in France:

- The receipt issued by the CNA (the ministry in charge of the environment or the relevant authorities of the overseas territories) to the applicant as the main outcome of the declarative scheme, for access to GR *in-situ* in a national park or access to genetic resources in the territories of the communities of inhabitants.
- The authorization, granted by the CNA (the ministry of the environment or the relevant authorities of the overseas territories) to the applicant for access to GR for utilization linked to some form of commercialization or for the utilization of aTK of the communities of inhabitants.

Exploring the contents of these decisions, the receipt issued by the CNA as the outcome of the declarative process and green light for the applicant to commence his activities is likely to be like all applications. However, it looks like the contents of the permit (authorization) which results from the authorization scheme may be different. This is because, as laid down in the 2016 law and the 2017 ABS implementing decree, the permits are likely to be structured on a case by case basis because they will reflect the negotiated ABS contracts and the consultation

[56] Art. L 412-9-I of the 2016 ABS law and Art. R 412-33 of the 2017 ABS Decree.

[57] Art. R. 412-29 to R. 412-33 of the 2017 ABS Decree.

[58] Art. R. 412-33 of the 2017 ABS Decree.

[59] Ibid.

process between the applicant, the CNA and the park management authorities on the one hand, or between the applicant and the moral authority representing the communities of inhabitants on the other. However, there are grounds to argue that perhaps the end point of these two types of access procedures in terms of the final document issued by the CNA may be similar after all. This is because, although the declarative process does not require negotiation of mutually agreed terms (benefit-sharing contract), in practice, all the permits that have now been published in the Clearing-House Mechanism (CHM) by France therefore becoming Internationally Recognized Certificates of Compliance (IRCC), mention MAT despite the fact that they are all linked to utilization of GR for non-commercial purposes by largely academic and research organizations.[60] Therefore, it appears that in the implementation of the law provisions on access for utilization of GR, be it through the declarative or authorization procedures, MAT are negotiated.

The declaration process starts with submission of an application form by the applicant to the CNA with the following information[61]:

- In the case of a natural person, his or her surname, first name and home address and, if it is a legal person, its name, legal status, address of its registered office as well as the quality of the signatory of the declaration.
- Description of the activities for which the Declaration is made and their purpose.
- Designation of the taxa concerned, with the best possible accuracy and the indication of the place of sampling or, if the material is in a collection, of the holding entity of the samples.
- Description of technical modalities for access to genetic resources and collection conditions.
- Detailed timetable of the planned activities.
- Confidential information which the declarant considers that disclosure could infringe trade secrets.

Provided the CNA is satisfied not only with all information but also with the reports of the consultations with the communities of inhabitants, it can issue a receipt as an approval of the declaration. This means collection activities can proceed.[62] At this stage, there is no additional condition that the applicant must fulfil before commencing the activities that are detailed in the declaration. Furthermore, there is no indication of any additional certificate e.g. in the form of a phytosanitary certificate, required by the applicant prior to the start of his activities.

Concerning the authorization scheme for access to genetic resources for their utilization, it begins with submission by the applicant of the application form to the CNA, with the following information:

[60] In France CHM, 38 IRCC were published and all mentioning MAT in non-commercial utilization of GR by largely academic and research institutions: https://absch.cbd.int/countries/FR. Accessed 29 May 2019.

[61] Art. R 412-13 of the 2017 ABS Decree.

[62] Art. R 412-14 of the 2017 ABS Decree.

- In the case of a natural person, his or her surname, first name and domicile and, if it is a legal person, its name, legal status, address of its registered office as well as the quality of the signatory of the application.
- Description of the activities for which the application is made, their objectives and their applications envisaged.
- Designation of the taxa concerned, with the best possible accuracy and the indication of the place of sampling, specifying whether it is within the geographical boundaries of a National Park defined in the Article L. 331-1 or, if the material is in a collection, the name of that collection.
- Description of technical modalities for access to genetic resources and collection conditions.
- Elements to assess the impact on the biodiversity of the activity or its applications in terms of the restriction of sustainable use or the risk of depletion of the resource for which access is sought.
- The planned timetable for the realization of the activities.
- The applicant's propositions for benefit-sharing, description of the applicant's technical and financial capabilities and, if it so wishes, reference to the time limit, exceeding the maximum period laid down in article R. 412-19, proposed to obtain an agreement on benefit-sharing.
- Confidential information which the applicant considers that disclosure could infringe trade secrets.

From the above, the main visible difference in the application between the declarative and the authorization scheme for access to GR for their utilization is in the additional benefit-sharing condition that is contained in the authorization scheme. This is a clear indication that the authorization scheme is the one through which utilization of GR leads to clearly identified benefits. These benefits must be shared between the user/applicant and the provider of the GR. There are additional differences, however. For one, once the CNA is satisfied with the application including the benefit-sharing contract and all the reports derived from the relevant consultations with the communities of inhabitants and the management of national park, it (the CNA) can grant the authorization and define its period of validity based on the activities described in the contract.[63] The CNA can also include additional conditions that the applicant/user must comply with in the course of his/her activities.[64] Such conditions may relate to compliance with the benefit-sharing mechanism, conservation of biodiversity and sustainable use of its components.[65] Although there is a separate list of conditions to be fulfilled by an applicant/user

[63] Art. R 412-22-II of the 2017 ABS Decree.

[64] Ibid.

[65] Art. R.412-25 of the 2017 ABS Decree.

that are specific to the utilization of aTK,[66] they are similar to those pertaining to the authorization/permit for access to GR for their utilization.[67]

It must be stressed, the two forms of permits or decisions (the declarative receipt or the authorization) discussed above are solely for the person who submitted and went through the application process. Additionally, the materials acquired through the process are only expected to be exploited as authorized, that is based on the information provided during the application process and in respect of any additional conditions that might have been prescribed by the CNA. The implication of these key elements is twofold. Firstly, if the primary applicant wants to pass the GR or the aTK to a third-party user, they must do so with all the obligations associated with their original decision. The primary applicant must also inform the relevant administrative authorities of the third-party transfer.[68] Secondly, any change of utilization of the GR or aTK either by the primary applicant or by the (new user) third party requires a fresh permit application from the user.

3.3 MAT, Enforcement, Alignment and/or Compliance of the Access Procedures with NP

The access procedures discussed above can be characterized as compliant with the letter and the spirit of the Nagoya Protocol. This is based on several elements such as the level of clarity and transparency of the procedures. The fact that the law and the implementing decree provide information on the application process, the office that administers the different permit schemes, the stakeholders involved and their roles including importantly the involvement and the role of the moral authority and the communities of inhabitants, align the domestic process with the prescriptions of Articles 6; 7; 12; 13.2 of the Nagoya Protocol. This alignment of the permitting schemes of the ABS regime of France with these provisions of the NP strengthens its credentials of legal certainty, which is an important consideration and an incentive for users to engage with any ABS regime.

The procedures described above make it clear that an ABS agreement is not required under the declarative scheme. However, for the CNA to grant an

[66] Art. R. 412-28 of the 2017 ABS decree. For example, the application form for the permit to utilize aTK requires additional information such as the detailed protocol for access to associated traditional knowledge, the description of the aTK and identification of the community or communities of inhabitants that are concerned and if access to the aTK requires access to the GR associated to it, then information on the origins of the samples of the GR will have to be specified.

[67] Arts R 412-33 and R.412-34 of the 2017 ABS Decree. Some of these additional prescriptions including the fixation of the validity of the permit; the guarantees that benefits will be shared fairly and equitably, that activities will not deplete GR but will rather contribute to their conservation and sustainable use and that benefits will be invested in project that would promote education, conservation and sustainability.

[68] Art. L. 412-17-III of the 2016 ABS law.

authorization/access permit either to GR or for the utilization of aTK one of the key conditions that the applicant/user must fulfil is to negotiate an ABS agreement and submit it to the CNA. The ABS regulatory framework of France provides two model contracts that stakeholders must use as templates for their negotiations. One is a model contract pertaining to access and the utilization of GR, which was promulgated as a separate implementing regulation,[69] and the other is the model contract pertaining to the utilization of the aTK, which is annexed to the May 2017 Decree implementing the 2016 ABS Law. These model contracts do not give details on the obligations of the suppliers or on the conditions of utilization of the GR or aTK. This section is intentionally left blank under article 2 of each of the model contracts because its content will derive from the negotiations. However, an important obligation that appears in the model contracts relates to the publication of research results based on the utilization of the GR and the aTK.[70] On publications, each of the model contracts stresses that scientific publications should mention the origin of the GR and aTK used and should indicate the registration number of the ABS contract that was negotiated.

On the enforcement measures, each of the model contracts promotes an amicable approach in solving any issues arising during contract implementation. However, these model contracts appear to put the onus on the user to submit a formal letter of complaint to the State (through the CNA) or to the moral authority representing the communities of inhabitants, highlighting any issues that affect the smooth implementation of the contract. The letter as submitted will be examined by the CNA (or the relevant authorities in the overseas territories) or the moral authority. Failure by the CNA or moral authority to come up with a decision after 2 months from the date of submission should mean rejection of the complaint. There is no additional recourse, although the model contracts do not have any provision preventing the user/applicant to resend this letter or to pursue the litigation route through the judiciary.

4 The Benefit-Sharing Approach of the ABS Regulatory Regime of France

Negotiations of the benefit-sharing terms are part of the overall MAT negotiations between the user and the State or between the user and the communities of inhabitants through their legal representative. Furthermore and importantly, the view of the permit issuing body on the benefit-sharing terms play a key role in the final decision for approval or refusal of the access authorization permit for GR and for the utilization of aTK. According to the 2016 ABS law, the authorization permit may be refused if from the perspective of the CNA, the negotiated terms do not reflect the

[69]The model contract is promulgated as the September 2017 ABS Implementing Regulation.

[70]Art. 4 of each of the model MAT for the utilization of GR and utilization of aTK.

technical and financial capacities of the user.[71] Another ground for refusal is the potential negative impacts of the activities and their applications to the conservation and sustainability of the resources that are requested.[72]

The model contracts referred to above provide more details on the benefit-sharing arrangements that should be negotiated by stakeholders. The two main types of benefits, non-monetary (non-financial) and monetary (financial) benefits are considered in the model contracts. The model contracts have broken down these two main types of benefits into five specific categories including:[73]

- Enrichment and preservation of biodiversity while ensuring its sustainable use
- Preservation of traditional knowledge associated with genetic resources through, for example, the development of TK databases with the participation and PIC of the communities of inhabitants
- At local level, contribution to local development through job creation and the development of value chains
- Collaboration, cooperation and contribution to research activities, education, training and capacity building
- Maintenance, conservation, management, supply or restoration of ecosystems services in a given territory
- Payment of financial contributions

Payment of financial contributions is just one category of benefits, which is listed last among all the categories considered, most of which are non-monetary benefits. This position of the monetary benefits in the list among the other categories reflects the 2016 ABS law's encouragement to stakeholders that they should prioritize the first four types of benefits during negotiations.[74] Stakeholders' participation in the benefit-sharing discussions is crucial and is included in both ABS model contracts. In the case of the model contract relating to the utilization of aTK, the communities of inhabitants who are the principal beneficiaries must be identified by their legal representative.

[71] Art. L 412-8-IV- 2 of the 2016 ABS law.

[72] Art. L 412-8-IV-3 of the 2016 ABS law.

[73] Art. 3 of each of the model contracts.

[74] Definitions Art. L-412-4-3 of the 2016 ABS.

5 Compliance Measures in the ABS Regulatory Framework of France

This chapter focused on the approach taken by the ABS regulatory framework of France in domesticating the user measures of the 2014 EU regulation and its 2015 implementing instrument.[75] These measures are laid down in articles L.412-18 through L.412-20 of the 2016 ABS law. They identify the relevant competent authorities and selected bodies such as funding bodies, the National Industrial Property Institute and the institutions that are responsible for market approval of products developed based on utilization of GR and aTK. Furthermore, in fulfilment of the requirement of article 6.1 of the 2014 EU Regulation, which requests EU member States to identify relevant competent national authorities, the 2017 ABS implementing decree designates the competent authorities that are responsible for checking user compliance, in respect of the exercise of the due diligence (DD) obligation.[76]

Specifically:

- The ministry in charge of research is the designated authority to receive a declaration from recipients of funding that they exercised due diligence when acquiring GR and aTK and in their subsequent utilization.[77]
- The ministry in charge of the environment is the designated competent authority where users shall declare that they exercised due diligence at the stage of final development of a product based on the utilization of GR or aTK.[78]
- In addition, the ministry in charge of research is the designated national authority responsible for the registration of collections pursuant to article 5 of the 2014 EU ABS Regulation.[79]

In relation to the approach set out by the EU regulation for proving that due diligence has been properly exercised, users are expected to either present to the relevant authorities, an IRCC as well as information on the content of the mutually agreed terms relevant for subsequent users.[80] However, where the IRCC is unavailable, users must record and provide to the relevant authorities, detailed

[75] Parties must fulfil three core obligations under the Nagoya Protocol. These are (1) obligations pertaining to access enshrined in Article 6, 7 and 12 NP; (2) obligations pertaining to fair and equitable sharing of benefits enshrined in Article 5 of the NP and obligations pertaining to compliance as well as monitoring and surveillance of utilization enshrined in Articles 15; 16; 17 and 18 of Nagoya Protocol. For a detailed discussion see Winter in this volume.

[76] Art. 4 of the 2014 EU ABS Regulation deals with the due diligence obligation of users of genetic resources and associated traditional knowledge.

[77] Art. D. 412-39-I-1 of the 2017 ABS Decree.

[78] Art. D. 412-39-II-1 of the 2017 ABS Decree.

[79] Art. D. 412-41 of the 2017 ABS Decree.

[80] Art. 4.3.a of the 2014 EU ABS Regulation.

information on the GR or aTK as requested by the EU Regulation.[81] The way the DD obligation is implemented in France is that, users of GR accessed from registered collections are considered to have fulfilled the DD obligation because the registration process of collections must have in itself gone through strict scrutiny and checks, which resulted in the collections' compliance with EU ABS Regulation. However, in case GR and/or aTK are not accessed from registered collections, recipients of funding must submit to the ministry in charge of research all the information required in article 4 of the EU ABS law.[82] In the event of development of final product based on the utilization of the GR and/or aTK, the information required in article 4 of the EU ABS Regulation are submitted to the national Industrial Property Institute or to the body responsible for products' market approval, depending on whether the user seeks patents or other IP protection over the product or market approval for the commercialization of final products.[83] The Industrial Property Institute or the market approval body is not expected to check this information and use it in its examination process. It should transfer it to the ministry in charge of the environment as the CNA responsible for checking that the user fulfilled the DD obligation.[84] In both instances, the role of the ministry in charge of research and the ministry in charge of the environment is to check the declarations received and to send them to the ABS Clearing-House, to the Commission and the ABS CNA.

The 2016 ABS law of France has made provisions on sanctions to users of GR and aTK who fail to use the resources as stipulated in the declaration or the authorization and to those who fail to abide by the DD obligation.[85] In effect, users are punishable by imprisonment and a fine of Euro 150,000 if they use GR and aTK without providing the information required under article 4.3 of the EU law to the relevant body, or for failing to search, record and transfer to subsequent users such pertinent information required under article 4.3 of the EU law. Fines may rise to Euro one million, in the event of commercial utilization of the GR and the aTK. Furthermore, providers of public research funds can include clauses in the funding decision dealing with reimbursement of the funds allocated should the recipient fail to deliver proof that DD was properly exercised.[86] The 2016 ABS law has identified several agencies and assigned them with the responsibilities to search and observe the offences committed by users. In addition to the CNAs that have been assigned responsibilities to oversee compliance with the EU user measures and DD,

[81] Art. 4.3.b of the 2014 EU ABS Regulation. Such detailed information includes but are not limited to date and place of access, description of genetic resources and aTK; presence of absence of rights and obligations relating to access and benefit sharing etc.

[82] Art. L. 412-18-II-1of the 2016 ABS law.

[83] Art. L.412-18-II-2 of the 2016 ABS law. Products marketed without the requirement of a market approval are not addressed, which could be a loophole.

[84] Ibid.

[85] Art. L.415-3-1 of the 2016 ABS law.

[86] Art. L.412.18-II-1 of the 2016 ABS law.

additional agencies include competition and fraud control and repression agencies, sworn agents designated by the ministry of defense, sworn agents designated by the ministry in charge of research, sworn agents designated by natural regional parks etc.[87] As it stands, the law has outlined the functions of these agents to include search and observation of offences committed but does not have any mechanism by which they can bring offenders to compliance before any formal enforcement action is pursued. Therefore, by only outlining the functions of these agents as to search and observe the offences committed, the law and the implementing instruments leave vague the link between offences committed and the mechanism by which corrective or punitive actions may be taken. Although the sanctions envisaged in the law sound particularly severe, it remains to be seen how they can be operationalized. Two possible enforcement routes may be explored by the relevant enforcement agencies in a bid to bring offenders to compliance. Firstly, general administrative law may be used to ensure that that the relevant enforcement agency has *per se* powers to order remedial action in case of unlawful behavior. Secondly the relevant agency may also rely on article 9 (6) Regulation 511/2014 which, because of the nature of a regulation, is directly applicable by national authorities.

6 Compliance with the Nagoya Protocol

An examination of the way the various instruments that make up the ABS regulatory framework of France encapsulate the key obligations of the NP shows close alignment with NP provisions under article 6.3. a, b, d; 7 and article 18.2,3. It is evident that France has chosen the path of requiring the PIC of the CNA for access to GR on the one hand, and the PIC and involvement of the communities of inhabitants for access to GR in their localities and for the utilization of aTK. As laid down in the 2016 ABS law, the access rules and procedures are not arbitrary, in terms of giving special privileges to a certain category of applicants and refusing such privileges to others. For example, the law does not appear to differentiate between applications from nationals and non-nationals.[88] Furthermore, the ultimate decision of approval through the declarative scheme and the authorization scheme is a written decision, which becomes an IRCC when submitted to the ABS Clearing-House. Because the roles and prerogatives of the different bodies[89] that intervene in the access procedures and access conditions are clearly enunciated in the law and implementing decree including clarification of the timeline of the decision making process, there is

[87] Art. 38 of the 2016 ABS Law.

[88] Art. R. 412-13 and Article R.412-18 respectively on the declarative and authorization schemes states that, anybody (toute personne) wishing to access GR…There is no distinction between applications from nationals and foreigners.

[89] Compliance with Articles 12 and 13 of the NP is reflected in the in the ABS regulatory framework. The functions and the roles of the CNA, the communities of inhabitants / or their representatives in the PIC process and MAT negotiations are clearly enunciated.

a substantial level of legal certainty, clarity and transparency in the ABS permitting process.[90]

Concerning the ABS framework's compliance with the Nagoya Protocol provisions on compliance with mutually agreed terms, each of the model contracts provides that: 'In the event of a dispute relating to the interpretation, execution or validity of this contract, and in the absence of amicable settlement, the dispute may be brought before the territorially competent administrative court'.[91] This is the last resort because bringing the dispute to the court should be considered only after failure of amicable settlement of the dispute[92] on the one hand, or if the contract has not been terminated by the State[93] or the moral person[94] representing the communities of inhabitants. Each of the model contracts states that the applicable law is French Law. None of the model contracts plainly prevents a foreigner or non-national from bringing the matter to court, in the event of failure of amicable resolution, which implies that access to justice in France is provided to foreigners.[95] One issue that has been left out in the dispute resolution regime is the utilization of alternative dispute resolution mechanisms such as mediation and arbitration, particularly mutual recognition and enforcement of foreign judgements and arbitral awards. The absence of these alternative dispute resolution mechanisms is a clear limitation of the dispute resolution mechanism of the ABS regime suggesting non-compliance with Articles 18.1.c and 18.3.b of the Nagoya Protocol.

7 Strengths and Weaknesses of the ABS Regulatory Framework of France

7.1 Strengths

The most obvious and arguably principal strength of the ABS regulatory framework of France lies in the level of clarity, not only of the ABS procedures, but also of the roles and functions of the different intervening agencies. These agencies include all CNAs including the ministry in charge of the environment responsible for processing the declaration or authorization permits, the ministry in charge of research and ministry in charge of environment as ÇNA to receive the DDDs and the checkpoints responsible for monitoring the utilization of the genetic resources pursuant to article 17 of the NP. The law sets out the circumstances in which access

[90] Table 1 below on linkages between domestic ABS provisions with articles 6.3 (c), (e), (f), (g), 12 of the NP.

[91] Art. 7 of the standard aTK contract and article 7 of the standard contract for access to GR.

[92] Art. 6 of each of the model contracts.

[93] Art. 5 Model ABS contract for access to GR annexed to the September 2017 ABS regulations.

[94] Art. 5 Model ABS contract for access to TK annexed to the 2017 ABS Decree.

[95] There is compliance with NP article 18.1 (a) and (b).

Table 1 Alignment of domestic ABS regulatory instruments with the Nagoya Protocol

Nagoya Protocol Provisions	Provisions ABS regulatory framework France			
	2016 ABS Law	2017 Decree	2017 Arete and annex to 2017 Decree	comments
Art. 6.3.a	√	√	√	A holistic reading of all the instruments of the France ABS regulatory framework suggest compliance with the NP Art. 6.3. a
Art. 6.3. b	Art. L 412-7 and L 412-8	Art. R. 412-12 and Art. R 412-18		
Art. 6.3.d	Art. L 412-7 and Art. L 412-8	Art R.412-14 and Art R.412-22		The decision can only be made after negotiations of the ABS contract
Art. 7	Art. L 412-7 and Art. L 412-8	Art. R 412-28 to Art. R 412-38		The communities of inhabitants are represented by the designated moral person pursuant to Art. L 412-10
Art. 18.2			Art. 7 annex 2017 Decree and Art. 7 Arete 2017	
Art. 18.3	√	√	√	In respect of compliance measures with MAT, the regulatory framework does not capture the mutual recognition and enforcement of foreign judgements and arbitral awards
Art. 6.3.c		Section 4 of the 2017 ABS decree		
Art. 6.3.e		Art. 412-16; Art. R412-24 and Art. R 412-37		
Art. 6.3.f		Art. R 412-30 to Art. R 412-34	Art. 1 to Art 3 annex 2017 decree, standard contract for Use of traditional knowledge	
Art. 6.3.g	Art. L 412-8; L 412-9	Sub section 2 and 3	√	
Art. 12	Paragraph 5	Sub section 3		
Art. 13	Art. L 412-8; L 412-9; L 412-15 and L 412-16	R 412-13; R 412-18; R 412-27; R 412-28; R 412-39 and R 412-41		

(continued)

Table 1 (continued)

Nagoya Protocol Provisions	Provisions ABS regulatory framework France			
	2016 ABS Law	2017 Decree	2017 Arete and annex to 2017 Decree	comments
Art. 18.1			√	The only weakness here is that, apart from the amicable avenue for solving disputes, the model ABS contracts fail to capture options for alternative dispute resolution, such as mediation or arbitration

Table is adapted from the author's interpretation of the compliance dimension of the ABS regulatory framework of France with the specific provisions of the NP that were identified to be of special interest to this research
Tick = Specific ABS instrument is generally in compliance with the specific article of the NP
No tick = No specific observation to highlight in respect of the specific compliance area.
Source: *Own illustration*

applications are required and outlines the conditions that the applicant must fulfil in the application form and its submission to the CNA. As discussed above, the law and the implementing decree outline the prerogatives of the administration of parks and the involvement of the communities of inhabitants through their legal representative in PIC and MAT negotiations. The regulatory framework has provided the timeline for the treatment/examination of the access application, be it through the declarative or the authorization scheme. The regulatory framework has also outlined the roles of the overseas territories in the administration of the access procedures. In the event of disputes arising from the execution of ABS agreement, each of the model contracts provides a dispute resolution scheme which, despite its limitations, is nonetheless clear to all parties and guarantees access to justice for foreigners. Overall, the way the regulatory framework is laid out delivers a significant level of *de jure* compliance with the NP on ABS and ensures that any actors interested in accessing GR and aTK for their utilization knows what to do, where to go, who to consult and how long it will take to have a decision.

Another strength of the regulatory framework is its strong emphasis on non-monetary benefits from the utilization of GR and aTK as compared to monetary benefits. Not only do non-monetary benefits outweigh the monetary types in the list of benefits, it is the purposes for which they will be utilized that make non-monetary benefits more interesting and important. The regulatory framework stresses on allocation of benefits for the purpose of biodiversity conservation and sustainable use, but equally for training, education and local/community empowerment and local development. Generally, developing countries' ABS regulations pursue monetary benefits from the utilization of their GR and aTK as priority, despite sometimes the non-established commercial value of the resources in question and whether commercialization *per se* may even occur.

7.2 Weaknesses

Alongside the strengths mentioned above, there are some weaknesses or limitations, which, if not addressed raise concern on the level of *de facto* compliance of the domestic ABS regime with the NP and can provide room for illegal ABS activities to take place unnoticed and without attracting any element of sanction.

The first limitation is the lack of description by the regulatory framework of the kind of activities that amount to commercial utilization of the GR and aTK. As discussed above, the principal demarcation between access procedures of the declarative scheme and the authorization scheme is that an applicant/user follow the declarative scheme if their subsequent handling/utilization of the resources has no commercial ends. However, what the law has not attempted to clarify is what amount to commercialization or commercial use of the genetic resources. Supposedly, it may be the intent of the law that commercial utilization and the prospective benefits will be discussed during the contract negotiations between the user and the CNA or the communities of inhabitants. Nevertheless, failure of the 2016 ABS law and the implementing decree to outline or provide an indicative list of activities/uses that can amount to commercial utilization of genetic resources may provide incentives to access applicant to avoid the authorization scheme and follow mainly the declarative route.

Another weakness is the absence in the law and its implementing regulations of the mechanism by which the agencies responsible for monitoring compliance with user measures, especially the DD obligation (the ministry in charge of research and the ministry in charge of the environment) as well as the agencies responsible for searching and observing general offences in the implementation of the law, can bring either corrective or punitive actions to the offenders. As indicated earlier, two possible routes for this are the application of general administrative law by the relevant competent authorities or their reliance on article 9 (6) of the 2014 EU Regulation that have direct effect at domestic level in France. In domesticating the 2014 EU Regulation in respect of the DD obligation, the 2016 ABS law of France expects users to fulfil their obligation under article 7.1 of the EU Regulation by filing the DDD to the ministry in charge of research when they receive public funding for their research. Similarly, users are expected to file DDD to the ministry in charge of the environment during the development of the final commercial product in fulfilment of article 7.2 of the EU Regulation. However, there are discrepancies between the way France has domesticated these obligations and the guidance provided by the EU in 2016. While the EU 2016 guidance document provides a fair amount of details on the timing for submitting the DDD, the ABS regulatory framework of France has domesticated them rather loosely. The legislation does not replicate the timing suggested in the EU Commission's guidance document,[96] leaving the timing of DDD submissions perhaps open to the discretions and sole judgment of users. In addition, with regard to DD obligation in relation to research funding, the EU

[96] EU Commission (2016).

guidance document states that: "The EU ABS Regulation does not make a distinction between public and private funding. Both types of funding for research are covered by the obligation to declare due diligence as provided for in Article 7(1)".[97] However, the 2016 ABS law of France seems to have narrowed DD obligation at the stage of research funding only to public funded research. These two discrepancies alongside a lack of mechanism to bring corrective or punitive actions in case of violation by users of these measures do not help to instill trust, from the perspective of supplier countries, in the ability of the regulatory framework to detect and take actions against illegally or inappropriately acquired materials that are utilized in France.

A third limitation or weakness is the failure of the ABS regulatory framework to address collection of bulk materials that may end up being utilized in the way utilization is defined under the Nagoya Protocol (i.e. in R&D). In fact, it is perfectly possible that bulk materials of biological and genetic resources of France, collected not as a result of submission of an access application in respect of the ABS regulatory procedures, can be utilized for non-commercial or commercial purposes without being noticed by the institutional apparatus responsible for the administration of the ABS framework. The 2016 law has provisions for change of intent, which requires the user to make a fresh/new application for an authorization permit or an application for a new declaration. This change of intent only works for already known ABS activities that were authorized through the permitting outlined in the law. Bulk collection is not mentioned in the law and could represent a significant loophole in the operationalization of the ABS regulatory framework.

The fourth weakness of the ABS regime of France is its arguable *de facto* compliance with the NP. Discussions above under the compliance section lead to the conclusion that by and large, the post Nagoya Protocol ABS regulatory framework of France is *de jure* compliant with the letter of the Nagoya Protocol on ABS. However, questions may be asked if the spirit of the NP is properly reflected in this ABS regime in the way that can deliver *de facto* compliance and properly ensure conformity with the regime by any entity engaged in GR collection and utilization in France. One of the grounds of limited *de facto* compliance is for example failure of the regime to cover the activities of intermediaries other than the scientific collections. To benefit from the facilitated approach under the declarative scheme, scientific collections must register with the Ministry in Charge of environment and are expected to update their registration information every year before 31st March. If they do not register, they are theoretically expected to follow the authorization scheme when they want to access GR. As it stands, the regulatory framework is completely silent on the activities of less formal intermediaries such as individual researchers, parataxonomists and private landowners. The implications of this include the fact that a great amount of GR collection and supply to researchers or R&D actors is likely to take place undetected by the institutional apparatus in charge of the administration and enforcement of the ABS regime.

[97] Ibid, p. 14.

Lastly, the ABS regulatory framework does not provide for alternative dispute resolution tools such as arbitration and mediation and reduces dispute resolution to two options: amicable approach and court proceedings if necessary. Alternative dispute resolution mechanisms are included in the NP among the choices available to actors and failing to accommodate them in the domestic ABS regime arguably has a number of implications in terms of e.g. fairness, equity and cost in the quest of justice. On the one hand, there may be an issue of power imbalance between parties to an ABS contract and the outcomes of an amicable resolution may well reflect that. On the other hand, although there are elements of costs in mediation or arbitration which are borne by parties to the dispute, costs of these alternative dispute resolution mechanisms are lower than the cost (time and money) of triggering and sustaining court litigation. In this respect, having other alternative dispute resolution mechanisms is important as it widens the choices available to actors and adds an element of fairness and equity in the quest of justice.

8 Conclusions

Like some members of the European Union *viz.* Spain, Bulgaria, Croatia and Malta, France has decided to regulate access to GR under the country's sovereignty and the utilization of aTK of the communities of inhabitants of the overseas territories of Guyana and Wallis et Fortuna. The domestic ABS regulatory framework on access complements the implementation by France of Regulation (EU) 511/2014, which regulates the utilization of genetic resources within the EU territorial jurisdiction. Having chosen the path to regulate access to GR and utilization of aTK, the ABS regulatory framework of France has clearly laid down the requirements and procedures depending on whether access is for commercial ends or not. The procedures are structured in two strands: the declarative strand for access with no commercial ends in sight and the authorization strand that applies to access linked to commercial utilization. For each of these strands, the roles of the CNAs in the PIC process, negotiation of the benefit-sharing agreements, the involvement of the communities of inhabitants and their legal representatives in the PIC and MAT negotiations processes are outlined. Furthermore, the regulatory framework sets out the powers of the administrations of the overseas territories in the implementation and administration of the access procedures.

Furthermore, the compliance regime especially in relation to the DDD is not strong enough to instill and establish trust of developing countries when their genetic resources are utilized in France. For example, while the requirement of the DDD may be seen as a proof of compliance by the EU framework to the user measures of the NP, the way it will be implemented begs the question as to whether users who do not make DDD will be caught at all and punished. France appears to have domesticated the DD obligation more loosely which creates discrepancies with the guidance provided by the EU Commission. Furthermore, the fact that the relevant CNA and other agencies are only responsible to search and observe offences with no link to any mechanism to trigger corrective or punitive actions raise questions as to whether the rather severe sanctions earmarked in the law will ever be enforced. From the perspective of supplier countries (largely developing countries) with limited technical capacities whose genetic resources are likely destined for utilization overseas in technologically capacitated countries like France, the ABS regulatory framework does not in practice help strongly enough to address use of illegally or inappropriately accessed resources from other parties.

Acknowledgments The author expresses sincere gratitude to all the experts who agreed to be interviewed in this case study. My heartfelt thanks to Dr. Evanson Chege Kamau, for kindly inviting me to be part of the team he put together for this research project and for reviewing the draft materials. I would also like to thank the other members of the team, whose comments shared during all the research meetings held in Bremen between 2017 and 2019, were very instrumental in fine tuning my case studies. I want to thank Dr. Kamau assistants at Bremen, who were always ready to help in any way, especially in relation to all the meetings that were held. I want to thank my family, my wife and children for putting up with me through all related travel.

References

Publications

Aubertie S (2017) L'APA Pas à Pas: Mise en Œuvre du Protocole de Nagoya et des Règlementations d'Accès aux Ressources Génétiques et aux Connaissances Traditionnelles associées et de Partage des Avantages issues de leur Utilisation (APA) dans le cadre des activités de recherche et de Développement. Fondation pour la Recherche sur la Biodiversité (FRB), Paris, France

Chiarolla C (2015) Commentary on ABS provisions of the draft biodiversity law of France. In: Coolseat B et al (eds) Implementing the Nagoya protocol: comparing access and benefit sharing regimes in Europe. Brill/Martinus Nijhoff, Leiden, pp 77–114

Other Documents and Websites

CBD and Nagoya Protocol Clearing House Mechanism France., https://absch.cbd.int/countries/FR (accessed 29 June 2019)

Convention on Biological Diversity (2011) Strategic plan for biodiversity 2011–2020, including Aichi Biodiversity Targets, https://www.cbd.int/sp/ (accessed 28 May 2019)

Ministère de l'Ecologie, du Développement durable des Transports et du Logement (2011) La Stratégie Nationale pour la Biodiversité 2011–2020, Paris France

Ministère de l'Ecologie, du Développement durable et de l'Energie (2014) Le cinquième Rapport National de la France a la Convention sur la Diversité Biologique, Paris, France

UICN France & MNHN (2014). La Liste rouge des espèces menacées en France 2008–2013, Paris, France

World Megadiverse Countries., https://www.worldatlas.com/articles/ecologically-megadiverse-countries-of-the-world.html (accessed 28 May 2019)

Legal Texts

France

Loi No2016-1087 du 08 Aout 2016 pour la Reconquête de la Biodiversité, de la Nature et des Paysages, Titre V: Accès aux ressources Génétiques et Partage Juste et Equitable des Avantages, codifié aux articles L412-3 à L412-20 du Code de l'environnement

Décret No 2017-848 du 09 Mai 2017 Relatif à l'Accès aux Ressources Génétiques et aux connaissances traditionnelles associées au partage des avantages découlant de leur utilisation. This decree provides for the designation of the relevant Competent National Authorities (CNA) responsible for the implementation of the monitoring of utilization and compliance provision of the 2014 EU ABS Regulation and for the registration of collections. It also includes a model contract for benefit sharing from the utilization of traditional knowledge associated with genetic resources

Arrêté du 13 septembre 2017 fixant le contrat type de partage des avantages découlant de l'utilisation de ressources génétiques prélevées sur le territoire national, mentionné à l'article R. 412-20 du code de l'environnement

European Union

Regulation (EU) 511/2014 of the European Parliament and of the Council of 16 April 2014 on compliance measures for users from the Nagoya Protocol on access to genetic resources and the fair and equitable sharing of benefits arising from their utilization in the union

Commission Implementing Regulation (EU) 2015/1866 of 13 October 2015 laying down detailed rules for the implementation of Regulation (EU) No 511/2014 of the European Parliament and of the Council about the register of collections, monitoring user compliance and best practices

EU Commission (2016) Guidance document on the scope of application and core obligations of Regulation (EU) No 511/2014 of the European Parliament and of the Council on the compliance measures for users from the Nagoya Protocol on Access to Genetic Resources and the Fair and Equitable Sharing of Benefits Arising from their Utilization in the Union. Official Journal of the European Union, C 313/01

Marcelin Tonye Mahop PhD, held academic research positions in the UK and the Netherlands, senior programme management position with ECCAS in Central Africa and he is currently a Research Fellow at the School of Earth and Environment (SEE), the University of Leeds. Marcelin was a freelance consultant working on various topics including Access and Benefit Sharing/Genetic resources policies broadly and BioTrade, IPRs, agricultural biotechnology, seed regulations and seed systems, forest law enforcement and trade, projects and programmes evaluations, project formulation. tonye2169@aol.com, or tonye2169@gmail.com.

Part III
Implementation Experiences and Lessons

Post Nagoya Protocol Experiences of Basic Research in Ecuador

Erwin Beck

Abstract The chapter briefly summarizes the political context, in which nature utilization and protection is realized in Ecuador, which is a member of the Community of Andean Nations and hence bound to the Regional Biodiversity Strategy for the Tropical Andean Countries. The chapter then describes the changes in the conditions under which non-commercial, basic biodiversity research could be performed during the past two decades, using as an example a German Research Unit engaged since 1997 in ecosystem studies in South Ecuador. Major changes of the research conditions originated from the legislation of a new Environmental Law in 2017 that led the way to the ratification of the Nagoya Protocol (NP) by Ecuador in the same year. Ecuador as a member to the NP obviously strives to make use of the possibility of a facilitated procedure (NP, Art. 8 (a)) resulting in a research permit for academic research instead of a MAT (mutually agreed terms) contract. However, transport of biological materials irrespective of the purpose of research requires special authorization and this holds also for export. The chapter concludes with examples of non-monetary benefit-sharing, e.g. by capacity building and joint publications.

Keywords Biodiversity · Non-monetary research · Research permit versus MAT · Nuevo código orgánico

---

E. Beck (✉)
Present Address: University of Bayreuth, Bayreuth, Germany
e-mail: erwin.beck@uni-bayreuth.de

E. Chege Kamau (ed.), *Global Transformations in the Use of Biodiversity for Research and Development*, Ius Gentium: Comparative Perspectives on Law and Justice 95, https://doi.org/10.1007/978-3-030-88711-7_17

1 Introduction: Ecuador as a Member of the Community of Andean Nations and as a Party to the NP

Ecuador belongs to the Community of Andean Nations (CAN, Fig. 1) and hence participates in the sub-regional common regime on access to genetic resources established in 1996 under the Andean Pact. In 2005 the country agreed to the Regional Biodiversity Strategy (RBS) for the Tropical Andean Countries (Decision 523 of the Andean Community, 2005), but only in 2011 adopted a national

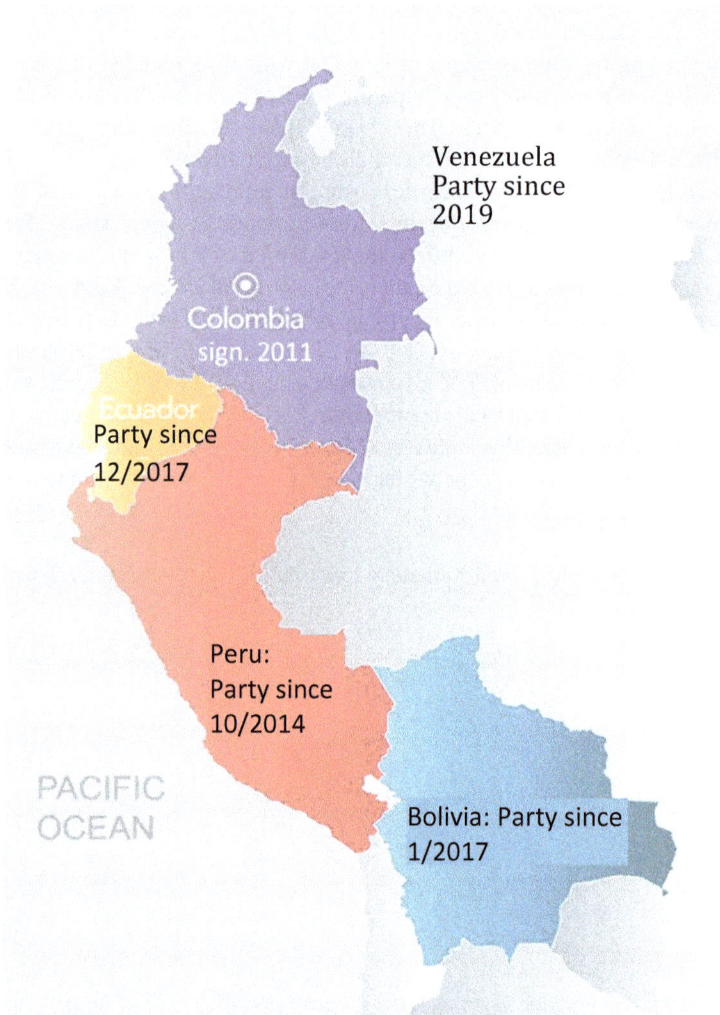

Fig. 1 The status of the Nagoya Protocol in the Andean States. **Source:** http://u.osu.edu/peek.9/files/2014/04/somosCAN_mapa2. Information supplemented by the author

regulation to implement the CAN legal framework. This regulation is provider-centered under the principle of sovereignty of states over their natural resources. It distinguishes between commercial and non-commercial research and establishes different procedures for access to genetic resources (GR), but under conditions that subject even basic research to the prior informed consent (PIC). Following the adoption of the RBS, a competent national authority was created and, in addition, "supportive institutions" that took part in the process of determination of access applications. Nevertheless, a number of instruments that had been conceptualized in order to operationalize the sub-regional regime in Ecuador finally remained unfulfilled.

Ecuador signed the Nagoya Protocol (NP) on 1 April 2011, but the examination of its compatibility with the country's Constitution, the development of the required administrative structures and the subsequent decision to ratify the protocol took more than 5 years (20 September 2017). Finally, on 19 December 2017 Ecuador became a party to the Protocol.[1] Immediately before the ratification of the NP, the Law on Environment and Natural Resources (La Ley de Gestión Ambiental y el Manejo de Recursos Naturales from 2004) was replaced by a new Codex (Nuevo Código Orgánico del Ambiente en Recursos Naturales, Energía e Infraestructura,[2] COA) which repealed several laws associated with the environment and biological diversity in the sense of the NP. Among others, the new Codex regulates the following issues:

- Environmental education
- Environmental Research
- Citizen participation
- Unique System of Environmental Information (SUIA)
- Funds for environmental management
- National System of Protected Areas
- National Forest Regime
- Unique System of Environmental Management (SUMA)
- Environmental Incentives

As article 1 of the New Codex states, its purpose is to guarantee the right of people to live in a healthy and ecologically balanced environment, as well as to protect the rights of nature for the realization and sustainability of good living. With respect to the NP the new Codex appoints the Ministry of the Environment and its provincial branches as the National Environmental Authority which logically is also the competent authority for the implementation of the NP.

[1] https://www.cbd.int/information/parties.shtml#tab=2.

[2] https://www.ambiente.gob.ec/wp-content/uploads/downloads/2018/01/CODIGO_ORGANICO_AMBIENTE.pdf.

2 Ecuador's Implementation of the NP: Consequences for Biodiversity Research

As in many countries with a developing economy and administration, the implementation of the Nagoya Protocol turned out as a difficult process that is still not yet completed. According to the author's experience there are a multitude of reasons for that: Because Ecuador has been and partly still is a country with remote regions in the jungle and the Andes, which are difficult to reach (e.g. only on horseback) and with a still uncompleted telecommunication net, its 24 provincial governments are practicing a relatively high level of autonomy, especially so with respect to biodiversity and nature protection issues. These have been handled individually by the provincial ministries of environment. Administrative supervision of particular projects of the author's research group that were working in several provinces were treated differently by the provincial branches of MAAE and so were the research permits.

With the implementation of the NP focal points were installed. The internet records 9 focal points related to biodiversity issues and the Nagoya Protocol, as well as 4 ABS-competent authorities (Tables 1 and 2). Centralization of the competences in a country like Ecuador is associated with many difficulties, e.g. reluctance to comply with a decrease of autonomy by the provincial administration on one side and still too few experts in federal ministries on the other. Biological research, as a field reaching into several administrative responsibilities is not easy to handle by the authorities.

Table 1 Ecuadorian administrative national focal points related to the CBD and the NP

N.F.P. Convention on Biological Diversity, Environment & Sustainable Development	Ministry of Foreign Affairs & Human Mobility
N.F.P. Convention on Biological Diversity, SBSTTA	Ministry of Environment and Water (MAAE)
N.F.P. Traditional Knowledge and Genetic resources	Ministry of Environment and Water (MAAE)
F.P. Resource mobilization	Ministry of Environment and Water (MAAE)
N.F.P. Traditional Knowledge	Servicio Nacional de Derechos Intelectuales (SENADI) and Ministry of Environment and Water (MAAE)
N.F.P. Protected areas	Ministry of Environment and Water (MAAE)
N.F.P. Marine and Coastal Biodiversity	Ministry of Environment and Water (MAAE)
N.F.P. Clearing House Mechanism, Protected Areas, Global Taxonomy Initiative (GTI) Global Strategy for Plant Conservation (GSPC)	Ministry of Environment and Water (MAAE)
N.F.P. Global Taxonomy Initiative (GTI), Plant Conservation & Wildlife	Ministry of Environment and Water (MAAE)

Table 2 Ecuadorian administrative authorities related to the NP

N.F.P. Access and Benefit Sharing	SENESCYT
Intellectual Property Rights	ABS Competent National Authorities
Education, Science, Technology & Research	SENESCYT & ABS Comp. Natl. Authorities
Environment	Ministry of Environment and Water (MAAE)
Biodiversity	Instituto Nacional de Biodiversidad (INABIO) & ABS Comp. Natl. Authorities

N.F.P. = National Focal Point; F.P. = Focal Point
Source: https://www.cbd.int/countries/nfp/?country=ec (accessed 11 November 2020)

For granting access to GR and permission for basic/academic/non-commercial research with biological materials three governmental authorities are relevant: The Secretaría Nacional de Educación superior, Ciencia, Tecnología e Innovación (SENESCYT), analogous to a Ministry of Science and Technology, the Ministry of Environment (Ministerio del Ambiente, Agua y Transición Ecológica del Ecuador, MAAE) and its provincial branches, and the National Biodiversity Institute (Instituto Nacional de Biodiversidad INABIO[3]). Competence for a research permit in Ecuador is principally with SENESCYT. However, with respect to biological materials, MAAE has to agree and has to decide whether the project contains commercially oriented aspects and thus will need a framework contract (Contrato marco) with prior informed consent (PIC), MAT and the International Certificate, or is only basic research which could make use of article 8 (a) NP allowing a facilitated procedure of granting research permission. This differentiation holds also for projects of Ecuadorian research institutions like Universities. For instance the Universidad Tecnica Particular de Loja, one of the counterparts of the author's research group, had to conclude such a framework contract because it is producing and marketing products of biological origin.

The implementation phase of the NP in Ecuador revealed a lack of expertise in the interface between biology and jurisdiction, as already mentioned. Therefore at present the authority of granting access to GR and issuing research permits is with MAAE, making use of counselling by INABIO. For the author's projects in South Ecuador ministerial supervision has been transferred to the provincial branches of MAAE. Nevertheless permission to transport biological materials between provinces requires support by INABIO. INABIO is a research institute that was established by the State of Ecuador in 2014, in the time span between signing the NP and its ratification. In spite of its written independence it is associated with MAAE. Its mission is to "generate knowledge and develop science, technology and innovation required by the Ecuadorian State to ensure the conservation of its natural heritage

[3] http://inb.ambiente.gob.ec/.

through the sovereign, strategic and sustainable use of biodiversity and its components for the consolidation of the society in a good living".[4]

INABIO encounters 12 scientific departments and is considered counterpart of research projects which request transfer of biological materials between provinces ('movilización'), and export. Application for mobilization and export is at MAAE, but needs written support from INABIO.

The latter is as well competent for activities in the scope of REDD^{+}.

3 Academic, Non-commercial Research Projects in Ecuador Using Biological Materials

In this contribution the legal situation of a German research program before and after ratification of the NP by the Ecuadorian Government is compared. The general aim of that biodiversity-related program is an understanding of the ecosystem "tropical mountain rain forest" and its man-made agricultural replacement ecosystems, mostly mountain pastures. A more recent expansion of the project is a comparison of the tropical mountain rain forest with the tropical mountain dry forest (the so-called Tumbesian dry forest on both sides of the southern Ecuadorian/Peruvian border). The author is since 1997 member of this program.

3.1 The Pre-Nagoya Protocol Phase

Research started in 1997, housed by the field station Estación Científica San Francisco (ECSF)[5] in the eponymous valley of the Rio San Francisco, which partly belongs to the provinces of Loja and Zamora-Chinchipe. In general, the research program of over 22 years can be characterized as ecological research comprising geo- and life-sciences. Starting with 6 research groups for the first 2 years, the program continued with varying main foci and up to 30 concomitantly working sub-projects in between, and at present is financed until end of 2020 with an extension for another three years. The main funding agency is the German Research Foundation (DFG) but also projects resourced by other organizations have joined our program. They have been working on separate research permits. The entire research is exclusively basic research, however application of the results to problems of land use, e.g. reforestation of abandoned agricultural areas and restoration of

[4] Ibid.

[5] Beck et al. (2020).

pastures is included. According to the definitions in the OECD Frascati Manual[6] the entire research fulfills the criteria of non-commercial.

Until December 2017, when Ecuador became party to the NP, research was carried out on the basis of research permits, initially issued for groups consisting of three sub-projects and later on for each subproject separately. In the course of time the research permits covered different numbers of years, but after 2006, annual reports and annual applications were obligatory and the permits were for one year only. Regarding the scientific substance, counterparts were the local universities on the basis of Memorandums of Cooperation. Responsibility to supervise the performance of the researchers was committed to one of these universities who had to give annual notice of a proper realization of the allowed research. Right from the beginning the German researchers had negotiated with MAAE to abstain from any kind of bioprospecting and in the unlikely case of stepping on a promising organism or compound, would leave it to the Ecuadorian counterpart. Fortunately this has not happened. Following this agreement during the whole period of 22 years, the group never had any problem with suspicion of biopiracy. Fruitful collaboration with several Ecuadorian Universities developed into a sustainable fundament of mutual support and trust.

Untiring efforts of 2 project managers who as German researchers set up home in Ecuador and are meanwhile well-known by the provincial and federal authorities turned out as a very important factor in creating trust.

Apart from Ecuadorian universities the group cooperates with non-academic counterparts. The most important of these is the foundation "Nature and Culture International" (Naturaleza y Cultura Internacional, NCI[7]), originally a US-American organization, which in the meantime has obtained some autonomy with respect to projects in Latin America. NCI is the owner of the above mentioned research station ECSF and another one in the Tumbesian dry forest. The rent of both well-equipped research stations is defined by contracts including the rental-fees. Activities of NCI in nature conservation, e.g. the implementation of several UNESCO Biosphere Reserves in South Ecuador were supported by the German researchers with scientific arguments. During the four years immediately preceding Ecuador's ratification of the NP research focused on knowledge transfer to non-academic users on the one hand, and on the development of early indicators for global change effects on the investigated ecosystems on the other. This phase of the program was initiated as a common German-Ecuadorian Research Platform which after international evaluation consisted of 12 German and 5 sub-projects from a consortium of several Ecuadorian universities. Unfortunately the Ecuadorian projects could only start with a delay of 4 years, at least partly due to the requirement of a framework contract between MAAE and the Ecuadorian Speaker University of the consortium.

Knowledge transfer aimed at reforestation of cleared former forested areas and of restoration of abandoned pastures. For that part, counterparts were landowners,

---

[6] OECD (2015).

[7] naturalezaycultura.org.

supported by cooperatives and non-governmental organizations (NGOs). Development of the global change indicators was in collaboration with NCI, including also the local universities. For these projects research permits ("Autorizaciòn de Investigación Científica") on an annual basis were granted by the provincial ministries of environment which also granted export of biological materials for in-depth analysis. At that time, the issue of transport permission (mobilization, "Movilización") of biological materials from one province to another came up. This could be considered analogous to an export of GR from a province and the formal requirements were somehow alike. Mobilization permission pertained not only to activities of non-Ecuadorian researchers, but also to those of the Ecuadorians and has to be applied at the Ministry of Environment and Water. A reference manual has been developed together with the forms, termed "Guìa de movilización de especímenes de flora y fauna silvestre" in 2015, that must be used since then. With the issue of mobilization control, part of the responsibilities of the supervising partner university was re-internalized by the Ministry of Environment. Other obligations for the research groups were similar to those under the new contract and will be assessed below.

Apparently, Ecuador was seeking to adapt its legislation to the requirements of the NP first before ratifying it. With the implementation of the NP, the situation of basic non-commercial biological research changed.

3.2 *The Post-Nagoya Protocol Phase*

3.2.1 A Research Program Spanning the Transition Phase Between Pre- and Post-Nagoya Protocol

By December 2017 a new project phase was started based on the knowledge elaborated during the precedent 20 years of ecological research. The new projects aim at a comprehensive model of the investigated ecosystems, further supported by complementing field measurements in order to update their actual states and assess their vulnerability by global change effects. Field measurements are partly *in-situ* and partly on materials transferred to Germany for identification and further in-depth analysis. The use of molecular-genetic methods was proposed by only one of the subprojects for a proper identification of collected insects.

From the outset it is clear that the entire research program, partitioned into 10 sub-projects is academic and non-commercial. Collaboration with Ecuadorian universities is continued, however, their responsibility for our NP-compliant performance has been taken over by INABIO under its framework treaty on access to GR of Ecuador with the Ministry of Environment. That framework treaty (contrato marco), termed "Genetic Biodiversity of Ecuador" (Biodiversidad Genética del Ecuador) was established in July 2016. Its main tasks are:

- Identification of floristic and faunistic elements of Ecuador;
- Analysis of the genetic flux and the connectivity of floristic and faunistic communities;
- Finding and evaluating morphological characters for the proper description and classification of Ecuadorian plant and animal species; and
- Analysis of biogeographical processes taking place in Ecuador's flora and fauna.

Encompassing 21 clauses, the framework treaty

- permits access (via INABIO) to and handling of biological and genetic resources of Ecuador **exclusively for academic research purposes** to increase basic knowledge of Ecuador;
- at the same time excludes all other kinds and aims of research on those resources, as well as occupancy and utilization for other purposes;
- excludes also access to traditional knowledge;
- encompasses 15 individual projects at the time of establishment.

In addition, INABIO is entitled to apply on behalf of the associated projects for mobilization and export of GR with the Ministry of Environment.

When applying for a research permit for the new German project negotiation with INABIO was recommended, which obviously is the institution where all basic biodiversity-related research should be brought together. An alternative would have been a separate framework treaty of the German group with MAAE, which, however, would have required time-consuming negotiations. Therefore the program including the descriptions of the individual contributions was submitted to INABIO which after some consultations proposed it in May 2018 as number 16 of its projects to MAAE. In June 2018 MAAE agreed and the research program was registered under the title "RESPECT: Environmental Changes in Biodiversity Hotspot Ecosystems of South Ecuador: Response and Feedback Effects" (Cambios Ambientales en Ecosistemas de Megabiodiversidad en el Sur del Ecuador: Efectos de Respuestas y Realimentación (RESPECT)) with a list of all principal investigators (PIs).

Thus in a legal sense the framework treaty of INABIO defined the rights and obligations and hence constituted the research permit, granting access to those GR which had been specified in advance in the projects. Under this contract material transfer agreements (MTAs) and mobilization permits could be applied for. The negotiations revealed a considerable mutual interest in closer collaboration between the German research group and INABIO and attendance at events, e.g. workshops and symposia organized by members of both groups witnessed this interest. Unfortunately the framework treaty expired by 31 July 2019 and therefore the question of a research permission arose again.

3.2.2 Obligations in the Initial Research Permits and Changes Caused by the Framework Treaty

In the following, the author refers to experiences in the context of his own projects "Water consumption and carbon capture by trees, as indicators of a smooth climate change in two mountain forests in South Ecuador" (2013–2018)—the project just before ratification of the NP by Ecuador—and "Plant physiological and structural properties as response and effect traits of biotic-atmospheric interactions in natural and anthropogenic eco-systems in South Ecuador" (2018–2020)—the project after ratification, realized as a member of the German team associated with INABIO.

The research permit for the first project stated apart from personal data the following counterparts: Two Ecuadorian universities, the foundation Naturaleza y Cultura Internacional in Loja, and the Director of the Ministry of Environment of the province (Zamora Chinchipe) where the research was carried out. It defined the time period of the validity of the document (1 year) and the dates for submission of the annual report and a potential new application. It excluded mobilization of plants, animals and microorganisms for which another permit by the provincial Ministry of Environment was necessary. The same held for export of biological materials from Ecuador for which support by INABIO or another authorized collection was required. Explicitly not allowed was the use of samples for bioprospecting or for access to genetic resources (unfortunately, "access to GR" was not defined in the document) or the use of results from the study for later studies requiring access to GR. In the latter case, permission by MAAE must be separately applied for.

Allowed was marking of a 100 m × 5 m plot in the forest, quarterly measurement of tree growth, measurement of photosynthesis in the crowns of 10 trees, and measurement of sap flow in the stems of 20 trees. These measurements required the use of towers, up to 35 m high in the forest, the construction of which was subject of a special permit.

Since the project was a subproject in the program of the German research group there were also points of general importance mentioned in the research permit:

1. A list of the approaches proposed by the group and agreed by the Ministry:
   - Identification, quantification, procurement and analysis of data on the fauna registered in mining concessions
   - Determination of the floristic and faunistic diversity- and similarity indices of the research area
   - Determination of the ecological importance of the various species
   - Identification and establishment of zones of special ecological importance
   - Suggestion of measures for conservation
2. A list of further obligations and limitations in the research permit
   - Indexing of samples of individuals or batches by the already given numbers prior to application for transportation or export
   - Collection and treatment of living samples is allowed, but not as parent material for use for commercial purposes

- Restriction of the use of species and numbers of individuals as listed in detail in the application
- Interferences in the field must be fair and responsible, and conservative methods must be applied
- Working on private areas requires permission by the owners
- Work in protected areas requires special governmental permission
- Fire arms, explosives and poisons are banned
- The same holds for plant material, animals and everything that threatens the integrity of the area

The document closes with articles for the renewal procedure of the research permit, articles for sanctions in case of violation of the conditions and one article on the costs (20 US $ for each subproject).

For transportation (mobilization) of collected materials a form has been developed which had to be submitted together with an application letter to the provincial Ministry of Environment. This form requested detailed information about the samples, their containers, the aims of the investigation, the itinerary of transportation, the car, the driver and the date. Of course, the mobilization permit was only valid in connection with the research permit.

Special attention was given to the numbers of samples and particular explanations were not only required for higher numbers, but also if smaller numbers than granted in the research permit should be investigated, transported or exported. The mobilization document could also be used for exportation if the German university was declared as the final destination of the samples.

Therefore the way to permit mobilization and export could be considered as analogous to concluding a MTA.

3.3 Research Under the Umbrella of INABIO: The Latest Research Project (2018–2020)

From the legal viewpoint the new projects of the German group are projects of INABIO under the framework treaty "Genetic Biodiversity of Ecuador" which, as said above, allows exclusively academic research. The application form requires details on the research area, on species and specimens used for research, and of the methods employed in fieldwork and the laboratory. Most importantly, the framework contract authorizes the named coordinators of the projects (in joint responsibility with INABIO) for the execution of collection, manipulation and access to biological resources. These coordinators are responsible for the compliant performance of all persons working in the projects, ensuring that these are familiar with the contract provisions. Any change of the coordinators must be immediately reported to MAAE. Detailed obligations concern data sharing with INABIO, deposition of original samples and duplicates, the format of reports about the project inclusive peer-reviewed publications, capacity building for the staff of the MAAE, information of the ministry in case of detection of new species, report to the relevant authorities if research is conducted in protected areas. The same holds for private areas or *ex-situ* collections. If contracts with third parties are necessary MAAE must be consulted prior to concluding a contract or an agreement. On the basis of the requests by its individual research groups and projects INABIO has applied access to and management of estimated numbers of samples, e.g. 1000 samples of plants, 300 batches of amphibia, 400 samples of birds etc. However, in the individual applications very detailed data are required about species, numbers, methods and place of analysis as shown in Table 3 for the project of the author.

In the presented case all potentially investigated species were not known in advance, so it was declared that "functional types" of trees shall be studied by their leaves, their wood and by a parameter characterizing capacity of photosynthesis (amount of the protein RubisCO). Fresh material must be analyzed in Ecuador whereas for specified analysis (e.g. stable isotope signals) dry samples are needed which can only be analyzed in Germany (University of Erlangen) and thus must be exported. To this end a mobilization permit was obligatory using the forms of the relevant provincial branch of MAAE and a MTA that was applied for by INABIO at the Ministry of Environment.

While the rights over the GR are principally under the sovereignty of Ecuador, the property rights of the authors over their research data is up to the present guaranteed by Ecuadorian legislation. However, the framework treaty and the research permits do not mention Digital Sequence Information (DSI), which obviously was not a political issue in 2016 when it was concluded. A new problem arose from the different termination dates of the framework treaty (July 2019) and the funding of the German projects (December 2020). Still relying on close collaboration INABIO proposed a simple research permit for the German projects by the provincial branches of MAAE, based on the consideration that access to GR was not applied for. Conversely this means that sequencing for taxonomic purposes was not judged as R&D that requires access. Research permits were granted by December 2019 for one year.

With that decision the Ecuadorian authorities made use of article 8 (a) NP which encourages provider countries to promote non-commercially oriented basic research on the countries' GR by a facilitated legal process of permission. Whether this politics might generally be expected will come up upon the forthcoming extension of the program for another three years.

4 Benefit-Sharing

Apart from clauses on capacity building for the staff of the Ministry of Environment, the framework treaty did not request activities other than sharing of results. Also not specifically requested was cooperation with universities or other counterparts. The reason for this might be seen in the fact that the framework treaty was negotiated by an Ecuadorian institution for seven Ecuadorian and only one external group. The

Table 3 Example of data required for application of a subproject in the scope of the framework treaty Biodiversidad Genética del Ecuador

Sample	Number of samples		Transport	Export outside Ecuador	Country	Analysis	Institution	Laboratory	Cooperation
	1st phase	2nd phase							
Leafs	1395	465	Dry	Yes	Germany	Isotope analysis of the leaf cellulose	University of Erlangen	Isotope Lab	Prof. Achim Bräuning
Leafs	90	30	Fresh	No	Ecuador	Morphological & anatomical leaf traits	Universidad National de Loja	Dendrochronology Lab	Dr. Darwin Pucha Cofrep
Leafs	630	210	Fresh	No	Ecuador	Amount of RubisCO[a]	Universidad Técnica Particular de Loja	Molecular Biology Lab	Dr. Juan Pablo Suarez Chacon
Leafs	75	25	Fresh	Yes	Germany	Amount of RubisCO	University of Osnabrueck	University of Osnabrueck	Prof. Renate Scheibe
Increment cores wood samples	1395	465	Dry	Yes	Germany	Isotope analysis of wood cellulose	University of Erlangen	Isotope Lab	Prof. Achim Bräuning
Increment cores wood samples	135	45	Dry	Yes	Germany	Wood anatomical analysis	University of Erlangen	Dendrochronology Lab	Prof. Achim Bräuning
Increment cores wood samples	1395	465	Fresh	No	Ecuador	Wood anatomical analysis	Universidad National de Loja	Dendrochronology Lab	Dr. Darwin Pucha Cofrep

Source: Compilation from several completed questionnaires by author
[a]RuBisCO is a special protein in the leaves which is important for modelling of photosynthesis

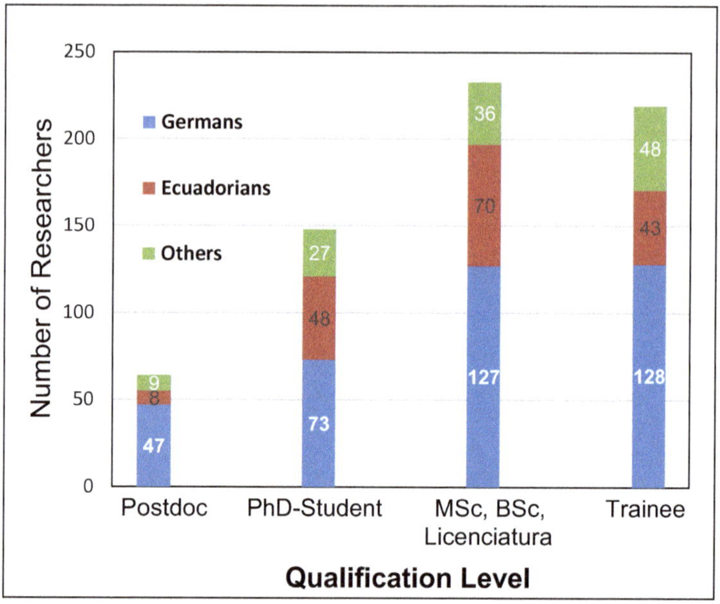

Fig. 2 Capacity building (2007–2020) in South Ecuador by the German research groups. **Source:** The German's group data warehouse (http://www.tropicalmountainforest.org) accessed December 2020 (communicated by J. Bendix and M. Dobbermann, University of Marburg)

author's group was the second external group which was affiliated with INABIO under the umbrella of the framework treaty. In future clauses for capacity building might be expected also for basic research projects from a user country. As a measure of benefit sharing collaboration and capacity building was always an obvious component of the German programs and it turned out as a real win-win situation for both (Figs. 2 and 3). Figure 2 shows the relevant data of personal capacity building in the cooperation of the German groups with four South Ecuadorian universities, their researchers and students. For an appreciation of the numbers in Fig. 2, it must be mentioned that Ecuadorian universities did not have the right of doctoral promotion. Therefore the Ecuadorian postdocs and doctoral students receive(d) their doctoral degrees from German universities.

Benefit-sharing in academic research projects is also documented by the author/co-authorship in peer-reviewed publications. Figure 3 shows not only a substantial number of such publications but also an increasing percentage of Ecuadorian authors and co-authors.

Data sharing is another quantifiable component of non-monetary benefit sharing. Due to the obligation to submit annual reports of activities and findings more than 8 million data have been already deposited in the data warehouse of the German program to which the Ecuadorian coworkers have access.[8]

[8]http://www.tropicalmountainforest.org.

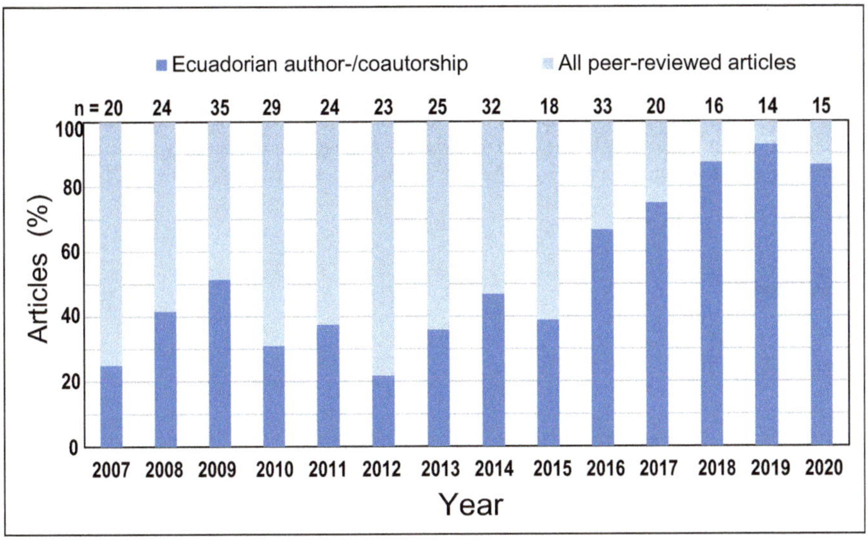

Fig. 3 Shared benefits: Peer-reviewed papers with Ecuadorian participation. **Source:** The German's group data warehouse (http://www.tropicalmountainforest.org) accessed December 2020 (communicated by J. Bendix and M. Dobbermann, University of Marburg)

5 Conclusion

The administrative conditions under which the described research is conducted in Ecuador may appear rather restrictive. As described in this paper, the conditions in the pre-Nagoya Protocol phase were gradually tightened and during the years when Ecuador had signed but not ratified the NP several of the applied regulations were an anticipation of the NP. Some twenty years of mutual experience between the German researchers, the Ecuadorian scientific counterparts and administrative authorities have resulted in a considerable portion of trust, and since both sides are playing with open cards, most of the projects' aims, sometimes with some delay, sometimes in an unexpected manner, could be achieved. Due to the long-standing funding of the work by the German Research Foundation (and others), a few of the German researchers, e.g. the author of this chapter, could experience how the spirit of the CBD and the NP gained more and more foothold in Ecuador's academic sector and state administration, but also how important non-governmental organization (NGOs) are for the protection and sustainable use of the goods and services provided by nature. Such NGOs together with an environmentally competent education lay the foundation stone for a responsible management of the outstanding biological wealth of Ecuador. As compared to the situation in the era before the implementation of the NP, the researchers' level of freedom has decreased. However, on the other hand the development of partnerships with the local universities opened a wealth of research facilities, which not only balances out the restrictions but strongly promotes the planned research.

Acknowledgements The author gratefully acknowledges ongoing information on the legal requirements and obligations during the past two decades by Dr. Felix Matt (Coordinator of the German research group) and Dipl. Geoecologist Joerg Zeilinger (Co-Coordinator). He also wants to thank the leaders of the third and later phases of the research group, Prof. Dr. Joerg Bendix and Prof. Dr. Nina Farwig (University of Marburg) for reliable guidance and friendship. Funding of the described research in Ecuador by the German Research Foundations is gratefully mentioned as well as granting of the research permits by MAAE in the provinces of Loja and Zamora-Chinchipe.

References

Books/Articles

Beck E, Paladines P, Paladines R, Matt F, Farwig N, Bendix J (2020) Alexander von Humboldt would have loved it: Estación Científica San Francisco. Ecotropica 21. https://doi.org/10.30427/ecotrop201909

OECD (2015) Frascati Manual 2015: guidelines for collecting and reporting data on research and experimental development, the measurement of scientific, technological and innovation activities. OECD Publishing, Paris. https://doi.org/10.1787/9789264239012-en. Accessed December 2020

Links

http://inb.ambiente.gob.ec/ (accessed 8 December 2020)

http://u.osu.edu/peek.9/files/2014/04/somosCAN_mapa2 [or https://www.uni-bremen.de/fileadmin/user_upload/fachbereiche/fb6/feu/FEU/ABS/ABS_PDFs/Experiences_of_non-commercial_research_with_ABS.pdf] – Presentation made by Erwin Beck and Lily Rodríguez during an ABS workshop at the University of Bremen in April 2017 on “Experience with non-commercial research and access to biological material” (accessed 8 December 2020)

http://www.tropicalmountainforest.org (accessed 7 December 2020)

https://www.cbd.int/information/parties.shtml#tab=2 (accessed 7 December 2020)

naturalezaycultura.org (accessed 7 December 2020)

Erwin Beck born 2 November 1937, Dr. rer. nat. (1963) in Plant Systematics (University of Munich), Associate Professor (1970–1975) University of Munich. Full Professor for Plant Physiology (1975–2006) at University of Bayreuth. Professor Emeritus (2007). Since 1979 research in the tropics. Academic services: Dean, Vice-president University Bayreuth, External Examiner: Universities of Nairobi and of Addis Ababa. President: German Biologists Association; German Botanical Society, Treasurer of IUBS. Honors: Order of Merit of the German Federal Republic, Professor Honoraria (UTPL, Ecuador), Dr. rer. nat. h.c. (TU Kaiserslautern, Germany); Ethiopian Academy of Science: Associate Member. bt250004@myubt.de.

Rights Over Genetic Resources and Ways of Monitoring the Value Chain. A Case Study from the Royal Botanic Gardens, Kew

China Williams

Abstract The access and benefit-sharing regime established by the CBD aims to ensure that benefits from the use of genetic resources are shared fairly and equitably with countries providing those resources. However, despite advances in implementation, provider countries face challenges in ensuring that benefits can be traced down the value chain of utilization, and shared fairly. This chapter focuses on the experiences of the Royal Botanic Gardens, Kew, and the range of measures it has put in place to ensure that the use of genetic resources can be tracked and benefits shared.

Keywords Convention on Biological Diversity · Nagoya Protocol · Access and benefit-sharing · Genetic resources · Royal Botanic Gardens, Kew

1 Introduction

Whilst the Convention on Biological Diversity (CBD) recognizes the sovereign rights of states over their biodiversity, its article 15 establishes the 'access and benefit-sharing (ABS) regime' that frames this relationship. Access to genetic resources is to be according to the national law of the provider State, with prior informed consent and on mutually agreed terms that ensure that benefits from the use of genetic resources are shared fairly and equitably. Ensuring that this bargain is kept has been a key challenge for both users and providers of genetic resources, particularly as use and actors change along the value chain as the resources are used in different ways. The Nagoya Protocol makes it compulsory for all parties to put in place compliance measures to monitor use and ensure compliance and benefit-sharing. This is set out in its article 17 and all parties must designate checkpoints to monitor compliance, ensure there are penalties for non-compliance, and use internationally recognized certificates of compliance, with their minimum

C. Williams (✉)
Royal Botanic Gardens, Kew, Surrey, UK
e-mail: c.williams@kew.org

E. Chege Kamau (ed.), *Global Transformations in the Use of Biodiversity for Research and Development*, Ius Gentium: Comparative Perspectives on Law and Justice 95, https://doi.org/10.1007/978-3-030-88711-7_18

information requirement, or their equivalent, to track use. The Protocol also establishes the ABS Clearing-House[1] as the central repository of information needed to complete these checks.

Although a basic system is established by the Nagoya Protocol, in practice there are challenges in implementation. These vary according to the kind of intermediaries involved and their roles. How exactly in practice should restrictions on use travel with the material, how is use tracked along the value chain given that the rights of the provider may extend to information created such as the results of research? Other questions relate to how future users are bound by the terms of what is essentially a bilateral agreement, how is this monitored and enforced? In this chapter we look at the experience of the Royal Botanic Gardens Kew.

2 Royal Botanic Gardens, Kew

The Royal Botanic Gardens, Kew ("Kew") is a botanic garden on the edge of London in the United Kingdom, a public attraction with over 1.7 million visitors a year and designated a UNESCO World Heritage Site in 2003. Kew as a historic, originally royal, pleasure garden is over 250 years old—the layout of the garden, as well as some of the living specimens, date from that time. Kew is also an active scientific research organization and employs over 900 staff in science and horticulture. To support this role, Kew houses vast global Science Collections, and these have been consistently added to for the past 170 years. They number over 8.5 million items (see Box 1) and, crucially, represent approximately 95% of the world's vascular plant genera and 60% of fungal genera.[2] They constitute a vast range of material which includes preserved plant collections—dried plants and fungi in the herbarium and spirit collections, a fungarium with over a million specimens, living material including over 2 billion seeds, a DNA and tissue bank, a microscopic slide collection and an *in vitro* collection. Kew also has an historic Economic Botany Collection which contains a range of samples and objects documenting the use of plants by people, and as well as this a huge Library of printed books and journals, Art collections, and Archives such as field notebooks, records, photographs and correspondence. These collections are added to at the rate of approximately 38,000 new specimens a year. While around 30% of these new additions are collected in the field by Kew scientists, the rest are sent to Kew from collaborating institutions around the world, either as part of an exchange agreement, or for further research and, crucially, identification and naming. Kew's collection represents a vast global resource that is unique and dynamic. It provides "an exceptional evidence base for addressing some

---

[1] absch.cbd.int/.

[2] Science Collections Strategy (2018).

of the critical challenges facing humanity – from climate change to food security and human health".[3]

Research undertaken at Kew focuses on enhancing understanding of plant and fungal diversity and is largely fundamental, exploratory research that does not have a commercial application. At its heart is accurate taxonomy to ensure species are accurately named, but spans research on seeds to better understand seed storage requirements and dormancy, comparative research looking at how plants interact with their environment, research on ecosystem services and the effects of climate change, propagation, and genomic studies to help conserve the diversity of genes and genomes at the population level. A small area of research focusses on the exploration of phytochemicals and their biological activity. All this work underpins the implementation of the CBD, informing its objectives of the conservation and sustainable use of biodiversity.

Box 1 Global Science Collections

Kew's Science Collections comprise of:

- the Herbarium—**7 million** pressed plant specimens
- the Spirit Collection—**76,000** specimens preserved in spirit
- the Fungarium—**1.25 million** specimens of dried fungi and over 1000 fungal cultures in liquid nitrogen
- the Economic Botany Collection—**100,000** samples documenting the use of plants by humans
- the Seed Collection at the Millennium Seed Bank—**86,000** collections and over 2 billion seeds
- the DNA and Tissue Bank—**48,000** samples of plant genomic DNA and 10,000 tissue samples
- the Microscope Slide Collection—**150,000** microscope slides of plant and fungal anatomy
- the In Vitro Collection—**6000** living plant and fungi cultures in jars of agar

Source: Science Collections Strategy (2018)

3 Development of an ABS Policy

Following extensive consultations with Kew staff and partners, and faced with the challenges of the developing ABS regime, Kew was one of the first botanic gardens to develop an Access and Benefit Sharing Policy[4] in 1998. The ABS Policy stated: "RBG, Kew will make reasonable efforts to clarify in a written agreement the respective roles, rights and responsibilities of RBG, Kew, the source country and

[3] Ibid.

[4] https://www.kew.org/about-us/reports-and-policies/conservation-and-sustainable-use.

relevant stakeholders in activities involving the collections of genetic resources and associated research, including prior informed consent. Where collecting permits do not clarify benefit-sharing or other obligations, RBG Kew may suggest that a material acquisition agreement (MAA) between the supplier and RBG Kew be used to clarify the terms on which RBG Kew acquires the genetic resources from the Supplier". At the same time, staff in Kew's newly created 'CBD Unit' began working with science staff to develop templates for more formal agreements with specific terms on the acquisition and supply of genetic resources. As identified in the first ABS Policy, agreements were being developed to cover the gaps and uncertainties raised by the adoption of the CBD, and the range of national legislation being adopted to implement it in countries where Kew had important partnerships and collecting programs.

4 Development of Agreements

In October 2019 Kew had over 140 active research agreements with a wide range of international partners in over a hundred countries. Formal agreements with international partners have been key to enabling Kew's conservation work, helping to build long term partnerships, enhance Kew's collections and develop global expertise in taxonomy and conservation both in the UK and partner countries. They ensure that Kew is working legally, within the national and international laws and regulations to which it is subject, both in the UK and the partner country. As partners are from countries at different stages of introducing and implementing national law on ABS these agreements have become a vital tool in ensuring agreed best practice is followed. On a practical level they ensure that all parties to an agreement have sat down and discussed what will be involved, and this ensures that the research is well planned, fits with partners' conservation priorities, any national legislation and procedures in place, and that the scientific benefit gained from the cooperation is shared and fed back through infrastructure, resources, capacity building and training. They also provide some clarity on terms and conditions of use of the material being transferred.

4.1 Memoranda of Collaboration

Memoranda of Collaboration (MoC) are the primary agreements used by Kew and its partners. These agreements are used where Kew is seeking to formalize an ongoing relationship with another similar institution (another botanic garden or research institute), or to set out the terms of a project that involves the transfer of material and associated information and data for purely taxonomic analysis. They are considered low risk from a legal point of view, due to the nature of the material being transferred and the use being made of it at Kew. They also serve as a way for Kew to set out its commitment to abide by international environmental conventions such as The Convention on International Trade in Endangered Species (CITES) and the CBD and to serve as a formal way of ensuring that partners are aware, and agree to, the use that will be made of material and data transferred to Kew. These agreements may address and set out the broad prior informed consent (PIC) and mutually agreed terms (MAT) requirements of the CBD, but specific terms will be set out in the permits and national legislation of the provider country. These broad agreements tend to recognize a longstanding relationship between the two institutions, as well as laying out a broad commitment to work together, develop future collaborations and build capacity. They also tend to include an Annex setting out a more specific framework of activities between the Parties. This might cover a specific project that has been funded for a specific purpose and for a specific amount of time. On the whole, commitments in MoC are not measurable and would be difficult to enforce in law. They are designed to represent a partnership and as a way of formulizing a bilateral relationship that exists in the context of developing and changing national law. In the future, as the legal systems solidify, we are likely to see a growing number of more enforceable bilateral agreements emanating from provider countries.

4.2 Access and Benefit Sharing Agreement

The model Access and Benefit Sharing Agreement (ABSA) developed by Kew following the inception of the Millennium Seed Bank Project (MSBP)[5] represents a more formal style of agreement that was developed to set out Kew's commitment to international conventions, in particular the Convention on Biological Diversity, and used in the case where material that is considered more 'high risk' (such as live plant material and seeds where there are more research and development options) or activities (DNA extraction etc.) were planned, to confirm that PIC had been obtained from the relevant authority and MAT negotiated. The MSBP was established in the 1990s, with the development of the CBD, and the concepts of sustainable use as a backdrop, and recognizing the central role of seed banking to successful conservation projects. When the MSBP was formally established in 2000, with the help of UK national lottery funding, the concept of sharing benefits from the use of the biodiversity was recognized as essential, and each project was carefully and individually crafted and negotiated with partners to work out what these benefits would be. Each project, and thus each ABSA, which reflects and frames the project, was individually designed to reflect not just the individual characteristics and needs of the partners themselves, but also the often evolving implementation of the CBD at the national level, the development of a national plan or strategy towards ABS, the introduction of draft laws, final decrees and interpretative statutes. Kew's ABSAs are

[5]https://www.kew.org/wakehurst/whats-at-wakehurst/millennium-seed-bank.

usually signed, or counter signed by the government authority responsible for granting access.

In addition to MoCs and ABSAs, Kew has developed a suite of other model agreements to be used to ensure that terms and conditions relating to the use of the plant material it has accessed, used or supplied, are monitored and passed on to third parties. So, for instance, Kew uses material supply agreements for all transfer of material to third parties and these pass on any restrictions on the use of the material to third parties. These supply agreements have standard terms prohibiting commercialization and sale, and include terms ensuring that Kew is acknowledged as the supplier of the material. They also ask that the recipient "shall maintain retrievable records linking the Material to these terms of acquisition and to any accompanying Data provided by Kew". In addition, the material supply agreement (MSA) contains the term "the recipient will contact Kew to request prior permission from Kew or, where appropriate, from the provider of the material to Kew for any activities not covered under the terms of this agreement". All these terms are a way of retaining control over use and movement of the material, and also, a way of monitoring the use of the material down the supply chain. In situations where Kew is donated material from a third party, and where that third party has not yet developed their own material transfer agreements, Kew uses 'donation letters' to clarify the terms and conditions of use. Kew's donation letters establish that the material has been legally obtained (and asks for permits and conditions to be attached), that the donor has permission to donate the material to Kew, and sets out standard terms of what will happen to the material once accessioned into Kew's collections. These 'standard use' (See Box 2) of material conditions ensure that the majority of material accessioned to Kew can be held and used on these standard terms. Ultimately though, monitoring use raises significant challenges for research institutes that exchange large amounts of material with third parties and who do not have the capacity to ensure that obligations passed on have been met or adhered to. This is an issue which has, to an extent, to be solved by compliance regimes in user countries.

4.3 Analysis of Key Terms in Model Agreements Used at Kew Relating to Monitoring the Value Chain

4.3.1 Use of the Material

To ensure that *consent* is *informed prior* to use, the standard Kew agreements outline the way in which material transferred to Kew will be used. As a working botanic garden with an active research program, many staff, visiting researchers and students, needing to track the use and supply of hundreds of thousands of specimens, it is vital, from a curation point of view, and to ensure that Kew can continue to actively use material collected for scientific research, that all material is acquired largely on the same terms and conditions. Kew's standard terms of use (Box 2 below) provide a starting point for the negotiation of mutually agreed terms and may be amended, added to or changed in individual bilateral agreements.

Box 2 Standard Terms of Use of Material by Kew
Kew may use the material as follows. It may be:

- Made available for scientific study to Kew staff and authorized visitors; and/or
- Used for the common good in the areas of education and public display; and/or
- Sampled for pollen, spores, DNA, anatomical or cytological preparations and/or chemicals for scientific research purposes; and/or
- Sent on loan, or further distributed, to other scientific institution(s) for further scientific research, on standard terms that prohibit commercialization; and/or
- Digitally imaged and published in botanical databases freely available on the internet.

Source: Own illustration

4.3.2 Commercialization

All Kew's standard agreements state that material acquired is for non-commercial use only, and that any commercial activity should be subject to a separate agreement between the parties to the standard agreement. Commercialization is defined as: "applying for, obtaining or transferring intellectual property rights or other tangible or intangible rights by sale or license or in any other manner, commencement of product development, conducting market research, seeking pre-market approval; and/or the sale of any resulting product". This definition of commercialization was developed by the Pilot Group[6] of botanic gardens working to develop common principles on access and benefit-sharing for the botanic garden sector.

4.3.3 Permissions to Collect, Transfer and Study the Material

Kew's standard agreements outline provisions for the collection and transfer of material. In some cases, where negotiated with the designated access authority, the agreements themselves lay out the PIC and MAT requirements as negotiated with the relevant in country authority. In others, they may outline the steps that need to be taken to get the proper permissions or permits.

[6] https://www.bgci.org/resources/bgci-tools-and-resources/principles-on-access-to-genetic-resources-and-benefit-sharing/.

4.3.4 Transfer to Third Parties

The standard clause in Kew's agreements is that material can be exchanged with third parties for scientific study. ("Kew may loan or supply the material or any derivatives from the material to other institutions for the purposes of scientific research or education, provided that such loan or supply is on terms which prohibit commercialization"). In some cases, this standard clause will be amended after negotiation between the parties, and the terms of supply to third parties is restricted totally or partly.

The most common additional restriction to be found in Kew's agreements relates to restrictions on the supply of the specimens collected to third parties. This is a clear way provider countries seek to control use of the material—a restriction on passing material on. However, from the point of view of an academic research institute this kind of restriction really undermines the value of specimen, and creates complicated, and thus expensive, curatorial burdens. The most common restriction is "no supply to third parties". In some cases, particularly if the other Party is also a scientific research organization, this has been softened through negotiation—to "no supply to third parties without PIC", or "notify the provider of any supply". However, any change leads to higher curation costs for Kew, as these specimens will need to be labelled and databased accordingly.

4.3.5 Benefit-Sharing

The agreements outline what benefits will be shared. These will be discussed by the parties to the agreement (that is Kew, the provider country Competent National Authority (CNA) and any local partners), but an indicative list is included in agreements as follows:

- Informing one another of the results of relevant scientific studies;
- Sharing specimen data and images, where appropriate;
- Providing one another with copies of relevant subsequent publications;
- Informing one another of relevant opportunities for formal or informal training and/or study by appropriate staff personnel at [insert name of counterparty] or at Kew; and
- Acknowledging [insert name of counterparty] as the origin of the Material in publications arising out of this collaboration.

Kew is undertaking an annual audit of projects to record benefits that have been shared as agreed.

4.3.6 Reporting Requirements

In addition to restrictions, another way provider countries can retain control over their material is through adding reporting requirements to the use of material. This establishes a mechanism by which provider countries can be updated with regard to activities undertaken using material transferred under the agreement.

4.4 Key Restrictions in Agreements

An analysis of Kew's agreements with African countries came up with the following list of key restrictions requested by partners on the use of material as set out in bilateral agreements:

1. No digitization
2. No third-party loan/supply
3. No third-party loan/supply without PIC
4. Notification of supply to third party
5. Do not extract DNA
6. Do not extract DNA without PIC
7. Notification of DNA extraction
8. Destroy progeny (Living collections)
9. Country retains ownership
10. Reporting requirements (annual/end of project)
11. PIC for transfer of threatened/confidential data/nationally important/listed material
12. Data transfer restrictions (such as no transfer, or only with PIC)
13. No publication without PIC
14. Restrictions on publication (such as only joint publication or only with PIC)

These restrictions reveal that provider countries do have concerns on how they can continue to monitor the use of plant material transferred and use a variety of clauses and restrictive terms in agreements as a way of being able to control and limit use of their material.

5 Overseas Fieldwork Process

Kew's science staff are involved in an active overseas collecting program, currently going on over sixty fieldwork trips a year. Material (herbarium specimens, seed, live material, DNA *in silica* gel, information, data etc.), is brought back to Kew and incorporated into its diverse collections according to terms and conditions of permits and agreements with provider countries.

To ensure that fieldwork is well planned and in accordance with Kew's Access and Benefit Sharing Policy, Kew has established an online registration system and reviewing committee for overseas fieldwork trips. This process ensures that all aspects of legal collection are prepared for well in advance. Staff who are planning fieldwork need to apply and give details of their trip to the cross-departmental Overseas Fieldwork Committee (OFC) three months before travel, to ensure:

- National laws and legislation are followed;
- Appropriate permissions are obtained and kept;
- Benefits are agreed and shared fairly;
- Staff work according to sectoral best practice standards and models;
- Staff are working safely and have adequate advice;
- Collections are curated appropriately afterwards, linked to permits and terms of use;
- A record of the countries in which Kew is working, and their requirements, is kept and updated.

The reviewing committee is representative of all Kew's Science departments and is led by the Policy Section, who ensure that all aspects of legal collection (CITES, Plant Health, ABS legislation and permitting) have been considered. When collected material is brought back to Kew (either with the staff member or sent later) it will not be incorporated into Kew's collections until all relevant permits and permissions have been received and inspected. Terms and conditions are recorded to ensure they have been met or will be met in the future. They are scanned and kept in databases linked to the collection number. Linking the collections to the OFC number ensures that the relevant permits and paperwork are linked to the material and future use and benefit-sharing can be tracked. This is a challenge for institutions and at Kew dealt with in a variety of ways. Firstly, policies and procedures are in place to ensure standard benefits are shared with countries of origin (see list of standard use box (Box 2) and standard benefit-sharing list (Sect. 4.3.5)). Secondly, staff procedures (annual reviews and project reporting obligations) help line managers keep track of conditions that need to be met. In addition, funded projects have a mid-way or end of project review to ensure that obligations have been met. One thing that is a challenge is the length of time these obligations run for—sometimes many years and in this case it is the standard policies and procedures that help to ensure conditions are met.

Thirdly, Kew undertakes an annual review of OFC trips to ascertain whether and how benefits have been shared as agreed and to audit the terms and conditions of permits and whether they have been met.

Box 3 RBG Kew Standard Procedures

Local counterparts: All Kew fieldwork is undertaken in collaboration with local counterpart(s), usually a botanic garden, herbarium, university or forestry department.

Distribution of specimens: Prior to collecting, it is usual practice for Kew to agree the number of sets of herbarium specimens, live plants and/or seeds to be collected with its counterpart and how the specimens will be distributed. It is customary for Kew and its counterpart to each retain one set of specimens collected. In the case of fungal material, there is not always sufficient material; where it has been possible to collect an extra set of specimens, this set will be left with the counterpart institution. Those specimens sent to Kew will be accessioned into the collections at Kew. Additional duplicates may be distributed to other institutions as mutually agreed by Kew and its counterpart.

Research at Kew: Plant and fungal material at Kew will be made available to Kew staff and authorized visitors for scientific research on systematics, ecology, conservation, genetics, horticulture, morphology, physiology and sustainable use. Such work may involve making anatomical and cytological preparations as well as sampling for pollen, spores, DNA, and/or chemicals.

Research results: It is usual practice for Kew to provide a copy of published research results to its local counterpart(s) and to acknowledge its counterpart(s) in any such publications.

Loans: Kew may loan plant and fungal material to other botanic, mycological, horticultural and scientific research institutions for further scientific research or for educational purposes subject to the standard "Conditions Governing Loans". These terms include that the specimen may only be used for non-commercial purposes.

Supply to third parties: Kew may supply plant and fungal material to other botanic, mycological, horticultural and scientific research institutions and/or to individual mycologists or botanists for further scientific research or for educational purposes, provided that such institution or individual signs a "Material Supply Agreement" with Kew, prohibiting commercialization of the plant or fungal material supplied.

Propagation and public display: Live plant material will be made available to Kew staff for propagation. Any plants grown from such propagation may be put on public display in the gardens at Kew.

Information and images: As a scientific institution it is important that Kew makes its collections as accessible as possible to its direct scientific counterparts and to the wider scientific and conservation community. This increasingly involves the digital imaging of plant and fungal material and of associated data, chemical or genetic analysis and publication of much of this information in databases freely available on the internet.

Source: Own illustration

6 Best Practices and Codes of Conduct

As well as having its own ABS Policy, Kew ensures that its internal policies and processes (Box 3 Standard procedures) are in line with sectoral best practice in the area. Through collaboration and information sharing with colleagues in the non-commercial research community Kew has helped to build a harmonized sectoral approach to ABS. Examples include the Common Policy Guidelines and Principles on ABS[7] for botanical institutions, and the Consortium of European Taxonomic Facilities (CETAF) package of policies and procedures[8] which has been formally recognized as a best practice model by the European Commission.

7 Conclusion

Kew has a wide range of international partners in over 60 countries worldwide. The fieldwork, research, scientific activities and sharing of information is undertaken in accordance with Kew's policies and procedures and these relationships are usually set out in specific bilateral agreements such as Memoranda of Collaboration and Access and Benefit Sharing Agreements which lay out a specific set of activities to be undertaken by the partners, and benefits to be shared.

Kew's collecting activities can be seen as a series of specific, bilateral and cumulative collecting events. The collecting event leads to the material coming back to Kew, research being done on that material, and specific benefits such as taxonomic identification and co-authored papers going back to the country of origin. Specific collection events usually involve the need to acquire a collecting permit or other written demonstration of prior informed consent (such as a material transfer agreement (MTA)) from the authorized provider. These documents may also include specific benefits that need to be shared following the acquisition and use of the material. However, it is also important to note that the effect of numerous historic bilateral collection events resulting in Kew's global and historic collections is also to create a global accumulation of benefits with its own added value, hugely important to achieving the objectives of the CBD.

So, for example, the collections stewarded by Kew become an international resource for taxonomy and identification, threatened species recovery programs, education and awareness raising, reintroduction and restoration programs, *in-situ* conservation, historic information on population shifts, on the effects on populations due to climate change/deforestation, and to support many other science and conservation studies. Collections can be also used to create identification models and tools

[7] https://www.bgci.org/resources/bgci-tools-and-resources/principles-on-access-to-genetic-resources-and-benefit-sharing/.

[8] https://cetaf.org/services/natural-science-collections-and-access-and-benefit-sharing.

for implementation of, for example, CITES legislation, for toxicology or medicinal plant identification.

All this creates great challenges for provider countries as they strive to assert their rights over their genetic resources and ensure that benefits are shared with them. Monitoring the use of genetic resources along the value chain is a challenge for both users and providers. Provider countries increasingly use terms in permits and agreements to try and either restrict the movement of material or to ensure they retain more control over movement of genetic resources. The development of Kew's ABSAs and MoCs, as well as the development of its ABS Policy have, essentially, been trying to fill this policy gap. They seek to set out, what is fair and equitable, standard setting and working with others in the sector to ensure there is an agreed understanding of what is 'best practice'.

8 Review and Recommendations

There is much work to be done to improve the ways in which tracking long term use can be improved to ensure benefit-sharing—the success of this being integral to the success of the CBD, for benefits incentivize countries to conserve and use their biological resources sustainably. Users and providers need to engage in discussion of strategies on how it is possible to keep track of third party use of the material to ensure benefit-sharing along the value chain. Currently there are gaps in the Nagoya Protocol compliance system—few countries have introduced legislation and, where they have, it needs to be clear, transparent and, crucially, uploaded on the ABS Clearing-House system. There is not yet enough information on what happens in the user value chain, and, in particular, the linkages between non-commercial research and utilization are unclear. This makes it difficult for administrative authorities to evaluate scientific research proposals submitted for ABS access requirements, or to design workable regimes.

More research is urgently needed to look at where, how and under what circumstances this change of use (from non-commercial research to commercialization, or an intention to embark on commercial research) takes place and specific indicators to assist compliance. In addition, there is still a lack of agreement as to who in the value chain is responsible for negotiating change in intent and use.

We need more detail on research patterns, in both the commercial and non-commercial sectors. This would assist provider countries wanting to develop access legislation that encourages biodiversity research, but with the security of knowing that checks are in place to ensure they are notified of change of use, or are in the position to predict this themselves. The non-commercial research community needs to work hard to develop policies and work-flows that ensure that conditions of access are followed and that the relevant authorities are informed or consulted in the case of a change of intent or use. They have started to do this individually, and on a sector level.

A key moment, as well as when change of use takes place within an institution, is at the point of transfer to a third party—either another non-commercial institution, or perhaps a commercial one. This is the point at which terms and conditions of use need to be clearly transferred and a decision needs to be made as to whether to inform the provider country, seek revised PIC and renegotiate MAT, and whose responsibility it is to do so. This may be set out in the terms of the original bilateral contract, permit or material transfer agreement. Alternatively, it may be clarified by national access legislation, which may set out exactly what steps to take in the event of a change of use (as suggested in article 8 NP), or national Nagoya Protocol implementing legislation will clarify this for users (for instance in the case of the Implementing Acts of the European Union ABS Regulation). In the absence of such information, sectoral guidelines are encouraged (Nagoya Protocol articles 19 and 20) and can provide a useful guide to ensure best practice is followed (e.g. CETAF). This is an especially crucial issue for *ex-situ* collections which hold huge and important collections that were acquired before the Nagoya Protocol, or CBD, came into force. Such trust building exercises could also help to resolve the continuing uncertainty about the retroactive applicability of the Nagoya Protocol (Article 5 (1)) in respect of new utilization of those resources.

References

Collections Strategy

Science Collections Strategy (2018), https://www.kew.org/sites/default/files/2019-02/Kew%20Science%20Collections%20Strategy%202018-2028%20Download.pdf (accessed 3 September 2020)

Links

absch.cbd.int/

https://cetaf.org/services/natural-science-collections-and-access-and-benefit-sharing (accessed 3 September 2020)

https://www.bgci.org/resources/bgci-tools-and-resources/principles-on-access-to-genetic-resources-and-benefit-sharing/ (accessed 3 September 2020)

https://www.kew.org/about-us/reports-and-policies/conservation-and-sustainable-use (accessed 3 September 2020)

https://www.kew.org/wakehurst/whats-at-wakehurst/millennium-seed-bank (accessed 3 September 2020)

China Williams is Senior Science Officer (Policy) at Royal Botanic Gardens, Kew, UK. c.williams@kew.org.

First Experiences in the Implementation of the EU ABS Regulation in Germany

Thomas Greiber and Ellen Frederichs

Abstract On 16 April 2014, the European Union adopted its Regulation (EU) No 511/2014 on the Nagoya Protocol on Access to Genetic Resources and the Fair and Equitable Sharing of Benefits Arising from their Utilization which applies as of 12 October 2014. In Germany, the European Union (EU) access and benefit-sharing (ABS) Regulation with its due diligence obligations is implemented by the Federal Agency for Nature Conservation since July 2016. First experiences with setting up the required compliance monitoring system show a number of open questions as well as practical challenges with regard to all components of the EU compliance system: Due diligence declarations, efficient and effective compliance checks, registration of collections, recognition of best practices, but also general awareness raising. The following chapter will provide an overview and preliminary analysis of the situation in Germany.

Keywords Nagoya Protocol · EU ABS Regulation · Due diligence · Risk-based compliance checks · Registered collections

1 Introduction to the EU ABS Regulation and the German Implementing Act

On 16 April 2014, the European Union (EU) adopted its Regulation (EU) No 511/2014 on the Nagoya Protocol on Access to Genetic Resources and the Fair and Equitable Sharing of Benefits Arising from their Utilization.[1] The EU access and benefit-sharing (ABS) Regulation which entered into force on 9 June 2014 and

[1] Herein after referred to as EU ABS Regulation.

T. Greiber (✉) · E. Frederichs
Division I 1.5 "Competent National Authority for the Nagoya Protocol", German Federal Agency for Nature Conservation, Bonn, Germany
e-mail: thomas.greiber@bfn.de; ellen.frederichs@bfn.de

E. Chege Kamau (ed.), *Global Transformations in the Use of Biodiversity for Research and Development*, Ius Gentium: Comparative Perspectives on Law and Justice 95, https://doi.org/10.1007/978-3-030-88711-7_19

applies as of 12 October 2014 (date of entry into force of the Nagoya Protocol) aims to implement the compliance pillar of the Protocol in the EU. It does not, however, regulate access to genetic resources and associated traditional knowledge (aTK) as well as related benefit-sharing in EU Member States as this is left at their discretion.[2]

The Nagoya Protocol contains a number of provisions which are intended to support ABS compliance:

- Articles 15 and 16 of the Protocol focus on compliance with requirements to obtain prior informed consent (PIC) when accessing genetic resources and to establish mutually agreed terms (MAT) for sharing the benefits arising from their utilization as foreseen in domestic ABS legislation or regulatory requirements of the contracting Party providing genetic resources, respectively the contracting Party where indigenous peoples and local communities holding aTK are located.
- Article 17 supports articles 15 and 16 as it obliges the contracting Parties to take measures to monitor the utilization of genetic resources, in particular by designating one or more checkpoints.
- Article 18 concentrates on compliance with the conditions set in MAT through the inclusion of contractual provisions on dispute resolution, the opportunity to seek recourse when disputes arise from MAT and access to justice.
- Finally, articles 19 and 20 promote the development of compliance support tools, i.e. model contractual clauses as well as codes of conduct, guidelines and best practices.

The EU ABS Regulation aims to implement all of these compliance provisions mainly through a so-called due diligence system, except article 18 which is not dealt with at the EU level but left to the individual Member States.[3] This due diligence system is composed of different measures and instruments: a general due diligence obligation of users of genetic resources and/or aTK (article 4 of the EU ABS Regulation); an additional obligation of users to file due diligence declarations at different points in time (article 7); an obligation of EU competent authorities to check user compliance with these due diligence obligations (article 9); and a voluntary EU register of collections (article 5) as well as the recognition of best practices by the European Commission (article 8) to support due diligence on the user side.

The Federal Republic of Germany is a Nagoya Protocol party since 20 July 2016. Germany does not specifically regulate access to its genetic resources and related benefit-sharing. Access to genetic resources collected *in-situ* within the national jurisdiction of Germany is therefore not subject to PIC and MAT requirements,

---

[2] EU's Interim National Report, response to question 7: "There are no access measures established at EU level. Some EU Member States have established access measures at national level whereas others decided not to introduce specific ABS access measures." For details on the EU ABS regime established by Regulation (EU) 511/2014 see Winter in this volume.

[3] For example, according to Germany's First Interim National Report recourse can be sought subject to the rules of international civil procedure, in particular Regulation (EU) No 1215/2012 on jurisdiction and the recognition and enforcement of judgments.

but solely subject to the general restrictions of public and private law. *Ex-situ* access to genetic resources via a collection located in Germany may, however, be linked to ABS rights depending on the provider country where the resources were originally collected *in-situ*.

In the implementation of the Nagoya Protocol, Germany thus concentrates on its compliance obligations. The EU ABS Regulation builds the basis for this and is supplemented by the German Act Implementing the Obligations under the Nagoya Protocol and Transposing Regulation (EU) No 511/2014 which entered into force on 1 July 2016. Accordingly, the Federal Agency for Nature Conservation (Bundesamt für Naturschutz—BfN) has been designated as the competent authority in charge of undertaking compliance checks, receiving due diligence declarations, handling applications for registration of collections as well as providing ABS advice to users in Germany. Furthermore, BfN has the mandate to cooperate with competent national authorities (CNAs) of provider states to ensure user compliance. The Act also regulates possible measures that can be taken by BfN in order to address situations of non-compliance with the EU due diligence obligations.

The following paper will provide an overview of the experiences of BfN in the application of the German Implementing Act and the EU ABS Regulation since its designation as competent authority in July 2016 until end of 2019. In doing so the BfN activities will be described, challenges identified, and the EU due diligence system critically analyzed.

2 Due Diligence Declarations

The key concept of the EU ABS Regulation is the general due diligence obligation under article 4. According to article 4.1 users shall exercise due diligence to ascertain that genetic resources and aTK which they utilize have been accessed in accordance with applicable ABS legislation or regulatory requirements, and that benefits are fairly and equitably shared upon MAT, in accordance with any applicable legislation or regulatory requirements.

The due diligence declarations foreseen by the EU ABS Regulation at two different points of the research and development chain—at the stage of research funding (Art. 7.1) and at the last stage of product development (Art. 7.2)—are to be understood as so-called checkpoints.[4] They are supposed to play an important role in demonstrating that the general due diligence obligation is met.[5]

[4]EU's Interim National Report on the Implementation of the Nagoya Protocol, response to question 26.

[5]Preambular para. 25 of the EU ABS Regulation.

2.1 *Understanding of Checkpoints*

Article 17.1 (a) of the Nagoya Protocol formulates the concept of checkpoints as follows:

(i) Checkpoints shall collect or receive relevant information related to PIC, to the source of the genetic resource, to the establishment of MAT, and/or to the utilization of genetic resources.
(ii) Contracting Parties shall require users of genetic resources to provide the information specified above at the checkpoints.
(iii) Such information shall be provided to relevant authorities, to the Party providing PIC and to the ABS Clearing-House.
(iv) Finally, checkpoints shall be relevant to the utilization of genetic resources or to the collection of relevant information at, *inter alia*, any stage of research and development, innovation, pre-commercialization or commercialization.

In the EU, it is the Member States' competent authorities that receive due diligence declarations from users of genetic resources at relevant stages of the research and development chain, namely at the beginning when a publicly or privately funded research project using genetic resources takes place (Art. 7.1 of the EU ABS Regulation), as well as at the end when a product is developed which also includes the sale of research results (Art. 7.2).

The declarations at both stages are mandatory in Germany. Due diligence declarations under article 7.2 of the EU ABS Regulation are obligatory in all EU Member States since 9 November 2015 when the Implementing Regulation (EU) 2015/1866 entered into force. Due diligence declarations under article 7.1 have to be requested by the competent authorities in EU Member States; this was done in Germany by a General rule[6] that BfN issued in April 2018. Since then due diligence declarations under article 7.1 are also obligatory in Germany.

As further specified in Annexes II and III of the EU Implementing Regulation, in their declarations users have to provide information on the genetic resources which were used, the subject matter of their research or name of the product, the country from which the utilized resources originated, and evidence of the necessary PIC (in the form of an internationally recognized certificate of compliance (IRCC) or a national permit). Furthermore, they need to declare that they are in possession of information on MAT.

Through the EU online system DECLARE[7] the submitted due diligence declarations are then forwarded to the ABS Clearing-House by BfN in collaboration with the Federal Ministry for the Environment, Nature Conservation and Nuclear Safety (Bundesministerium für Umwelt, Naturschutz und nukleare Sicherheit—BMU) as Germany's national focal point for the Nagoya Protocol. On the ABS

[6] General rule on submitting a due diligence declaration according to Art. 7 para. 1 of Reg. (EU) No. 511/2014.

[7] See https://webgate.ec.europa.eu/declare (accessed 14 Sept. 2020).

Clearing-House the due diligence declarations are converted into checkpoint communiqués and published as such.[8] A copy of the communiqué is also submitted to the provider State.

While the EU due diligence declaration system thus fulfills all necessary checkpoint criteria foreseen by the Nagoya Protocol, different understandings of the checkpoint concept currently exist amongst Nagoya Protocol parties. This became clear from an international Nagoya Protocol compliance workshop which was jointly organized by the Secretariat of the Convention on Biological Diversity (CBD) and BfN in 2019. As indicated in the official report the discussions held during the workshop supported the EU's checkpoint approach and understanding.[9] At the same time, the report shows that a number of countries have misunderstood the concept as they have set up their national checkpoints in a way to monitor the use of their own genetic resources in other countries instead of the utilization of foreign resources within their own territory.[10] Such monitoring activities can surely generate important compliance related information which supports the collaboration between CNAs of provider and user countries. However, they are not directly envisaged by the checkpoint provisions of the Nagoya Protocol.

2.2 *Value of Due Diligence Declarations for Compliance Monitoring*

Whereas the EU due diligence declaration system appears to be compliant with the Nagoya Protocol requirements under article 17.1 (a), another question regards the actual value of this system for compliance monitoring in practice.

The first due diligence declarations filed in the EU were submitted to BfN in mid-2018.[11] Afterwards these declarations became the first checkpoint communiqués worldwide to be published on the ABS Clearing-House. As of December 2019, a total of 11 declarations were submitted to BfN and 10 out of 17 checkpoint communiqués on the ABS Clearing-House came from Germany.

So far all due diligence declarations submitted in Germany fall under article 7.1 of the EU ABS Regulation whereas declarations under article 7.2[12] have not been filed yet. One of the reasons for the lack of article 7.2 declarations could be that the EU

[8] See https://absch.cbd.int/search/nationalRecords?schema=absCheckpointCommunique (last checked 14 Sept. 2020).

[9] Report on the Global Capacity-building Workshop on Monitoring the Utilization of Genetic Resources, para. 34 and 36.

[10] Ibid.

[11] A critical step that triggered this development was the general rule by BfN in April 2018 which changed the filing of due diligence declarations under Art. 7.1 of the EU ABS Regulation from voluntary to mandatory.

[12] Due diligence declarations under Art. 7.2 of the EU ABS Regulation have been obligatory EU-wide since 9 Nov. 2015 when the Implementing Regulation (EU) 2015/1866 entered into force.

ABS Regulation only applies to genetic resources which were accessed after 12 October 2014. The final development of products from such resources might take longer than 5 years. Furthermore, products can also still be developed on the basis of "old" material which falls out of the temporal scope of the Regulation.

After receiving a due diligence declaration BfN checks the completeness and timeliness of the declaration. Furthermore, a plausibility check of its content will be undertaken which may lead to further queries. Such plausibility checks, however, can be challenging as ABS permits and agreements are not internationally standardized and therefore differ substantially from one country to another in terms of their form and content. Furthermore, ABS documents are generally issued in local languages which may create further comprehension problems. As a result, it is sometimes difficult to assess whether a document or permit number submitted as proof of legal access provides for an actual ABS permit/agreement or not. The first due diligence declaration submitted to BfN and afterwards published on the ABS Clearing-House led to such confusion. While BfN had no reason to question the ABS permit number conveyed by the user in his due diligence declaration, the CNA of the provider State faced initial problems with relating the number mentioned in the checkpoint communiqué to an existing ABS permit issued in the past. This example confirms the great importance of further ABS harmonization to which IRCCs can contribute in the compliance monitoring context. IRCCs are a critical tool to create transparency and clarity between provider and user authorities as their content is standardized, they are given a unique identifier in the ABS Clearing-House and can only be published by Nagoya Protocol authorities. Such lack of harmonization is of course not only an issue when due diligence declarations are submitted and checked, but also during general compliance controls by the competent authorities as described below.

Once a due diligence declaration is converted into a checkpoint communiqué and thus submitted to the ABS Clearing-House as well as the provider State, a bilateral communication between BfN and the provider state authorities can be launched. If the provider State as the recipient of the communiqué reports irregular practices or ABS concerns, BfN may become active again in order to shed light on the allegations. Such bilateral communication also helped to bring further clarification in the case of the first due diligence declaration/checkpoint communiqué mentioned before.

Nevertheless the question arises if this system of due diligence declarations is in practice an effective tool to monitor Nagoya Protocol compliance or is only an apparent monitoring instrument leading in the end to increased paper work of limited practical value, i.e. bureaucracy. As a user can only submit a declaration together with information on the necessary IRCC or national permit, it is highly likely that those making a submission are convinced that they obtained the needed ABS documents and have nothing to hide. The EU checkpoints will therefore rather be a means to confirm ABS compliance or to detect cases of negligent non-compliance. In contrast, users who deliberately and intentionally violate the legislation of a provider country in order to avoid ABS procedures and relevant obligations will generally refrain from filing a due diligence declaration as they would obviously not be in a position to provide the necessary ABS documents and thus directly disclose their non-compliance. First experiences with the due diligence declaration system therefore seem to indicate that the ability of the declaration system to monitor and detect the most severe and urgent cases of ABS non-compliance is rather limited.

3 Compliance Checks

While from the point of view of the EU competent authorities the established checkpoints/due diligence declarations can be considered as more "reactive" tools (i.e. triggering their compliance actions), the EU due diligence system also comprises a "proactive" compliance monitoring component.[13] Article 9 of the EU ABS Regulation obligates competent authorities of EU Member States to undertake compliance checks in order to verify whether users fulfill their due diligence obligations, including the general due diligence obligation under article 4 as well as the obligations to file due diligence declarations under article 7.

Article 9.2 requires that such user compliance checks are

- effective,
- proportionate,
- dissuasive as well as
- fit to detect non-compliance of users with their due diligence obligations.

In order to better understand the meaning of these qualifiers, they need to be read in conjunction with other provisions of the EU ABS Regulation. Effectiveness but also proportionality can be related to the two triggers of user compliance checks foreseen in article 9.3: (a) periodically reviewed risk-based plans and (b) substantiated concerns. Both are built on the assumption that users are generally willing and eager to comply with ABS requirements. Consequently, these triggers do not require all users to be checked but only those where risks, i.e. threats of due diligence violations, are likely. The EU approach building on risk-based plans as well as substantiated concerns thus provides the competent authorities with the necessary discretion to focus their compliance checks and adapt the checks to changing situations as well as new information.

The criterion of effectiveness can be further interpreted in view of the obligation of competent authorities under article 9.6 to demand remedial action by users in case of non-compliance. Such remedial action aims at rectifying ABS violations that have occurred in the past, and thus provide a tool to ensure that ABS measures of provider states are followed, even when this may only happen retroactively. Making ABS work could thus be seen as one of the main goals of EU compliance check processes.

[13] See preambular para. 32 of the EU ABS Regulation.

A key for dissuasiveness again is the obligation of EU Member States to lay down penalties in their national implementing laws to address infringements of due diligence obligations (Art. 11.1).

In Germany, the Implementing Act further specifies the rights of BfN as the competent authority in the context of compliance checks. It confirms the rights already established by the EU ABS Regulation, i.e. possible response measures to address situations of non-compliance:

- orders to remedy breaches of ABS laws (s. 2.1 of the Implementing Act),
- prohibition of utilization (s. 2.2),
- seizure and confiscation of unlawfully utilized genetic resources (s. 2.2 and 2.3).

Furthermore, it stipulates the necessary penalties in case of due diligence violations: administrative fines of up to 50,000 Euros as well as skimming of financial profits (s. 4.3).

3.1 Risk-based

As indicated above, covering all potential users is not necessarily envisaged by the EU ABS Regulation and also not realistic in practice, as will be further explained below.

In 2018, BfN started with its first round of compliance checks after developing a first risk-based control plan. The starting point for this plan was the assumption that due to the novelty of the EU ABS Regulation and the subject matter overall no tried and tested risk criteria yet existed. The questions which specific institution, which type of institution or at least which sector would be more risky in terms of due diligence non-compliance could not be answered without any previous monitoring experiences. In order to gain such experience as well as a better insight into the prevailing Nagoya Protocol awareness in the very diverse sectors potentially affected by the Protocol it was decided to check the same number of institutions (10) in each of the following 8 sectors: pharma, cosmetics, biotechnology, food and feed, biocontrol, plant breeding, animal breeding and academic research.

A risk-based approach was applied in the process leading to the selection of the institutions to be checked. From a previously developed list of companies and institutions potentially affected by the Nagoya Protocol (approx. 2600 institutions in total) 20 institutions per sector were selected on a random basis. All 20 institutions went through a severe background analysis to establish a "potential user profile". For this, annual reports containing data on research and development (R&D) activities, the geographic location of R&D company sites, R&D expenses and strategies, databases including publications and patents related to genetic resources, as well as company homepages and other internet sites comprising information on products as well as bio-innovation processes were evaluated in view of ABS relevance. The 10 institutions with the most Nagoya Protocol relevant profiles were then selected for the actual compliance checks.

Before the end of 2019, 50 companies from 5 different sectors (pharma, cosmetics, biotechnology, food and feed as well as biocontrol) were contacted by mail and requested to fill in a questionnaire. 30 additional institutions from the plant breeding, animal breeding as well as academic research sectors followed in 2020 to close the first round of compliance checks.

The questionnaire, which was similar for all sectors, was comprised of two parts, containing a set of general Nagoya Protocol questions and a set of specific ABS compliance questions. The general questions were focused on the identification of responsible ABS persons as well as ABS policies, procedures and best practices which were adopted by an institution. The underlying idea was to raise and at the same time test internal Nagoya Protocol awareness. Furthermore, it provided an opportunity to review existing internal policies in order to suggest possible improvements and thereby foster due diligence preparedness for the future.

The specific ABS compliance questions addressed the issues of access to and utilization of genetic resources in order to find out whether an institution was within the scope of the EU ABS Regulation or not, and if yes, whether it fulfilled its due diligence obligations. Based on the answers received more concrete BfN investigations were triggered, additional queries posed, or even on-the-spot checks undertaken to clarify open questions on both sides. By the end of 2019 this written procedure had led to on-the-spot checks in 4 companies from different sectors.

3.2 *Substantiated*

In addition to risk-based compliance checks, article 9.3 (b) of the EU ABS Regulation foresees checks based on substantiated concerns. Information leading to such substantiated concerns can be collected by the competent authority itself or provided by third parties. Third parties may include provider states, but possibly also competitors reporting ‘free riders’ or even non-governmental organizations focusing on the issue of biopiracy.

Such controls have a comparative advantage in the sense that they must be based on some sort of evidence, meaning more concrete substance than a mere risk analysis. As a consequence, checks based on substantiated concerns might have a higher probability of identifying institutions which utilize genetic resources, fall within scope of the EU ABS Regulation or even violate due diligence obligations. However, it is also clear that their occurrence is much less predictable and rather *ad hoc*. The emergence, collection or reception of such ABS relevant information cannot be planned ahead in the same way as risk-based compliance checks.

By the end of 2019, in Germany a total of 3 companies as well as 1 institution from the academic research sector had been checked based on substantiated concerns. In all cases BfN collected the concerning information itself. Triggers were e.g. biopiracy accusations published in a press article which led BfN—after contacting the CNA of the relevant country of origin—to a specific patent based on a genetic resource which was accessed without ABS documents. As in the case of risk-based compliance checks a questionnaire was sent to the selected institutions before more concrete questions were asked about the particular case at hand.

3.3 Practical Challenges

As mentioned before, until the end of 2019 BfN had started compliance checks with a total of 54 institutions: 50 based on the previously adopted risk-based plan and 4 based on substantiated concerns. In 47 cases, the compliance checks could already be closed because it was clarified that the institutions' R&D activities were not or not yet in scope of the EU ABS Regulation.

While at first glance such figures might be interpreted as indicating low effectiveness of BfN's compliance checks, the absence of relevant ABS/user data must be considered here. No databases of ABS players yet exist (neither in the different commercial sectors nor in the academic research community) which could facilitate an *a priori* selection and reliable assessment of institutions falling within scope of the Regulation regarding their activities. Datasets of submitted due diligence declarations could potentially provide necessary lists of such companies and research institutes at some point in the future. However, the slow submission process leading to low declaration numbers poses a big question mark in this regard. Besides, as mentioned before, the value of due diligence declarations to detect actors that are intentionally avoiding ABS obligations is rather limited. Consequently, a focus on declaration submitters would rather lead to additional work for those who have already gone through often cumbersome ABS permitting processes and most likely complied anyway.

Furthermore, the temporal, geographical and material requirements that all need to be fulfilled at the same time for the EU ABS Regulation to apply provide a plethora of options for institutions to find their way out of its scope. For instance, in several compliance checks it could be determined that companies had undertaken relevant R&D on genetic resources. However, it was then claimed by the institutions that the resources were all accessed before 12 October 2014, at a point in time before a country had ratified the Nagoya Protocol, or it was argued that resources were accessed before and after 2014 but only the old samples were the subject of R&D while the new samples had not been used yet. Putting the burden of proof on the competent authorities the EU Regulation makes the determination of its applicability therefore a real difficulty.

The highly differentiated, small-sized interpretation of what is to be considered utilization and what not, adds additional complexity to compliance checks and limits the scope of the Regulation. For example, R&D activities were sometimes declared as simple screening activities which would be out of scope. Verifying such statements proved to be challenging and time consuming as information about the concrete research subjects and approaches had to be gathered from the companies, compared with research results already published in the academic literature and analyzed in detail in order to take an informed decision whether new knowledge on

Table 1 Summary of reasons for out of scope conclusions

N° of cases[a]	Reason for out of scope conclusions
11	Temporal scope (access before 12.10.2014)
5	Material scope
3	– Genetic resource out of scope (human, synthetic, digital sequence information,
21	cut-off)
6	– Derivatives without connectivity
	– R&D without utilization (e.g. formulation, screening, contract research)
	– Innovation claimed/assumed but no R&D
5	Geographical scope
7	– Not Party to Nagoya Protocol
	– No ABS legislation in place (or not applicable)

Source: Own illustration
[a]Total number of cases is higher than 47 as for some institutions more than one reason is applicable

the genetic or biochemical composition of genetic resources was developed or not. Deciding whether the concept of contract research applies or not, or whether connectivity is given between a derivative and a genetic resource are further examples for the great complexity of determining the scope of the Regulation in practice. Due to the limited amount of information available to BfN it proved to be difficult to differentiate between institutions outsourcing their R&D activities to service providers (which is still within scope) and those simply purchasing R&D results as final products (which is out of scope if the product is only used for formulation purposes without further R&D). The same was the case for derivatives bought on the market and those received from collaboration partners. Clarity could only be achieved here by disclosing the specific relationships between business partners, which of course discomfited several institutions.

Last but not least, first experiences with compliance checks raise the general question whether bio-innovation is in many cases used rather as a marketing than a research strategy. In a number of cases it was observed that companies promoted themselves as highly active and innovative in bio-based R&D (e.g. on company websites, in advertisements or reports) while their true research and innovation was limited to packaging and marketing which had nothing to do with research on genetic resources. Consequently, the development of "natural" products and brands combined with innovation prices and research budgets do not necessarily serve as reliable indicators for utilization in the meaning of the EU Regulation.

BfN's compliance checks therefore show that clarifying the applicability of the EU ABS Regulation is not only challenging but an integral part of the control process which cannot be anticipated. Final confirmation whether an institution falls within scope of the Regulation or not can only be achieved through the compliance check itself as taking the decision requires the collection of detailed information which can only be gathered through direct contact and exchanges with the monitored institutions (Table 1).

Despite these challenges, BfN's compliance checks have led to the detection of due diligence violations in 2 cases (1 risk-based compliance check and 1 based on substantiated concerns). These violations have triggered administrative processes (still ongoing) requiring remedial actions as a first step. Orders were adopted requesting the institutions to get the necessary ABS documents retroactively. At the same time, the institutions were prohibited to use, publish or otherwise share their research results until the breach of the ABS rules of the provider states is rectified. As a second (future) step, BfN will sanction the institutions through administrative fines. The degree of these penalties will depend on whether the institutions succeed in correcting their ABS non-compliance, i.e. get the necessary ABS permits and agreements.

4 Registration of Collections

According to article 3.9, the term collection is defined as "*a set of collected samples of genetic resources and related information that is accumulated and stored, whether held by public or private entities*". The mere collection/sampling and storing of genetic resources is not considered as utilization under the EU ABS Regulation and thus does not fall within its scope.

However, collections are major suppliers of genetic resources as well as aTK and therefore may play an important role in ABS processes.[14] As the EU ABS Regulation is not only focused on monitoring ABS compliance through checkpoints and controls but also aims to support users with implementing their ABS due diligence obligations, the Regulation foresees a register of collections within the European Union (Art. 5.1). Users obtaining a genetic resource from a registered collection shall be considered to have exercised due diligence as regards the seeking of information listed in article 4.3.

The EU register is an internet-based tool[15] that contains the information referred to in article 2 of the EU Implementing Regulation for each registered collection or registered part of a collection: e.g. registration code, name, contact details, short description. Collection holders may voluntarily apply for registration of their collection to the respective Member State in whose territory the collection is located (Art. 5.2).

4.1 Registration Process

Until the end of 2019, in Germany (and the whole EU) only 1 collection (the Leibniz Institute DSMZ—German Collection of Microorganisms and Cell Cultures GmbH)

[14] See preambular para. 28 of the EU ABS Regulation.

[15] See http://ec.europa.eu/environment/nature/biodiversity/international/abs/pdf/Register%20of%20Collections.pdf (accessed 14 Sept. 2020).

was registered under the EU ABS Regulation. The process of this first registration comprised different steps which can be summarized as follows:

- **Preparation**

The process started with an informal meeting between DSMZ and BfN where the expectations on both sides (collection as well as competent authority) were exchanged and a first set of open questions was discussed. In particular, the procedural path, the registration requirements but also the form and content of the application itself which is only indicated in Annex I of the EU Implementing Regulation had to be clarified. Afterwards a draft application was submitted leading to a number of feedback rounds in which BfN made suggestions for improvement.

- **Application**

After DSMZ had finalized its preparations and considered the necessary documentation as ready and complete, the official application was submitted to BfN. This triggered a compulsory assessment whether the requirements set by article 5.3 of the EU ABS Regulation were fulfilled. After minor inquiries and suggestions for additional changes BfN undertook an on-site visit to inspect the collection. During the visit DSMZ was given the opportunity to present the measures undertaken to comply with the registration requirements. Furthermore, BfN was able to clarify the last remaining questions before an official (positive) decision on the application could be taken and communicated.

- **Notification of the European Commission**

After BfN had given notice of its positive decision DSMZ had 4 weeks to object the decision. Only then the European Commission was informed and the necessary information according to article 2 of the EU Implementing Regulation was submitted which is a prerequisite for the inclusion of a collection in the EU register. As a final step the European Commission listed DSMZ in its online register and issued a registration number.

Although this first application process was successful, no other German collection has followed and applied for registration yet. An unofficial, voluntary survey amongst German collections (undertaken by BfN) seems to indicate the reasons for this slow start may be manifold ranging from general unawareness about the instrument of registration and its legal requirements, to lack of financial resources, fear of increased bureaucracy, and lack of clarity about the opportunities and threats in practice.

As a consequence, in 2018 BfN organized a series of ABS awareness-raising and capacity-building activities for German collections. The benefits of registration were described as follows:

- From a user's point of view, the EU register of collections can be an important tool to bring ABS transparency and legal certainty into R&D chains and thus help reducing administrative burden and mitigating risk of non-compliance where the intended use is in scope of the PIC and MAT provided by the collection.

- From a collection's point of view, the entry in the EU register will make a collection more ABS trustworthy and visible to users. It might attract users seeking safe access to genetic resources and aTK and therefore provide a comparative advantage (if its business is in supplying outside users).

4.2 Registration Requirements

Apart from the formal application process, the substantive registration requirements also needed to be clarified. Based on the first experiences with DSMZ it can be noted that in order to be eligible for registration a collection needs to demonstrate its capacity to

- apply standardized procedures for exchanging or supplying genetic resources and related information: i.e. technical, internal workflows specified for example in standard operating procedures (SOPs), quality management systems or general terms and conditions;
- keep records of all samples of genetic resources and ABS related documentation supplied to third parties in order to enable a later re-identification: in particular name of the resource, number assigned by the collection, possibly assigned numbers of previous collections/keepers, originally assigned strain/resource name, country of origin as well place of sampling;
- establish and use unique identifiers for genetic resources supplied to third persons: e.g. in the form 'collection acronym followed by sequential number'; and
- use tracking and monitoring tools for exchanging genetic resources and ABS related documentation with other collections: this makes a registered collection responsible for the recording of existing data (received and self-generated data) for traceability purposes which is the core of the obligations of a registered collection.

Should a registered collection have to undertake a substantive ABS review of the genetic material received, held and shared with its clients? Collections do not necessarily sample themselves but also accept material from external depositors under certain conditions. Such depositors are the ones initially collecting genetic resources *in-situ* and therefore in charge of applying for ABS documents (if needed) in the first place. Furthermore, under normal circumstance collections do not have full information about the specific R&D activities planned by its clients. To what level a registered collection should (and in practice could) check ABS compliance of the material held, the correctness of the ABS documentation attached to the deposited material, or even going beyond, the R&D undertaken further downstream was therefore another issue to be clarified.

A full ABS compliance review is not directly required by article 5 of the EU Implementing Regulation. Instead, the criteria listed in article 5.3 indicate the need for a risk management system containing procedural as well as documentation steps related to ABS transparency. Amongst others, the following steps should (and actually could also in practice) be applied:

- Regarding the ABS situation in the country of origin, the status of a country as a party to the Nagoya Protocol and the contact details of the national focal point and competent national authority(ies) can be assessed based on the information published on the ABS Clearing-House.
- The information on ABS measures provided on the Clearing-House however cannot be fully relied upon as experience shows that new measures are not always promptly displayed and the parties do not always keep their entries up to date.
- In case an internationally recognized certificate of compliance is presented by a depositor, the name, number and identity of the certificate should be compared using the ABS Clearing-House database.
- Insofar as a depositor claims that application for ABS documents is not feasible (achievable), the collection should request information from the national focal point of the provider country. However, the obligation to present ABS documents remains with the depositor himself so that the deposit should be rejected if documents are needed but their submission is not possible.
- In the case of well-known provider countries or those where ABS processes are known to be difficult, increased awareness and diligence is required when material is deposited.
- Recipients of material should be notified that the material can only be used for the specific R&D activities mentioned in the ABS documents. This is particularly important in the case of clients from industry who receive material with PIC/MAT attached which only foresees non-commercial utilization. However, collections are not required to investigate and if necessary enforce compliance of their clients with PIC/MAT obligations.

As the substantive ABS compliance monitoring and enforcement is thus limited, users who obtain genetic resources from a collection listed in the EU register shall be considered to have exercised due diligence only as regards the seeking of information (Art. 4.7 of the EU ABS Regulation). However, keeping and transferring the information and most importantly complying with PIC and MAT, as well as filing due diligence declarations remain as users' obligations. Compliance with these due diligence obligations will still be subject to compliance checks under article 9 of the EU ABS Regulation.

5 Recognized Best Practices

Another instrument envisaged by the EU ABS Regulation to support users in ABS compliance are best practices which have been recognized by the European Commission or adopted by the Nagoya Protocol Parties on the basis of article 20.2 of the Protocol.[16]

Under article 8 of the EU ABS Regulation associations of users or other interested parties may apply for the legal recognition of certain procedures, tools or mechanisms that have been developed to facilitate user compliance with the general due diligence obligation as well as the obligation to submit due diligence declarations. The overall idea behind such instruments is that compliance may be promoted through due diligence measures which are adapted to the needs and particular situations of users from specific sectors. At the same time, the development and adoption of best practices are considered as possible ways to reduce compliance costs and create legal certainty.[17]

The recognition process under the EU ABS Regulation foresees an application to be sent to the Commission including a description of the particular procedures, tools or mechanisms as well as other supporting information and documentation which is further specified in Annex IV of the EU Implementing Regulation. Before the Commission decides on an application, the competent authorities of EU Member States are given the opportunity to provide comments which shall help the Commission in its assessment whether best practice standards are met (Art. 8.4 of the EU Implementing Regulation).

By the end of 2019, 3 applications had been submitted to the Commission. So far only 1 of them, by the Consortium of European Taxonomic Facilities (CETAF), could finally be approved after going through several rounds of revision. The CETAF Code of Conduct and Best Practices are now published as the first and so far only recognized best practice under article 8 of the EU ABS Regulation.[18] The other 2 applications which were submitted by different associations from the cosmetics sector were not pursued further by the applicants after the first round of commenting by the EU Member States and the Commission which requested significant improvements.

One of the issues that turned out to be problematic in all 3 applications was the development of clear step by step handouts that would serve as sound and helpful tools for users applying the best practices. In the case of CETAF, this difficulty seemed to originate (at least to some extent) in the great heterogeneity of institutional setups in the target group. CETAF members differ widely in view of the objects held,

---

[16] By the end of 2019 no best practices were proposed for recognition under the Nagoya Protocol. Furthermore, no mechanism and process for such recognition have been agreed yet.

[17] See preambular para. 21 of the EU ABS Regulation.

[18] See http://ec.europa.eu/environment/nature/biodiversity/international/abs/pdf/Register%20of%20Best%20Practices.pdf (accessed 14 Sept. 2020).

their collection sizes as well as the overall management frameworks in place.[19] The proposed tools therefore had to meet the right balance between on the one hand a certain level of generalization to be applicable by all CETAF members, and on the other hand the necessary concreteness to be still useful in order to guide the institutions' compliance efforts. This could be achieved by the identification of common management principles in CETAF collections and the development of flexible instead of one-size-fits-all tools and deliverables.[20] The recognized best practice in the form of the CETAF Code of Conduct and Best Practices now includes:

- the CETAF Code of Conduct itself which comprises the agreed governance principles;
- the Best Practices—as the core—which are recommendations for policies and processes related to data management, curation, staff training, fieldwork, utilization, third party transfer, etc. to implement these principles;
- a Statement of use of Biological Material which outlines the uses of biological resources;
- a glossary with an explanation of key terms;
- a section on non-monetary benefit-sharing;
- another section with practical advice on implementing the Code of Conduct and Best Practices;
- Material Transfer Agreements (MTAs); and
- the Data Use Statement.

Furthermore, concerns about the obligation of applicants to "oversee" the recognized best practice procedures, as stated in the Regulation, raised questions about potential liability for the represented target group. Associations normally lack the legal authority but also the practical means to check compliance of their members with best practice procedures, tools and mechanisms. Thus, they have only very limited capacities to fulfill oversight and supervision functions. Such concerns could be addressed during the CETAF application process by clarifying that the recognition of a best practice does not lead to and correspond with auditing or certification obligations. Being responsible for the development of and application for a recognized best practice does not trigger legal liability for its implementation by the represented users. In the end it was agreed that regular awareness raising activities combined with internal communication processes to update CETAF member institutions about future revisions and improvements of the best practice are sufficient to fulfill the "oversee" criteria.[21]

In this context it has to be noted that the recognition of a best practice does not automatically lead to the fiction that users who belong to the represented group and apply the best practice have acted with due diligence. Instead, they still need to fulfil

[19] Neumann (2018), pp. 71, 72.

[20] Ibid.

[21] Ibid.

all due diligence obligations. However, if a recognized best practice is applied by a user group, the risk of non-compliance of its members can be estimated as lower, which again may have consequences for risk-based approaches in compliance checks.

By the end of 2019 Germany had no practical experiences with the application of CETAF's recognized best practice by relevant users and its usefulness to promote ABS compliance. This is due to the fact that the best practice was only recognized relatively recently, in May 2019. Furthermore, compliance checks in the German public research sector (to which CETAF members belong) only started in the course of 2020. Nevertheless, the CETAF recognition process once more showed the complexity of ABS approaches for heterogeneous groups of users. It could therefore be recommended to focus more on smaller user groups and more specialized procedures and mechanisms when developing further recognized best practices in the future.

6 Awareness Raising

In order to support the implementation of the EU ABS Regulation, article 13 foresees a number of complementary measures to be taken by the European Commission and the EU Member States. Since its designation as competent authority for the Nagoya Protocol in Germany BfN has put an emphasis and has undertaken great efforts to promote and encourage information, awareness raising and training to help stakeholders to understand the Protocol, ABS and the EU due diligence system, as envisaged under article 13 (a). It is estimated that the different awareness raising activities described below have absorbed about half of BfN's time up to the end of 2019.

As a first step, different information material was developed including a Nagoya Protocol flyer, a poster and a brochure to introduce stakeholders to the topic, clarify the new legal framework and spread BfN's contact details among potential users as well as associations. BfN's Nagoya Protocol website was fully revised and constantly improved offering also more practical tools, such as frequently asked questions (FAQs), recommendations for action, training materials, or background information on provider states' ABS frameworks.

To complement and make use of these information materials 2 surveys were undertaken: one online survey among a total of 2630 potential users, and a second survey among German collections. The outcomes of both surveys guided BfN's further awareness raising activities. At the same time, BfN entered into direct contact with ABS stakeholders by offering bilateral meetings with and presentations at affected institutions. Since 2016 around 100 meetings, conferences and workshops were attended to speak about the new legal framework and provide ABS guidance. In parallel, stakeholders could send their particular ABS questions to a centralized Nagoya Email address (nagoya-cna@bfn.de) which has most frequently been used to get clarity on the scope of the EU due diligence system or the ABS situation in Germany or third countries.

However, throughout the years concerns were raised that not all stakeholders feel comfortable in asking BfN as the competent authority for compliance checks for ABS advice, which in their eyes might create the risk of self-incrimination. As a consequence, in 2019 BfN launched a new awareness raising and support project[22] developing an independent Nagoya Protocol information platform for the academic research sector. In order to create the necessary trust in this platform, the project is being implemented by DSMZ together with different partners: the Consortium of German Natural Science Collections (Konsortium der Deutschen Naturwissenschaftlichen Sammlungen, DNFS), the German Life Sciences Association (Verband Biologie, Biowissenschaften und Biomedizin in Deutschland, VBIO) and the Leibniz Research Alliance for Biodiversity (Leibniz-Forschungsverbund Biodiversität, LVB).

Finally, BfN tries to keep interested stakeholders informed about recent ABS developments through a news section on its website as well as a biannual electronic newsletter.

7 Conclusion

Five years after the entry into force of the Nagoya Protocol as well as the EU ABS Regulation, and three and a half years after the entry into force of the German Implementing Act, mixed and rather sobering conclusions can be drawn from the first implementation experiences in Germany. Indeed, it is questionable whether the objectives of the EU ABS Regulation, in particular those set out in preambular paragraphs 9, 10, 23 and 25, are met in practice.

Accordingly, a clear and sound framework for implementing the Nagoya Protocol is sought that contributes to the conservation of biological diversity and the sustainable use of its components, as well as the fair and equitable sharing of benefits. At the same time, it is envisaged that opportunities for nature-based research and development activities in the EU are enhanced, legal certainty is improved, due diligence implementation costs are achievable at affordable cost and trust between Nagoya Protocol Parties and ABS stakeholders is increased.

First of all, it is currently unknown if or to what extend ABS frameworks and ABS interactions contribute to conservation and sustainable use. Appropriate baselines as well as indicators for related monitoring and measuring are still lacking in most countries around the world, an issue which is becoming more and more clear in the current CBD discussions on the Post-2020 Global Biodiversity Framework.[23]

[22]The German Nagoya Protocol help desk for academic researchers, HuB.

[23]See e.g. the Recommendation adopted by the Open-Ended Working Group on the Post-2020 Global Biodiversity Framework.

The development of possible monitoring approaches is highly complex for technical reasons (e.g. different understandings of scope amongst Parties, fragmented institutional ABS competencies in provider States, confidentiality issues on the user side), but also challenging in terms of capacity needs (resources for ABS implementation are already very scarce, a situation which will be further worsened through additional monitoring and reporting obligations).

Such complexity and challenges are further aggravated by a continuous lack of more holistic ABS approaches at the international as well as national levels. Monetary benefits are still distinguished from non-monetary benefits although the latter require investment and therefore bring indirect monetary benefits as well. Both types of benefits are also linked since non-monetary benefits build the basis for future international R&D collaboration and thus the generation of benefits by provider State actors themselves. The role of ABS related R&D in and its importance for conservation and sustainable use are not sufficiently acknowledged either. As a result, the issues are still too often approached separately instead of better linking and making them part and parcel of integrated R&D, economic and environmental strategies.

Similarly, it is far from clear that the implementation of the Nagoya Protocol and the EU ABS Regulation has facilitated R&D activities in the EU or at least created legal clarity and thereby certainty. Feedback received by BfN in various exchanges with both industry and public research seems to indicate the opposite. Indeed, the EU due diligence system proves to be challenging for all sides with not only users who have difficulties in analyzing whether they fall within the scope of the EU ABS Regulation, but also compliance authorities of EU Member States who lack concrete data on the exact R&D activities undertaken by users. Apart from difficulties in finding out the concrete ABS situation (i.e. obligations and procedures to follow) in provider countries, the greatest challenge in determining whether the EU ABS Regulation applies or not relates to its small-sized, highly complicated distinction between utilization activities on the one hand and activities which are pre- or post-utilization on the other hand. Implementation of the EU due diligence system in practice demands detailed and time-consuming investigations by all actors and therefore goes beyond a "low cost" documentation verification exercise.

Needless to say, CNAs in provider States also struggle with understanding the EU compliance system and consequently with building trust. The perception seems to prevail that the scope of the EU ABS Regulation offers too many loopholes and thus limits the effectiveness of the EU compliance system. Trust relationships between provider states, users and compliance authorities are still heavily loaded which is indicated by the often very slow responsiveness (sometimes even low response rate) of NFPs and/or CNAs in many provider states. Cumbersome and lengthy ABS permitting processes in provider states are related to limited institutional capacities of NFPs and CNAs, but also to complex and often over-protective ABS legal frameworks. Slow governmental progress in developing or updating ABS instruments contributes to further legal uncertainty. Limited understanding of many provider states is also detected in relation to the international compliance system established by the Nagoya Protocol, in particular the importance of the ABS

Clearing-House, its critical role in informing users on ABS obligations and procedures, the added value of IRCCs and checkpoint communiqués, or even the need for and benefits of international collaboration between CNAs on the provider and user side.

Despite such criticism and the evident deficiencies in the ABS systems on both EU and provider states' sides, it also has to be acknowledged that while the Nagoya Protocol and the EU ABS Regulation have already been in force for five years, they are still fairly new legal instruments which require more time and implementation experiences to reach their full operationalization. First experiences gained in Germany have revealed a number of challenges and open questions which are not only due to the novelty of the EU Regulation but also due to the extraordinary subject matter that it deals with. ABS and particularly ABS monitoring is marked by several complexities: transboundary situations and the objective to uphold third country legislation; very diverse sectors involved in highly innovative research and development activities; complicated and not always transparent supply chains; etc.

Nevertheless, some positive developments triggered by the start of the EU ABS Regulation implementation can be noted already at this stage. For example, suppliers of genetic resources seem to face increasing pressure by institutions that aim to use their resources for R&D activities and therefore require ABS transparency. Within companies and research institutions ABS and due diligence frameworks seem to be developing. Clarifying internal ABS responsibilities or developing policies and operation procedures are initial but very important steps in this context. Furthermore, the detection of first non-compliance cases will hopefully send a signal to users from all sectors that due diligence is not implemented by institutional policies alone but requires greater efforts in practice.

Finally, it can be observed that ABS awareness is increasing which is the key prerequisite for ABS due diligence and Nagoya Protocol compliance in the future.

Acknowledgements The authors would like to acknowledge the great efforts and crucial contributions made by Klara Nestvogel and Sebastian Jank, both Nagoya Protocol Compliance Officers at the German Federal Agency for Nature Conservation, in implementing the first compliance checks under the EU ABS Regulation in Germany.

References

Articles

Neumann D (2018) Application for acknowledgement of best practice under Article 8 of Regulation (EU) No 511/2014: The CETAF Code of Conduct & best practice. In: Feit U, Greiber T, Karger E (eds) Second Meeting of the European competent national authorities implementing the Nagoya Protocol and the corresponding EU Regulation - Final Report, BfN-Skripten 515

Studies, Reports and Presentations of State Officials

Secretariat of the Convention on Biological Diversity (2019) Report on the Global Capacity-building Workshop on Monitoring the Utilization of Genetic Resources under the Nagoya Protocol, Bonn, 30 September – 2 October 2019. CBD/NP/CB/WS/2019/1/2, https://www.cbd.int/meetings/NP-CB-WS-2019-01 (accessed 14 September 2020)

Recommendation adopted by the Open-Ended Working Group on the Post-2020 Global Biodiversity Framework, CBD/WG2020/REC/2/1, Preparation of the post-2020 global biodiversity framework: conclusions of the Work Group at its second meeting (accessed 14 September 2020)

Interim national report on the implementation of the Nagoya Protocol from the European Union, 30 October 2017., https://absch.cbd.int/pdf/documents/absNationalReport/ABSCH-NR-EU-238628/2 (accessed 14 September 2020)

First Interim national report on the implementation of the Nagoya Protocol from Germany, 27 October 2017., https://absch.cbd.int/pdf/documents/absNationalReport/ABSCH-NR-DE-238614/2 (accessed 14 September 2020)

Legal Texts

Regulation (EU) No 511/2014 of the European Parliament and of the Council on compliance measures for users from the Nagoya Protocol on Access to Genetic Resources and the Fair and Equitable Sharing of Benefits Arising from their Utilization in the Union. OJ L 150, 20.5.2014

Commission Implementing Regulation (EU) 2015/1866 of 13 Oct 2015 laying down detailed rules for the implementation of Regulation (EU) No 511/2014 of the European Parliament and of the Council as regards the register of collections, monitoring user compliance and best practices. OJ L 275/4, 20.10.2015

General rule on submitting a due diligence declaration according to Art. 7 para. 1 of Reg. (EU) No. 511/2014 (Allgemeinverfügung über die Abgabe der Sorgfaltserklärung nach Artikel 7 Absatz 1 der Verordnung (EU) Nr. 511/2014) from 19 April 2018, Notice of the Bundesamt für Naturschutz in the Federal Gazette from 9 May 2018

German Act Implementing the Obligations under the Nagoya Protocol and Transposing Regulation (EU) No 511/2014, Federal Law Gazette 2015, part I No 47, published on 2nd December 2015

Regulation (EU) No 1215/2012 of the European Parliament and of the Council of 12 December 2012 on jurisdiction and the recognition and enforcement of judgments in civil and commercial matters regarding international jurisdiction. OJ L 351, 20.12.2012

Thomas Greiber is Head of Division I 1.5 "Competent National Authority for the Nagoya Protocol", German Federal Agency for Nature Conservation. thomas.greiber@bfn.de.

Ellen Frederichs is Deputy Head of Division I 1.5 "Competent National Authority for the Nagoya Protocol", German Federal Agency for Nature Conservation. ellen.frederichs@bfn.de.

Due Diligence and the Regulation of Transnational Economic Activity: Regulation (EU) No 511/2014 Compared to Other EU Due Diligence Schemes

Christine Godt and Markus Burchardi

Abstract The Due Diligence obligation is a central component of the ABS Regulation No. 511/2014, which transposes the Nagoya Protocol into European Union law. This chapter explores the concept of 'due diligence' and the particular ways in which it got implemented by the Regulation. Built on the idea of a 'self-standing duty', the regulation absorbed the regulatory spirit of neighboring regulations of autonomous EU law. It is against this background, that Reg. 511/2014 has to be interpreted. We argue that the transposition into binding EU law transforms the standard of care into an objective one, leaning towards the standard established by due diligence in international business law ('what ought to be done needs to be done'). Concurrently, the subjective dimension is strict ('what the user ought to know') and correlates with the informational infrastructure stipulated by the respective due diligence regime. As far as the allocation of responsibilities between private and public actors is concerned, we identify Reg. 511/2014 as a rather peculiar due diligence regime compared to other EU Regulations.

Keywords Due diligence · Self-regulation · Risk management · Intermediaries · Compliance

C. Godt (✉) · M. Burchardi
Carl von Ossietzky University of Oldenburg, Faculty II, Institute for Legal Studies, Oldenburg, Germany
e-mail: christine.godt@uni-oldenburg.de; markus.burchardi@uni-oldenburg.de

E. Chege Kamau (ed.), *Global Transformations in the Use of Biodiversity for Research and Development*, Ius Gentium: Comparative Perspectives on Law and Justice 95, https://doi.org/10.1007/978-3-030-88711-7_20

1 Introduction

The European Union's (EU) implementation of the Nagoya Protocol rests on the concept of 'Due Diligence' (DD). The central article 4.1 Regulation (EU) No. 511/2014[1] (ABS-Regulation) stipulates:

> Users shall exercise **due diligence** to ascertain that genetic resources and traditional knowledge associated with genetic resources which they utilize have been accessed in accordance with applicable access and benefit-sharing legislation or regulatory requirements, and that benefits are fairly and equitably shared upon mutually agreed terms, in accordance with any applicable legislation or regulatory requirements.

In contrast to early thoughts, which envisioned a reciprocal linkage of provider and user state regulation by allowing a direct enforcement of provider State rules by user state countries,[2] the EU Regulation stipulates a duty of care regarding lawful access as an EU *sui generis* obligation. This is not evident, since modern law encounters various forms of 'import' of foreign and 'export' of domestic law.[3] Yet, any enforcement of 'foreign' law in the realm of administrative law was felt as an encroachment on the territoriality principle, restricting user states' sovereignty. Therefore, the European legislator of Reg. 511/2014 opted for a scheme, which we call 'regulatory hinge joint'. The term describes a regulatory program, which allows for a *translation* of provider state regulation into a self-standing rule under user state law, thus mitigating potential territoriality concerns. Beyond that, member States remain free to implement their own access regulations.[4]

At the time when the draft of Reg. 511/2014 was being deliberated in the European Parliament, the concept of DD was broadly discussed in UN-fora,[5] the

---

[1] Regulation (EU) No 511/2014 of the European Parliament and of the Council on compliance measures for users from the Nagoya Protocol on Access to Genetic Resources and the Fair and Equitable Sharing of Benefits Arising from their Utilization in the Union (2014) *Official Journal L150, p. 59.*

[2] Barber et al. (2003). This is standard practice in private international law, see Kegel and Schurig (2004), pp. 135ff. For the CBD, it was conceptualized under the principle of 'common but differentiated duties' as acknowledged in International Environmental Law, and implemented in the UNFCCC, see Godt (2003).

[3] Analyzed from the angle of both, public and private law. Public lawyers stress the "perforated sovereignty", e.g. Meyer (2018), pp. 66, 98 (analyzing standards and enforcement in transnational meat markets); Private lawyers focus on market regulation, e.g. Joerges (2013), p. 409 (proposing a concept of a "three dimensional conflict of laws").

[4] Some have and some have not. As of August 2020, only five countries have installed member state access regulations: France, Spain, Bulgaria Malta and Croatia, for further details see Winter in this volume.

[5] After John Ruggie as Special Representative had delivered his "Framework"-Report to the Secretary General in 2008, the UN Human Rights Council adopted the UN Guiding Principles on Business and Human Rights, UN Doc. A/HRC/17/31 (21 March 2011), on the relationship of DD-considerations and UN-principles on businesses and HR, see ILA Study Group on DD (2016), p. 28.

Organization for Economic Cooperation and Development (OECD)[6] and in academia.[7] At the same time, the EU Timber Reg. 995/2010,[8] which explicitly embraces a DD-system, was on its way to practical implementation.[9] Before that, DD had already become a well-established concept in business administration, routinely used in mergers and acquisitions,[10] and evolved into a tool to combine accounting rules and social responsibility (Directive 2014/95/EU[11]). The concept expanded, putting 'risk management' for all kinds of policy goals center stage. Today, various regulations subscribe to DD as architectural design (below, Sect. 3.3); covering a wide range of subjects. Reg. 2015/757,[12] for example, regulates carbon emissions from large vessels. The General Data Protection Regulation (Reg. 2016/679)[13] addresses the processing of personal data. And Reg. 2017/821 is about minerals and metals originating from conflict areas.[14]

The following analysis is interested in the precise content of the DD obligations contained in article 4.1 and article 4.5 Reg. 511/2014: What exactly do they require? The fulfillment of said DD obligations is to be declared when research funding is granted and when a product has reached the marketing stage ('monitoring'), and will be enforced through administrative supervision ('checks'). Considering that DD as a regulatory instrument is quite new, it is most likely that it will be the Court of Justice of the European Union (CJEU) that will determine the standard of care under article 4 Reg. 511/2014. Therefore, we ask: To what extent does article 4 Reg. 511/2014 borrow from earlier conceptualizations of DD in public international law and

---

[6] OECD Guidelines for Multilateral Enterprises, adapted on 25 May 2011 by OECD members as well as Argentina, Brazil, Egypt, Latvia, Lithuania, Morocco, Peru and Romania, http://oecd.org/daf/inv/mne/48004323.pdf (last visited 14.3.2019).

[7] At that time, the International Law Association (ILA) had already installed a 'Study Group on Due Diligence in International Law' (ILA Study Group on DD). The study group issued two reports, the first on 7 March 2014, the second in July 2016.

[8] Reg. (EU) No 995/2010 (2010) *Official Journal* L295, p. 23 (applicable since 03.03. 2013); complemented by Commission Implementing Reg. (EU) No. 607/2012 (2012) *Official Journal* L177, p. 16 and Commission Delegated Reg. (EU) No 363/2012 (2012) *Official Journal* L115, p. 12.

[9] Reg. 995/2010 is also referred to by the European Commission in its Guidance Document (2016) *Official Journal* C313/1 [hereinafter: EU Commission (2016)] at p. 11.

[10] Ibid.

[11] Dir. 2014/95/EU (2014) *Official Journal* L 330, p. 1 lays down the rules on disclosure of non-financial and diversity information by large companies, amending the general accounting Dir. 2013/34/EU (2013) *Official Journal* L182, p. 19. It requires companies to publish reports on their policies on environmental protection, social responsibility, treatment of employees, human rights, anti-corruption and diversity.

[12] Reg. (EU) 2015/757 of the EP and the Council on the monitoring, reporting and verification of carbon dioxide emissions from maritime transport (2015) *Official Journal* L123, p. 55, also abbreviated 'MRV-Regulation'.

[13] Reg. (EU) 2016/679 (2016) *Official Journal* L119, p. 1.

[14] Reg. (EU) 2017/821(2017) *Official Journal* L130, p. 1; complemented by Delegated Reg. 2019/429 (2019) *Official Journal* L75, p. 59.

international business law (Sect. 2)? Does the implementation in directly enforceable EU law give rise to an *autonomous* EU concept of DD, distinct from its predecessors? In order to identify the structures of such a unique European concept, we compare five European regulations which arguably rest on a DD scheme (Sect. 3). Subsequently, we scrutinize the DD scheme of Reg. 511/2014 in more detail (Sect. 4). At the end, we draw conclusions from the preceding inquiries (Sect. 5).

2 Conceptual Background of Due Diligence

The preconditions of what could be meant by 'Due Diligence' (DD), in fact, derive from two conceptual parents, namely DD in public international law (Sect. 2.1) and DD in international business law (Sect. 2.2). This legacy begs the question as to how to interpret DD regulations stipulated under EU law (Sect. 2.3).

2.1 *Public International Law*

In public international law, DD is conceived as a concept to mediate interstate relations when there is significant change. It has been traced back to Grotius.[15] In this context, it defines the yardstick against which to assess, whether a responsible state has complied with "certain obligations and standards".[16] Hence, also in international law DD refers to conduct.[17] It grew out of the debate on state responsibility.[18] The context is harm prevention and the addressee is the state deemed to be accountable for actual or future harm. The yardstick refers to traditional categories of human responsibility for harm, fault and reasonable care.[19] The reason is that in

[15]Kulesza (2016), p. 3. According to Freeman, as noted in his 1955 lecture at the Hague Academy of International Law, due diligence requires "nothing more nor less than the reasonable measures of prevention which a well-administered government could be expected to exercise under similar circumstances" (cited after ILA Study Group on DD, 2014, p. 3).

[16]Koivurova (2010).

[17]French, cited by ILA Study Group on DD (2016), p. 1.

[18]The ILC engaged in codifying the Law of state responsibility since the 1950, first focusing on wrongful acts, since the 1970s opening a second track on injurious consequences of acts not prohibited by international law. In 1996 draft Articles and Commentary dealing with prevention, co-operation, and strict liability for damages were proposed. In 2001, the ILC (International Law Commission) submitted Draft Articles on Prevention of Transboundary Harm from Hazardous Activities (the 'Prevention Articles'), *Int'l Law Comm'n, Rep. to the General Assembly on the Work of Its Fifty-third Session, U.N. Doc. A/CN.4/SER.A/2001/Add.1 (Part 2)*, and in 2006 Principles on the Allocation of Loss in case of Transboundary Harm Arising Out of Hazardous Activities ('Loss Allocation Principles'), *Int'l Law Comm'n Rep. to the G.A, U.N. Doc. A/CN.4/SER.A/2006/Add.1*; ILA Study Group on DD (2016), p. 24; in depth: Kulesza (2016), p. 115ff.

[19]ILA Study Group on DD (2016), p. 2.

international law causal (non-fault) responsibility was not embraced as a legal concept for accountability.[20] While few ‘certain obligations and standards’ exist, the key point of reference is the ’do-no-harm-principle’, which is acknowledged as a principle of customary international environmental law.[21] The required standard of care is to be determined based on the circumstances of the individual case.[22] Because of this flexibility, the level of care varies. While it is questionable to what extent subjective conditions like the economic stage of development, effective control (over a territory) and technological expertise are to be taken into account,[23] the specific duties of a given country depend on a balance test, which takes due regard of the level of effort required (reasonableness[24]) and the degree of harm (as a correlation between probability and grandeur of damage).[25] Scholars have described the DD principle as an instrument for transforming the duty ‘not to do harm’ into an obligation to install procedures which prevent significant transboundary damage.[26] This line of reasoning is explicit in cases involving environmental impact assessments.[27]

Since the beginning of the twenty-first century, international law has been moving towards binding multinational companies into the fabric of global governance and accountability. By and large, the Ruggie Report (2008[28] and 2011[29]) was seen as a turning point in this regard.[30] Earlier, the International Law Commission’s (ILC) Principles of Loss Allocation of 2006 had already referred to the responsibility of private actors, resulting in a shared responsibility of industry (prevent harm, bear the costs, provide third party cost coverage by insurance or pooling) *and* states (to regulate).[31] Yet, with international legal subjectivity of multinational companies

---

[20] Ibid.

[21] Takano (2018), p. 2.

[22] ILA Study Group on DD (2014), p. 2; ILA Study Group on DD (2016), p. 15.

[23] ILA Study Group on DD (2016), p. 11 (on control) and pp. 14–16 (on economic differentiation in various contexts).

[24] “The golden thread”, ILA Study Group on DD (2016), p. 8.

[25] ILA Study Group on DD (2016), pp. 12, 21, 23.

[26] Dupuy and Viñuales (2015), p. 60. McIntyre (2011), pp. 136, 143 clarifies that the International Court of Justice in the *Pulp Mills* case (Argentina v. Uruguay, 20.4.2010) as the respective landmark decision conceptualized procedural rules as preventive, but substantial rules as separate obligations, yet complementary. The Seabed Disputes Chamber of the International Tribunal for the Law of the Sea in its Advisory Opinion “Responsibilities and Obligations of States Sponsoring Persons and Entities with Resprect to Activities in the Area” (case no. 17, 1.2.2011), para. 120 concurs and suggests that the DD obligation is a conceptual bridge between harm prevention and precaution, cautioning: Brunnée (2016), p. 7.

[27] Bremer (2017).

[28] Known as “Framework for Business and Human Rights” (2008), published in *UN Doc. A/HRC/8/ 5*, p. 1.

[29] Known as “Guiding Principles on Business and Human Rights”, above n. 5.

[30] Lambooy (2010), p. 429.

[31] ILA Study Group on DD (2016), p. 26. An even earlier precedent is the IMO-created fund for oil spills.

remaining an open question, responsibilities have evolved on various levels, eventually crystallizing in statutory requirements and private liability rules in state law.

Thus, in international law, DD has become a bifurcated concept. Primarily interstate, it defines the duty of states to anticipate harm and install a legal regime for prevention, sic. regulation. Secondly and independent of state obligations, DD refers to the (general) duty of industry to respect human rights.[32] The duty as such is not conditioned by size and not restricted to the confines of one undertaking. The central idea is that DD relates to the supply chain. It therefore became a corner stone principle of Corporate Social Responsibility. With regard to the scope of accountability, Principle 19 (b) (ii) of the UN Guiding Principles on Business and Human Rights (2011) installs "leverage" as a central term, defined as the "ability to effect change in the wrongful practices of an entity that causes harm".[33] The OECD-Guidelines explicitly refer to a business responsibility to "seek ways to prevent or mitigate adverse human rights impacts [. . .] even if they do not contribute to those impacts".[34] Thus, under international law, DD obligations with regard to third parties' actions appear to be similar for states and businesses.[35] With regard to the specific content, the duty is flexibly conditioned by subjective factors such as resources, time and size, and objectively by the nature and context of the operations, as well as the likelihood and severity of the risk.[36] Yet, the principle of DD does not define the substance/the result/the outcome—neither for states, nor for multinationals. It is a process-oriented, open-ended standard. Kulesza stresses that the principle of DD is not binding law.[37] It is a normative instrument to determine the appropriateness of state regulation (respectively its absence) *ex post*, thus 'good governance'.

[32] ILA Study Group on DD (2016), p. 29 recognizes that the terminology 'due diligence' was deliberately referenced to both, human rights law and business management practice, indicating this as "cleverly". Yet, the report cautions as to the misleading potential. With regard to the intention of finding a language familiar for both, human rights and the business community, see also Martin-Ortega (2014), p. 50.

[33] Wood (2012), p. 63 at 65 argues that leverage applies also to businesses which have not contributed to a harm and coin this type of accountability "leverage-based negative responsibility".

[34] Sec IV OECD Guidelines on Multinational Enterprises (2011 version), p. 31, para 3; p. 33, para 43.

[35] ILA Study Group on DD (2016), p. 32.

[36] ILA Study Group on DD (2016), pp. 34–35—(at p. 34 reminding that Guiding Principle 17 explains that business enterprises conducting appropriate human rights due diligence "should not assume that, by itself, this will automatically and fully absolve them from liability for causing or contributing to human rights abuses").

[37] Kulesza (2016), pp. 115ff.

2.2 International Business Law

In International business law, DD has been a long-standing practice, going back to US regulation of financial markets after the Great Depression of the 1930s.[38] It was a means to reduce liability. Over time, it expanded to include any investigation relating to the acquisition of companies or assets in the commercial context, risk analysis in finance, and pre-contractual inquiries in general.[39] DD became an ubiquitous and globally applied (not limited to the US) business management tool that can be equated with the defense requirement in article 38 United Nations Convention on Contracts for the International Sale of Goods (CISG) "to give timely notice".[40] Today, the standard example of DD in various industries is the process by which a potential acquirer evaluates a target company or its assets for acquisition.[41] More recently, it was integrated into corporate sentencing laws.[42] The original business theory behind DD holds that performing this type of investigation contributes significantly to informed decision making by: (a) enhancing the amount and quality of information available to decision makers and (b) ensuring that this information is systematically used to deliberate in a reflexive manner on the decision at hand and all its costs, benefits, and risks.[43] The legal rationale is driven by the 'caveat emptor principle' (buyer beware), and thus by the risk of the purchaser to lose the right to claim contractual remedies if the defect was evident to him/her at the time the contract was performed. In other words, the seller will only be strictly liable for so-called hidden defects. Again, the question of how much effort is reasonable depends on the circumstances.[44] The notion that the seller will only be liable for hidden defects stands in opposition to both, the fault-based causation principle, which underlies DD in public international law, and the contractual liability under

[38] § 11 US Securities Act of 1933, see Bainbridge and Anabtawi (2017), p. 255. Acknowledging these US roots: DiMatteo (2009), p. 292; the same is reported in the 'human rights' literature, see: Martin-Ortega (2014) at p. 51, referring to Spedding (2005).

[39] For structuring the process of the various types of DD in the acquisition process see Bainbridge and Anabtawi (2017), pp. 255–263.

[40] Art. 38 sec. 1 CISG reads: "The buyer must examine the goods, or cause them to be examined, within as short a period as is practicable in the circumstances." It might be worth mentioning that equivalent norms do exist in national law, e.g. in Germany and Austria. However, in these two countries this rule is limited to contracts between two professional traders. On the related CISG-case law, see Schwentzer (2019), para. 6.

[41] Hoskisson et al. (2004), p. 251.

[42] In these concepts, due diligence serves as a defense against liability (examples: sec. 90.1 UK Financial Services and Markets Act 2000, sec. 7.2 UK Bribery Act 2010, sec.11 (b) (3) (A) US Securities Act 1933) or to reduce a sentence (example: § 8 B2.1. US Sentencing Commission's Federal Sentencing Guidelines Manual (2018), available at https://www.ussc.gov/sites/default/files/pdf/guidelines-manual/2018/CHAPTER_8.pdf, last visited 15.3.2019).

[43] Chapman (2006).

[44] Literature conceptualizes four stages of 'effort standards' on a continuum ranging from least to greatest efforts: "commercially reasonable efforts", reasonable efforts", "all reasonable efforts"/ "reasonable best efforts" to "best efforts", Bainbridge and Anabtawi (2017), p. 273.

general national laws.[45] Additionally—in contrast to most national systems of contractual liability—CISG liability is always limited in scope to 'direct' damages: Pure economic damages are not covered.[46]

To summarize, DD in the corporate world is qualified by the following characteristics: As a matter of principle, the exercise is undertaken in the firm's own interest and—in most cases—voluntarily.[47] The procedure is understood as multidisciplinary, not driven by law. It is structured as a risk assessment exercise and as a 'two party undertaking'. Both parties first agree on the scope, the timeline and the milestones. This approach highlights the nature of DD as industry self-regulation. The selling party is expected to submit documents, which the purchasing party then examines. A professional and strict 'out to know' standard applies; the addressee of this standard is the buyer, not the party which sets the risk in the first place. Still, it should be born in mind that in international business law, DD duties may also rest with the economic actor that 'sets the risk' in the first place (supportive/ residual duties). A seller, for example, needs to make sure that all pertinent information is disclosed to the potential buyer in the documents he or she submits.[48]

2.3 *EU Corporate Social Responsibility (CSR) Regulation*

It is questionable whether the European legislator took one of the above predecessors as a clear model when drafting article 4 Reg. 511/2014. While legislative material suggests that the European Parliament was inspired by the debate in public international law,[49] the European Commission seems to have been influenced more by the debate in international business law.[50] The EU Commission explains DD in its Communication of 2016[51] as follows:

- Due diligence refers to the judgment and decisions that can reasonably be expected from a person or entity in a given situation. It is about gathering and using information in a systematic way. As such it is not intended to guarantee a certain outcome or aiming at perfection, but it calls for thoroughness and best possible efforts.

---

[45] Just as example, §§ 437, 280, 281 283, 311a German Civil Code (BGB).

[46] In a previous legal analysis (Godt and Burchardi 2018, p. 60), we inferred that a DD liability in Art. 4.5 Reg. 511/2014, which would encompass the liability for registered collections for the stop of downstream utilization, is not in line with the origins of due diligence liability.

[47] An element stressed by McCorquodale (2009), p. 392.

[48] Bell (2001), pp. 125f.

[49] In a resolution of the EU Parliament made in 2010 in context of CSR in international trade agreements, the model of public international law is clearly articulated, cp Resolution P7_TA(2010) 0446, p. 11 Sec. 26 lit. e.

[50] In document (COM SWD 2012 0292, p. 31), the Commission explicitly refers to an ICC document, which takes reference to existing DD processes inside firms.

[51] EU Commission (2016), p. 10.

- Due diligence goes beyond the mere adoption of rules and measures; it also entails paying attention to their application and enforcement. Inexperience and lack of time have been held by the courts not to be adequate defenses.
- Due diligence should be adapted to the circumstances — e.g. greater care should be applied in riskier activities, and new knowledge or technologies may require adaptation of previous practices.

This language resonates more with DD in business administration than with DD in public international law. Article 4 Reg. 511/2014 stipulates the following duties: "to ascertain", "to keep, seek, and transfer", "to obtain" and "to establish" (sic. "produce compliance") "or discontinue". These duties only make sense against the background of a standard situation in which a firm did not originally access the resource but purchased it further down the chain. However, in contrast to what the business community might have wished for (and the Commission Notice insinuates[52]), article 4 Reg. 511/2014 does not primarily stipulate a purchaser's defense (being fulfilled in the party's own interest), an 'estoppel'. Instead, it stipulates a legal duty—not vis-à-vis the provider state directly, but vis-à-vis the state of utilization.[53] Yet, what does this imply? When analyzing the adjudication of the International Court of Justice (ICJ) with regard to the 'no-harm-obligation', Brunnée found that at the level of substantive law, the court distinguishes between a "duty to take diligent steps to *prevent* harm"—which it then treats as a separate procedural obligation—and the "duty to take diligent steps not to *cause* harm", which the judges consider to be a duty that cannot be violated simply by a failure to act diligently.[54] She cautions that this conceptualization contradicts the take of international environmental law scholars on the concept. The majority considers violations of procedural obligations to be in breach of the duty to prevent harm, regardless of whether harm occurred or not. If one mirrors this juxtaposition with the first part of article 4.5 Reg. 511/2014 (duty *to produce* compliance with provider states laws) and the last part ("or discontinue utilization"), one might speculate whether the CJEU will follow the ICJ's interpretation and interpret the "*or*" of the last part of the sentence as equally procedural in nature.[55] This interpretation will be discussed in more depth below under Sect. 4.1.

Moreover, at the time when the ABS-regulation was negotiated, other EU DD Regulations on the management of supply chains were already being deliberated. They focus more on DD as a business management tool than on either of the

---

[52]EU Commission (2016), p. 10: "If a user — no matter at which step in the value chain — takes reasonable measures in seeking, keeping, transferring and analyzing information, the user will be compliant with the due diligence obligation under the EU ABS Regulation. This way the user should also avoid liability vis-à-vis subsequent users, although this aspect is not regulated by the EU ABS Regulation." To be clear: The Commission speaks about the contractual liability towards the purchaser, not the statutory obligation under Art. 4.5 Reg. 511/2014.

[53]A proposed 'triangular duty to comply' (a diagonal duty to the provider state, enforced vertically by the user state) was rejected in the legislative process. Instead, Art. 4 Reg. 511/2014 installs a vertical (user state) 'duty to *produce* (diagonal) compliance' vis-à-vis the provider state.

[54]Brunnée (2016), p. 7.

[55]In contrast to respecting the residual substance (sic. 'compliance with provider state laws').

described precedents in public international law and international business law. Their focus is on 'risk management'. It is therefore necessary to inquire which rationale these 'parallel' Regulations pursue and whether they implement elements that characterize them as an autonomous EU concept of DD. We focus on the following questions: How is the risk management process structured? In what way do intermediaries (such as registered collections or brokers) contribute to self-governance? How is the flow of information organized in the various EU DD schemes?

The observation that context matters more than precedents would be in line with modern legal system theory, which holds that 'legal transplants' will not simply be adopted.[56] Inversely, it is the legal environment, which will transform the original instrument so that a peculiar novelty will emerge.[57] While literature has emphasized that the decisive transforming contextual forces are legal cultures, power structures, and actors with specific interests,[58] we submit that for DD, the decisive factor is the legal environment, here EU law (below, Sect. 3). Therefore, our observations would better match with what Gunter Teubner more modestly called (again with different connotations) "legal irritant".[59]

3 EU Schemes of Regulatory Due Diligence (DD)

Before comparing five selected EU DD regimes (Sect. 3.3), we must first introduce the relevant elements and actors of EU DD (Sect. 3.1) as well as the EU regulatory environment into which the concept of due diligence was transplanted (Sect. 3.2).

3.1 Elements and Actors of EU Due Diligence

The terms 'due diligence' and 'risk management' are often equated with one another. As a matter of EU regulation however, DD is the overarching concept whereas risk management is one of its central components (see Fig. 1). Consequentially, a risk management exercise can be found in all EU DD schemes. The standard configuration consists of three interrelated steps:

[56]Legrand (1997) thereby rejects the notion of Watson (1974) to whom the tern 'legal transplants' is attributed.

[57]Watson (2000) responded to Legrand's critique that both approaches to legal transplants are not very far removed from each other. More recent: Grozev (2012); Short (2016).

[58]Winter and Kalichava (2019), p. 41; Siems (2014), pp. 139–145 identified five categories, which influence the implementation of the transplant in the novel environment; on legal transplants in the modern context of private international law: Siems (2018), pp. 236–238, and pp. 256–260.

[59]Teubner (1998), p. 11; similar: Chen-Wishart (2013), p. 1.

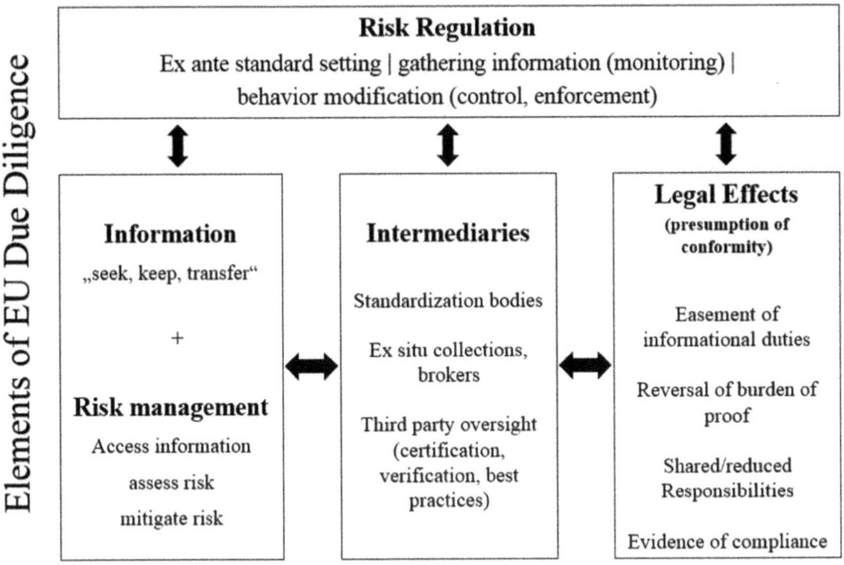

Fig. 1 Elements of EU due diligence. **Source:** Own illustration

(1) Access relevant information and document it,
(2) analyze the information, perform a risk assessment, and
(3) if necessary, use the available options of risk mitigation.

Yet, the structure and detailedness of this exercise varies considerably across regimes (below, Sect. 3.3). Most likely being the result of trade-offs in the EU lawmaking process, this divergence has a direct impact on the extent of user compliance measures (below, Sect. 4.4). In the system of EU DD, risk management is complemented by risk regulation. Risk regulation can also be understood as a three-pronged approach:[60]

(1) Ex ante standard setting
(2) Gathering of information (monitoring)
(3) Behavior modification (controls, enforcement)

As a business management tool, risk management is usually performed by private entities. Risk regulation on the other hand, is usually attributed to the state. However, as operational programs, risk management and risk regulation are open to both. The regulator, for example, may take a risk management approach when carrying out checks on target compliance via national competent authorities.[61] And regulatory targets (e.g. importers, manufacturers) may engage in risk regulation as 'classical' business self-regulation.

[60] Hood et al. (2001), p. 21.

[61] See e.g. Art. 9 Sec. 3 lit. a Reg. 511/2014.

Information fuels both risk management and risk regulation.[62] All EU DD regimes oblige the addressees of regulation to install internal systems and processes that help to seek, keep and transfer information. The regulatory goal of DD is essentially understood to be information management ("it is about gathering and using information in a systematic way").[63] In risk management, information is transferred between (economic) actors along a (supply-) chain pursuant to the requirements of internal information policies. At times, this may include disclosure vis-à-vis the public for the sake of accountability. When information is transferred to national regulators instead, it serves as valuable feedback about the effectiveness of regulation (=regulatory information). Information regarding transnational supply chains and economic conduct abroad is of particular importance, since it is not immediately accessible for either the regulator or consumers.[64]

With regard to transnational economic activity, the capacities of a state to regulate—i.e. formulating, implementing and enforcing rules—are typically cut-short by territorial borders.[65] The regulator may therefore decide to enlist the help of knowledgeable third parties—intermediaries—[66] in order to overcome some of those limitations.[67] Because of its dynamic, informational and situational nature, the concept of DD is well suited to accommodate specialized intermediaries. EU DD is no exception here—the regimes we analyzed all include intermediaries in one form or another. The reasons for doing so can be found in the operational capacity, expertise, independence, and legitimacy that some intermediaries possess.[68] Intermediaries, too, are accessible to both the regulator and the targets of regulation. As service providers, they may help firms to comply with their DD requirements, typically by providing third-party certification or verification. They may also assist the regulator in achieving its regulatory goals, for example by monitoring target behavior. When intermediaries are "orchestrated"[69] to work "directly or indirectly in

[62] Gellert (2015), p. 9.

[63] *Inter alia*, see Art. 5.2 Implementing Reg. 607/2012: "In applying their due diligence system operators shall be able to demonstrate how the information gathered was checked against the risk criteria provided for in Art. 6.1.b. of Reg. 995/2010, how a decision on risk mitigation measures was taken and how the operator determined the degree of risk."

[64] According to Haufler, mechanisms revolving around "information politics" have come to dominate transnational business regulation, Haufler (2018), p. 116.

[65] The disparities between global production and national regulation demarcate a "critical governance gap", which many scholars regard as the main driver behind the promulgation of corporate social responsibility and industry self-regulation, see Haufler (2018).

[66] *Inter-* means "between, among"; therefore, an intermediary denotes "someone who moves back and forth between two sides – a 'go-between'," https://www.merriam-webster.com/dictionary/intermediary (last visited: 15.04.2019).

[67] Abbott et al. (2017), p. 18. This way, EU public actors "increase their capacity to regulate global value chains", Partiti (2019), pp. 95, 105.

[68] Abbott et al. (2017), p. 20.

[69] Orchestration describes a form of public action that "mobilizes the voluntary participation of intermediaries in the regulatory process in order to address a target in the pursuit of public goals", Partiti (2019), p. 95f.

conjunction with a regulator to affect the behavior of a target", they become "regulatory intermediaries".[70] In its most dynamic form, this relationship can be expressed in the formula $\underbrace{R \leftrightarrow I \leftrightarrow T}$.[71] The arrows and the bracket denote the multidirectional flow of regulatory information[72] in what can be described as the 'best case scenario'.[73] By contrast, classical principal-agent relationships, where "regulation is prescriptive and compliance is based primarily on deterrence and sanctions" are marked by unidirectional flows of regulatory information.[74] While the EU DD schemes we analyzed by and large follow the dynamic RIT-model described above, not all the intermediaries contained therein are 'proper' regulatory intermediaries. This may indicate structural shortcomings in certain regimes (below, Sect. 3.4 f.).

Finally, yet importantly, it must be borne in mind that all three actors—regulators, intermediaries and targets—pursue their own agenda and that divergence of goals may render some intermediaries a less perfect choice than others in certain regulatory settings.[75] Intermediary 'capture' by either the regulator or the target[76] may occur as well. This would likely reduce the effectiveness of any DD scheme. Independence and impartiality of the intermediary are therefore basic requirements for proper decision-making in intermediary bodies.[77] Doubts along these lines may be counterbalanced by accreditation[78] or other forms of formal recognition.

3.2 *EU Regulatory Environment*

The fact that EU DD is broader than the ubiquitous risk management exercise can be attributed to the fact that DD as a concept fell on fruitful ground when it was transplanted into EU law. In 1985, the EU had already started to promote industry co-regulation in the area of technical harmonization in order to speed up the removal of technical barriers to trade and complete the EU Single Market before 1992.[79] A new European standardization policy was introduced, based on adherence to voluntary industry standards and conformity assessment by accredited certification bodies,

---

[70] Abbott et al. (2017), p. 19.

[71] R = regulator, I = intermediary, T = target (of regulation). Seminal: The RIT-Model by Abbott et al. (2017).

[72] Abbott et al. (2017), p. 26f.

[73] Below, Sects. 3.4, 4.4.

[74] Abbott et al. (2017), p. 17. With its procedural and open nature, due diligence is more or less the opposite.

[75] Abbott et al. (2017), p. 19f.

[76] Abbott et al. (2017), p. 28ff.

[77] Ensthaler et al. (2016), p II.

[78] Abbott et al. (2017), p. 23; Ensthaler et al. (2016), p. XVII.

[79] Lachaud (2018), p. 253.

the results of which would be mutually recognized by the EU Member States. Standard setting, i.e. norm building, was delegated to private European standardization bodies.[80] With its increased trust in business to regulate itself without compromising public health and security, the so-called "New Approach"[81] marked a paradigm shift in EU product safety law.[82] The trust afforded to industry was epitomized by the granting of presumptions of conformity. If a company used harmonized industry norms and standards or provided third party certification, then there would be a (rebuttable) presumption of conformity in favor of the company that its products comply with the basic regulatory requirements and could thus be placed on the market without any additional state approval being required.[83] Annex II of the Council Resolution of 7 May 1985 on a new approach to technical harmonization and standards, listing the "Means of proof of conformity and effects" under B. V. and VIII., read:

> Member States shall accept that the products [. . .] are considered to be in conformity [. . .] when their conformity is demonstrated by [. . .] certificates and marks of conformity issued by a third party; results of tests carried out by a third party; [. . .].

Verified compliance with the standards would not exonerate the manufacturer from its liability for defective products vis-à-vis consumers.[84] Nevertheless, the knowledge that the efforts to produce compliance were sufficient, would increase. The New Approach also defined a new role for the state. It was supposed to become a "regulatory gorilla in the closet"[85] that would largely refrain from regulation and only intervene in case a hazard relating to an identified product had surfaced. When the New Approach was revised and expanded via the "New Legislative Framework" in 2008,[86] its basic regulatory features were kept. However, the legal effects of the presumption of conformity were watered down.[87] Article 19.1 para 3 Regulation

[80] Ibid.

[81] Council Res. on a new approach to technical harmonization and standards (1985) *Official Journal* C136, p. 1.

[82] Schucht (2017), p. 46.

[83] Schucht (2017), pp. 47, 49.

[84] Council Dir. 85/374/EEC (1985) *Official Journal* (1985), L210, pp. 29–33. The European Product Liability Directive 85/374/EEC is based on causal (no-fault) liability.

[85] Verbruggen (2013). This is sometimes also referred to as the "shadow of hierarchy", Héritier and Eckert (2008).

[86] The 'New Legislative Framework' is formed by Reg. (EC) 764/2008, Reg. (EC) 765/2008 and Decision 768/2008, see Reg. (EC) No 764/2008 laying down procedures relating to the application of certain national technical rules to products lawfully marketed in another Member State and repealing Decision No 3052/95/EC *Official Journal* L218, pp. 21–29; Reg. (EC) No 765/2008 setting out the requirements for accreditation and market surveillance relating to the marketing of products and repealing Regulation (EEC) No 339/93 (2008) *Official Journal* L218, pp. 30–47; Decision No 768/2008/EC on a common framework for the marketing of products, and repealing Council Decision 93/465/EEC (2008), L218, pp. 82–128.

[87] Schucht (2017), p. 49.

765/2008 setting out the requirements for accreditation and market surveillance relating to the marketing of products stipulates:

> Where economic operators present test reports or certificates attesting conformity issued by an accredited conformity assessment body, market surveillance authorities shall take due account of such reports or certificates.

Taking due account of something (= being an indicator of proof) is not the same as accepting it as proof.[88] As Schucht reminds us, the implementation of the privileging presumption of conformity in specific harmonization legislation and in non-compliance proceedings has never really lived up to its intended role as evidence of compliance.[89] The Commission's 2002 Action Plan[90] and the 2003 Inter-Institutional Agreement on better law making[91] (both preceding the New Legislative Framework) undergird an understanding of co-regulation that is still very much top-down; something that is complementary rather than an alternative to legislation.[92] The result is "conditioned self-regulation", situated on a continuum between 'pure' self-regulation on the one hand and command and control legislation on the other.[93] That has consequences for both regulators and the regulatory targets, which are supposed to be in a position to regulate themselves effectively:

- National competent authorities may not have transparent criteria for the execution of their powers of intervention in case of an assumed breach of compliance.[94] The question becomes: *When to intervene?*
- A reliable, content-based means of providing compliance is not readily available for business.[95] Here: *When are efforts to produce compliance exhausted?*

Since manufacturers *and* market surveillance authorities could not maintain their roles as originally envisaged, both need to take risk-based approaches in their day-to-day activities as a consequence. Authorities are required to cooperate with economic operators when engaging in corrective action in order to prevent or reduce risks.[96] The idea is to afford economic operators the possibility of mitigating the risk (s) themselves.[97]

In the following section we will examine five selected EU Regulations, and re-encounter the identified elements, issues and limitations. It will become evident how the experiences made with the 'New Approach', in particular with regard to the

88 Portalier (2017), p. 9; Schucht (2017), p. 49.

89 Schucht (2017), p. 49f. See also Senden (2005), p. 11.

90 EU Commission (2002).

91 Inter-Institutional Agreement on better law-making (2003) *Official Journal* C321, p. 1.

92 Best (2003), p. 3. Senden (2005), p. 12.

93 Senden (2005), p. 12; Lachaud (2018), p. 254.

94 Schucht (2017), p. 49f.

95 Ibid.

96 Art. 19.2 second sentence Reg. 765/2008—"well hidden" and sometimes overlooked, Schucht (2015), p. 656.

97 Schucht (2015), p. 658f.

orchestration of private intermediary actors,[98] influenced the configuration of EU DD schemes decades later.

3.3 EU Due Diligence Schemes in Comparison

Despite the fact that the following five DD Regulations tackle very different subjects, some general observations can be made: The ABS Reg. 511/2014, the Timber Reg. 995/2010 and Reg. 2017/821 on conflict minerals and metals set out obligations regarding supply chains in commodities.[99] All three have in common that the term 'due diligence' literally appears in them. In contrast, Reg. 2015/757 on the monitoring of CO_2 emissions from maritime transport and the General Data Protection Reg. 2016/679 neither regulate supply chains in commodities, nor are they specifically labelled as DD regimes. Yet, the elements and actors of EU DD outlined above are all present, thus allowing for a direct comparison along a continuum of self-regulation. When juxtaposed, it becomes apparent that there are many features, which set the ABS Reg. 511/2014 apart from its relatives. Most importantly, three of the above Regulations seek to 'export' regulation, while Reg. 511/2014 and its direct conceptual precursor, the EU Timber Reg. 995/2010, aspire to 'import' regulation.[100]

3.3.1 Reg. (EU) 995/2010 Laying Down the Obligations of Operators Who Place Timber and Timber Products on the Market

Reg. 511/2014 and Reg. 995/2010 both contain a general obligation to 'exercise due diligence'. However, Reg. 995/2010 also contains a general prohibition of placing of illegally harvested timber on the internal market.[101] This prohibition was incorporated at the urging of the European Parliament,[102] which also tried to include a general prohibition of using illegally acquired genetic resources in Reg. 511/2014—but ultimately failed.[103] Thus, Reg. 995/2010 contains both procedural (DD) *and*

---

[98] Partiti (2019), p. 103f.

[99] Note: As the Guidance Document on the scope of application and core obligations of Regulation (EU) No 511/2014 makes clear, genetic resources fall outside the scope of the regulation when they are traded and exchanged as commodities and no research and development is taking place, EU Commission (2016), p. 7. With that caveat in mind, we decided to keep the term for purposes of comparability.

[100] For an overview on the global reach of EU law, see Scott (2014), who prefers to speak of 'action forcing contingent unilateralism' instead of 'exportation' (of regulation). See also Fishman and Obizdinski (2014) on the wider trade law implications of Reg. 995/2010.

[101] Art. 4.1 Reg. 995/2010; European Commission (2016).

[102] Levashova (2011), p. 294.

[103] European Parliament (2012), pp. 28, 54.

substantive (prohibition) obligations.[104] Meeting the DD requirements does not automatically shield a timber importer from subsequent findings of illegality by a national competent authority.[105] Instead, it is suggested that records, which show that a timber importer was duly diligent, may have a moderating influence on the calculation of penalties or fines.[106]

What stands out in Reg. 995/2010 is the dual system of privileged and unprivileged timber imports.[107] For imports that fall under either the CITES-regime (Convention on International Trade in Endangered Species of Wild Flora and Fauna)[108] or the FLEGT-scheme (Forest Law Enforcement, Governance and Trade Action Plan)[109]—and are thus registered during border controls—a presumption of conformity applies ("shall be considered to have been legally harvested for the purposes of this Regulation"). This so-called "green lane" exempts importers entirely from their DD obligations.[110] All other imports will trigger the general DD requirement of article 4.2 Reg. 995/2010 which mandates the use of a DD system. Its compulsory elements are outlined in a concise and clearly structured manner in article 6 Reg. 995/2010, further concretized by Implementing Reg. 607/2012. Alternatively, timber importers may use a DD system established by an accredited private monitoring organization (MO).[111] In that case, the obligation to maintain and regularly evaluate the DD system will fall to the MO, which will also need to verify its correct use by the importer and report any failure to do so to the competent national authorities.[112] Thus, MOs are directly involved with the national competent authorities, who control them.[113] Yet, hierarchical accountability[114] of MOs is based on formal recognition by the EU Commission, not national accreditation. On a final

[104] For a discussion of procedure vs. substance in Reg. 511/2014, see below under Sect. 4.1.

[105] Unwin (2011), p. 4f.

[106] Ibid.

[107] This is not dissimilar to the system of registered and non-registered collections under Reg. 511/2014. Users obtaining genetic resources from unregistered collections will not benefit from Art. 4.7 Reg. 511/2014.

[108] Reg. (EC) No 338/97 transposing the Convention on International Trade in Endangered Species of Wild Flora and Fauna [CITES] (1997) *Official Journal* L61, p. 1.

[109] Reg. (EC) No 2173/2005 (2005) *Official Journal* L347, p. 1. Licensing scheme within the FLEGT-Regulation. The respective licenses can only be issued by countries that have a Voluntary Partnership Agreement (FLEGT-VPA) with the EU in place, confirming that the timber products at hand were logged in full compliance with the laws of the exporting country. FLEGT is the acronym for the EU's Forest Law Enforcement, Governance and Trade Action Plan, which was established in 2003. EU FLEGT Facility, Import Procedures for FLEGT-licensed products, available under http://www.flegtlicence.org/eu-import-procedures-for-flegt-licences (last accessed 28.9.2019).

[110] Importers who first place such timber on the market can assume that the respective wood is marketable without constraints (EU Commission, 2016, 10A, 11).

[111] Art. 4.3 Reg. 995/2010.

[112] Art. 8.1 lit. b and c Reg. 995/2010.

[113] 'Supervisory accountability', Moser and Leipold (2019), p. 7. Cp Art. 8.4 Reg. 995/2010, Art. 6 Reg. 607/2012.

[114] Ibid.

note, it must be borne in mind that while MOs engage in operational burden-sharing in support of their clients,[115] timber importers retain full legal responsibility for using the DD system properly.[116]

3.3.2 Monitoring, Reporting and Verification (MRV) of CO_2 Emissions from Large Vessels—Reg. (EU) 2015/757 (as Amended by Delegated Regulation 2016/2071)

Since January 2018, large merchant vessels calling ports in the European Economic Area (EEA) must carry on board valid documents of compliance with the requirements of Reg. 2015/757. Compliance, here, is directed towards information about the total amount of CO_2 exhausts from a given ship ("disclosure")[117] and not towards total emissions caps.[118] The MRV Regulation 2015/757 obliges ship owners (usually companies) to accurately monitor their ships' CO_2 emissions and fuel consumption based on a previously drawn up monitoring plan and submit a verified emissions report to the EU Commission before the end of each reporting period. Verification itself is performed by accredited third-party verifiers. Their first task is to review ship owners' monitoring plans for, *inter alia*, completeness and accuracy.[119] When the verifier concludes that the monitoring plan is in conformity with the respective requirements of Reg. 2015/757, it will use the monitoring plan as a basis when assessing a ships annual emissions report.[120] Eventually, the verifier will issue a verification report, stating that the emissions have been determined in accordance with the Regulation and the monitoring plan. Finally, the verifier will issue timely limited documents of compliance based on the verification report and submit them to the EU Commission and the respective flag states (which are not necessarily EU Member States).[121] During inspections and in cases where emission report data is not available from Commission publications, the competent authorities of EU Member States will view these documents as (rebuttable) evidence of compliance. This is the presumption of conformity contained in article 19.1 Reg. 2015/757.

In Reg. 2015/757, the focus of the risk assessment (performed by the intermediary) is on identifying potential risks related to the monitoring process, such as misstatements and non-conformities with the monitoring plan and the respective requirements of the Regulation.[122] Here, the verifier may conduct on-site visits,

[115] Ibid., 'market accountability'.

[116] Unwin (2011), p. 12f.

[117] Fedi (2017), p. 383.

[118] Deane et al. (2017), p. 2.

[119] Art. 6 Reg. 2015/757.

[120] Art. 13.2 Reg. 2015/757.

[121] Art. 17 Reg. 2015/757.

[122] Arts 5 and 11 Commission Delegated Reg. (EU) 2016/2072 (2016) *Official Journal* L320, p. 5.

enquire with relevant staff, inspect documents and engage in active observation in order to verify the proper implementation of the (self-) monitoring and reporting system. Intervention by the regulator is limited to penalizing the failure to comply with monitoring and reporting obligations set up by the Regulation. The broad regulatory authority and margin of discretion afforded to the verifier[123] mandate high levels of expertise, professional independence and impartiality. Consequently, article 16 Reg. 2015/757 stipulates that verifiers shall be accredited for their activities by a national accreditation body pursuant to Reg. 765/2008.[124]

3.3.3 Reg. (EU) 2016/679 on the Protection of Natural Persons with Regard to the Processing of Personal Data and on the Free Movement of Such Data

The General Data Protection Regulation 2016/679 (GDPR) aims at protecting natural persons with regard to the processing of personal data and lays down rules relating to the free movement of such data as well. The obligatory risk management exercise can be found in article 24.1 GDPR. It mandates the controller (as the entity that collects data from the data subject) to *"[...] implement appropriate technical and organizational measures to ensure and to be able to demonstrate that processing is performed in accordance with this Regulation [...]"*. This is where regulatory intermediation comes in: Data protection certification mechanisms may be used for demonstrating compliance with the Regulation. Pursuant to article 42.2 GDPR, certification may also be used to demonstrate the existence of appropriate safeguards provided by processors of personal data located in third countries (outside the EU and EEA areas). However, this presumption of conformity is narrower than its wording might suggest. If one looks at the specific provisions dealing with the responsibilities of controllers and processors,[125] it becomes apparent that certification mechanisms, as referred to in article 42 GDPR, may merely be used "as an element" by which to demonstrate compliance. They do not serve as evidence of compliance[126] and can be a "snapshot" of compliance at most.[127] There is not even a reversal of the burden of proof in the GDPR.[128] The privilege afforded by article 42 GDPR is rather geared towards the calculation of administrative fines for infringements, where certifications can have a mitigating effect pursuant to article 83.2 lit. j GDPR.

---

[123] Fedi (2017), p. 405f.

[124] Commission Delegated Reg. (EU) 2016/2072 further details the accreditation of verifiers, which are subject to, *inter alia*, annual surveillance and extraordinary assessments by the national accreditation bodies.

[125] E.g. Arts 24.3 and 28.5 GDPR.

[126] Lachaud (2016), pp. 814, 820ff.

[127] Eckhardt (2017), para 45.

[128] Art. 82.3 GDPR.

GDPR data protection certifications can be issued, renewed and withdrawn either by (private) certification bodies or by the competent supervisory authorities themselves.[129] In the GDPR, national supervisory authorities remain very much at the helm of most regulatory functions, such as monitoring, enforcement, industry guidance and investigations of non-compliance.[130] Moreover, they approve the certification criteria developed by certification bodies and may order a certification body to withdraw a certification or not to issue one, in case the requirements for certification are no longer met. Certification bodies are also obliged to inform supervisory authorities about the reasons for granting or withdrawing certifications.[131] Finally, supervisory authorities are responsible for the accreditation of certification bodies.[132] As far as the distribution of responsibility among regulatory actors is concerned, it should be noted that certification does not shield the data processor from liability or reverse liability by shifting it to the certifier. Burden sharing is not an issue either: As a consequence of the "empty" presumption of conformity described above, the duties as stipulated remain with each participant (regulator-intermediary-target).[133]

3.3.4 Reg. (EU) 2017/821 Laying Down Supply Chain Due Diligence Obligations for Union Importers of Tin, Tantalum, Tungsten, Their Ores, and Gold (3TG) Originating from Conflict-Affected and High-Risk Areas[134]

Reg. 2017/821 seeks to curtail opportunities for armed groups to trade in 3TG minerals and metals by holding EU importers of those commodities publicly accountable via the mandatory disclosure of information on their supply chain. It therefore establishes a system of supply chain DD. When compared with the EU DD schemes analyzed above, it becomes apparent that Reg. 2017/821 takes the broadest conceptual approach yet: Beyond risk management-, information management- and intermediary audit obligations, it incorporates "structural obligations",[135] which assign responsibility for the supply chain DD policy directly to the senior management of Union importers.

In order to determine whether their supply chain DD practices are in conformity with the requirements of Reg. 2017/821, Union importers have two options. Either

---

[129] Art. 42 GDPR.

[130] See Art. 58 GDPR for a list of investigative and corrective powers of the supervisory authorities.

[131] See Art. 43.5 GDPR.

[132] Pursuant to Art. 43.1 GDPR, accreditation may also be carried out in conjunction with national accreditation bodies named in accordance with Reg. 765/2008.

[133] This is even provided for in the Regulation itself, see Art. 42.4 GDPR.

[134] For the international background of Reg. 2017/821, see Martin-Ortega (2014), pp. 62ff; also Partiti and van der Velde (2017).

[135] Heße and Klimke (2017), p. 449.

they carry out audits of their own supply chain DD practices via an independent third party, or they demonstrate that they sourced their minerals and metals exclusively from smelters and refiners that have been "white-listed" by the EU Commission.[136] In the latter case, a privileging presumption applies, exempting Union importers from the obligation to carry out third-party audits.[137] In Reg. 2017/821, there is only an indirect flow of regulatory information from the intermediary to the regulator, when importers submit their mandatory third-party audit reports to the national competent authorities. The latter are the ones carrying out *(ex-post)* checks of Union importers by, *inter alia*, conducting on-site inspections and by examining the documents and records that demonstrate compliance with the supply chain DD obligations.[138] Third parties operate almost exclusively as service providers for Union importers. Hitting a low point of intermediary accountability, Reg. 2017/821 does not require accreditation or formal recognition of third-party auditors by a national accreditation body, a competent national authority, or the EU Commission. That certainly raises questions of credibility and intermediary capture.[139] It is interesting to note that like Reg. 511/2014, Reg. 2017/821 does not contain a general import prohibition. The obligations contained in Reg. 2017/821 are purely procedural in nature. What stands out is the frequent referral to the standards laid down in the OECD DD Guidance for Responsible Supply Chains of Minerals from Conflict-Affected and High-Risk Areas (2013),[140] especially with regard to the risk management obligations of Union importers. One example is article 5.3 Reg. 2017/821, obliging Union importers to design their risk mitigation strategies by relying on the measures suggested in Annex III of the OECD DD Guidance. The OECD DD Guidance, however, is a (non-binding) soft law instrument. Relying on that instead of listing the risk management duties of importers fully within the Regulation itself, has been criticized as hardly harmonious and possibly even violating the principle of determinacy.[141]

3.4 Interim Conclusion

When one compares the level of detail that Reg. 511/2014 and its immediate relatives—Reg. 995/2010 and Reg. 2017/821—afford to the risk management

[136] Art. 9 Reg. 2017/821.

[137] Art. 6.2 last sentence Reg. 2017/821. This is reminiscent of the 'green lane' under Reg. 995/2010 (see above).

[138] Arts 3.2 and 11 Reg. 2017/821.

[139] Business representatives (just like NGOs) were involved at every stage of the EU negotiation process that led to Reg. 2017/821, see Partsch (2018). The regulatory markup of Reg. 2017/821 suggests that they were successful in achieving regulatory concessions.

[140] OECD Due Diligence Guidance for Responsible Supply Chains of Minerals from Conflict-Affected and High-Risk Areas (Second Edition, OECD, 2013).

[141] Teicke and Rust (2018), p. 43.

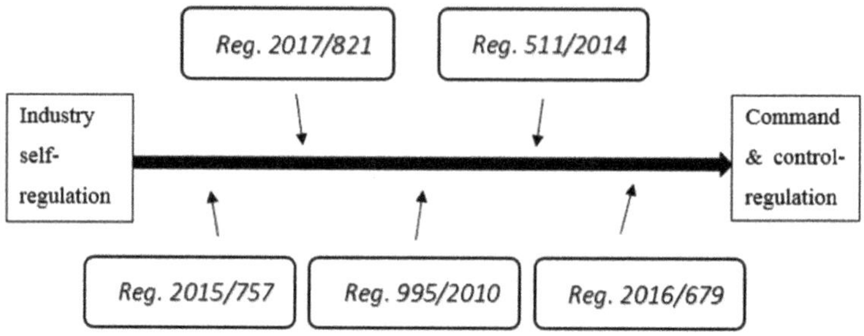

Fig. 2 The (self-) regulatory continuum of EU due diligence. **Source:** Own source

exercise, the EU Timber Reg. 995/2010 stands out in terms of clarity and comprehensiveness. Its risk management obligations are highly (pre-) structured and outlined further in Implementing Reg. 607/2012. The elements of risk management in article 6.1 Reg. 995/2010 are labelled accordingly[142] The procedural (DD) and substantive (import prohibition) obligations are separated and recognizable as such.[143] And finally, the addressees are clearly identifiable ('importers').That clarity was not duplicated in article 4 Reg. 511/2014 on user obligations: Risk mitigation is referred to only indirectly in article 9.1 Reg. 511/2014 (recognized best practices), Article 4.5 Reg. 511/2014 demands risk assessment procedures without using that term, obtaining genetic resources from registered collections pursuant to article 4.7 Reg. 511/2014 is not labelled as a risk mitigation procedure, and identifying 'users' of genetic resources will not be as straightforward as identifying those who import timber and put in on the market. In Reg. 2017/821 the risk management process is, again, more structured, and the criteria for risk mitigation are "surprisingly far reaching."[144]

When we now add Reg. 2015/757 and Reg. 2016/679 to the picture, we can go one step further and conceptualize a (self-) regulatory continuum of EU DD. The extreme points of this continuum are the command-and-control model of regulation on the one hand, and industry self-regulation without any state involvement on the other. Thus, we can qualify the Regulations analyzed as follows (Fig. 2):

Compared to DD in the corporate world, it becomes evident that none of the EU Regulations delegates full self-regulatory discretion to industry. In the MRV-Reg. 2015/757, the intermediary takes over a good amount of regulatory functions, with the regulator more or less staying in the "regulatory closet".[145] Yet still, the Regulation meant a switch from voluntary self-monitoring to a compulsory control

[142] Procedures providing access to information, risk assessment procedures and risk mitigation procedures in Art. 6.1 lit. a-c Reg. 995/2010.

[143] For a discussion of substantive elements in Reg. 511/2014 see below, Sect. 4.1.

[144] Teicke and Rust (2018), p. 41. Those criteria are largely contained in soft law, which is also problematic, see the discussion above in the preceding paragraph.

[145] Verbruggen (2013).

system.[146] Likewise, none of the Regulations is purely based on command and control mechanisms. Even in the GDPR, certification is an "arrangement mixing public and private involvement".[147] Still, the GDPR affords the least 'trust' to self-regulation.[148] Regulatory cooperation (and therefore wiggle-room for industry) is replaced by oversight, which is why Lachaud suggests calling certification under the GDPR "monitored self-regulation" instead of co-regulation.[149] Remaining self-regulatory features such as monitoring and enforcement via certification[150] are designed as substitutable and can be provided by either the certification bodies or the supervisory authorities.

If we look at the respective presumptions of conformity, we see that the benefits afforded by them—and thus the level of legal certainty—decrease, the further we move towards the command and control end of the spectrum (above, Sect. 3.3.3). Towards the other end, article 19.1 Reg. 2015/757 obliges Member States to regard the documents of compliance issued by verifiers as "evidence of such compliance," thus bearing close resemblance to how presumptions of conformity were originally envisioned under the New Approach (above, Sect. 3.2). The presumptions of conformity contained in Reg. 2017/821 and Reg. 995/2010 are exempting the regulatory target from the need to carry out third-party audits or DD altogether. Both are significant alleviations of burden. However, they are the result of something that can be attributed to the regulator (who registered 'green lane' wood or 'white-listed' certain smelters). They are not a consequence of industry co- and self-regulation, such as adherence to harmonized norms or third-party certification (above, Sect. 3.2). The Regulations dealing with supply chains in extractive[151] industries (timber, genetic resources, conflict minerals) are nestled somewhere in between the two extremes, with the ABS-Reg. 511/2014 leaning more towards the command-and-control end of the spectrum. Here, the presumption of conformity *is* a result of intermediation—but it is limited in scope, only alleviating the user from having to access the relevant information him-/herself, not more. Likewise limited is the functional scope of the intermediary itself. The absence of any enforcement, monitoring or reporting duties on the part of registered collections is a deviation from how the concept of DD is arranged in the other Regulations. Registered collections do not function as risk-absorbers for users (e.g. by monitoring behavior

---

[146] Fedi (2017), p. 410.

[147] Lachaud (2018), p. 252.

[148] This may also be due to the 'delicate' subject matter—personal data—and the accountability principle that Reg. 2016/679 is built on, see Art. 5 Reg. 2016/679. As of October 2019, there are still not approved private GDPR certification mechanisms, let alone accreditation bodies.

[149] Lachaud (2018), p. 251. Where oversight is not possible, e.g., when personal data is transferred to third countries, certification must be backed up by binding and enforceable commitments (e.g. contractual or otherwise) of the controller or processor in the third country.

[150] Spindler (2016), p. 409.

[151] We use this term with the proviso, naturally, that environmental impacts are different in the case of genetic resources (collecting/sampling) when compared to timber (logging) and minerals (mining).

for compliance or by helping to interpret rules). There is also no formal regulatory feedback (e.g. via reporting duties) from registered collections to the regulator. Depending on their position and tasks, regulatory intermediaries may provide authorities with feedback on rule deployment and by pointing out boundaries to effective enforcement and/or monitoring.[152] This feedback appears particularly important for dynamic rule development[153] within transnational regimes marked by enforcement gaps. The only other regime where there is no regulatory feedback from the intermediary to the regulator is Reg. 2017/821. We therefore do not consider registered collections and third-party auditors to be regulatory intermediaries in the classical sense. They are 'functional'[154] intermediaries whose job is the procedural safeguarding of information for the downstream supply chain, without having any supervisory authority.[155]

The aforementioned limitations translate into a strong role for national competent authorities in Reg. 511/2014. Users might expect something else when reading the term 'due diligence'.[156] At the same time, the risk for industry to fail with their own duties seems markedly higher than in the other Regulations. The more risk, the higher the duties.[157] It is worth discussing in this regard, if other actors can be considered as regulatory intermediaries.[158] The Guidance Document speaks of "specialized companies or organizations" with a "similar function" to registered collections.[159] It goes on to note that intermediaries such as registered collections and brokers "are best placed to inform the user about the legal status of the material they hold".[160] Currently, associations of users and other interested parties (such as e.g. commercial bio-prospectors or brokers[161]) can apply for the recognition of 'best practices' under article 8 Reg. 511/2014.[162] Yet, monitoring and enforcement duties

152 Abbott et al. (2017), p. 24.

153 Abbott et al. (2017), p. 23.

154 "Driven by functional considerations", see Abbott et al. (2017), p. 19.

155 Below, Sect. 4.2.

156 Especially if one follows the 'logic' that new forms of co- and self-regulation (especially in the international sphere) are indicative of a power shift to the private sector at the expense of public authority, Partsch (2018), p. 482; Green (2013). Also, Reg. 511/2014 does not use words like 'audit', "certification" and 'verification' which relate to third-party quality control (as a common feature of industry self- and co-regulation).

157 Bell (2001), p. 124.

158 Brès et al. (2019) developed a typology along the axis 'formalized vs. unformalized' and 'unofficial vs. official'.

159 EU Commission (2016), p. 5.

160 Ibid.

161 Commercial bio-prospectors/brokers had already engaged in developing codes of conduct in the past. For examples see Godt et al. (2020), § 3 VIII, 1–3.

162 Until today, applications under Art. 8 Reg. 511/2014 were only successful for associations of collections, thus mostly complementing Art. 5 Reg. 511/2014. The first 'best practices' acknowledged on 10.5.2019 is the CETAF Code of Conduct, see https://cetaf.org/sites/default/files/documents/leaflet-a4_codeofconduct_hd.pdf (last accessed 06.03.2020), for background Neumann

on part of associations establishing best practices are too limited (article 8.1 Reg. 511/2014 mentions only "oversight"[163]), and there is no regulatory feedback.[164] Second, if brokers or associations of users were to be assigned regulatory functions beyond article 8 Reg. 511/2014, notions of accountability (and upstream "trust") would have to be addressed by e.g. accreditation or formal recognition.[165]

4 Lessons to Be Learnt for Articles 4.1 and 4.5 Reg. 511/2014

Against the background of both preceding sections, we can draw conclusions as to how article 4 Reg. 511/2014 is to be interpreted. First, it must be understood as a compromise. The fuzzy concept of DD has spoken to various audiences. Environmental lawyers found the prospect of enhanced responsibility appealing, the business community was attracted by the prospect of more self-regulation and practices familiar to them. This ambiguity served to organize the necessary majorities. Therefore, the concrete content of the DD standard under article 4 Reg. 511/2014 neither derives from the model forerunners, nor from any particular legal family (civil law or common law), nor is it simply a management tool. But some legacies of the predecessors will remain. Second, Reg. 511/2014 must be understood as autonomous Union law. When DD was incorporated into Reg. 511/2014, the requirement to be duly diligent was stripped off its (inter-) governmental context and became a *behavioral* standard for firms—directly applicable, aiming at enforcement.[166] With the choice of the specific measure of a Regulation under article 288 sec. 2 TFEU, DD shifted from soft law to hard law.[167] Article 4 Reg. 511/2014 stipulates an *ex ante* standard of care (as in international business law), not a liability rule where the standard of care is evaluated *ex post* (as in international public law). DD was embedded in a set of parallel Regulations, which differ from each other due to sectorial specificities. However, together they define a new type of Regulation, aiming at risk regulation in transnational (supply-) chains.

(2018). Two other applications from the cosmetics industry have remained (until today, Oct. 2020) unsuccessful.

[163] Unfortunately, this is not really elaborated on. Annex IV to Commission Implementing Reg. (EU) 2015/1866 merely stipulates that associations and interested parties provide a description of how the overseeing will be carried out when applying for the recognition of best practices.

[164] There is passive regulatory feedback (per Commission request) from associations establishing best practices in cases of non-compliance by users implementing those practices, see Art. 12 Commission Implementing Reg. (EU) 2015/1866.

[165] In case of user associations, that would mean an overlap of the intermediary and the target, Abbott et al. (2017), p. 26.

[166] Lambooy (2010).

[167] Martin-Ortega (2014), p. 52.

Considering this new mixture, the following section will identify what exactly the legislator stipulated.

4.1 Procedural or Substantive?

The duty under article 4.1. Reg. 511/2014 "to ascertain that [. . .resources] have been accessed in accordance with applicable access and benefit-sharing legislation or regulatory requirements" is explained as "to keep, seek, transfer" in article 4.3 Reg. 511/2014 and coupled with "to obtain", "to establish" (sic. "produce compliance"), "or discontinue" under article 4.5 Reg. 511/2014. The question is, if there is a reason for the CJEU to interpret the "or" as a legislative means to put the obligations before and after the "or" on the same doctrinal footing. That would mean interpreting the duty "to discontinue" as procedural. The consequence of this reading is that the discretion of when "to discontinue" would rest with the user. The duty to do so would exist, but it could essentially not be enforced. This argumentation would be in line with the procedural DD-reading of the preceding models in international public law and in international business law. It could also be inferred from the rejection of a transnational implementation of article 15 NP, which would have substantially linked user measures to provider measures.[168] Such a reading would imply two further consequences. First, the search for remedying an incompliant situation could be infinite; the agency would have no authority to order a shutdown of operations. Secondly, if conceptualized as procedural, one could infer from article 4.7 Reg. 511/2014 that recipients of genetic resources from a [registered] collection [. . .] are preempted from liability. The argument would be that article 4.7 stipulates a real privilege for the user and shields him/her from a shutdown under article 4.5 Reg. 511/2014.[169]

This reading has to be opposed for the following reasons: With due diligence becoming part of an enforceable Regulation, it was stripped off its intergovernmental *and* purely business self-regulatory contexts. As binding and directly enforceable EU law, article 4 Reg. 511/2014 directly defines the standard of firms' conduct. While the legislator of Reg. 511/2014 rejected to install a substantial obligation to 'comply with foreign law' (more precisely, to link an [actual] domestic legal consequence to a [past] violation of foreign law), it did install an (actual) "duty to ascertain that the *resource* utilized *was* accessed in compliance with provider states laws". This rule *mediates* the substantive 'obligation to obey the foreign law' by a duty to check the legal status *of the object*. The technique resembles the concept of

[168] The consequences of the EU-rejected model for possible scenarios of transnational enforcement are discussed by Godt et al. (2020), § 2 II 5 b-f.

[169] Art. 5.7 Reg. 511/2014 reads: "Users obtaining a genetic resource from a collection included in the register of collections within the Union referred to in Article 5(1) shall be considered to have exercised due diligence as regards the seeking of information listed in paragraph 3 of this Article".

liability in private law, which prescribes a (non-enforceable) substantive obligation for everyone 'not to do harm' (lat. '*neminem laedere*'). Yet, the legal consequences of a violation of this substantive norm—such as injunctions and damage claims—carry conditions, which go beyond the causal link of result and behavior (lat. '*caveat emptor*').[170] With due respect to national differences and therefore generalizing: Injunctions require that the *result* can be (normatively) qualified as 'unlawful'—conceptualized as an objective standard.[171] In most jurisdictions, additional damage claims require 'fault'; meaning that the preceding *behavior* can be qualified as 'sub-standard'—conceptualized as a subjective standard.[172]

In the common law context, the tort of negligence consists of two parts: The objective part asks for the 'standard of care', the subjective part asks, if the specific individual in the case at hand could act as required (again more or less standardized). The standard examples here are children or freshmen surgeons.

As a *legal transplant,* DD changes its context when implanted into *statutory law*. The juxtaposition between substance and procedure cannot be the same when compared to the jurisprudence in public international law. Before the (substantive) appraisal that a resource was accessed in violation of foreign law can trigger a legal action, additional conditions must be met. In this context, DD defines the standard of care which triggers the legal action (injunction; shutdown of operations). The substantive standard 'compliance with foreign law' remains the normative objective, the general obligation. The Commission clearly stated that it understands the "duty to discontinue" as an "obligation of result".[173]

Thus, the legal action is conditioned. In contrast to the current status in public international law (above), the European legislator substantively *qualified* the 'standard of care' (article 4.5 Reg. 511/2014) by the word "or discontinue". This language links procedure with substance. It requires, first, "efforts to remedy",[174] but does not

[170]This is exactly how national legislators transposed the norm into national laws, e.g. for Germany, see § 2 of the German Law transposing Reg. 511/2014. It reads: (1) The competent authority adopts the necessary orders to remedy breaches of the legal acts designated in section 1 paragraph (1). (2) Should a user fail to comply with an order in accordance with paragraph (1), the competent authority may in individual cases seize the unlawfully-utilized genetic resource or prohibit specific utilization activities. This shall be particularly considered if the user is unable to submit the information required in accordance with Article 4 paragraph 3 of Regulation (EU) No. 511/2014.

[171]A situation which is "simply" unlawful does not give rise to individual obligations. Modern rule of law implies that preceding related conduct is in some way "blameworthy" in order to trigger legal consequences.

[172]This is what in German doctrine is conceptualized as "outer care". This does not equate with "guiltiness" which is "inner care" in the German doctrinal tradition. In the fault based German tort tradition, it is the inner care which provides for the basis of a single general clause of § 823 German Civil Code (BGB) which imbeds both, intent and negligence. For a profound critique, see Brüggemeier (2004), pp. 76ff.

[173]EU Commission (2016), p. 11.

[174]Art. 4.5 Reg. 511/2014 stipulates "When the information in their possession is insufficient or uncertainties about the legality of access and utilization persist, users shall obtain an access permit or its equivalent and establish mutually agreed terms, or discontinue utilization".

stop there.[175] Efforts might not last indefinitely. This qualification has two consequences: First, a normative reading—and one that deliberately avoids to simply equate compliance with substance—of the wording "or discontinue" requires an action plan with qualitative milestones and quantitative timelines.[176] It does leave space for discretion.[177] Second, single procedural steps cannot escape the normative substantive requirement ("or discontinue").[178] Therefore, the acquisition of material from a registered collection cannot escape the injunction commanded by article 4.5 Reg. 511/2014, as will be explained in the next section.

4.2 A Shift of Responsibilities?

The comparison with the other regulations clarifies that Reg. 511/2014 does not integrate a "regulatory intermediary" which might increase legal certainty for users. Classical elements of industrial self-regulation are scaled down. Registered collections (article 5 Reg. 511/2014) secure information, but there is no information flow to the regulator, no coordination or any other third-party quality control. In the broader context, registered collections and recognized best practices (article 8 Reg. 511/2014) appear rather peculiar. It is therefore, that we conceptualize them as purely informational ('functional') intermediaries (above, Sect. 3.4). And while the function of registered collections is to secure the important information about the 'country of origin' for the downstream supply chain, 'recognized best practices' are a tool to help users to comply with their DD obligations. However, article 4.5 Reg. 511/2014 has no connection to the "legality of access and utilization"; article 4.7 Reg. 511/2014 only says: "shall be considered to have exercised due diligence as regards the seeking of information". These stipulations are even missing with regard to 'recognized best practices'. Thus, Reg. 511/2014 does not provide for a legal assumption of lawful acquisition.[179] Indemnity is not granted. It narrowly relates to seeking information, and therefore can only help to provide evidence that the standard of care was met (with regard to 'seeking information'). This legal concept

[175] From the perspective of the competent authority, this triggers a conditioned program of actions. A first step will see the CA order the user to submit ABS documentation.

[176] From the perspective of the Competent Authority that means, if papers are not produced upon request, the CA will order the discontinuation of utilization until the papers are produced.

[177] In principle, utilization without ABS-documents has to be stopped. However, a respective order might be unproportionate. Recent experience has shown, that political turmoil or other political reasons result in 'no-answers' of competent authorities in provider states. In those cases, the CA of the user state may provide for a temporary allowance.

[178] Third, if it turns out that the approval of use is either impossible, or the user is unwilling to engage in action, the CA is entitled to prohibit utilization (including the publication of results) *ad infinitum*. The German law provides for the genetic resource(s) to be confiscated and destroyed.

[179] Godt and Burchardi (2018), p. 64f.

of article 5 Reg. 511/2014 does not prevent state held collections from doing more.[180]

In line with the preceding DD regulation on tropical timber, one might have expected that the ABS-regulation would install something akin to a monitoring organization, which develops standards as a means of self-regulation, evidenced by certification.[181] Yet, this element of industrial self-regulation was not taken on board in Reg. 511/2014.[182] As substitutes, 'registered' collections were envisioned. Yet, they fulfill a different function, and do not define the professional standard for users. The only regulatory effect is that the acquisition of resources from registered collections reverses the burden of proof regarding the question whether all pertinent information pursuant to article 4.3 Reg. 511/2014 was accessed. Thus, they have an upstream function for users,[183] but they do not belong to that group.

4.3 The Objective Standard of Care: Or "What Ought to Be Done?"

The standard of care as developed in tort law consists of two facets, an objective and a subjective one. The objective facet defines the standard of care of 'what ought to be done'. The subjective facet is composed of two different questions and will be dealt with in the subsequent two sections (4.4 and 4.5). Under the forerunner models of DD, the standard of care derives from an equation of likelihood (risk) with the extent of the harm. What, however, is the expected harm when the user of a resource is not able to establish compliance and the situation remains 'ABS-incompliant'?

In principle, and in contrast to the predecessors of EU DD, a lack of ABS-compliance neither results in a direct environmental damage, nor is a domestic

---

[180] The first registered collection (May 2018) is the German Collection of Microorganisms and Cell Cultures DSMZ. This collection pursues a specific service agency-approach for German research. It provides a legal check for access compliance, when requested by connected institutions; for background: Hartmann-Scholz (2018). In February 2020, the French Collection for Plant-associated Bacteria (CIRM-CFBP) became the second registered collection; in April 2020, the French 'Pierre Fabre Research Institute Library of dry ground plant parts' became the third registered collection, https://ec.europa.eu/environment/nature/biodiversity/international/abs/pdf/Register-of-Collections.pdf (last accessed 22.09.2020).

[181] Originally, the EP had proposed certification of the economic operators by the monitoring organizations. However, certification is not considered as ultimately necessary for due diligence. Yet, independent third-party auditing is mentioned as an element in the OECD DD Guidance and the UN-Expert Group Guidelines, as acknowledged by Martin-Ortega (2014), p. 73.

[182] It seems that the efficacy of certification was doubted in international fora (for the emerging conflict mineral regime, see Martin-Ortega (2014), p. 71; for diamonds see Ekmen (2011). In the realm of genetic resources, it has been argued that the affected industries are too dispersed as to come to grips with certification.

[183] Something, which was reiterated explicitly for conflict minerals in Recital 12 of Regulation 2017/821.

rule (contractual promise; financial regulation) or an international human right violated.[184] Only foreign administrative law is not complied with. This is a significant deviation from earlier models, and most probably the central source of misunderstanding and the lack of acceptance. As a matter of principle, the very reason for territoriality is the breaking down of sovereign governmental power into spatial pieces. It is for democratic reasons that we do not owe respect to foreign law. This is exactly the reason why the European legislator rejected the direct binding force of foreign law and stipulated a self-standing DD obligation. Yet, modern law is familiar with both, accepting the violation of foreign domestic law (tax evasion is the simplest example), and the importation of foreign law. The central questions here are: What is being imported, why and under which conditions? Importation of norms is standard in private international law. In public law, it is accepted when negotiated under an international treaty.[185] Here, the only basis is the sovereignty of the provider state, its accession to the Nagoya Protocol and the implemented ABS-Regulation. In combination with the underpinnings of DD in business administration, it is sometimes purported that the DD duties are to be determined fully by the user.

We submit that the listed duties must be read autonomously with the legislative intent of the EU lawmaker in mind (see already above, Sect. 2.2). The legislator had rejected to bind the user directly to 'compliance with provider state law' (which would ultimately result in instant discontinuation). The listed duties "to ascertain, to keep/seek/transfer, to obtain, to discontinue" describe an exercise of (legislative) translation (comparable to the discipline of private international law). The listed duties fall into two distinct categories:

The first set of procedural duties (to ascertain, to keep/seek/transfer, to obtain as stipulated in article 4.3 Reg. 511/2014) submit the user to undertake "efforts to obtain" regarding 'compliance with foreign law'. These efforts might differ among sectors, but they are of an objective nature. The chapeau-term 'due diligence' does not mean that the user has the authority to determine the duties. Article 4 Reg. 511/2014 duties are not primarily in the interest of the user.[186] Due to the fact that industry did not engage in norm building, the element of 'objective determination' of what 'ought to be done' remained somewhat under-conceptualized in the EU Regulation (above, Sect. 3.2). This gap will only subsequently be filled by the

[184] Unless we qualify the matter in terms of the dissemination of the benefits from science with regard to the human right to science, which is provided for in Art. 27 Universal Declaration of Human Rights (1948) G.A. Res. 217A (III) *UN Doc. A/810*, p. 71 and Art. 15 International Covenant on Economic, Social and Cultural Rights (1966) G.A. Res. 2200A (XXI) *United Nations Treaty Series 993*, p. 3; see also Morgera (2018/19), p. 55.

[185] Cf. the United Nations Framework Convention on Climate Change 'UNFCCC' (1992) G.A. Res. 48/189 *United Nations Treaty Series 1771*, p. 107, the principle of origin under the Treaty on the Functioning of the European Union 'TFEU' (2012) *Official Journal* C326, p. 47, EU Reg. 995/2010.

[186] It is in the interest of the user to be given support by the user state; this is the guiding implementation principle of the Dutch competent agencies, see van Winkoop (2018).

Guidance Paper(s) exercise,[187] to be orchestrated by the implementing agencies. The consequence for the interpretation of these duties is that they are to be determined according to the respective industrial sector (only in this regard "subjectively"). Therefore, the efforts under article 4 Reg. 511/2014 are double-conditioned: At the beginning, the user is only submitted to the duties under Reg. 511/2014, provided that the provider state is a party of the Nagoya Protocol and has enacted respective ABS-regulation. Only then, the Regulation is applicable. Yet still, in order to know if the Regulation applies in the first place, the user has to inquire the origin of the resource, as acknowledged in the Commission's Notice.[188]

The nature of these duties is determined by the respective supply chains of a given sector. Being the 'downstream' user in one country, one has to ascertain that the resource was legally acquired. That means that a user has to inquire if preceding 'upstream' owners/users complied with the respective statutory laws of an 'upstream' provider country (therefore going beyond the mere 'adoption of rules and measures'). This is the peculiar feature of article 4 Reg. 511/2014: What 'ought to be done' (objectively) is essentially determined by provider state legislation,[189] ergo foreign law (which is only enforceable in that given country). Yet, it is the user state law, which determines the objective standard of care ('thoroughness and best possible efforts') with regard to the resource used. If doubts remain, it is up to the purchaser to refuse the resource or to contact the provider in order to 'produce compliance'. In constitutional terms, the proprietary freedom of the user is burdened with a duty to inquire possible legal restrictions originating in a foreign country. If compliance cannot be established, the use must be discontinued. The central question is, when are 'efforts to produce compliance' exhausted? As long as 'efforts are not yet exhausted', the user has a valuable defense against a shutdown order. It shifts the burden of argumentation regarding the exhaustion of efforts to the competent agency. The evaluation of exhausted means of efforts is the precondition for an order to discontinue.

In comparison, the duty to discontinue under article 4.5 Reg. 511/2014 is of a different nature than those duties stipulated under article 4.3 Reg. 511/2014. Besides being a substantive duty (above, Sect. 4.1), the duty to discontinue is to be determined by the user state—neither by the provider state nor by the user. Said duty mirrors—at a different regulatory level—the international duty of the EU and its

[187] After the horizontal Guidance Paper (EU Commission 2016) eight vertical Guidance Papers (3rd version April 2017) were drafted for the following sectors: animal breeding, biotechnology, cosmetics, foods and feeds, plant breeding, pharmaceuticals, the biocontrol and bio-stimulants sector, and for ex-situ collections. After deciding not to adopt them as official guidance papers, competent authorities will move forward with singular examples and (published) explanations.

[188] EU Commission (2016), p. 12.

[189] Notwithstanding the fact that respecting foreign law is limited by restrictions, such as the temporal scope (cut-off date 12.10.2012), the material scope ('commodities', strongly disputed 'derivatives', 'digital sequence information'), the interpretation of the term 'utilization' (discussed is the exemption of activities defined as 'pre-' or 'post-utilization' which would fall under 'utilization' defined by provider state laws).

Member States as parties to the Nagoya Protocol. The competent authorities will execute said international duty by issuing an order to discontinue. The decision is at the discretion of the user state. A consequence of this is that said state could also decide to tolerate the incompliance with foreign law, e.g. if a permit cannot be achieved in wartimes or when the provider state rejects co-operation. Then, it is for the user state to decide if it accepts that efforts have been exhausted without the possibility to achieve compliance.

The point in time when to discontinue will crystallize from various sources, while the administration enjoys wide discretion to determine said point in time. It primarily depends on whether the procedural part has come to an end. The respective behavioral standard differs from sector to sector and is distinct as far as professions are concerned (researchers, collections (registered/non-registered), industry). The practical thing to do will be to come up with an implementation plan that is structured according to timescales and includes milestones and a cut-off day.

4.4 The Subjective Standard of Care or "What Ought to Be Known?"

The subjective standard is part of the DD forerunner models. In public international law, it comes in as a category of fault, linked to the duty to prevent harm. In international business law, the subjective standard is part of the purchaser's defense. In both models, 'knowledge' and 'ought to know' are equated. It is closely connected to 'what ought to be done', since more knowledge might have prompted more activity. In both preceding DD models, the question of 'what ought to be known' is determined *ex post*. In contrast, article 4 Reg. 511/2014 installs an *ex ante* duty. This is a significant requirement. The rule of law mandates that the law has to inform the addressee in advance about what he/she needs to know in order to comply. The classical public law requirements of foreseeability and determinacy are *ab initio* in tension with DD. What did the legislator mean?

As seen in Sect. 3, the user has to install a reasonable risk management system. This system determines what the user knows and can know. This correlation of knowledge and information management is linked to the correlation of the normative duty and informational infrastructure: the more risk, the higher the duties; the stronger the informational architecture, the lower the risk of industry to fail with their own duties. Since Reg. 511/2014 installed a rather weak informational architecture, industry itself will need to engage in establishing a suitable management system. 'What should have been known' will then depend on 'what should have been done' in terms of the management system. The importance of the correlation becomes evident when the user only becomes aware of the incompliant use of a resource after he/she had filed a DD declaration and the competent agency raised concerns. In such a case, the relevant point in time when the DD efforts should have started will not be the moment in which the user acquired positive knowledge, but the moment in which the user *could have known* that the use is non-compliant. This is an important element of the business legacy of the concept of DD (above, Sects. 2.2, 2.3). In addition, the standard of care depends on the respective business *usance*, not on the actual knowledge. While the DD defense might be brought forward, it is up to the agency to determine if reasonable efforts were exhausted in the given time period. As identified, this evaluation is not at the discretion of the user under Reg. 511/2014. Thus, the point in time when the user 'ought to have known' will affect the decision for the shutdown of operations. This may be a surprising result for some corporate compliance officers. Consequently, we may see disputes as to when efforts are objectively exhausted—companies might argue that further steps can still be undertaken and that time would not yet have elapsed.[190] It is up to the agency to identify if all *reasonable* measures were taken and the efforts therefore exhausted.

4.5 *The Professional Standard: A Firm's Capacities*

How much do subjective capacities matter, which traditionally have an impact on fault in private law? The Commission Notice[191] explicitly refers to the strict business standard: "Inexperience and lack of time are traditionally not seen as valuable defenses as to due diligence". Both are typical defenses rejected by civil law courts.[192] In administrative law, subjective categories principally play no role. They are categories of criminal law which affect the length and nature of a sentence. In negligence literature, the capacity estoppel is by and large rejected. The only exceptions are children and, in some jurisdictions, mental sickness. Thus, in the given context of Reg. 511/2014, the estoppel of individual capacities can clearly be rejected. However, it seems that the Commission, here, refers to the procedural duties. With regard to the substantive duty (to discontinue), it makes sense to take the attributes of the user and the surrounding situation into account, such as the size of a business, its role as a global player in a certain field, prior dealings of the firm with the domestic Competent Authority in the provider state, or respective negotiations between provider and user state governments.

---

[190] This is exactly the opposite rationale, when compared to the due diligence concept in international business law. Here, it is the buyer who argues that all efforts have been exhausted.

[191] EU Commission (2016), p. 10.

[192] See Landgericht Stuttgart, Germany 31 August 1989, IPRax 1990, 317; District Court Roermond (*Fallini Stefano v. Foodik*), Netherlands 19 December 1991, available under http://cisgw3.law.pace.edu/cases/911219n1.html; Supreme Court (*Trekking shoes case*) Austria 27 August 1999, available under: http://cisgw3.law.pace.edu/cases/990827a3.html.

5 Conclusion

Only in broader terms can DD be understood as a "single concept" of transnational governance. The two conceptual parents provided the EU legislator with a toolbox to choose from. In the context of EU DD, a notion of hybrid public-private risk regulation materialized. Differences emerge due to sectorial adaptations, which gave rise to a distinct category of "orchestration" (sic. the mix of public-private engagement[193]). It is against this framework that specific articles of the Regulations have to be interpreted.

In comparison with the other Regulations, the DD concept of Reg. 511/2014 turns out to be much more "public". Despite the absence of a clear prohibition—such as the import prohibition in the Timber Reg. 995/2010—and despite the non-existence of mandatory reporting duties, the procedural duties in Reg. 511/2014 are not 'open ended', and they are not at the discretion of the user. Since organized industry did not take part in norm production, there is very little articulation of the private side (sic. industrial self-organization) in this Regulation. That raises the relative power of the state.[194] It is this analytic framework, which guides the interpretation of the contested rule of article 4.5 Reg. 511/2014. The 'duty to discontinue' is of a substantive quality—however, it is an integrated part of a 'program of duties'. The effect is that it proceduralizes a straightforward prohibition and provides leeway for agency discretion. If non-compliance is detected, a firm may take efforts to bring its use into compliance. In other words, DD in this Regulation gives room to 'produce compliance'. If compliance cannot be produced, a closing order has to follow. This is the Regulation's unique technique to install the ABS-regime in a transnational fashion, in fact 'importing' provider state regulation. The user is submitted to 'his/her' home country administration. Once it is determined, that a provider state is not communicating any more (or not willing or able to examine the case), the user is freed from the hassle to deal with a foreign government enforcing the 'sovereign property rights' in its resources. It is up to the user state to assert the situation. This is of a double advantage for industry. Technically, the elements of corporate DD are melted down to a defense against the order to discontinue. Due to the underdeveloped informational infrastructure in Reg. 511/2014, the agencies discretion requires forensic sensitivity regarding the gathering of information necessary to evaluate the user's compliance. This implies the possibility of random checking. The heavy burden for the competent national authority is counter-balanced by article 9.3 Reg. 511/2014, which says that agencies should take a risk-based approach when checking on users.

---

[193] Partiti (2019).

[194] In this regard see also Bartley (2014), p. 102ff., who discusses how the transnational timber regime (including Reg. 995/2010) "re-centers" the state and reduces the relative space for private authority (by prioritizing the legality of timber over sustainability); very similar Partiti (2019), p. 103: "Orchestration, however, is a two-way street. It also affects the roots of public authorities' understanding and operationalization of concepts such as [. . .] due diligence".

The analysis displays Reg. 511/2014 as a fine example for Pierre Legrand's skepticism towards legal transplants, and supports Teubner's reflection of 'legal irritants'. While two different models inspired the law-making process, we see the EU regulatory environment having a strong impact on the newly implemented legal institution. Despite the fact that DD came around as a novel, innovative legal institution, preceding and following Regulations shape a regulatory concept of EU due diligence. The strict form of an EU Regulation has an impact on the content, but more importantly, the compromises enshrined in legal language make due diligence under Reg. 511/2014 a unique, EU-autonomous legal instrument.

References

Books and Chapters

Bainbridge SM, Anabtawi I (2017) Mergers and acquisitions. A transactional perspective. Foundation Press, St. Paul

Barber CS, Johnson S, Tobin B (2003) Options for developing measures in user countries to implement the access and benefit-sharing provisions of the Convention on Biological Diversity, 2nd edn. United Nations University Press, Yokohoma

Bartley T (2014) Transnational governance and the re-centered state. Sustainability or legality? Regul Gov 8(1):93–109

Brüggemeier G (2004) Common principles of tort law. A pre-statement of law. British Institute of International and Comparative Law, London

Chapman CE (2006) Conducting due diligence. Practicing Law Institute, New York

DiMatteo LA (2009) Law of international contracting, 2nd edn. Kluwer Law International, Alphen aan den Rhijn

Dupuy PM, Viñuales JE (2015) International environmental law. CUP, Cambridge

Godt C, Burchardi M (2018) Strict liability for "registered collections"? – Assessing Reg. (EU) 511/2014. In: Feit U, Greiber T, Karger E (eds) Second Meeting of the European Competent National Authorities Implementing the Nagoya Protocol and the Corresponding EU Regulation, BfN-Skripten, No 515. Bundesamt für Naturschutz, Bonn, pp 57–70

Godt C, Šušnjar D, Wolff F (2020) Umsetzung des Nagoya Protokolls in EU- und nationales Recht – Ein Alternativvorschlag zur Umsetzung der EU VO 511/2014/Transposition of the Nagoya Protocol into EU and national law – an alternative proposal for the transposition of EU Reg. 511/2014. Nomos, Baden-Baden

Green JF (2013) Rethinking private authority: agents and entrepreneurs in global environmental governance. Princeton University Press, Princeton

Hartmann-Scholz A (2018) First registered collection in the EU. In: Feit U, Greiber T, Karger E (eds) Second Meeting of the European Competent National Authorities Implementing the Nagoya Protocol and the Corresponding EU Regulation, BfN-Skripten, No 515. Bundesamt für Naturschutz, Bonn, pp 71–78

Hood C, Rothstein H, Baldwin R (2001) The government of risk – understanding risk regulation regimes. Oxford University Press, Oxford

Hoskisson RE, Hitt MA, Ireland RD (2004) Competing for advantage. South-Western/Thomson Learning, Mason

Joerges C (2013) Perspektiven einer kollisionsrechtlichen Verfassung transnationaler Märkte. In: Falke J, Joerges C (eds) Handelsliberalisierung und Sozialregulierung in transnationalen Konstellationen. Nomos, Baden-Baden, pp 391–433

Kegel G, Schurig K (2004) Internationales Privatrecht, 9th edn. Beck, München
Kulesza J (2016) Due diligence in international law. Nijhoff/Brill, Leiden/Boston
Meyer K (2018) Grenzen und Entwicklungsmöglichkeiten des Souveränitätsprinzips in transnationalen Handelsbeziehungen. Mohr Siebeck, Tübingen
Neumann D (2018) Application for acknowledgement of best practice under Article 8 of Regulation (EU) No 511/2014: the CETAF code of conduct & best practice. In: Feit U, Greiber T, Karger E (eds) Second Meeting of the European Competent National Authorities Implementing the Nagoya Protocol and the Corresponding EU Regulation, BfN-Skripten, No 515. Bundesamt für Naturschutz, Bonn, pp 71–78
Schwentzer I (2019) Art. 38 (Commentary). In: Schlechtriem P, Schwentzer I, Schroeter UG (eds) Kommentar zum UN-Kaufrecht (CISG): Übereinkommen der Vereinten Nationen über Verträge über den internationalen Warenkauf, 7th edn. C.H. Beck, München, pp 615–628
Short JL (2016) Transplanting law in a globalized world: private transnational regulation and the legal transplant paradigm. In: Bignami F, Zaring D (eds) Comparative law and regulation: understanding the global regulatory process. E. Elgar, Cheltenham, pp 430–444
Siems M (2014) The curious case of overfitting legal transplants. In: Adams M, Heirbaut D (eds) The method and culture of comparative law. Hart, Oxford/Portland, pp 133–146
Siems M (2018) Comparative law, 2nd edn. Cambridge University Press, Cambridge
Spedding LS (2005) Due diligence and corporate governance. Butterworth-Heinemann, Oxford
van Winkoop A (2018) Evaluation of user checks in 2017 in the Netherlands. In: Feit U, Greiber T, Karger E (eds) Second Meeting of the European Competent National Authorities Implementing the Nagoya Protocol and the Corresponding EU Regulation, BfN-Skripten, No 515. Bundesamt für Naturschutz, Bonn, pp 31–34
Watson A (1974) Legal transplants: an approach to comparative law. Scottish Academic Press, Edinburgh

Journals and Articles

Abbott KW, Levi-Faur D, Snidal D (2017) Theorizing regulatory intermediaries: the RIT model. Ann Am Acad Pol Soc Sci 670(1):14–35
Bell R (2001) Basic due diligence. Environ Claims J 13(4):123–140
Best E (2003) Alternative regulations or complementary methods? Evolving options in European Governance. Eipascope 1:1–10
Bremer N (2017) Post-environmental impact assessment monitoring of measures or activities with significant transboundary impact: an assessment of customary international law. RECIEL 26(1):80–90
Brès L, Mena S, Salles-Djelic ML (2019) Exploring the formal and informal roles of regulatory intermediaries in transnational multistakeholder regulation. Regul Gov 13:127–140
Brunnée J (2016) Procedure and substance in international environmental law: confusion at a higher level? ESIL 5(6):1–7
Chen-Wishart M (2013) Legal transplant and undue influence, lost in translation or a working misunderstanding? ICLQ 62(1):1–30
Deane F, Huggins A, Karim MS (2017) Measuring, monitoring, reporting and verification of shipping emissions: evaluating transparency and answerability. RECIEL Special Issue:1–10
Ekmen P (2011) From riches to rags – the paradox of plenty and linkages to violent conflicts. Goettingen J Int Law 3(2):473–493
Fedi L (2017) The monitoring, reporting and verification of ships carbon dioxide emissions: a european substantial policy measure towards accurate and transparent carbon dioxide quantification. Ocean Yearb 31:381–417

Fishman A, Obizdinski K (2014) European Union Timber Regulation: is it legal? RECIEL 23(2):258–274
Gellert R (2015) Understanding data protection as risk regulation. J Internet Law 2015:3–15
Haufler V (2018) Producing global governance in the global factory: markets, politics, and regulation. Glob Policy 9:114–120
Héritier A, Eckert S (2008) New modes of governance in the shadow of hierarchy: self-regulation by industry in Europe. J Public Policy 28(1):113–138
Heße D, Klimke R (2017) Die EU-Verordnung zu Konfliktmineralien: Ein stumpfes Schwert? EuZW (Europäische Zeitschrift für Wirtschaftsrecht) 12:446–450
Lachaud E (2016) Why the certification process in the General Data Protection Regulation cannot be successful. CLSR 32(6):814–826
Lachaud E (2018) The General Data Protection Regulation and the rise of certification as a regulatory instrument. CLSR 34(2):244–256
Lambooy T (2010) Corporate due diligence as a tool to respect human rights. NQHR 28(3):404–448
Legrand P (1997) The impossibility of "legal transplants". MJECL 4(2):111–124
Levashova Y (2011) How effective is the new EU Timber Regulation in the fight against illegal logging? RECIEL 20(3):290–299
Martin-Ortega O (2014) Human rights due diligence for corporations: from voluntary standards to hard law at last? NQHR 32(1):44–74
McCorquodale R (2009) Corporate social responsibility and international human rights law. J Bus Ethics 87(2):385–400
McIntyre O (2011) The world court's ongoing contribution to international water law: the Pulp Mills Case between Argentina and Uruguay. Water Altern 4(2):124–144
Morgera E (2018/19) Fair and equitable benefit-sharing in a new international instrument on marine biodiversity: a principled approach towards partnership building? MarSafeLaw J Special Issue on Ocean Commons 5:48–77
Moser C, Leipold E (2019) Toward "hardened" accountability? Analysing the European Union's hybrid transnational governance in timber and biofuel supply chains. Regul Gov 2019:1–18
Partiti E (2019) Orchestration as a form of public action: the EU engagement with voluntary sustainability standards. ELJ 25(1):94–117
Partiti E, van der Velde S (2017) Curbing supply -chain human rights violations through trade and due diligence: possible WTO concerns raised by the EU Conflict Minerals Regulation. JWT 51(6):1043–1068
Partsch L (2018) The new EU Conflict Minerals Regulation: normative power in international relations? Glob Policy 9(4):479–488
Schucht K (2015) Kooperation zwischen Wirtschaftsakteuren und Marktüberwachungsbehörden – auch ein Compliance-Instrument im Produktsicherheits- und Technikrecht. BB (Betriebs-Berater) 12:654–659
Schucht K (2017) 30 Jahre New Approach im europäischen Produktsicherheitsrecht – prägendes Steuerungsmodell oder leere Hülle? EuZW (Europäische Zeitschrift für Wirtschaftsrecht) 2:46–51
Scott J (2014) Extraterritoriality and territorial extension in EU law. Am J Comp Law 62(1):87–126
Senden JL (2005) Soft law, self-regulation and co-regulation in European law: where do they meet? EJCL 9(1):1–27
Spindler G (2016) Selbstregulierung und Zertifizierungsverfahren nach der DS-GVO Reichweite und Rechtsfolgen der genehmigten Verhaltensregeln. ZD (Zeitschrift für Datenschutz) 9:407–414
Takano A (2018) Due diligence obligations and transboundary environmental harm: cybersecurity applications. Laws 2018 7(36):1–12
Teicke T, Rust M (2018) Gesetzliche Vorgaben für Supply Chain Compliance – Die neue Konfliktmineralien-Verordnung. CCZ (Corporate Compliance) 1:39–43

Teubner G (1998) Legal irritants: good faith in British law or how unifying law ends up in new divergences. MLR 61(1):11–32

Verbruggen P (2013) Gorilla in the closet? Public and private actors in the enforcement of transnational private regulation. Regul Gov 7(4):512–532

Watson A (2000) Legal transplants and European private law. EJCL, vol 4.4. https://www.ejcl.org//44/art44-2.html. Accessed 15 Mar 2019

Winter G, Kalichava K (2019) Rechtstransfer und Eigendynamik in Transformationsländern: Das Beispiel der Verwaltungsrechtsentwicklung in Georgien. ZaöRV (Zeitschrift für ausländisches öffentliches Recht und Völkerrecht) 2:273–424

Wood S (2012) The case of leverage-based corporate human rights responsibility. Bus Ethics Q 22(1):63–98

Proceedings and Conference Papers

Godt C (2003) IPRs and environmental protection after Cancún. Paper submitted to the Int'l Conference "Moving forward from Cancún - The Global Governance of Trade, Environment and Sustainable Development" organized by Ecologic – Institute for International and European Environmental Policy, Berlin, 30/31 Oct 2003, https://uol.de/fileadmin/user_upload/wire/fachgebiete/eurowr/download/Texte/IPRs%20and%20Environmental%20Protection%20after%20Canc%C3%BAn.pdf (accessed 29 October 2019)

International Law Association – ILA Study Group on Due Diligence in International Law (2014) First Report (French, D [Chair] and Stephens, T. [Rapporteur]), https://www.ila-hq.org/index.php/study-groups (accessed 7 November 2019) (cited as 'ILA Study Group on DD')

International Law Association – ILA Study Group on Due Diligence in International Law (2016) Second Report – Working Session Report Johannesburg (Marboe, I [Chair]), https://www.ila-hq.org/index.php/study-groups Accessed 07 Nov 2019 (cited as 'ILA Study Group on DD')

Online Publications

Eckhardt J (2017) Art. 42 DS-GVO (Commentary). In: Wolff HA, Brink S (eds) Beck Online-Kommentar Datenschutzrecht, 21st edn. Beck, München

Koivurova T (2010) Due diligence. In: Wolfrum R (ed) Max Planck encyclopedia of public international law, http://opil.ouplaw.com/abstract/10.1093/law:epil/9780199231690/law-9780199231690-e1034?prd=EPIL (accessed 7 March 2019)

Other

Ensthaler J, Gesmann D, Zink KJ (2016) Evaluierung der deutschen Akkreditierungsstruktur. Project No. 25/15 on behalf of the German Federal Ministry for Economic Affairs and Energy, https://www.bmwi.de/Redaktion/DE/Publikationen/Studien/evaluierung-der-deutschen-akkreditierungsstruktur.pdf?__blob=publicationFile&v=4. Accessed 29 Oct 2019

Grozev C (2012) Dynamics in (auto)-acculturation of legal transplants. IMADEC University, https://www.academia.edu/6661543/Dynamics_in_Auto_-acculturation_of_Legal_Transplants. Accessed 29 Oct 2019

Portalier P (2017) Myths and realities of the presumption of conformity. Background paper for the Action 8 working group under the Joint Initiative on Standardisation initiated by the European Commission, https://www.orgalim.eu/insights/myths-and-reality-presumption-conformity. Accessed 28 Nov 2019

Unwin E (2011) What does the exercise of due diligence under the Timber Regulation require? Client Earth Legal Briefing. https://www.documents.clientearth.org/wp-content/uploads/library/2011-10-01-what-does-the-exercise-of-due-diligence-under-the-timber-regulation-require-ce-en.pdf. Accessed 29 Oct 2019

Policy Documents

European Commission (2002) Commission Communication of 05.06.2002. Action plan: simplifying and improving the regulatory environment. C (2002) 278 final

European Commission (2016) Commission Notice of 27.08.2016. Guidance document on the scope of application and core obligations of Regulation (EU) No 511/2014 of the European Parliament and of the Council on compliance measures for users from the Nagoya Protocol on Access to Genetic Resources and the Fair and Equitable Sharing of Benefits Arising from their Utilization in the Union. *2016/C 313/1*

European Parliament (2012) DRAFT REPORT on the proposal for a regulation of the European Parliament and of the Council on Access to Genetic Resources and the Fair and Equitable Sharing of Benefits Arising from their Utilization in the Union. C (2012) 576 final

OECD (2011) OECD guidelines on multinational enterprises. OECD Publishing, Paris

OECD (ed) (2013) OECD due diligence guidance for responsible supply chains of minerals from conflict-affected and high-risk areas, 2nd edn. OECD Publishing, Paris

Cases

District Court Stuttgart (*Shoe case*), Germany 31 August 1989, Praxis des Internationalen Privat- und Verfahrensrechts (IPRax) 1990, 317

District Court Roermond (*Fallini Stefano v. Foodik*), Netherlands 19 December 1991, Weekblad voor privaatrecht, notariaat en registratie (WPNR) 1994, 564

Supreme Court (*Trekking shoes case*), Austria 27 August 1999, Österreichische Zeitschrift für Rechtsvergleichung (östZRVgl) 31

Case concerning pulp mills on the river Uruguay (*Pulp Mills* case), Argentina v. Uruguay, judgement on the merits, ICGJ 425 (ICJ 2010)

Christine Godt is Full Professor for European and International Economic Law, Private Law, and Hanse Law School Director of the Carl von Ossietzky University of Oldenburg since 2010. She was guest professor at UCLA (2019), the Center of Excellence of the University of Constance (2015), University Institute Florence (2011), LMU Munich (2008/2009). She 'habilitated' 2007 with a thesis on patent law and theory, her PhD (s.c.l.) 2007 covered environmental liability (both, University of Bremen). Studies of law in Berlin and Washington DC, State Exams (1997/1991). Her research focusses on issues of biodiversity, (comparative) property, intellectual property, the emerging data constitution, transnational governance schemes. Publications posted under: https://uol.de/eurowr/cgodt-publikationen. christine.godt@uni-oldenburg.de.

Markus Burchardi LL.M. (Glasgow/Leuphana) is a Ph.D. candidate and research associate at the Carl von Ossietzky University of Oldenburg, Germany (Jean Monnet Chair of European and International Economic Law, Professor C. Godt), where he researches the transposition of the Nagoya Protocol into EU law. His doctoral work focuses on the interplay between big data biodiversity, open science and the fair and equitable sharing of benefits. markus.burchardi@uni-oldenburg.de.

Part IV
Unresolved Issues and Solutions to Implementation Challenges

Digital Sequence Information on Genetic Resources and the Convention on Biological Diversity

Christopher H. C. Lyal

Abstract There is disagreement whether international agreements over genetic resources (GR) under the Convention on Biological Diversity, including the Nagoya Protocol, cover Digital Sequence Information (DSI). DSI is data on the composition of GR, which may be downloaded from databases and used by entities who have no contractual relationship with the country of origin of the original GR from which the data were derived, and with no applicable legal requirement to agree on benefit-sharing with that country. Proposals for inclusion of DSI under the CBD are intended as changing this situation and increasing benefit-sharing. However, inclusion of DSI under the CBD is contested, and there is a legal basis for both inclusion and exclusion, so any eventual decision must be based on policy agreement by parties. To date there has been little explicit basis for a definition of DSI; here a distinction between 'inclusive' and 'exclusive' definitions is discussed, the latter using the scope of the CBD to support the concept to include only Nucleotide Sequence Data. Irrespective of inclusion under the CBD countries are including DSI in national ABS legislation, through inclusion in PIC and MAT on newly accessed GR, stating sovereign rights over DSI already generated, and through article 5 (1) of the Nagoya Protocol. These approaches are discussed, and the need for further consideration of the practical implications of article 5 (1). Ultimately a strong focus of future discussions should be benefit-sharing and the ethical position of users.

Keywords ABS · Nagoya Protocol · CBD · DSI · Digital Sequence Information

C. H. C. Lyal (✉)
Department of Life Sciences, Natural History Museum, London, UK
e-mail: c.lyal@nhm.ac.uk

E. Chege Kamau (ed.), *Global Transformations in the Use of Biodiversity for Research and Development*, Ius Gentium: Comparative Perspectives on Law and Justice 95, https://doi.org/10.1007/978-3-030-88711-7_21

1 Introduction

1.1 Background

The topic of 'Digital Sequence Information on genetic resources' (DSI) is currently dominating discussions of access and benefit-sharing (ABS).[1] However, these discussions have been mired in a swamp of misunderstandings, confused terminology and conflicting legal views. Do the sovereign rights of a country extend to DSI derived from genetic resources (GR) accessed in its territory? If they do, does this mean that DSI should be within the scope of the Convention on Biological Diversity and the operation of its Nagoya Protocol? Can or should the Protocol be applied when the DSI is held on public databases outside the country of origin of the original GR? What exactly is DSI?

The scope of the Convention on Biological Diversity (CBD) is set out in article 4 and refers explicitly to items under the national jurisdiction of contracting parties: components of biological diversity, and processes and activities carried out under national jurisdiction. Bearing in mind the recognition of sovereign rights in article 3 of the CBD, it would appear that the intended meaning is that the components of biological diversity[2] referred to come under those sovereign rights. This is reinforced by article 15 CBD, which commences "Recognizing the sovereign rights of States over their natural resources, the authority to determine access to GR rests with the national governments and is subject to national legislation" and explains that "For the purpose of this Convention, the genetic resources being provided by a Contracting Party, as referred to in this article and articles 16 and 19, are only those that are provided by Contracting Parties that are countries of origin of such resources or by the Parties that have acquired the genetic resources in accordance with this Convention." This relationship of the scope of the CBD to sovereign rights is a key to interpreting DSI and assessing to what extent it might come within scope of the Convention.

The CBD and its Nagoya Protocol on ABS (NP) were developed in a conceptual framework in which access to physical genetic resources was followed by research and development which generated monetary or non-monetary benefits. These benefits were to be shared fairly and equitably by the users with the country providing those genetic resources. This simple linear model bears little relationship to the complex interactions that often characterize supply and value chains of physical genetic resources, let alone the modalities of sharing 'DSI'.

Since agreement of the CBD in 1992, and the adoption of the Nagoya Protocol (NP) in 2010, there have been considerable advances in DNA sequencing

[1] Medaglia (2020).

[2] The term 'components of biological diversity' used in CBD article 4 on jurisdictional scope (see above) can be understood through the uses of terms set out in article 2 of the CBD, in particular 'biological diversity', 'biological resources', 'genetic material' and 'genetic resources'.

methodology and biotechnology.[3] The three large public databases of the International Nucleotide Sequence Database Collaboration (INSDC) currently hold over 216 million nucleotide (DNA and RNA) sequences[4] (including human DNA, which is out of scope of the CBD). Such sequence data can now be downloaded from a database, its properties predicted if not known, and synthetic DNA constructed, thus obviating the need (at least in some circumstances) for examining the original genetic resources.[5] In this situation there is a break in the legal chain between the source of the physical genetic resources and the user of information derived from those genetic resources.[6]

This has focused attention on the potential value of data and information produced through utilization (if utilization is taken to include sequencing the DNA or RNA), and caused concern to some provider countries and NGOs.[7] They perceive that use of such information downloaded by a third party bypasses requirements for benefit-sharing,[8] some viewing it as 'digital biopiracy'.[9] As a result, there is a call from developing countries in particular for the Parties to the CBD to agree the inclusion of DSI within the ambit of the CBD and its Nagoya Protocol and thus subject to the same requirements for compliance and benefit-sharing as physical genetic resources. However, views on whether or not DSI is, or even can be, covered by the CBD differ sharply.[10] The CBD makes no mention of genetic or sequence information, and the 'Use of Terms' statements in article 2 appear to restrict the scope of the Convention to physical biological material.[11] However, the opposite viewpoint also has support, with the assertion that such a restriction does not apply and that a teleological interpretation of the Protocol (and the CBD) suggests that benefits generated through the use of DSI must be shared "by the terms of the Nagoya Protocol".[12]

As in other realms, laws and policies affecting ABS do not keep up with technology.[13] One possible response is to argue that the definition of genetic resources as containing functional units of heredity (Art. 2 CBD) needs to be dynamic in order to meet overall ABS objectives.[14] However, because a 'dynamic

---

[3] Soplín and Muller (2009), Schei and Tvedt (2010), van Dijk et al. (2014), Hu et al. (2018) and Baltz (2018).

[4] NCBI (2019).

[5] Boles et al. (2017) and Park (2017).

[6] Lawson et al. (2020).

[7] Bagley (2015).

[8] Lawson et al. (2020).

[9] Bagley (2015), ICSWGSB (2016) and Servick (2016).

[10] SCBD (2018a).

[11] Spranger (2017).

[12] Sollberger (2018).

[13] Sollberger (2018) and Wynberg and Laird (2018).

[14] Schei and Tvedt (2010) and Tvedt and Schei (2013).

interpretation' is necessarily not fixed, it could give rise to legal and contractual uncertainty.

If DSI were not to be considered as equivalent to genetic resources, might it be considered as a 'derivative'? Although proposals were made during the negotiation of the Bonn Guidelines that information as a product developed from GR could be included under the concept of derivatives,[15] this was not explicit in the final Guidelines.[16] The term 'derivative' as first defined in the Nagoya Protocol was limited to "a naturally occurring biochemical compound ...", thus omitting information.[17] Australia, in a recent submission to the CBD[18] considered that DSI (or any other such terminology) are not 'derivatives' as defined under article 2 of the Protocol.

Discussions on whether or not DSI falls under the CBD rely in part on agreement on just what DSI comprises. The term is rarely used outside the CBD process and its meaning is undefined and disputed. The term 'DSI' was developed in the CBD process for the purposes of that process. Scientific advice can give options for its meaning (see table developed from Ad Hoc Technical Expert Group meeting below (Table 1)), and from those options an understanding of the implications of different choices of the meaning of DSI can be assessed. Nevertheless, ultimately a decision on scope of 'DSI' can only come from negotiations under the CBD.

The key issue for provider countries is whether benefit-sharing results from the use of DSI. This is more than a legal issue, but also an ethical issue, and a practical matter of how scientific and commercial relationships can be maintained and prosper with mutual benefit. Increasingly countries are adding DSI to their national ABS legislation,[19] although with different or unstated definitions. Ultimately the conclusions around DSI and the CBD will determine whether benefit-sharing from the use of DSI is governed by legal, contractual or ethical constructs, or a mixture of all three.

1.2 Policy Process

The issue of 'Digital Sequence Information' is not new,[20] and was raised for example under the International Treaty on Plant Genetic Resources for Food and Agriculture (ITPGRFA) in 2015 when the 6th meeting of the AHTEG to Enhance the Functioning of the Multilateral System of ABS was asked to consider genetic information. The concern over DSI was formally brought before the CBD in 2016 at

[15] SCBD (2002).

[16] Sollberger (2018).

[17] Spranger (2017).

[18] Australian Government (2019).

[19] Submission by the African Group of Negotiators in SCBD (2017b) and Bagley et al. (2020).

[20] Soplín and Muller (2009).

Table 1 Groups proposed by the AHTEG on DSI in 2020 to clarify scope of DSI

	Information related to a genetic resource			
	Genetic and biochemical information			Associated information
Group reference	Group 1	Group 2	Group 3	
High-level description of each group	DNA and RNA	Group 1 + proteins + epigenetic modifications	Group 2 + metabolites and other macro-molecules	
Examples of granular subject matter	• Nucleic acid sequence reads; • associated data to nucleic acid reads; • non-coding nucleic acid sequences; • genetic mapping (e.g. genotyping, microsatellite analysis, SNPs, etc.); and • structural annotation.	• Amino acid sequences; • information on gene expression; • functional annotation; • epigenetic modifications (e.g. methylation patterns and acetylation); • molecular structures of proteins; and • molecular interaction networks.	• Information on the biochemical composition of a genetic resource; • Macro-molecules (other than DNA, RNA and proteins); and • cellular metabolites (molecular structures).	• Traditional knowledge associated with genetic resources • information associated with DSI Groups 1, 2 and 3 (e.g. biotic and abiotic factors in the environment or associated with the organism) • other types of information associated with a genetic resource or its utilization.

Source: Copied from table in the AHTEG report (SCBD 2020)

COP 13/MOP 2 in Cancun, with a proposal that it be considered within scope of the Nagoya Protocol. After considerable discussion the CBD parties decided that information should be sought on the implications of including DSI within the Convention, and asked for an Ad Hoc Technical Expert Group (AHTEG) to be set up.[21] In response to this decision stakeholders' views were submitted,[22] consultants reported on terminology,[23] and the AHTEG provided a report.[24] The subsequent text from SBSTTA was almost completely bracketed,[25] indicating little agreement on the way forward.

Further negotiations at COP 14/MOP 3 delivered no agreement on whether or not DSI should fall under the CBD, but did produce decisions seeking more information.[26] This included information providing conceptual clarity on the nature of DSI through a "science- and policy-based process", and a further AHTEG. Four peer-reviewed studies were commissioned, covering the concept and scope of DSI, how DSI is currently used, traceability of digital information, public and private databases of DSI, and how domestic measures address benefit-sharing arising from commercial and non-commercial use of DSI.[27] The AHTEG considered this information and its report[28] has been submitted for consideration by a meeting of an open-ended working group supporting the preparation of the post-2020 global biodiversity framework prior to COP 15.

The disagreements about inclusion of DSI within the CBD have been so profound that some countries suggested that they would withdraw from the NP if its scope were to include DSI, while others have threatened to withdraw from the post 2020 process if it does not. Similar disagreements in the ITPGRFA at the last Governing Body (December 2019) led to failure of negotiations on benefit-sharing under the treaty.[29] The implications of including DSI within the NP are certainly considerable, and some stakeholders predict a negative impact on non-commercial biodiversity research that supports CBD implementation,[30] as well as activities with a more commercial aim.[31] There has also been a suggestion that inclusion might derail current work on implementation of the Nagoya Protocol, which is already suffering from lack of capacity in at least some countries.[32]

---

[21] SCBD (2016).

[22] SCBD (2018a, b).

[23] Laird and Wynberg (2018).

[24] SCBD (2018c).

[25] SCBD (2018d).

[26] SCBD (2018e, f).

[27] Bagley et al. (2020), Houssen et al. (2020) and Rohden et al. (2020).

[28] SCBD (2020).

[29] Medaglia (2020).

[30] SCBD (2018a).

[31] Bagley (2015, 2017).

[32] ICC submission in SCBD (2018a).

At the same time that CBD is discussing DSI the same topic is also under consideration in other fora: the ITPGRFA, the Commission on Genetic Resources for Food and Agriculture,[33] the UN Convention on the Law of the Sea (UNCLOS) and UN General Assembly discussions on an international legal instrument for Biodiversity Beyond National Jurisdiction (BBNJ), the World Health Organization (WHO) Pandemic Influenza Preparedness Framework (PIP Framework) and World Health Assembly, and the World Intellectual Property Organization (WIPO).[34] It would be challenging if different solutions were found in the different contexts.

1.3 National Responses

As Parties to the CBD and Nagoya Protocol develop or revise their ABS legislation some are including DSI within it in some form.[35] There are at least two (non-exclusive) ways in which this is expressed: (i) to include potential DSI in Mutually Agreed Terms (MAT) applying to the genetic resources from which the DSI are subsequently generated; (ii) to assert rights over DSI downloaded from a database (either in the provider country of the genetic resources or anywhere else in the world), at least sometimes irrespective of the conditions that were placed at the time of access of the original genetic resources. In the latter case, the country may treat DSI as equivalent to GR and require PIC and MAT from anyone using it, or alternatively not seek to regulate access (although some registration procedure may be required), but require benefit-sharing on utilization; this may arise from a stated equivalence to GR, or through (perhaps implicit) invoking of article 5 (1) of the NP and treating DSI as being covered by persistent rights; the difference may not be apparent in legislation. The range of approaches will pose problems in implementation, particularly when users may not know or never visit the country concerned, or even be aware of such legislation (see also Sect. 5). Moreover, because different countries are defining DSI independently, users cannot assume they know what concepts apply in any given case.

1.4 Main Questions

Three underlying questions need resolution to enable progress in discussion:

[33] Commission on Genetic Resources for Food and Agriculture (2019a).

[34] SCBD (2018g).

[35] Bagley et al. (2020).

1. What is Digital Sequence Information?
2. Does DSI fall under the sovereign rights of the provider country of the genetic resource?
3. Can the use of DSI fall under the CBD/Nagoya Protocol?

The following sections will address these points.

2 What Is DSI?

2.1 Terminology

Before there can be any meaningful discussion on how to manage DSI in the context of ABS there must be clarity on what it encompasses. However, the discussions on DSI are taking place in a context of scientific change, when hitherto unconsidered possibilities rapidly become the norm. Terminology and definitions used in agreements need to be sufficiently dynamic and flexible to encompass future changes, but also sufficiently precise so that users and regulators can determine with an appropriate degree of legal certainty whether regulations apply in any given case.[36] Insufficient precision on the scope of the term 'Digital Sequence Information' has already led to documents submitted to the CBD based on very different concepts, and consequently with different policy and legal implications.

The term 'Digital Sequence Information' appears to be largely limited to the CBD process, being brought into the mainstream of discussion in CBD COP decision XIII/16. COP decision XIV/20[37] notes "the term "digital sequence information" may not be the most appropriate term, and ... it is used as a placeholder until an alternative term is agreed". The WHO PIP framework is discussing the issue under the terms 'genetic sequences' and 'genetic sequence data' (GSD)[38] and 'Digital Sequence Data'. The ITPGRFA has used both DSI[39] and 'Genomic Information'[40] and the UNCLOS process on BBNJ is using 'genetic sequence data', 'digital sequence data' and in silico.[41] Other terms used in various sectors include genetic sequence data, genetic sequences, genetic sequence information, genetic resource sequence data, digital sequence data, genetic information, dematerialized genetic resources, intangible genetic resources, in silico utilization, information on nucleic acid sequences, nucleic acid information, nucleotide sequence data, natural

[36] Tvedt and Schei (2013).

[37] SCBD (2018e).

[38] WHO (2018).

[39] Welch et al. (2017).

[40] Manzella (2016).

[41] United Nations (2019).

information and sequences;[42] this is by no means an exhaustive list, and few of the terms, where they can be defined, are precise synonyms.[43]

Each of the words in 'Digital Sequence Information' has been the subject of detailed analysis and interpretation by a variety of stakeholders.[44] The three words cover different aspects:

'Digital'—the means of storage (also 'in silico'). There seems to be a general agreement that the term 'digital' is an unnecessary qualification since, while digital storage and transmission is the current prevalent means of information transmission, the term might exclude either novel systems or more traditional paper-based information media (see submissions to CBD[45]). The key understanding is that the subject is available for use without recourse to the original genetic resource or a physical derivative of it.

'Sequence'—the data/information that is available for use instead of the physical genetic resource (also 'genetic sequence', 'genetic resource sequence', 'nucleic acid sequences' 'nucleic acid', 'nucleotide sequence'). The term 'sequence' appears to refer only to the order of base pairs in nucleic acid (DNA and RNA) ('Nucleotide Sequence Data'), and possibly of amino acids in a protein. Although this term would seem to preclude wider interpretations, the subject of this term assumed in submissions to the CBD and other documents can include much broader concepts. This is discussed in the next section.

'Information'—Nucleotide sequences are usually referred to by researchers as 'data'.[46] 'Information' is a wider term, which may be distinguished from 'data', 'data' being observations lacking meaning while 'information' emerges through cognitive (or other) processing of data,[47] although this terminology is not clearly understood in law.[48] The use of 'information' in DSI terminology implies a wider scope than the use of 'data'.

2.2 *Scope of the Concept*

As noted, while the term 'DSI' is used in CBD discussions it is not used anywhere else, at least in the scientific world, and the CBD process has not so far defined it. The term has been applied without restriction, and attempts to discover what it means outside the CBD context are necessarily futile. Two possible approaches to understanding the meaning of DSI may be considered as 'inclusive' and

---

[42] Brazil in SCBD (2017a), Augusto and Gutiérrez (2014) and Laird and Wynberg (2018).

[43] Houssen et al. (2020).

[44] Laird and Wynberg (2018), SCBD (2018a) and Houssen et al. (2020).

[45] SCBD (2019).

[46] Laird and Wynberg (2018) and SCBD (2018a).

[47] Baškarada and Koronios (2013) and Rowley (2007).

[48] Bygrave (2014).

'exclusive'.[49] The inclusive approach has been the norm so far, and has been used to include almost any product of utilization that is stored electronically. Variations of scope using this methodology are effectively *ad hoc*, and justification for them is often unclear. In contrast, the exclusive approach asks "If DSI were to be agreed to fall under the CBD as an intangible equivalent to genetic resources, what could it include (and what would have to be excluded)?" This approach has been followed implicitly or explicitly by some of the recent submissions to the CBD.[50]

The inclusive approach was adopted by the CBD AHTEG on DSI in 2017, and led them to identify a list "relevant to the utilization of genetic resources".[51] Costa Rica's recent submission to the CBD[52] repeats this list to explain the concept of DSI as they understand it, although later in the submission focus more on amino acid sequences and metadata.

The 2020 AHTEG did not attempt to define DSI but instead, making use of the study by Houssen et al.,[53] provided a set of increasingly inclusive concepts. These did not refer to the logic of an exclusive concept proposed above but, as discussed below, can be related to that.

The broadest proposal for scope of 'DSI' in the submissions to the CBD[54] is probably "natural information", defined by Ruiz Muller[55] as "any non-uniform expression, difference or distinction produced by nature".[56] The African Group of Negotiators also favored this term in their recent submission,[57] but stated "the more relevant focus should be on utilization and not terminology, as precise terminology is not critical for this issue. This is because if information results from the utilization of genetic resources, it is within the scope of the CBD and Nagoya Protocol (in particular article 5.1) and subject to benefit sharing".[58] The implications of the use of article 5 (1) are discussed below. India, in the 2019 submissions to the CBD[59]

---

[49] Lyal (2020).

[50] CETAF, NHM-RBGK-RBGE, SPHNC, ICC in SCBD (2019).

[51] SCBD (2018c).

[52] SCBD (2019).

[53] Houssen et al. (2020).

[54] SCBD (2018a, 2019).

[55] Ruiz Muller (2018).

[56] The concept of 'natural information' as a basis for benefit sharing has been expounded also by Vogel (1994) and Vogel et al. (2011) and others. The concept is far broader than 'genetic resources', and in the context of the CBD is used in concert with 'Bounded openness', which is the conceptual foundation which stresses that natural information flows freely for R&D, until commercial success of an innovation at which time the innovation protected by intellectual property is obligated to share monetary benefits (Vogel et al. 2018; Ruiz Muller 2018). In this model the scope of DSI is unimportant, and could include information on any derivative; the important condition is that there should be benefit sharing at the commercial end of a process that makes use of a biological resource.

[57] SCBD (2019).

[58] African Group of Negotiators submission in SCBD (2019), p. 2.

[59] SCBD (2019).

suggested a very broad scope, although did not use the term 'natural information'. They suggest DSI "may include any information derived out of these nucleotide or amino acid sequences, their biochemical composition, gene locations, genetic maps, chromosome maps, localization details, functional expression details, genome sequences and their annotations, and applied aspects in decoding the sequence information. The scope of DSI thus would inter alia cover information on genetic or biochemical composition including nucleic acid sequence gene reads and associated data, sequence assembly, its annotation and gene mapping describing the genome, individual genes or fragments thereof, organelle genome, gene expressions, macromolecules and cellular metabolites." Many other submissions used a much more restricted scope.

Concepts proposed or assumed in submissions to the CBD and in the first AHTEG report may be arranged along two rough axes: content and human intervention. In list 'A' there is progression from close to DNA to more distant; in list 'B' increasing amount of human input (processing), although in neither is the progression simple. This method was used by Houssen et al.[60] in their 2020 report to the CBD and agreed in principle by the AHTEG.[61] The two axes are:

A. Genetic/Biochemical:

1. Nucleic acid sequence—the order in which different nucleotides occur in a strand of DNA or RNA;
2. amino-acid sequence of proteins produced from gene expression (i.e. derivatives);
3. molecular structures of gene products and derivatives (cell metabolites etc.);
4. epigenetic heritable elements. Not all heritability is encoded in the genes themselves, but some heritable factors have been found elsewhere in the cells. Mexico and Argentina in the 2019 submissions on DSI mentioned one of these, Methylation patterns.[62]

B. Human intervention

1. Nucleotide sequence of an organism as read from that organism (raw sequence). It does not include information connected with or resulting from the analysis or further application, e.g. sequence assembly, sequence annotation, genetic maps, metabolic maps, three-dimensional structure information or physiological properties related to it.
2. Technical aspects of sequencing: sequencing libraries, preparation techniques and data files associated with a nucleotide sequence. This is the 'associated data' referred to by the AHTEG in its report.[63]

[60] Houssen et al. (2020).

[61] SCBD (2020).

[62] SCBD (2019).

[63] SCBD (2018c).

3. Metadata of collection of the genetic resource: taxonomic name, information on ecological relationships, abiotic factors of the environment.
4. Aligned nucleotide sequences, with information on sequence assembly etc.
5. Structural annotation of genomic elements (ORFs and their localization;[64] gene structure; coding regions; location of regulatory motifs).
6. Functional annotation of genomic regions (biochemical function; biological function; involved regulation and interactions; expression).
7. Function, such as behavioral observations.
8. Structure of organism, including morphological observations and phenotype.
9. Molecular structures of gene products and derivatives (cell metabolites etc.).
10. Patentable discoveries and inventions (modalities of use)

The items on these lists can be combined into groups that reflect the degree of biological processing and the proximity to the underlying genetic resource.[65] The 2020 AHTEG, using this approach, proposed three cumulatively inclusive groups (Group 2 includes all elements of Group 1, and Group 3 contains all elements of Groups 1 and 2) (Table 1).[66]

The groups were proposed as an aid to negotiators so that there can be clarity on the topic of their discussions and considerations on what might be encompassed in the concept 'Digital Sequence Information'; the proposal was not intended to deliver an opinion on this scope itself. Consideration of each of the three groups in terms of what impact a choice would have on the CBD implementation, use outside the CBD, and benefit sharing, will inform any decision. The AHTEG made a start on considering such impacts. Notably, associated information (other than structural annotation of nucleotide sequences) was proposed not to be DSI itself; Associated Traditional Knowledge was also excluded from DSI although the AHTEG noted its importance.

Using the inclusive approach all of the contents of Groups 1–3 (and, indeed, the associated information) might be termed 'DSI'. Ethically, there are arguments for benefit-sharing when novel use is made of any of groups 1–3. However, the point of the negotiations within the CBD is not purely ethics, but legal guidance and the scope of the CBD itself.

Since DSI is a concept that arose from the CBD and is being used to decide a specific application of the CBD it can only be defined in the context of the CBD. The main basis for accepting DSI as coming under the CBD and NP is the (disputed) proposition that DSI is the 'intangible equivalent' of a physical genetic resource and as such falls under the sovereign rights of the country from which the original genetic resource was accessed (see Sect. 3.1 below). Taking this position as a starting point an exclusive concept of DSI can be identified. This is based on what tangible resources fall within the scope of the CBD and therefore what their 'intangible equivalents' are.

[64] An ORF is a continuous stretch of codons that contain a start codon.

[65] Houssen et al. (2020).

[66] SCBD (2020).

When a genetic resource is accessed in-situ in a provider country, all that is acquired is an organism or set of organisms. These lack an identity, and there is no 'attached' information; any information has to be provided through research or pre-existing knowledge of the person accessing that organism or the subsequent researcher. Discussion of "DSI" in the context of sovereign rights would logically be an analogue of what is accessed as a material entity, focusing on the genetic resource element of that entity.

Article 15 of the CBD refers specifically to genetic resources (GR). The definition of a GR in article 2 of the CBD is "genetic material of actual or potential value" where "'Genetic material' means any material of plant, animal, microbial or other origin containing functional units of heredity". "DSI", consequently, would contain an intangible analogue of "functional units of heredity" if it is the intangible equivalent of a GR. Functional units of heredity were generally considered to be genes when the CBD was negotiated, but the term was never defined officially and practical and legal meanings of the term have not been agreed or established.[67] The clearest and most generally-agreed concept is to consider DNA or RNA a necessary component of any functional unit of heredity. Consequently, for an intangible analogue a DNA or RNA sequence[68] must be included.

DNA or RNA sequences may be termed 'Nucleotide sequence data' (NSD)[69] or 'Genetic Resource Sequence Data (GRSD).[70,71] These are the actual genetic sequence data of a specific genetic resource. They may include sequences of parts of genes, DNA 'barcodes', entire genes, full genomes of organisms, clades (pangenomes) or environmental samples (metagenomes). While some previous discussions have included consideration of different types of DNA and whether they all represent 'functional units of heredity' (e.g. non-coding vs. coding DNA sequences, regulatory sequences, conserved sequences, genes that encode specific traits, 'junk' DNA etc.), with increasing understanding of gene functionality these distinctions are becoming unhelpful and all may deliver information of value on analysis.[72] Consequently, the closest functional analogy between a genetic resource and an intangible equivalent is with NSD. A number of countries have apparently adopted this concept.[73] 'GSD' (Genetic sequence data) is used in the WHO PIP

---

[67] Tvedt and Young (2007) and Tvedt and Schei (2013).

[68] DNA and RNA—nucleic acids—are made up of long chains of connected 'nucleotides'. The four types of nucleotides in DNA and RNA are Adenine, Thymine (Uracil in RNA), Guanine, and Cytosine, which are usually abbreviated to A, T, U, G and C. The order in which these nucleotides occur in a strand of DNA or RNA is the Nucleotide Sequence.

[69] CETAF submission in SCBD (2019).

[70] This is a more restricted concept than NSD, excluding information on sequence assembly etc., but the differences in detail are not significant for the purposes of this paper.

[71] ICC submission in SCBD (2019).

[72] Schei and Tvedt (2010) and Wynberg and Laird (2018).

[73] Submissions to the CBD in SCBD (2019) from South Africa, Malawi, Canada, EU and its Member States, Japan, Republic of Korea, Switzerland, USA.

Framework and has the same meaning. This is the Group 1 of the latest AHTEG.[74] If negotiators cannot agree if this Group falls within the scope of the CBD there can be no agreement that the other two groups do, since they both include these elements.

The additional elements included in Groups 2 and 3 are more problematic for inclusion using the approach of equivalence to GR, and include proteins, epigenetic heritable factors and metabolites. There are different views regarding proteins and the amino acids of which they are comprised. Some submissions to the CBD (Argentina, Costa Rica, India, Iran, Mexico, Third World Network) include amino acid sequences in proteins in their concept of DSI.[75] Others implicitly or explicitly exclude these (South Africa, Malawi, Canada, EU and its Member States, Japan, Republic of Korea, Switzerland, USA). Since amino acid sequences are not themselves units of heredity inclusion is questionable given the logic of equivalence to genetic resources, although they might be considered as the intangible equivalent of derivatives.

The 2020 AHTEG did not determine any difference between 'data' and 'information', although a distinction is mentioned above. In the DSI context 'data' are analytically derived from naturally-occurring genetic resources. However, the genetic resource itself when accessed has no intrinsic 'information'. Thus the properties of the genes are not discoverable in the same way that the sequence of nucleotides in that gene is discoverable. The data need to be analyzed and researched upon before such information can be developed. Since 'information' is developed through analysis of 'data', it is potentially covered by intellectual property rights (IPR). For example, while patents cannot be granted on DNA segments simply because they have been isolated from the surrounding genetic material (at least under some jurisdictions), when those data are coupled with the discovered properties of the gene (and perhaps its 'use' in an anthropocentric context) this may be patented.[76] Sharing of IPR is identified as an example of both monetary and non-monetary benefits in the Nagoya Protocol Annex, and is often a component of ABS negotiations;[77] IPR are a result of research and not under sovereign rights of a country, unlike genetic resources. This point is raised by Belarus in its recent submission to the CBD.[78]

Although at the time of negotiation of the CBD concepts of heritability were focused on genes (hence the terms 'genetic resources' and 'genetic material') there has been more recent explorations of epigenetic inheritance (i.e. heritable factors not contained in genes), such as methylation patterns.[79] Some submissions to the CBD, such as from Argentina, Mexico and the Third World Network, refer explicitly to

[74] SCBD (2020).

[75] SCBD (2019).

[76] Cole (2015), WIPO (2016) and Nicol et al. (2019).

[77] WIPO (2018).

[78] SCBD (2019).

[79] Bird (2002) and Trerotola et al. (2015).

methylation patterns.[80] Whether or not these fall within current CBD understanding of 'functional units of heredity' is unclear; they are explicitly not 'genetic resources'. As with other elements of Group 2, discussion could usefully be deferred until the status of Group 1 has been decided.

If the outcome of the current negotiations is to bring DSI within scope of the CBD, because it is agreed that it is the intangible equivalent of GR and falls under national sovereignty, it is difficult to see how the concept can be extended beyond the current understanding of GR within the Convention and its Nagoya Protocol, and include subjects beyond the reach of sovereign rights.

As mentioned above, a priority for developing countries is establishing fair and equitable benefit-sharing protocols for DSI. This might not rely on the explicit inclusion of DSI (in whatever sense) within the CBD, but a sui generis system might be developed.

3 DSI as a Component of ABS

3.1 Sovereign Rights, the CBD and the Nagoya Protocol

The sovereign right of countries to their natural resources is expressed in UNGA resolution 1803 (XVII) and reaffirmed in the context of the CBD in its preamble ("Reaffirming that States have sovereign rights over their own biological resources"). Article15 (1) CBD sets out the applicability of these rights to ABS ("Recognizing the sovereign rights of States over their natural resources, the authority to determine access to genetic resources rests with the national governments and is subject to national legislation.") This text treats GR as a subset of natural resources, but while explicitly recognizing States' rights to determine access to GR, it does not diminish their rights to determine access to other natural or biological resources, since the right stems from the sovereign rights and is not in addition to it. However it does, in its use of the term 'genetic resources' as opposed to 'biological resources' (which is used elsewhere in the text eleven times), set out the scope of ABS.

As discussed in Sect. 1 above, article 15 CBD, and the preambular text reaffirming the sovereign rights of states over their natural resources, provide the only rationale (other than ethical considerations) in the text for the inclusion of genetic resources within the CBD for the application of ABS. The arguments for including DSI within CBD are based on the premise that it is the intangible equivalent of a GR, and thus falls under sovereign rights of states.

This argument, if accepted, would change the status of DSI from something that currently is discussed under the CBD as a benefit or a tool into a subject of the CBD.

[80] SCBD (2019).

If DSI were to be recognized as the equivalent of a genetic resource and falling under the sovereign rights of countries then it would be addressed within the third objective of the CBD—the fair and equitable sharing of benefits arising out of the utilization of genetic resources. The Nagoya Protocol is a Protocol to the CBD, thus is subsidiary and must use the same concepts and rationale. Article 1 of the Nagoya Protocol gives its objective as fair and equitable sharing of benefits arising out of the utilization of genetic resources and, if DSI is the equivalent to genetic resources, potentially DSI would fall under the provision of the Nagoya Protocol. However, the Parties to the CBD may consider that many of the provisions of the NP are inapplicable operationally to DSI, because of factors such as the means of storage in huge databases held outside the counties where the original GR were accessed, the high transaction rate and the potential for users to sample perhaps thousands of genes simultaneously. In such a case there could be a decision to handle DSI within the CBD but not the NP, perhaps through setting up a dedicated multilateral benefit-sharing mechanism (and thus avoiding other issues surrounding article 10 of the NP that might delay its implementation) or even using a sui generis system.

3.2 Does DSI Fall Under the Sovereign Rights of the Provider Country of the Genetic Resource?

Clearly, many countries take the position that DSI does fall under their sovereign rights, although this precise formulation has not necessarily been used. However, it is questionable whether data derived from the study of natural resources are themselves natural resources, in the context of the 1962 UN resolution. Outside the debate around DSI and whether genetic data (or information) is equivalent to genetic resources the question does not seem to have been raised and there seems no general assumption that data and natural resources are equivalent; no definition of natural resources has been found that includes 'information' or 'data'. Rather, definitions such as "Natural resources are natural assets (raw materials) occurring in nature that can be used for economic production or consumption" (OECD) seem to exclude data, since data are generated through human action and do not occur in nature.

An alternative possibility is that sequence data fall under IPR of the provider state. However, Deplazes-Zemp[81] argues that genetic resources/genetic information cannot itself be IPR without human intervention. This argument is also used by WIPO, which states that "GRs as encountered in nature are not creations of the human mind and thus they cannot be directly protected as intellectual property (IP)".[82]

---

[81] Deplazes-Zemp (2018).

[82] http://www.wipo.int/tk/en/genetic/.

Since there are no obvious routes to determine whether DSI can a priori be considered natural resources, arguments relating to its falling under the CBD should be considered for illumination.

3.3 *Can DSI Be Within Scope of the CBD?*

There are strongly contrasting views about whether DSI can be within scope of the CBD (and thus the Nagoya Protocol).

Although the Convention does not explicitly cover genetic data, Brazil[83] asserts that "Most Parties understand that intangible genetic resources are defined within the Convention". India[84] expresses a similar view, identifying the use of the word 'value' in the definition of genetic resources as supporting application to intangible elements. Part of Brazil's argument rests on a definition of the word 'material' in the Oxford English Dictionary[85] (OED) to include information or ideas (as in 'genetic material') and contrasting this with 'matter', although the OED also defines 'material' as "the matter from which a thing is or can be made".

Several stakeholders have made the point that the intent of the CBD and Nagoya Protocol would be frustrated if DSI were to be excluded.[86] Their position is that since the intent of ABS in the original negotiation of the CBD was that provider countries would benefit in a fair and equitable fashion from any utilization of genetic resources that fall under their sovereign rights, this intent should be persistent when novel means of enabling that utilization are developed—in this case, using the product of utilization of GR rather than the original GR. Sollberger[87] argues that, considering the spirit and purpose of the Nagoya Protocol, it cannot make a difference whether the product of research and development is based on utilization of physical genetic resources or is based on a synthetic substance which was created making use of information gained when sequencing the original genetic resource. He also does not believe it to be relevant whether the product is entirely developed by the entity who accessed the original GR, or by actors who are connected indirectly only via their use of a publicly accessible database.

In contrast, information has been asserted not to be material but immaterial,[88] and consequently could not fall within a definition of 'genetic material'. Spranger[89] in his analysis for the German Federal Ministry of Education and Research, notes that in article 2 of the CBD 'genetic resources' fall within 'biological resources, which

---

[83] SCBD (2017a).

[84] SCBD (2017b).

[85] Onions (Ed) (1973).

[86] Sollberger (2018); Ethiopia on behalf of the African Union, Brazil in SCBD (2017b).

[87] Sollberger (2018).

[88] E.g. European Seed Association in SCBD (2017a).

[89] Spranger (2017).

are defined as including "genetic resources, organisms or parts thereof, populations, or any other biotic component of ecosystems with actual or potential use or value for humanity". He notes that the phrase "or any other biotic component of ecosystems" appears to exclude anything that is not a biotic component of ecosystems, so intangible data cannot be included, and there is no flexibility to include them subsequently. Others assert more broadly that the scope of the CBD does not cover information,[90] although this blanket view might be challenged by the inclusion of traditional knowledge associated with genetic resources in the scope of the Nagoya Protocol.

Some correspondents to the CBD[91] note that detail of scope of terms used in the Nagoya Protocol were discussed in its preparation, and recommend that these be considered when deciding whether DSI was already within scope.[92] As pointed out by Schei and Tvedt,[93] under the Vienna Convention on the Law of Treaties, article 32, "recourse may be had to ... the preparatory work of the treaty and the circumstances of its conclusion" if, among other things, the meaning of a term is obscure, or leads to a "result which is manifestly absurd or unreasonable".[94] Indeed, information related to genetic resources and sequence data was discussed in the development of the Bonn Guidelines (as a possible derivative) and the Nagoya Protocol.[95] However, these discussions did not lead to a clear introduction of the concept to the Guidelines, and no documents have been seen that suggest that this was an agreed intention.

Article 31 (1) of the Vienna Convention provides another relevant tool to help understand whether DSI should be considered as equivalent to genetic resources in the context of the CBD. It states: "A treaty shall be interpreted in good faith in accordance with the ordinary meaning to be given to the terms of the treaty in their context and in the light of its object and purpose." In the context of the CBD the disagreement as to the 'ordinary meaning' of genetic resources has been discussed above. In academic literature there is also disagreement. For example, "in the case of genetic resources, the genetic information is the resource being used and valued",[96] "genetic resources as natural information"[97] and "the gene as a form of information".[98] However, others take a contrary view, referring to genes as 'carrying' or 'containing' information. For example "Genes are not information",[99] "chromosomes were identified as the carriers of genetic information" and "genome as a

[90] Chartered Institute of Patent Attorneys in SCBD (2017a).

[91] E.g. Brazil in SCBD (2017a).

[92] SCBD (2002) and Schei and Tvedt (2010).

[93] Schei and Tvedt (2010).

[94] United Nations (1969).

[95] SCBD (2002, 2008).

[96] Deplazes-Zemp (2018).

[97] Vogel (1994), Soplín and Muller (2009), Vogel et al. (2011) and Ruiz Muller (2015).

[98] Dawkins (2008).

[99] Rossi (2014).

physical entity encoding the information",[100] and "Each genome contains all of the information".[101]

There is no conclusive decision provided by the texts, and legal opinions can be found supporting both alternatives—that DSI should be (is) covered by the CBD, and that it is not and cannot be so covered. Ultimately it is a policy decision.

4 Implementation of National Legislation Covering DSI[102]

The NP in general sets only minimum rather than maximum obligations.[103] Consequently, in developing domestic access legislation for genetic resources, some national governments have chosen to include greater detail and requirements than has been agreed in the NP. Thus, irrespective of whether or not DSI falls under the CBD and Nagoya Protocol, some countries are already explicitly including DSI or an equivalent in their national access legislation. CBD COP noted "that some Parties have adopted domestic measures that regulate the access to and use of digital sequence information on genetic resources as part of their access and benefit-sharing frameworks".[104] Such domestic legislation has been reviewed in the CBD-commissioned study by Bagley et al.,[105] who also point out that some countries have taken the decision not to regulate DSI. Inclusion in legislation or regulations takes two non-exclusive forms; (i) the explicit statement in MAT covering GR that benefit-sharing from DSI generated from those GR forms part of the access requirements, and (ii) that the country's ABS rights extend to DSI that has already been generated and is held in a database elsewhere. Relevant to the understanding of both of these options is the application of article 5 (1) of the NP, as identified by the African Group of Negotiators on Biodiversity-Ad Hoc Group on Digital Sequence Information in their submission to the CBD.[106] As pointed out by Bagley et al.[107] some countries require benefit-sharing without putting access requirements in place.

Different nations are not constrained to a particular scope, concept or terminology for DSI, leading to these differing between different countries. Terms used (where there is an explicit inclusion) include genetic information, genetic heritage,

[100] Goldman and Landweber (2016).

[101] Goldman and Landweber (2016).

[102] Considerably more information and analysis is present in the 2019 Study commissioned by the CBD - Fact-finding Study on How Domestic Measures Address Benefit-sharing Arising from Commercial and Non-commercial Use of Digital Sequence Information on Genetic Resources and Address the Use of Digital Sequence Information on Genetic Resources for Research and Development.

[103] Bagley (2015).

[104] CBD COP decision 14/20, para 5.

[105] Bagley et al. (2020).

[106] SCBD (2019).

[107] Bagley et al. (2020).

intangible components, gene sequences, sequence information, information, and information of genetic origin.[108] In some cases original text that was intended to cover Traditional Knowledge has been implicitly or explicitly expanded to include DSI, such as in Peru and Kenya.[109] Often the concepts are not described in detail, complicating interpretation of national legislation by users.

4.1 Inclusion in PIC and MAT When GR Are Accessed

PIC and MAT provisions may govern whether sequence data can be published following sequencing, some countries, such as Panama, prohibiting publication on databases such as those in the INSDC.[110] Others allow publication, but place conditions on this. For example Malawi states that "ABS contracts indicate that in any digital publication of sequence data or any other digital expression of the genetic resources or results from the genetic resources, full acknowledgment is to be given to the government of Malawi, the project enabled by the contract, and the following clause shall be enclosed in the digital publication: "The government of Malawi has commercial rights or other further use rights in products or processes developed based on the research results or this DSI, and any use requires a contract of use with the Government of Malawi."[111] Colombia has a similar requirement (Bagley et al. 2020). Such contractual relationships seem to be the main tool used by countries to regulate conditions of use and benefit-sharing obligations on DSI.[112]

Such conditions might provide management of resulting DSI by the provider country. However, once any such information is made available on a public database that control is lost regarding third parties. Costa Rica apparently addresses this through powers granted to its national ABS authority (CONAGEBIO), which has the authority to impose restrictions on the dissemination or deposit of genomic information (e.g. sequence data) in public databases, where this information or data results from utilization of genetic resources accessed under a permit. Restrictions or bans on publishing DSI, if general, would lead to a stalling of the population of the international public databases, at least from the countries concerned. These databases de facto form a part of the Global Taxonomic Information System requested by CBD COP decisions on the Global Taxonomy Initiative[113] and, indeed, are used by every country in the World. As a model, placing such restrictions is less attractive in many ways than seeking benefits by other means.

---

[108] Bagley et al. (2020).

[109] Bagley et al. (2020).

[110] Bagley et al. (2020).

[111] African Group of Negotiators in SCBD (2019).

[112] Bagley et al. (2020).

[113] Submission by CETAF in SCBD (2017b).

4.2 Stating Country's Sovereign Rights to DSI That Has Already Been Generated and Is Held in a Database

In 2015 Brazil passed Law 13,123/2015 and implementing Decree 8772/16. The latter clarified that Brazil's genetic heritage is deemed to encompass information held in public genetic sequence databases such as GenBank.[114] The requirement of the legislation is that foreign researchers are only able to access[115] Brazilian genetic heritage if associated with a public or private Brazilian research institution, which will take responsibility for registering the activity. This is expected not only for material acquired directly in Brazil but also for DSI held outside the country.[116] However, registration is not required on acquisition, but before commercialization, so that benefit-sharing may be agreed. Brazilian Law 13,123/2015 defines genetic resources (or genetic heritage) as the genetic information from plants, animals, and microbial species, or any other species, including substances originating from the metabolism of these living organisms (Art. 2, I). The legislation is not clear what comprises in silico genetic heritage, but Brazil's submission to the Commission on Plant Genetic Resources for Food and Agriculture[117] defines DSI as "data information contained in a digital file with a precise order of nucleotides or amino acids". There have been a number of access registrations of in silico genetic heritage in Brazil's SisGen.[118]

Other countries require PIC to be sought for DSI from genetic resources that originated under their sovereignty. Malawi has developed guidelines on access to biological resources and benefit-sharing based on its Environmental Management Act[119] including such a provision. Similarly, the Ethiopian revised draft law covering access to genetic resources, which defines GR as 'any material of biological resource containing genetic information having actual or potential values for humanity and includes derivatives and digital sequence information', includes a requirement for PIC to be sought for sequence information from Ethiopian genetic resources prior to access. However, currently there is no clarity on how this would be obtained when the data are available in publicly and privately accessible databases hosted outside of Ethiopia. Nor is it clear how non-compliance with the PIC requirement for such sequence information would be tracked, or how benefit-sharing obligations would apply.[120]

Of these examples only Brazil has a clear statement of what comprises DSI (although not in its legislation). All of the examples require bilateral agreements at

[114] da Silva and de Oliveira (2018).

[115] 'Access' in Brazilian legislation, is equivalent to 'utilization' in the Nagoya Protocol use of terms.

[116] da Silva and de Oliveira (2018).

[117] Commission on Genetic Resources for Food and Agriculture (2019b).

[118] For a detailed discussion of the Brazilian case see Mozini in this volume.

[119] Government of Malawi (2019).

[120] Ethiopia in AU submission to CBD in SCBD (2019).

some stage, but none have a clear model of how their legislative requirements might be transmitted to users of databases outside their country.

If the DSI is contained within a database within the provider country itself there seems little reason to doubt that the state can exercise control and set conditions for its use, depending on its laws.[121] The emerging concept of data sovereignty, although related more to national security issues,[122] may prove to have relevance here. Extending this right to information held outside the country seems more tenuous and, indeed, falls under the same issue of data sovereignty. Spranger[123] notes that article 15 (3) of the CBD refers to "genetic resources ... provided by Contracting Parties that are countries of origin of such resources or by the Parties that have acquired the genetic resources in accordance with this Convention", and argues that in the case of DSI (provided and to the extent that it falls under the CBD) if stored in databases (which he equates with ex situ collections), the power of disposal does not lie with the country from which the original genetic resources originated, but with the country where the database is located where this is different. Such a position has not been stated by major databases in the field; indeed, the INSDC databases are data hosts and not owners—data ownership remains with the original data provider.[124] If such data providers are publishing the data under PIC and MAT there is then a potential route for transfer of obligations.

The various legislations mentioned above do not necessarily refer to a temporal scope; the implication is that downloading a sequence is equivalent to accessing a GR. The date at which the GR from which the DSI was derived was accessed appears not to be a consideration for the countries concerned.

There is no requirement under the Nagoya Protocol for countries where use of DSI is taking place to monitor such use or take action when benefits are not shared, although it might be assumed that compliance measures of countries that consider DSI to be covered by the concept of GR would assert this in compliance. Bagley et al.[125] did not report any such arrangement, although they noted that India and South Africa applied the consideration to their own DSI. While article 18 of the NP requires Parties to ensure there is an opportunity to seek recourse in the case of a dispute on compliance with MAT, this does not extend to cases where there are no MAT between provider and user, and no requirement under the Protocol for the user to seek them.

[121] Argentina comment on Laird & Wynberg in SCBD (2017a).

[122] Polatin-Ruben and Wright (2014).

[123] Spranger (2017).

[124] Karsch-Mizrachi et al. (2017).

[125] Bagley et al. (2020).

4.3 Use of Article 5 (1)

The African Group of Negotiators state that "if information results from the utilization of genetic resources, it is within the scope of the CBD and Nagoya Protocol (in particular Article 5.1)".[126]

This approach requires a little more discussion on the scope of the CBD (and hence NP). DSI is derived from utilization of GR, and is considered by many to constitute a benefit in its own right. This is the strong implication of the peer-reviewed study prepared for the CBD by Laird and Wynberg,[127] and benefits conferred by open access to the data were supported by Japan and Costa Rica in the study by Bagley et al.[128] Sharing of such data is frequently included in MAT for access to genetic resources. Although benefit-sharing as a process can be considered as within the scope of the Convention (Arts 4 (2), 5 (7) CBD) the benefits themselves are distinguished from genetic resources in the text and would not appear to fall under sovereign rights. Consequently, they would not be covered by CBD article 4. Although article 3 of the NP ("This Protocol shall apply to genetic resources within the scope of article 15 of the Convention and to the benefits arising from the utilization of such resources") seems to contradict this understanding, since the objective of the Protocol is effective benefit-sharing, the benefits themselves cannot logically fall within sovereign rights of states. Thus, while the NP can apply to benefits (e.g. as set out in article 5) it does not confer rights over those benefits.

Article 5 (1) of the Nagoya Protocol states "benefits arising from the utilization of genetic resources as well as subsequent applications and commercialization shall be shared in a fair and equitable way" (my underline). Rather than conferring sovereign rights over products of utilization this addresses persistence of rights to benefit-sharing beyond utilization. This might apply to cases where processing has taken place to the extent that genetic material is no longer present in the product.[129] Greiber et al.[130] take the view that the provisions of article 5 (1) regarding benefit-sharing "are pertinent" outside access requirements and agreements with the person accessing the genetic resources. However, they stop short of stating that there are legal obligations under the NP in such cases, and do not consider the specific case of DSI.

The application of article 5 (1), although not explicit, perhaps underlies Malawi's approach. They propose that users of online information would be bound by the standard clause (see Sect. 4.1) a "standard online accept-condition" which requires certain information about the user and acceptance of a contract with Malawi. They further propose a clickwrap agreement that would automatically send a notification to Malawi about access to the information, although with no clarity how this might

[126] SCBD (2019).

[127] Laird and Wynberg (2018).

[128] Bagley et al. (2020).

[129] Sollberger (2018); Brazil in SCBD (2019).

[130] Greiber et al. (2012).

be implemented.[131] Laird and Wynberg[132] discuss the use of clickwrap for DSI in their report to the CBD.

The relevance of article 5 (1) to DSI (or, indeed, tangible genetic resources) has not yet been discussed in detail. However, there is an important legal disconnect between the researcher who developed the DSI and the entity who downloads it from a public platform with the intent of using it. The former has PIC and MAT with the provider country, and is potentially subject to compliance scrutiny by the competent authorities in the country where utilization takes place. The latter has no agreement with the provider country of the genetic resource unless they have voluntarily sought one or there is a clickwrap condition on downloading the sequence as described by Malawi. In either case, these are contractual matters and not currently covered by the Nagoya Protocol.

5 Moving Forwards

The reason that DSI is an issue for many countries is that they believe benefit-sharing should take place in a fair and equitable manner when DSI is used. They see no distinction between utilization of GR and use of DSI and, irrespective of the legal discussion on whether DSI already falls under the CBD or not, benefit-sharing for its use is in the spirit of the agreement. This seems at very least a strong ethical position.

Overall, the benefit-sharing discussion is focused about the delivery of monetary benefits from commercial use. Although benefit-sharing has been discussed largely without this qualification, a negotiating position of many countries has been to exclude non-commercial research that delivers non-monetary benefits from additional requirements. Many submissions to the CBD in 2017[133] stressed the importance of the use of DSI to the implementation of the Convention. Hindering such activities impedes implementation of all three objectives of the CBD. Another area which has become very important recently is the rapid dissemination of the COVID-19 virus sequence and the use of this to develop test kits and vaccines, and track the evolution of different strains. The results of this work are of global significance and value.

As seen, some countries are putting provisions in their domestic legislation covering use of DSI, but such laws are only applicable within the jurisdiction of the country concerned. This is reminiscent of the situation regarding utilization of genetic resources prior to the Nagoya Protocol, when it was agreed that an internationally agreed compliance mechanism was required to monitor utilization and hence increase benefit-sharing. The purpose of countries wishing to have DSI included under the CBD and Nagoya Protocol is to generate a similar change and ensure that

---

[131] SCBD (2019).

[132] Laird and Wynberg (2018).

[133] SCBD (2017b).

benefit-sharing on the use of DSI does take place. One practical question is whether the available compliance mechanisms enshrined in the legally-binding Nagoya Protocol, and the potential to address benefit-sharing (Arts 5 (1, 3), 10 and 18 of the Nagoya Protocol), might be appropriate to DSI, given the attendant complexities for providers, regulators and legislators.

These practical challenges are not inconsiderable. The CBD and Nagoya Protocol are predicated on a simple linear model in which a user accesses a single genetic resource from a provider country, utilizes it and develops a product; benefit-sharing can take place at appropriate points in this value chain. However, even with a physical genetic resource that is no longer an accurate depiction of the real-life process in many cases. With DSI the user is acquiring the data from a third party and does not need to visit the provider country of the original genetic resources. The user may make use of many hundred DNA sequences from different organisms and even species, and also use them to build a novel artificial sequence that has the desired properties. Databases are increasingly being queried by computer not individuals,[134] increasing speed and the number of queries possible. The main data source for DSI (at least for sequence data, but also other data), is the INSDC; this has tens of thousands of consumers monthly and tens of millions of 'hits' monthly (from every country in the world).[135] Rapidity of response is extremely important to users, a factor very apparent in the COVID-19 pandemic, where health requirements made rapid access to data vital. The modality of use of the databases seems incompatible with the simple bilateral model of the Nagoya Protocol and, at the very least, the volume of transactions would not be serviceable for competent authorities who in many cases currently lack the capacity to address access requests for genetic resources.

The three databases of the INSDC have a uniform policy of "free and unrestricted access to all of the data records their databases contain" and "will not attach statements to records that restrict access to the data, limit the use of the information in these records, or prohibit certain types of publications based on these records. Specifically, no use restrictions or licensing requirements will be included in any sequence data records, and no restrictions or licensing fees will be placed on the redistribution or use of the database by any party".[136] These strong limitations, based in each case on the policies of the countries or regions in which the databases are based (including the USA, a non-Party) seem incompatible with requirements for a model that includes registration or tracking users. Moreover, many of the sequences included are of human DNA (and thus out of scope of the CBD) or from GR accessed in countries that do not have ABS requirements,[137] so a system such as

[134] Wilkinson et al. (2016).

[135] To better understand the use of these databases, the reader is invited to look at the real-time map of use of one of the INSDC databases, EBI-EMBL, at www.ebi.ac.uk/about/our-impact (accessed 24 June 2020).

[136] Brunak et al. (2002), also at http://www.insdc.org/policy.html (accessed 24 June 2020).

[137] Rohden et al. (2020).

registration to address DSI benefit-sharing would not be appropriate to apply to all users for all contents. One major database that does include registration, GISAD, also includes in its policies restrictions on use and distribution of data.[138] This provision has, as discussed in a recent meeting,[139] caused some challenges for researchers making use of sequence data from both GISAID and INSDC. Transmission of conditions automatically may be possible,[140] but if there is a requirement to take action on each sequence or data element prior to use this would slow research dramatically. Overall, there is a trade-off between rapidity and ease of use of database contents and the introduction of a registration or other tracking system, and loss of speed and ease will impact researchers, environmental managers etc. in every country of the world.

Developing a benefit-sharing mechanism seems both ethically sound and pragmatic. If no such system is developed provider countries are likely to increase restrictions on generation of DSI, and even of access to GR; signs of this are already to be seen. A mechanism that requires bilateral agreement prior to use is in most cases unlikely to be practicable. The alternative to this would be a multilateral system,[141] and to preserve and enhance societal benefits from the availability to and use of DSI for all parties this seems preferable. A multilateral system for DSI might fall under the Global Multilateral Benefit-Sharing Mechanism foreseen under article 10 of the Nagoya Protocol. However, this is subject to other concerns and negotiations are ongoing. Tying a solution to benefit-sharing for use of DSI to article 10 implementation could slow progress considerably. However, a separate system might be set up to address DSI in particular. There is no requirement *a priori* that a benefit-sharing mechanism should be under the Nagoya Protocol or, indeed, the CBD. Given that similar discussions are taking place in other fora (WHO, ITPGRFA, UNCLOS) it might be appropriate to develop a benefit-sharing mechanism that can support all three.[142]

How payments are put into such a system would need to be determined. This might be through a subscription or licensing model, which might have a scale of contributions according to the type of user or their geographic placement, or a model which requires contributions on commercialization (and in this reflects the 'bounded openness' model discussed by Vogel and Ruiz Miller). The submission of the African Group of Negotiators to the CBD,[143] although focusing on inclusion of DSI within scope of the CBD, does contain a message about negotiating

---

[138] GISAID EpiFlu™ Database Access Agreement, para 2(d) "Subject to applicable law, You agree not to distribute Data to any third party other than Authorized Users as contemplated by this Agreement." (accessed 24 June 2020). GISAID was set up to service the WHO PIP framework not the CBD.

[139] Under Chatham House Rules, so unattributable.

[140] Wilkinson et al. (2016) and Jacobsen et al. (2020).

[141] ABS Capacity Development Initiative (2019).

[142] Lawson et al. (2019, 2020).

[143] SCBD (2019).

differentiated benefit sharing rates for different classes of 'natural information'. One benefit of a subscription model is that resources would be available far earlier than if the model is built on payments on or subsequent to commercialization. However, there may be a perception that such a system would not be as fair or equitable as one related to value of the final product. One might consider a mixed model, with a basic subscription to access DSI and a subsequent payment related to commercialization. Similarly, how funds are disbursed, for example pro rata for geographic origin, or on application for a project implementing the CBD, is also a matter for consideration.

A further point is whether, should a benefit-sharing mechanism for DSI be implemented, compliance requires to be monitored and, if it does, how this might be achieved. The mechanisms developed for compliance under the Nagoya Protocol[144] do not lend themselves to multilateral systems or DSI.

Irrespective of whether DSI is ultimately brought into the scope of the NP, the development of agreed modalities of benefit-sharing from use of DSI, coupled with clarity on the scope of DSI, would be a welcome and necessary next step.

Acknowledgments I would like to thank many people who have shared or contributed ideas, particularly Kate Davies, Peter Geire, Evanson Chege Kamau, Paul Kersey, Alicja Kozlowska, Thomas Greiber, Dirk Neumann, Anne Nivart, Alan Paton and China Williams. Marcelin Tonye Mahop, Gerd Winter and Evanson Chege Kamau provided very helpful comments and suggestions on an earlier version of this chapter. Interpretations and opinions are entirely my own unless stated otherwise.

References

Books and Journal Articles

Augusto C, Gutiérrez C (2014) Governing synthetic biology in the light of the access and benefit sharing regulation (ABS). Rev derecho y genoma Hum 41:63–87

Bagley MA (2015) Digital DNA: the Nagoya Protocol, intellectual property treaties, and synthetic biology. Wilson Center

Bagley MA (2017) Towering wave or tempest in a teapot? Synthetic biology, access & benefit sharing, and economic development. In: Frankel S, Gervais D (eds) The internet and intellectual property: the nexus with human and economic development. Victoria University Press

Baltz RH (2018) Synthetic biology, genome mining, and combinatorial biosynthesis of NRPS-derived antibiotics: a perspective. J Ind Microbiol Biotechnol 45:635–649. https://doi.org/10.1007/s10295-017-1999-8

Baškarada S, Koronios A (2013) Data, information, knowledge, wisdom (DIKW): a semiotic theoretical and empirical exploration of the hierarchy and its quality dimension. Australas J Inf Syst 18:5–24. https://doi.org/10.3127/ajis.v18i1.748

Bird A (2002) DNA methylation patterns and epigentic memory. Genes Dev 16:16–21. https://doi.org/10.1101/gad.947102.6

[144] For details see Winter in this volume.

Boles KS, Kannan K, Gill J et al (2017) Digital-to-biological converter for on-demand production of biologics. Nat Biotechnol 35:672–675

Brunak S, Danchin A, Hattori M, Nakamura H, Shinozaki K, Matise T, Preuss D (2002) Nucleotide sequence database policies. Science 298(5597):1333

Bygrave LA (2014) Information concepts in law: generic dreams and definitional daylight. Oxf J Leg Stud 35:91–120. https://doi.org/10.1093/ojls/gqu011

Cole P (2015) Patentability of genes: a European Union perspective. Cold Spring Harb Perspect Med 5:1–12. https://doi.org/10.1101/cshperspect.a020891

Dawkins R (2008) River out of Eden: a Darwinian view (Paperback reprint of 1995 ed.). Basic Books

Deplazes-Zemp A (2018) 'Genetic resources' an analysis of a multifaceted concept. Biol Conserv 222:86–94. https://doi.org/10.1016/j.biocon.2018.03.031

Goldman AD, Landweber LF (2016) What is a genome? PLoS Genet 12. https://doi.org/10.1371/journal.pgen.1006181

Greiber T, Moreno SP, Åhrén M et al (2012) An explanatory guide to the Nagoya Protocol on access and benefit-sharing, https://portals.iucn.org/library/efiles/documents/EPLP-083.pdf. Accessed 22 May 2020

Hu H, Scheben A, Edwards D (2018) Advances in integrating genomics and bioinformatics in the plant breeding pipeline. Agriculture 8:75. https://doi.org/10.3390/agriculture8060075

ICSWGSB (2016) Synthetic biology and the CBD Five key decisions for COP 13 & COP-MOP 8. 8 pp

Jacobsen A, de Miranda AR, Juty N et al (2020) FAIR principles: interpretations and implementation considerations. Data Intell 2(2020):10–29. https://doi.org/10.1162/dint_r_00024

Karsch-Mizrachi I, Takagi T, Cochrane G (2017) The international nucleotide sequence database collaboration. Nucleic Acids Research, 2018, vol 46, Database issue Published online 28 November 2017. https://doi.org/10.1093/nar/gkx1097

Lawson C, Humphries F, Rourke M (2019) The future of information under the CBD, Nagoya Protocol, Plant Treaty, and PIP Framework. J World Intellect Prop 22:103–119. https://doi.org/10.1111/jwip.12118

Lawson C, Rourke M, Humphries F (2020) Information as the latest site of conflict in the ongoing contests about access to and sharing the benefits from exploiting genetic resources. Queen Mary J Intellect Prop 10:7–33. https://doi.org/10.4337/qmjip.2020.01.01

Lyal CHC (2020) Current situation on Digital Sequence Information (DSI). In: Kamau EC (ed) Regulating access and benefit-sharing under the Nagoya Protocol - selected case studies and themes, BfN Script 564, https://www.bfn.de/fileadmin/BfN/service/Dokumente/skripten/Skript564.pdf. Accessed 29 Dec 2020

Medaglia JC (2020) Digital Sequence Information (DSI) and benefit-sharing arising from its use: an unfinished discussion. GRUR Int 0:1–2, doi:https://doi.org/10.1093/grurint/ikaa057

Nicol D, Dreyfuss RC, Richard Gold E et al (2019) International divergence in gene patenting. Annu Rev Genomics Hum Genet 20:1–23. https://doi.org/10.1146/annurev-genom-083118-015112

Onions CT (ed) (1973) The Shorter Oxford English Dictionary. Clarendon Press, Oxford

Park N (2017) 5.Digital Sequence Information: a new but inevitable challenge to the Nagoya Protocol. J Int Relations Area Stud 46:41–47

Rossi J (2014) Genes are not information: rendering plant genetic resources untradeable through genetic restoration practices. Geoforum 55:66–75. https://doi.org/10.1016/j.geoforum.2014.05.001

Rowley J (2007) The wisdom hierarchy: representations of the DIKW hierarchy. J Inf Sci 33:163–180. https://doi.org/10.1177/0165551506070706

Ruiz Muller M (2015) Genetic resources as natural information: implications for the Convention on Biological Diversity and Nagoya Protocol. Routledge, London, New York

Ruiz Muller M (2018) Access to genetic resources and benefit sharing 25 years on: progress and challenges. International Centre for Trade and Sustainable Development (ICTSD), Geneva

Schei PJ, Tvedt MW (2010) "Genetic resources" in the CBD: the wording, the past, the present and the future. Montreal

Servick K (2016) Rise of digital DNA raises biopiracy fears. Science (80-). https://doi.org/10.1126/science.aal0395

da Silva M, de Oliveira DR (2018) The new Brazilian legislation on access to the biodiversity (Law 13,123/15 and Decree 8772/16). Brazilian J Microbiol 49:1–4. https://doi.org/10.1016/j.bjm.2017.12.001

Soplín SP, Muller MR (2009) The development of an international regime on access to genetic resources and fair and equitable benefit sharing in a context of new technological developments. Res Doc Andean Amaz Initiat Prev Biopiracy IV:16 pp

Trerotola M, Relli V, Simeone P, Alberti S (2015) Epigenetic inheritance and the missing heritability. Hum Genomics 9:17. https://doi.org/10.1186/s40246-015-0041-3

Tvedt MW, Schei PJ (2013) The term "genetic resources": flexible and dynamic while providing legal certainty? In: Oberthür S, Rosendal GK (eds) Global governance of genetic resources: access and benefit sharing after the Nagoya Protocol. Routledge, London; New York, pp 18–32

Tvedt MW, Young T (2007) Beyond access: exploring implementation of the fair and equitable sharing commitment in the CBD. IUCN, Gland, Switzerland

van Dijk EL, Auger H, Jaszczyszyn Y, Thermes C (2014) Ten years of next-generation sequencing technology. Trends Genet 30:418–426. https://doi.org/10.1016/J.TIG.2014.07.001

Vogel JH (1994) Genes for sale. Oxford University Press, New York

Vogel JH, Álvarez-Berríos N, Quiñones-Vilches N et al (2011) The economics of information, studiously ignored in the Nagoya Protocol on access to genetic resources and benefit sharing. LEAD J (Law, Environ Dev Journal) 7:52

Vogel JH, Ruiz M, Angerer O et al (2018) Inside views: ending unauthorised access to genetic resources (aka Biopiracy): bounded openness. IP-Watch, 6 April 2018, www.ip-watch.org/2018/04/06/ending-unauthorised-access-genetic-resources-aka-biopiracybounded-openness/

Welch EW, Bagley MA, Kuiken T et al (2017) Potential implications of new synthetic biology and genomic research trajectories on the International Treaty for Plant Genetic Resources for Food and Agriculture (ITPGRFA or 'Treaty')

WHO (2018) Fact Sheet. Nagoya Protocol and Public Health. 1–3

Wilkinson D, Dumontier M, Aalbersberg I et al (2016) The FAIR guiding principles for scientific data management and stewardship. Scientific Data 3, Article number: 160018

WIPO (2016) What is Intellectual Property? http://www.wipo.int/edocs/pubdocs/en/intproperty/450/wipo_pub_450.pdf

WIPO (2018) A guide to intellectual property issues in Access and Benefit-sharing Agreements, https://www.wipo.int/edocs/pubdocs/en/wipo_pub_1052.pdf. Accessed 22 May 2020

Wynberg R, Laird SA (2018) Fast science and sluggish policy: the Herculean Task of Regulating Biodiscovery. Trends Biotechnol 36:1–3. https://doi.org/10.1016/j.tibtech.2017.09.002

Legal and Policy

Vienna Convention on the Law of Treaties (Concluded 23 May 1969, entered into force 17 January 1980). United Nations, Treaty Ser 1155:331. https://doi.org/10.1007/978-3-642-19291-3

United Nations (2019) Draft text of an agreement under the United Nations Convention on the Law of the Sea on the conservation and sustainable use of marine biological diversity of areas beyond national jurisdiction. Advance unedited version

Reports, Submissions and Decisions

ABS Capacity Development Initiative (2019) First Global Dialogue on Digital Sequence Information on Genetic Resources, http://www.abs-initiative.info/fileadmin//media/Events/2019/6-8_November_2019__Pretoria__South_Africa/Report-First-Global-DSI-Dialogue-SouthAfrica-201911.pdf. Accessed 21 May 2020

Australian Government (2019) Digital Sequence Information on Genetic Resources: Submission of views and information and call for expression of interest to undertake studies Submission by Australia, https://www.cbd.int/abs/DSI-views/2019/Australia-DSI.pdf. Accessed 21 May 2020

Bagley MA, Karger E, Muller M et al (2020) dsi-ahteg-2020-01-05-en Fact-finding study on how domestic measures address benefit-sharing arising from commercial and non-commercial use of digital sequence information on genetic resources and address the use of digital sequence information on genetic resources and address the use of digital sequence information on genetic resources for research and development, https://www.cbd.int/doc/c/428d/017b/1b0c60b47af50c81a1a34d52/dsi-ahteg-2020-01-05-en.pdf. Accessed 21 May 2020

Commission on Genetic Resources for Food and Agriculture (2019a) "Digital Sequence Information" on genetic resources for food and agriculture and its relevance for food security

Commission on Genetic Resources for Food and Agriculture (2019b) Submissions by members and observers on "digital sequence information" on genetic resources for food and agriculture

Government of Malawi (2019) Guidelines on access to genetic resources and the fair and equitable sharing of benefits arising from their use, http://ead.gov.mw/storage/app/media/Resources/Guides/Guidelines ABS final copy 16-09-2019.pdf

Houssen W, Sara R, Jaspars M (2020) dsi-ahteg-2020-01-03-en Digital Sequence Information on genetic resources: concept, scope and current use, https://www.cbd.int/doc/c/fef9/2f90/70f037ccc5da885dfb293e88/dsi-ahteg-2020-01-03-en.pdf. Accessed 22 May 2020

Laird SA, Wynberg RP (2018) Fact-finding and scoping study on digital sequence information on genetic resources in the context of the Convention on Biological Diversity and the Nagoya Protocol, Montreal

Manzella D (2016) The Global Information System and Genomic Information: Transparency of Rights and Obligations. ITPGRFA. Background study paper No 10

Rohden F, Huang S, Dröge G, Scholz A (2020) dsi-ahteg-2020-01-04-en Combined Study on Digital Sequence Information on Public and Private Databases and Traceability, https://www.cbd.int/doc/c/1f8f/d793/57cb114ca40cb6468f479584/dsi-ahteg-2020-01-04-en.pdf. Accessed 22 May 2020

SCBD (2002) UNEP/CBD/COP/6/INF/40: Access and Benefit-Sharing as related to Genetic Resources. Compilation of submissions by experts on the use of terms

SCBD (2008) UNEP/CBD/WG-ABS/7/2. Compilation of Submissions By Parties, International Organizations, Indigenous and Local Communities and Stakeholders on Concepts, Terms, Working Definitions and Sectoral Approaches

SCBD (2016) Digital sequence information on genetic resources, decision XIII/16. 1–2

SCBD (2017a) Peer review of the fact-finding and scoping study, https://www.cbd.int/abs/dsi-gr/ahteg.shtml. Accessed 26 Aug 2018

SCBD (2017b) Submissions from Parties, other Governments, relevant organizations and stakeholders, https://www.cbd.int/abs/dsi-gr/ahteg.shtml

SCBD (2018a) CBD/SBSTTA/22/INF/2: Synthesis of Views and Information on the Potential Implications of the Use of Digital Sequence Information on Genetic Resources for the Three Objectives of the Convention and the Objective of the Nagoya Protocol. Montreal

SCBD (2018b) CBD/SBSTTA/22/INF/2/Add.1: Case studies and examples of the use of digital sequence information in relation to the objectives of the Convention and the Nagoya Protocol. Montreal

SCBD (2018c) CBD/SBSTTA/22/INF/4; CBD/DSI/AHTEG/2018/1/4: Report of the Ad Hoc Technical Expert Group on Digital Sequence Information. Montreal

SCBD (2018d) CBD/SBSTTA/REC/22/1: Recommendation adopted by the Subsidiary Body on Scientific, Technical and Technological Advice 22/1. Digital sequence information on genetic resources. Montreal

SCBD (2018e) CBD/COP/DEC/14/20. Decision Adopted by the Conference of the Parties to the Convention on Biological Diversity 14/20. Digital sequence information on genetic resources. 3

SCBD (2018f) CBD/NP/MOP/DEC/3/12. Decision Adopted by the Parties to the Protocol on Access and Benefit-Sharing 3/12. Digital sequence information on genetic resources. 1

SCBD (2018g) CBD/SBSTTA/22/INF/2/Add.2: Digital sequence information on genetic resources in relevant ongoing international processes and policy debates. Montreal

SCBD (2019) Submissions of views and information on Digital Sequence Information on Genetic Resources. In: SCBD website, https://www.cbd.int/abs/dsi-gr/2019-2020/submissions.shtml. Accessed 10 June 2019

SCBD (2020) CBD/DSI/AHTEG/2020/1/7: Report of the Ad Hoc Technical Expert Group on Digital Sequence Information on Genetic Resources

Sollberger K (2018) Digital Sequence Information and the Nagoya Protocol. Legal expert brief on behalf of the Swiss Federal Office for the Environment (FOEN)

Spranger TM (2017) Expert opinion on the applicability of the Convention on Biological Diversity and the Nagoya Protocol to digital sequence information. Submitted on behalf of the German Federal Ministry of Education and Research, Berlin

Conference Proceedings

Polatin-Ruben D, Wright J (2014) An Internet with BRICS Characteristics: Data Sovereignty and the Balkanisation of the Internet. In: Paper presented at the 4th USENIX Workshop on Free and Open Communications on the Internet (FOCI 14), San Diego, July 7, 2014. 10 pp

Website Links

NCBI (2019) GenBank and WGS Statistics. In: Website, https://www.ncbi.nlm.nih.gov/genbank/statistics/. Accessed 22 Oct 2019

OECD OECD Glossary of statistical terms. In: OECD. https://stats.oecd.org/glossary/index.htm

Christopher H. C. Lyal was a researcher at London's Natural History Museum for 46 years, and has published extensively on insect taxonomy. In 2001–2003 he was acting Programme Officer on the Global Taxonomy Initiative at the CBD Secretariat, and is UK GTI Focal Point. He has worked on policy and implementation aspects of the Nagoya Protocol, and was instrumental in developing the CETAF best practices, currently the only ones recognised under the EU ABS Regulation. He chaired the Informal Advisory Committee to the ABS Clearing House from its inception until 2018, and was co-chair of the 2020 AHTEG on Digital Sequence Information. C.lyal@nhm.ac.uk.

Lessons from Writing Binding and Enforceable ABS Contracts. A Contract Solution to Digital Sequence Data in ABS

Morten Walløe Tvedt

Abstract This chapter gives a summary of long developed theoretical and practical analysis of how functional contracts regulating biological resources in research and development can be drafted. The overall perspective is how to use contract language with precision to capture what is happening in biotechnological research. A core challenge is how to draft the subject matter of the contract to capture the value of biological samples. The aim is to contribute to well-working non-monetary benefit-sharing, which is often referred to as the most important part of 'access and benefit-sharing'. It outlines basic guidance on how to draft a contract that has been neglected in drafting ABS contract. Applying these 'golden rules' of contract drafting will increase precision and thereby the chance of achieving the objectives of the contract. Also the text outlines proposals for a contract-approach to the challenge of digital sequence data.

Keywords ABS contracts · Contract law · Standard Material Transfer Agreement · International Treaty on Plant Genetic Resources · Access and benefit-sharing · Multilateral System

1 Stocktaking on Implementation

On the path towards the 15th Conference of Parties (COP) to the Convention on Biological Diversity (CBD) there are some overall topics and some more practical small questions that are dominating the discussions in all fora dealing with genetic resources at a global policy level. Here, attention is given to the practical implementation of access and benefit-sharing (ABS) to become a functional legal tool for meeting the obligations in the CBD and its Nagoya Protocol (NP).

M. W. Tvedt (✉)
Molde University College (HiMolde), Molde, Norway
e-mail: Morten.W.Tvedt@himolde.no

E. Chege Kamau (ed.), *Global Transformations in the Use of Biodiversity for Research and Development*, Ius Gentium: Comparative Perspectives on Law and Justice 95, https://doi.org/10.1007/978-3-030-88711-7_22

Implementation of international law obligations happens by legal tools that are binding in national jurisdiction. Countries have been in the process of developing legislation for three decades, but on the way to alter behavior among the actual users these international principles need to become binding in a private law manner. A contract is a legally binding instrument between two or more parties. The use of contract is the practical legal tool that can make access and benefit-sharing work in practice, providing legal certainty and sufficient flexibility to cater for each individual situation. The contract must have clear terms and conditions to become binding and enforceable on the user. It must also have a format recognized across borders. A contract will *per se* be binding in other jurisdictions, *i.e.* that of the user if it is well drafted and if it contains binding and enforceable provisions. An ABS contract must be comprehensive and regulate the relationship between the parties in detail. Generally, in contract law, there are established practices and background law, which can be used to support their interpretation. For ABS contracts, this is not the case; so the contract must regulate all relevant aspects. The user country's legislation, which is established as part of the implementation of the NP, does not directly resolve the core challenges relating to contracts. An ABS contract can expect little interpretative support by other sources of law and must govern all necessary aspects of the transaction.

Based on experience in developing ABS capacity building, the experience from the provider countries is that many users approach them applying for access for research without immediate commercial application, typically by academic or research institutes. Many forms of research without immediate commercial purpose have the potential to produce useful findings that can be used for commercial purposes and this contract tool takes this into consideration. Experience has shown that provider countries can draw non-monetary benefits from securing their rights to findings even when the academic or research institution does not declare an immediate commercial objective.

The first section of this chapter looks into two main challenges for ABS implementation: (1) How to regulate access for so-called non-commercial research? Especially, how should access be regulated when the objective is to make resources available for other researchers? (2) How ABS regime can relate to the use of biological resources in bio-trade that is not aiming at exploring or exploiting the genetic resources or biochemical. Secondly, this section discusses the big challenge of all international fora—how to regulate the transformation from biology to digital format of genetic information. The next section goes through the golden rules for contract drafting. Then this chapter moves on to discuss the manner to draft the subject matter on how a contract can capture the value of biological samples in binding definitions of the subject matter of a contract. Then the next section is looking into how to draft contract language establishing well-functioning non-monetary benefit-sharing clauses. The last topic of this chapter is to outline a contract solution to the challenge of digital sequence data.

2 Fundamental Challenges to ABS

ABS covers a broad spectrum of value chains. For some types of value creation the biological material accessed will be identical or close to identical to the product on the market. One example is where bulk biological material is refined to a medicine or a cosmetic product. On more complex value chains there is a longer time lag. The material is first accessed, then researched, and developed into a novel product or product that can be launched into a market. Common for this spectrum is that there is a time-lag between access and value creation.[1] The time-lag in ABS contracts is unique to other contracts where both parties fulfill their obligations at the same time. The contracts in ABS need to capture this common challenge regarding the time-lag, and seek to cater for both rapid and more research-consuming value chains.

Use of material is a dynamic process where the material or parts thereof are transformed into new formats. One example is when biological samples of the material results in digital information or synthetic form of material (more on this topic in Sect. 6). The contract needs to foresee these types of changes of the expression of genetic information.

Research and development are dynamic processes and can take a variety of different paths. The text of a contract is static when it has been agreed to. A core challenge is to draft language in a manner that is able to capture the benefits to be shared regardless of when these benefits will be created or materialize along the value chain. Value creation and harvest-points in value chains are diverse.

An ABS contract needs to outline solutions to all these points in the chains of development even before access takes place. This calls for standardized manners to approach core parts of the contract.

2.1 Access for Non-commercial Purpose When the Objective Is to Make the Resource Available for Others

Academic institutions like universities, research institutions and gene banks have as one objective to make genetic material, information and related knowledge available to a larger research community and business community for research and development. This objective is pursued by the global and regional collections also in the plant sector where plant material is collected for the purpose of being made accessible. These activities are based on the rationale of bringing research and thus society forward. Academic research institutions that are accessing valuable samples want to publish their genetic code and information in open access databases.

Since the negotiations that led to the Nagoya Protocol there has been a notion of 'non-commercial' use or users. Experience in analyzing a number of contractual

[1] Tvedt et al. (2016).

situations with research institutions is that down the chain of research valuable findings by 'non-commercial' users tend to be picked up into commercial product development chains. So-called 'non-commercial' and academic actors are serving a public policy objective which leads to private harvest from research. Investments are done in these important steps to promote advances in science and the development of new products to solve the challenges of the future. This manner of translating academic research into commercially interesting products often benefits society. From an ABS perspective these practical experiences teach an important lesson that an access contract need to foresee the possible paths into product development and establish clear legal consequences regarding the non-monetary benefit sharing.

The ABS challenge is how to capture or reap public benefits from these steps of progress for the least affluent part of the global population. How will these advantages in science contribute to the three objectives of the CBD? Until there is a statutory system in place to promote the third objective of the CBD of fair and equitable benefit-sharing in place, contracts is the only functional legal tool for securing the public benefits.

A question for reflection is if access is done for pure non-commercial purposes, why should the academic institution have any problems of accepting that the property rights to a commercially interesting research result enabled by that accessed material should belong to the provider?

2.2 *Potential Loop-Holes: Bio-trade and Digitalization*

ABS has two scoping challenges. First, it is not aimed at regulating trade in biological material. A sale of bulk of commodities of biological material was probably never meant to trigger benefit-sharing obligations under the CBD. From pure bulk sales to clear cut transfer of samples for genetic research there is a series of different purposes of accessing biological samples. A special situation is what is often referred to as bio-trade where the trade is with biological bulk but the purpose of the trade is closer to the use of genetic or biochemical uses. If this kind of transfer is done without an ABS contract, a change in use down the line of development will not be covered by an obligation even if the product later developed should have been a typical ABS situation.

A second loophole is the digitalization of genetic information. There is a legitimate need for share and publishing digital sequences data in research and to spur innovation. Thus, a conflict is created between the open and unconstrained use of the genetic information, which stands against the legitimate fulfilment of the three objectives of the CBD. How can the benefit-sharing mechanisms from digitalized use of genetic resources have a contractual and private law solution? In this article, I submit that a contract approach provides a potential solution to cut the political Gordian knot of 'digital sequence data'.

One thesis explored in this chapter is that the solution is to regulate scenarios of commercialization in any type of ABS contract, including those contracts regulating bio-trade and use of digitalized published genetic information. Any non-commercial ABS contract should regulate scenarios of what happens in the case of commercial success. This does not necessarily have to be complex but creates more predictable legal conditions.

3 Golden Rules for Contract Drafting

3.1 Contract Drafting: An Interdisciplinary Challenge Between Biology and Law

The part of the contract that requires the most attention at the drafting stage is the substantive clauses—these are the obligations and rights to be specified in the contract. There has been a tendency in the past to be very brief and use general formulations in the substantive clauses. Experience has shown that this approach has not led to functional ABS contracts.

When defining contractual obligations, precision is a vital virtue.[2] In legal drafting we can learn from language theory. A legal term needs to reflect reality as concisely as possible. Between a legal term (intermediary) and the real world (reality), there is the human understanding (mind) of the term. There is no one-to-one-to-one relationship between these three: the 'legal term', the understanding or associations the legal term gives and the 'phenomenon' in reality. This is the case also in law and for legal drafting. Law is an area of specialized linguistics, thus modern philosophy of language brings insight to better understand these relationships. While ambiguous *legal terms* may help to reach an agreement in the first place, they carry a greater risk of not creating a legal rule that grasps the *reality* in a manner that *all minds* actually agreed to. The contract may not be fulfilled, underlying conflicts of interests may simply be postponed, a future court may not be able to identify the agreed upon obligation. Discrepancy or ambiguity between a *legal term* and the *normal, biotechnological interpretation* of that term will easily increase the degree of conflict in the contractual relationship.

A specific, clear and precise legal term which largely corresponds to the reality in the area where this legal source is meant to have normative effect has a better chance of succeeding. Contract drafting needs to use recognized biotechnological language and concepts to provide clarity. The higher the degree of correspondence between the *legal term* and the *phenomenological* occurrence is in the real *biotechnological world*, the more clarity the contract provides. Therefore, the first practical lesson is that a term with no reference to actions or phenomena in the real world is not useful in contract drafting.

[2] Tvedt (2013).

3.2 Be Clear, Specific and Dynamic, But Not Narrow

Contract obligations must be **clearly** defined in **specific** language stating when the obligation will be triggered, the content of the obligation, methods for calculating payments, and specify how each obligation shall be fulfilled. The contract needs to specify the consequences and actions that will be taken when one of the parties fails to meet any of its obligations.

Drafting **dynamic** clauses: At the same time as the obligations are made specific, they need to be dynamic so that they can embrace new and unpredictable research results. One feature of such contractual obligation is to formulate them in a technological neutral manner. And to choose wording that will grasp a plurality of research results, also those that are not easily predicted at the point of time of access.

In ABS, samples are accessed before the user conducts the research and development. The challenge for ABS contracts is to negotiate before utilization occurs, i.e. before anything is done with the resources, which could possibly result in value and potential benefits. An ABS contract must move beyond regulating obligations that are clearly foreseen and disclosed at the time of access. Its obligations need to anticipate a broad specter of potential outcomes from the utilization of the resource.

To overcome this major challenge, it is necessary for the contract to build in scenarios for research and development based on the said samples. Here, 'scenario' means typical pathways that the user can probably be expected to follow according to the relevant sectors and its associated trends and practices. The substantive obligations and the manner in which they will be fulfilled will build on these scenarios anticipated in the contract. The contract must also be able to cover unforeseen events, research results and the development of different types of products, meaning that obligations cannot be formulated in a *narrow* manner.[3] At the same time, the language of the contract must be clear and specific so that a judge will be able to ascertain whether the obligations have been complied with or not according to the facts of the case.[4]

3.3 Opt for the Positive Regulation Rather Than Attempt Prohibitions

In the past, ABS agreements have often included an obligation "not to commercialize", "not to seek intellectual property rights" or "not to access traditional knowledge". A common feature of these obligations is that it is problematic to prohibit and enforce such prohibition in a contract. Contractual obligations that seek to prohibit actions are generally impossible to enforce in the courts unless these are spelled out

[3] Tvedt (2022).

[4] Tvedt (2019) and Young and Tvedt (2017).

with clear and corresponding legal consequences, including verifiable trigger points and enforcement mechanisms that a court can objectively refer to.

When outcomes arising out of research and development lead to a product or process that is eligible for patent protection and all patent requirements are met, one cannot prevent a patent from being granted based on illicitly or breach of a contract. A contractual obligation is not a reason *per se* to prevent the granting or invalidating of a patent.

What is needed in a contract is to clearly regulate actions by stipulating clear and enforceable consequences should these actions take place. Instead of prohibiting a set of actions, opt for positive consequences that would be triggered. By establishing obligations in the contract on what happens when there is a patent, the contract would play along existing invention- and power-structures. For example, rather than drafting a contract saying that commercialization is prohibited, specify the payment obligations in case there is a commercial outcome. Then, the obligations can be connected to non-monetary and financial achievements or results.

3.4 *Regulate the Whole Scenario in the Original Contract*

Many first-generation ABS agreements have been weak in terms of defining clear obligations and trigger points at different stages of contract execution. In non-commercial ABS contracts, it has often been stipulated an obligation of coming back to negotiate a new contract in certain situations. Often these 'come-back clauses' have required that the user, in the event of commercialization, shall come back to the provider and negotiate terms and conditions for commercial activities. This obligation typically kicks in for unforeseen research results at the time of concluding the initial non-commercial contract. The problem with this approach is that this type of come-back clause does not lead to enforceable obligations. It is legally and practically impossible to impose an obligation on someone to come back to the negotiating table and make agreements on a given utilization or application that was not foreseen in the initial contract. One cannot be obliged to reach a new agreement.

Therefore, obligations that are not formulated in a contract from the onset have almost no chance of later becoming part of a contractual relationship. If these obligations and possible scenarios are not foreseen or governed from the onset, they remain voluntary and cannot be enforced as binding contractual obligations. The wording of the obligations must therefore be such that all relevant scenarios are stipulated in a clear and unambiguous manner. Thus, one core issue to have in mind is how contract clauses can cover relevant scenarios and translate them into binding language.

4 How to Define Biological Material into Written Subject Matters of Contractual Obligations

One of the most challenging elements in drafting the binding obligations is how to capture biological samples as subject matter of a contract. This is at the heart of establishing the respective rights and obligations that are incumbent to both the provider and the user.

The subject matter in the wording of the CBD and NP is 'genetic resources', as defined in article 2 CBD.[5] Everyone with a small interest in ABS should be able to describe in a few precise and objective terms what is meant by the definition. Wording that depends on subjective factors that cannot be applied without insight in cognitive observations in either of the parties cannot provide the clarity required for a contractual obligation to be certain.

The basic question to ask when drafting the text of a contract is: "Would a regular judge, trained in law, and not in biology, be able to interpret and apply the contract language pertaining to the definition of the subject matter in a concrete case?" More specifically: "Will the judge be able to interpret and apply the definition in article 2 CBD to the concrete case?" If the answer to this question is 'no', the wording of the contract must be redrafted to be more specific and self-explanatory so as to clearly define the subject matter that should be covered under the contract. This term does not provide the specificity or certainty required by a contract. A contract using such imprecise terminology cannot be expected to create binding and enforceable obligations on the user.[6]

To contrast the definition in article 2 CBD, the box below (Box 1) offers an example of a manner to draft the language defining the subject matter of the contract. Discussing concrete contract clause examples makes the challenges and their solutions in contract drafting more tangible.

Box 1 Turning Biological Samples into Written Subject Matters of Obligations

A plant samples [includes/covers/reaches on to] any seeds or propagating or multiplication material of any kind.

A plant sample includes/covers any part of plant,

(a) the vegetative material, including leaves, roots, straw, flowers, fruits, or any other, including but not limited to any extract in any form.

(continued)

[5] CBD article 2 "*Genetic resources*" means genetic material of actual or potential value.", and further on "*Genetic material* means any material of plant, animal, microbial or other origin containing functional units of heredity".

[6] Young and Tvedt (2017).

Box 1 (continued)
(b) any bioactive [elements/compounds] or organic composition (molecules, amino acids, proteins, lipids, glucides, [enzymes] or any other bioactive element in or [isolated/generated] from the vegetative material.
(c) every sub-cell element/parts, including DNA, RNA, or any other element carrying information.

A sample covers any information in the material, or which is extracted from it by any method, or genetic information in any format, or data generated therefrom, including but not limited to genetic information, biochemical content, or bioactive, or any other information in or extracted or analysed from the samples.

A samples also includes any micro-organisms in or on the plant, including but not limited to parasites or pathogens on or in the plant material.

Source: Own illustration, Tvedt 2021

This wording is an example clause which is exhaustive in its style. The article is a result of a linguistic translation of biology into legal language from longstanding dialogue with users and providers; academics and lawyers. It illustrates all aspects of what kind of subject matters that one sample of plant contains. In a specific ABS transaction such an exhaustive definition of the subject matter is a functional manner to build up the substantive clauses. It is a question of agreeing to the terms and conditions in the contract whether there shall be obligations connected to one or more of these subject matters. This comprehensive wording meets the criteria of clear, specific and dynamic. It avoids being too narrow. The additions ("but not limited to" or "of any kind" etc.) render it technologically neutral grasping non-known manner to use biological samples. It is drafted in a manner that grasps the dynamic application of technology to the samples, and forms a functional point of departure for the obligations to share benefits.

5 The Most Important Non-monetary Benefit-Sharing Obligation: *Free Use of Any Result*

In many situations, the most valuable right for the provider is to have a right to the research results and any commercial or non-commercial use and application of them. When a 'non-commercial' user develop something useful from the samples accessed in another country, the occurrence of that biological material in that country makes the research results particular interesting for that country. The contract should therefore establish a (non-exclusive) use right of the research results to the providing country (may be geographically limited to the state territory). It could also establish a right to be provided with the products developed and produced following the use of the resource ("access") under preferential terms. If a pesticide is developed from plants in providing country X, that country can be expected to have a particular strong use-interest in that newly developed product. This kind of right to use the results is a respond to the 'non-commercial' user; if no commercial use is foreseen, any such result would be fair to leave with the provider.

The empirical body of contracts studied and reviewed fails to include this crucial non-monetary sharing obligation. For example, if a pathogen is taken from a country, there is a good chance that the bacteria or virus causes more damage in that given country than elsewhere. For the provider country, free or low-cost access to any product developed based on these biological samples would be one of the most valuable benefits; giving out samples for research without a clause requiring this type of access to any product or process that may arise out of the use of the samples.

Box 2 Use-Right—Non-monetary Sharing of Benefits

[The Provider Country] has an unconditional right, without any charge, to experimental, commercial or developmental use of any result, product or process enabled by the samples under this Contract.

This right exists independently of any intellectual property rights that may be obtained from the use of the samples and includes an unconditional right to non-exclusive licenses for any outcome arising from this project. Any utilization of samples including knowledge or results achieved from any activities described in the Project description [the said project], shall be made available to [the Provider Country] without any charge, whether or not these results are subject to IPRs.

The property rights of the government of [the Provider Country] extend and include rights to any diagnostics, treatment or [INSERT ANY PRODUCT FROM THE RESEARCH OF PARTICULAR IMPORTANCE FOR THE PROVIDER] or any vaccine, medicine or any other treatment. The user shall ensure that any vaccines or [medicines or] techniques are made available to the population of [the Provider Country].

Source: Own illustration, Tvedt 2021

This draft example clause is one manner to formulate this core non-monetary benefit-sharing obligation. The brackets are there to illustrate that there are parts in the wording of a standard clause that will need to be filled in with the facts of a particular contract relationship. Such an obligation meets the criterion of 'fair and equitable' as it targets getting access back to what the provider country contributed.

6 Research Results from Screening or Scanning: Digital or Synthetic Use of Information

Technological changes entail challenges for law, but should not be a big surprise to anyone in the area of biotechnology and law. Screening and mapping the genome are typical activities conducted with samples of biological material. The information in the genome, DNA, RNA, proteins, molecules, atoms or alike, constituents of the samples is transferred into digital form. The transcription of genetic information opens a new set of possible applications (or uses) of the genetic information, genetic sequence data, knowledge or synthetic expression of the DNA molecule.

The concept 'digital sequence data' is in the eye of the storm of the political discussions in any international forum where 'genetic resources' as a subject matter is discussed. There are ongoing discussions in various international fora on whether digital sequence data, immaterial data or synthetic use of DNA, RNA, proteins and/or atoms in biochemical compounds fall within the scope of the respective international instruments. These discussions are taking place in the CBD and Nagoya Protocol (NP), the Food and Agriculture Organization both in the Commission on Genetic Resources and in the International Treaty for Plant Genetic Resources for Food and Agriculture (both parts of the Food and Agriculture Organization, FAO), in World Health Organization and in the World Intellectual Property Organization. These discussions at the international level are important in their own merit.

As early as at the Cali meeting of the negotiations leading to the Nagoya Protocol, in Colombia in 2010, the CBD Secretariat launched a report where one of the legal conclusions was that '*in silico*' use of genetic material was included in the concept 'genetic resources' in the CBD.[7] The main interpretative factor for that conclusion is that the definition in the CBD is technology neutral and focuses on the use with origin in the genetic material. The discussion of how to regulate digital sequence data is politicized. There are practical solutions to regulate these subject matters under the current regime by a contractual approach.

From the point of view of contract drafting, the manner the contract regulates these subject matters is more important as they can create sharing obligations independently of the result of the global political discussions. A contract depends on the mutual acceptance from its parties and not on the scope of or definition in any international law instrument. The global discussions and disagreement regarding whether these subject matters are covered or not, has made it even more important that ABS contracts regulate and thereby set the global standard for Digital Sequence Information (data) and synthetic, digital or any other technological uses.

The freedom of contract leaves flexibility for drafting contractual wording that captures property rights and sharing of benefits from these digital subject matters with the provider of the samples. When using the freedom to contract, one must

[7] Schei and Tvedt (2010) and Tvedt and Schei (2014).

recall that technology progresses at a fast rate. Therefore, contractual obligations must be written as 'technology neutral' as possible. The freedom of contract concerns all aspects, like the subject matter and the obligations to share research results or revenue from any kind of commercial use. In the current technological situation, screening and scanning of the DNA and alike are highly probable manners to use biological samples. The choice of the words 'screening and scanning' is precise. However, it can easily become too narrow and requires a swiping technology neutral wording as an addition of 'or any other manner'.

Technological development is moving fast which challenges the link between transfer of material in and property rights to research results enabled by them. It becomes important to seek to establish contractual obligations that are technology neutral. It is becoming easier to share information and thus the most valuable asset in the samples of biological material. One other important point is that, although technologies and processes such as genome editing and synthetic biology are still relatively new and at an early stage of development, they are evolving rapidly. This means that in the future, the time needed between generating knowledge in laboratories and the transformation of this knowledge into marketable products will continue to decrease. The main rationale for including these aspects in the contract is that the use of this type of data or related information has clear consequences for the scope of benefit-sharing obligations on the part of the user, and as mentioned in Sect. 2, this is currently a loophole in the benefit-sharing implementation.

There are existing research contracts between, e.g., a university and a company, which provide detailed rules on digital sequence data or synthetic use of DNA. As said, a contract depends on the mutual acceptance from its parties and not on the scope of or definition in any international law instrument. This body of existing contracts shows that it is possible to regulate digital aspects of genetic information in contracts.

Sequence data, genetic information or a synthetically made DNA or other synthetic genetic expression imply contract-drafting challenges concerning publication and sharing since digitalized information and data stored on a computer can easily be shared with a large number of users. One manner to regulate digital publication can look like illustrated in Box 3.

Box 3 Digital Publication of Research Results

In academic research projects there is often an obligation to make sequence data or any other genetic information digitally available online.

Making research or results available online in digital or other format shall only be allowed under this contract by a prior notification to [institution] in [Provider country]. Such notification shall include the web address(es) where the information is made available to the public.

The property rights retained by the Provider to the samples includes digital sequence data of genome, DNA, RNA, proteins, molecules, atoms (e.g. in

(continued)

Box 3 Digital Publication of Research Results (continued)
biochemical compounds) or similar sources. Screening or scanning or applying any other method to transcribe biological or genetic information into digital or other forms of information storage shall not alter the property rights of the [Provider country].

In the event that the User identifies a potentially commercially interesting application, the User shall immediately make any digital sequences data or information available to the provider by means of transferring digital information.

Making research results available online requires restricting access to the information behind an electronic 'wall' or a cookie or other electronic manner to register consent. Such a wall, cookie or alike shall include a log-in or any other manner to create identification by the person accessing the information. The accessing person shall be required to give consent to be bound by a standard clause respecting the economic interests of [Provider country]. To this end different technological solutions exist, including but not limited to a 'click-wrap' or any other standard online accept-condition. The log-in behind the electronic 'wall', cookie or alike shall be conditioned to online registration, including registering information about "name, affiliated institution, interest in accessing the digitalized information and the intended use". This online acceptance shall include an acceptance of the requirement in the Regulation of the [Provider country] of having a contract if embarking on using the information.

The click-wrap shall send a cookie notifying the [Provider country] about the entity accessing the information.

In any digital publication of sequence data or any other digital expression of the samples or results from the samples, full acknowledgement is to be given to the Government of [Provider country], the project enabled by the contract, and the following clause shall be enclosed in the digital publication:

"The government of the [Provider country] has commercial rights or other further use rights in products or processes developed based on the research results presented in this publication. Any use requires a contract of use with the Government of [Provider country]."

Source: Own illustration, Tvedt 2021

This contractual approach entails using well-known tools in data handling. This is similar to what happens when accepting conditions of access to online publications or governing software utilization purchased online. It has become a daily affair for anyone using the web to click and accept a set of standard contract clauses. For the digital gene bank to include this login and cookie approach, does seem as a small and insignificant change in their database computer system. The system, however, needs to be designed in a way that it generates a message back to the provider of the biological material.

In order to enforce property rights to digital sequences, data and associated information, these should only be made public behind a click-and-accept system that is binding on the person accessing the data. Lessons can be drawn from 'open source' computer software. Open-source software is a type of computer software in which source code is released under a license in which the copyright holder grants users the rights to study, change, and distribute the software to anyone and for any purpose. Open-source software may be developed in a collaborative public manner, meaning that any use of the software must also be made open on the same terms and conditions. Digitalized genetic information is more comparable to software than the biological samples. Access to digital genetic sequence data could introduce a system for sharing back information on any use or research based on these digitalized resources. Open source software is based on a principle of contributing back to the state of the art software. Applying such a contractual multilateral approach with recognition of a benefit-sharing mechanism will contribute to a higher level of fairness in the global scientific and research community.

7 Conclusions

In the body of contract discussion there are many opinions on how to draft a contract. In developing ABS regulation we need to be asking the question: how will this clause be working in a contract context? This chapter proposes a change in the manner we approach contract drafting in ABS. The level of details needs to be more hands on and in line with the use of language already in use on the research community. Therefore, bridging the linguistic gap between legal language and the language of research in biotechnology will become an important step in making ABS work on the ground, also in the time of data technology in applied research and development.

Acknowledgements The theoretical background for a contact analysis is the comprehensive monography by Tomme Young and Tvedt on how to draft successful access and benefit-sharing contracts (2017). This chapter draws on the research that has been provided under the collaboration with the ABS Capacity Building Initiative for me as an independent researcher and resources person for building competence among the providers of biological material as important input to research and development.

Deep-felt gratitude goes to partners in Bhutan, in particular Tashi Yangzome Dorji and Chencho Dorji at NBA and Singay Dorji currently at UNDP. I would like to extend my thanks to the Environmental Affairs Department in Malawi and in particular Mphatso Martha Kalemba, Chifundo Chinyama and Lilian Chimphepo for our close collaboration and for the insight gained on workable contracts through ABS implementation and negotiations.

Gratitude goes to Tomme R. Young with whom I wrote the monograph on Drafting Successful ABS Contracts published in 2017. Thanks to Olivier Rukundo for great collaboration in theory development and in teaching contract drafting over a decade.

The research behind this article forms part of the project Suitable Seeds for Food Security in Fragile States funded by the Norwegian Research Council in project number 274519. The research has also been developed to inform the ABS Capacity Development Initiative and Bio-Innovation

Africa. Thanks to the dean at my faculty for making it possible to conduct international research alongside teaching at Molde University College (www.himolde.no). Tvedt may be contacted at Morten.W.Tvedt@himolde.no, mwtvedt@gmail.com or by post at Molde University College, P.O. Box 2110, 6402 Molde, Norway.

References

Book Chapters

Tvedt MW, Schei PJ (2014) The term 'genetic resources': flexible and dynamic while providing legal certainty. In: Oberthür S, Rosendal KG (eds) Global governance of genetic resources. Routledge, London, pp 18–32

Young TR, Tvedt MW (2017) Drafting successful access and benefit-sharing contracts. In: Morgera E, Buck M, Tsioumani E (eds). Brill/Nijhoff, Leiden

Journal Articles

Tvedt MW (2013) Disentangling rights to genetic resources illustrated by aquaculture and forest sectors. Law Environ Dev J 9(2):129–140

Tvedt MW, Eijsink V, Steen IH, Strand S, Rosendal KG (2016) The missing link in ABS - the relationship between resource and product. Environ Policy Law 46(3–4):228–223

Studies and Others

Schei PJ, Tvedt MW (2010) The concept of "genetic resources" in the Convention on Biological Diversity and how it relates to a functional international regime on Access and Benefit-sharing, https://www.cbd.int/doc/meetings/abs/abswg-09/information/abswg-09-inf-01-en.pdf. Accessed 17 Nov 2020

Tvedt MW (in collaboration with Rukundo O) (2019) The ABS contract tool: Version 2.0. The ABS Capacity Development Initiative, Bonn, 40 pp

Tvedt MW (in collaboration with Rukundo O) (forthcoming 2022) The ABS contract tool: Version 3.0. The ABS Capacity Development Initiative, Bonn

Morten Walløe Tvedt is associate Professor at the University College in Molde. He has published extensively in the area of contract law and biotechnology (see app.cristin.no/persons/show.jsf?id=14465 for a complete list of publications). A review of the Standard Material Transfer Agreement of the Plant Treaty is one of the latest contributions to the contract-law understanding of ABS (2021). In 2018 Tvedt published a monography on ABS contracts with Tomme Young on how to draft successful ABS contracts. He works closely with the ABS Capacity Building Initiative giving legal advice to governments. Post contacts: Molde University College, P.O. Box 2110, 6402 Molde, Norway. Morten.W.Tvedt@himolde.no.

Cases and Questions in Application of ABS Regimes

Gerd Winter and Evanson Chege Kamau

Abstract The law in the books on access and benefit-sharing (ABS) has steadily reached the practice of research and development. Many researchers, research institutions, commercial organizations and intermediary enterprises have been confronted with requirements for access, utilization, and benefit-sharing. This has often resulted in problematic situations where the applicable law was unclear and users posed questions to competent authorities and experts in the field. The article presents real life as well as construed cases, and suggests answers to them. The answers exclusively reflect the personal opinions of the authors. Some of the issues raised concern provider State perspectives, but most of them reflect the view of a user country. Touched upon are questions of geographical, temporal and material scope of access and compliance regimes, the regularization of access which is not compliant with provider State conditions, research on biological resources previously bought as commodities, unavailability of information about provider State regimes, screening and testing genetic resources described in publications, monitoring powers of competent authorities, etc.

Keywords Access and benefit-sharing · ABS practice · Cases and questions in ABS · Nagoya Protocol · Solutions in ABS

G. Winter (✉)
Research Center for European Environmental Law, University of Bremen, Bremen, Germany
e-mail: gwinter@uni-bremen.de

E. C. Kamau
Faculty of Law, University of Bremen, Bremen, Germany
e-mail: echege@uni-bremen.de

E. Chege Kamau (ed.), *Global Transformations in the Use of Biodiversity for Research and Development*, Ius Gentium: Comparative Perspectives on Law and Justice 95, https://doi.org/10.1007/978-3-030-88711-7_23

1 Provider State Cases

1.1 Cooperation of Internal and External Researchers

Researcher A who works as a microbiologist at a university in provider state P is invited by a research team of a university in a European Union (EU) State to cooperate in a research project on drought tolerant plants. Her task shall be to liaise with the authorities of P concerning prior informed consent (PIC), mutually agreed terms (MAT) and permits required according to other laws concerning ABS. As such she shall be research cooperator. A is offered remuneration per day spent on her efforts. Should she agree to the offer?

It is true that some provider States require cooperation with an internal researcher as a precondition for access of a foreign researcher to genetic resources (GR). Such cooperation must be understood to mean cooperation in a substantial and not just an organizational sense. A should therefore insist in a substantial division of labor, that is specified as precisely as possible.

1.2 Multiple Permits

Inspector D who works for the authority of country P and is competent for ABS matters receives an application from a researcher A of a Dutch university for PIC and MAT for the sampling of extremophile bacteria in a nature park lake. The law of P provides that the sampling would, in addition to ABS PIC, require permits of the nature conservation agency, the research ministry, and the ministry for international trade. The ABS PIC and MAT shall only be issued after submission of the other permits. Can D facilitate the procedure in any way?

Article 8 (a) of the Nagoya Protocol (NP) suggests the simplification of procedures for non-commercial research. In that light D is indeed asked to accelerate the procedure. If the law of P endows the executive with powers to organize administrative procedures D involves the other authorities himself/herself rather than waiting for A to file the relevant applications and enter in the pertinent negotiations. A's access application could be taken to cover all other applications. D may by agreement with the other agencies be made responsible to serve as proxy also for them. Alternatively, streamlining procedures may be established by explicit legislative order.

1.3 Publication of Research Results

Inspector D of the competent authority of country G negotiates MAT with researcher A from a Polish university for access to specified plant material. The plant genome shall be sequenced and certain traits shall be researched as to their functions within the plant system. A intends to publish the research results. He does not aim at applying for a patent for the information. D and A agree that any results must be shared with G's research center R that shall be involved on the basis of a research contract to be concluded between A and R. D is not sure how to treat the question of publication of research and development (R&D) results.

D should reflect on the possibility that after publication of the information anyone can make use of it for developing products. This fact alone should not be a hindrance to grant access if the provider country is able to ensure the sharing of the benefits from the sales of the products. D should however insist that any publication must be made with the proviso that commercial benefits obtained from products developed via the genetic resources must be shared with the provider country, and that this requires prior *bona fide* negotiation of the product marketer with the provider State. D must nevertheless be aware of the fact that other than traditional publication media databases may not be prepared to flag such conditions with the genomic information. It is a question of policy of the provider State if such practice shall be accepted. At the same time, it should be recalled that access to GR and/or associated traditional knowledge (aTK) is always a prerequisite for benefit-sharing. Therefore, if the country's policy is to restrict access to its GR/aTK, this will have impacts on future benefit-sharing opportunities.

1.4 Consent of Indigenous Communities

Lawyer E, practicing in provider State F, is approached by a cosmetics company C concerning the collection of plant material of a specific variety in the Amazon forest. The targeted area is part of the territory of State F and is inhabited by indigenous communities. From the plant substances concerning interesting new fragrances shall be expressed and mixed into solutions for a perfume. C wants to take advice from some communities concerning interesting fragrances. What advice should E give C?

E would first of all have to find out if the extraction of substances is at all utilization of GR under the ABS regime of provider State F because only in that case the ABS regime applies. This would be answered in the positive, for example, if State F's law uses the same definition of utilization as included in article 2 of the NP. According to article 2 utilization includes biotechnology which again includes "any technological application that uses [...] living organisms, or derivatives thereof, to make or modify products". As a second step E needs to deal with the fact that traditional knowledge about the plants' fragrance properties will be used. In that case the consent of the indigenous communities must be obtained. In relation to the utilization of the plant itself E needs to research if F's legislation endows indigenous communities with rights to determine access to their genetic resources. If that is the case, consent of the communities must be obtained. Prior consent of the competent national authority might be needed in addition depending on the provider State's ABS regime. Further, E needs to advise on the number of communities that must be approached for consent. This may depend on the spread of the plants and the borders between communities. Finally, E must examine available sociological studies about customary rules of the communities on procedures, services and remuneration. He/she may also offer services to approach the communities and negotiate consent with them.

Modification: C already knows the plant varieties that are interesting for extracting fragrances.

In that case no prior consent concerning the related traditional knowledge is needed. It is however nevertheless needed if the local community is asked for additional information such as about processing the material. Consent by the competent national authority may also be needed.

Further modification: F's ABS legislation provides that the traditional knowledge is a national heritage. Its acquisition and use shall require PIC and MAT of the competent national authority. No consent of local communities is required.

Under national law of F as it stands C does not need to approach the local communities. However, F's national law is in violation of article 8 (j) of the Convention on Biological Diversity (CBD) and article 7 NP which prescribe consent of local communities although softening this by somewhat circularly referring to domestic law. If F's constitutional law adopts a monist conception concerning the application of international law, meaning that international law sets aside any contradicting national law, C may suggest a related improvement of F's law or even feel empowered to nevertheless require PIC of the local communities (in addition to PIC and MAT of the competent national authority).

2 User State Cases

2.1 *Temporal Scope*

2.1.1 Utilization Before Entering into Force of the EU Compliance Regime

In 1996 Company B in Germany isolated a microbial strain from a soil sample which was collected in country M in 1995. M became a party to the CBD in 1994 and to the NP in 2013. The NP was implemented by national law in 2017. In 1999, B isolated a compound from the said strain. In 2018, B found an interesting activity in the compound and would like to develop a product from it. Is it necessary from the legal point of view of German and M's law to apply for PIC and negotiate MAT with M?

According to Regulation (EU) 511/2014, article 2 (1), the European Union (EU) compliance regime applies to GR and associated traditional knowledge (aTK) that were accessed after the entering into force of the NP, i.e. 12 October 2014, and after the provider State became a NP party with access legislation. In conclusion, the described activities do not require PIC and MAT as the GR/aTK is out of temporal scope of the EU law. That means the relevant German enforcement authority cannot be expected to take measures if M claims that its ABS regulation was violated. Anyway, to be on the safe side in case of administrative checking it is advisable for B to have proof of the facts in its possession (exact access date and country of origin). However, B needs also to respect M's law even if that is not enforced by the EU Regulation. Assumed M's Act of 2017 requires that PIC and MAT must be obtained for previously accessed GR or aTK if "(a) there is a new use of the biological resource or traditional knowledge associated with the biological resource; or (b) there is development of a new product arising from the biological resource or traditional knowledge associated with the biological resource", M embraces a retrospective approach for biological resources, thus affecting actions taken prior to its coming into force. B's activity is a development of a new product thus belonging to category (b). Therefore, M would expect B to enter into a benefit-sharing agreement.

2.2 *Geographical Scope*

2.2.1 US Based R&D on GR Accessed in Nagoya Countries

Researcher R at a United States (US) university applies R&D on a plant specimen the sample of which she collected in provider State P, which is a contracting party to the CBD and NP and has an ABS access regime. She did not obtain PIC or MAT before collecting. R wants to involve researcher A at an Italian university in the research project.

There are no specific ABS compliance related obligations for R under US law.[1] Still, R has to comply with the laws of the provider country, although the US government will not check this. Further, A who is subject to the EU ABS regime must perform due diligence in finding out if P requires PIC and MAT. As that is the case, A must obtain PIC and MAT for her R&D.

2.3 *Material Scope*

2.3.1 Screening a Number of Plants

Company C bought a large number of herbs at local markets in various countries that have an ABS regime in force. C plans to sequence the herbs' genomes and search for specific genetic traits. These plants shall further be studied in depth. C is willing to comply with ABS requirements concerning the selected plants. C wants to know if already screening is subject to PIC and MAT.

The EU Commission in its revised Guidance Document on Regulation (EU) No. 511/2014 draws a line between screening as such, screening for taxonomic purposes alone, and screening for further research. The first two cases are classified as no utilization.[2] We share this classification, but with the qualification that screening shall also be considered as utilization if the data are uploaded to a public database because this would open up any further R&D. However, different provider countries may treat the question differently. Therefore we would advise C to consult with relevant authorities of individual countries.

---

[1] It may be considered that the Lacey Act (16 USC Ch. 53: Control of illegally taken fish and wildlife) is applicable in this case. The Act *inter alia* prohibits the importation of wild animals or plants that were taken in violation of any foreign law.

[2] See Commission Guidance document on the scope of application and core obligations of Regulation (EU) No 511/2014 …, OJ 2021/C 13/01, para. 2.3.3.1. Hereinafter cited as Commission Guidance.

2.4 *Information Stored at and Taken from Public Databases or Print Media*

2.4.1 MAT Conditions Concerning Published Data

Researcher A working for Company B in France reads a published scientific article that refers to genomic data uploaded to a public database. The article informs that the relevant genomic data from a GR accessed from country C is subject to prior consent for the commercialization of products based on the GR. A synthesizes some strain and inserts it into an organism the sales of which yield monetary revenue. Is B obliged to share the revenue with the provider State?

In our understanding A was first of all obliged to ask C for consent for the commercialization.[3] This would have been an opportunity to negotiate benefit-sharing. Having failed to do so A or B should reveal their mistake to the competent authority of the provider State and offer negotiations. The percentage of shared revenue would among other factors be influenced by the relevance of the inserted trait for the product.

Modification: A took the genome data from a public database which only informs about the source organism and the country—country C—where the pertinent sample was taken. C has legislation in force that defines genetic resources to extend to information on the genome, and requires PIC and MAT for any access, be it material or informational, and benefit-sharing.

It is subject to a vivid debate whether the ABS regime of the CBD and NP only relates to GR as material or also to GR as information. In the latter case this would mean that the sovereign rights of provider States extend to the information thus constituting a kind of intellectual property right in "wild information". This would however overstretch the meaning of article 15 (1) of the CBD. The provider State may nevertheless establish such right independently of the pertinent international legal situation. This would however only cover activities under C's jurisdiction. As C is flagged as the country of origin A is advised to ask C for consent for the commercialization if he/she wishes to respect C's regulation. He/she is however not obliged to file a due diligence declaration under the EU ABS Regulation because the EU regime does not apply to access to non-material information.

Further modification: C's legislation does not extend the term GR to information but C has a practice whereby MAT establish that the commercialization of R&D results based on C's genetic resources requires C's prior consent.

It is a largely accepted opinion that a provider State may direct the informational level of GR by PIC and MAT conditions. According to article 4 of Regulation (EU) 511/2014 the user of accessed GR has to practice due diligence concerning compliance of access with the provider State regime. A will in principle have to inquire about the access requirements of the provider State. However, he/she would not be bound by EU law because he/she is not the person who originally acquired the sample, neither was he/she, per viral clause, drawn into

[3]Contrastingly, according to the EU Commission's understanding the EU ABS regime does not apply as the researcher only used sequence data without having access to the original material, Commission Guidance, Annex 2 para. 6.1. See further modification of the case, below.

the PIC and MAT conditions. In conclusion, A would be freed from any benefit-sharing duties.

2.4.2 MAT Conditions Concerning Publication of R&D Results

MAT between company B and provider State C establish that genomic data of the accessed GR shall not be published neither in print media nor public databases. B shall be obliged to share a certain percentage of revenues from products or intellectual property rights (IPRs) developed on the basis of the data. Any comment?

In terms of general policy the agreement is against the objective laid down in NP article 8 (a) that states shall create conditions "to promote and encourage research which contributes to the conservation and sustainable use of biological diversity". Such objective can best be reached on the basis of publicly accessible research results. But article 8 (a) NP is not a conclusive statement in that regard. Therefore B and C are free to conclude the said MAT. They might nevertheless consider to keep secret only information on the final stages of product development or patentable information. They could also agree that C's claim to benefit-sharing shall be flagged with any publication of data that have a market value.

Researcher A of a user State university finds out that a gene of a certain plant codes for a certain scientifically interesting function. She took the investigated plant specimen from a small plant collection of the university the origin of which is not documented, and offers her results for publication in a journal. The journal asks her to provide evidence on the country of origin of the specimen and, as the case may be, PIC and MAT of the provider State.

In principle journals are free and should even be encouraged to check ABS compliance of submissions. But they should not go further than the EU compliance regime demands. This means they should apply a standard of due diligence and acquiesce if there is no possibility at the university to identify the origin of the specimen. This fact should however be noted in the publication.

2.5 Contributions or Not to Product Development

2.5.1 Research "on" and "with" GR

A uses a bacterium accessed in a NP party country for the screening and characterization of chemically synthesized compounds. Is this contributing to the development of a final product and thus a trigger for benefit-sharing?

If the microorganism is known to be simply usable in that way it is not to be regarded as GR that contributes to the development of a product. The situation is different if further research is needed to find out about the properties of the microorganism. For such cases it has been proposed to distinguish between research "on" GR as being within scope and research "with" GR as being out of scope.[4] This distinction appears to be rather artificial. It is

[4]Commission Guidance, para. 2.3.3.2.

suggested to regard such contribution as causal contribution. The relevance of the same would have to be considered when the amount of benefit-sharing is calculated considering fairness and equity.

2.5.2 Research Without Modification

What if the strains are discovered and are to be used without modification to further develop an improved assay for selecting the chemical compounds?

If the strain is modified this is a clear case of being within scope. If it is researched to find out about its potential to further develop an assay, this is also within scope. If its useful property was known before, there is no utilization in the sense of the NP and thus no access for utilization (which would be the trigger for requiring PIC and MAT).

2.6 *Bulk Uses or Utilization?*

2.6.1 Change of Intent from Consumption to R&D

Company A buys on a German market fruit pulps of the yellow shower tree that were imported from country I. A explores the purgative properties of the plant with a view to use the results for developing a drug. Does A have to seek PIC from and MAT with I's competent authorities?

While the consumption of the fruit would be out of scope, access to GR is generally understood to include cases where a previously purchased consumable commodity is made object of R&D. A therefore needs to contact I's authorities and seek to obtain PIC and establish MAT.

2.6.2 Known Properties and New R&D

Company A uses an (unmodified) bacterium for producing an enzyme that the bacterium naturally produces. Is that utilization of the bacterium?

The use of the bacterium is to be classified as biotechnological application. Utilization of GR is defined as R&D on the genetic/biochemical composition of GR, including "through" the application of biotechnology. The word "through" links R&D and biotech application. Simple biotechnological application alone would not qualify as R&D.[5] Therefore the answer is "no" if the potential of the bacterium is a well-known fact, and "yes" if research was undertaken to find out.

[5] Commission Guidance, para. 2.3.4.

2.6.3 Propagation and Cultivation of GR

Company C produces cosmetic products. At a local market of State S, C's researcher R acquires roots of Ginseng which is known for its multiple properties, including an anti-oxidative effect. C confirms the anti-oxidant effect in various prototype cosmetic product formulations and finally produces a new finished cosmetic product. The Ginseng ingredients needed for the product are taken from the cultivation of the roots. Does C have to share its revenue from the product with S?

Two activities could be considered to be utilization: the formulation of the product and the cultivation of the roots. But neither of them qualifies as R&D. The formulation is just a new combination of well-known properties and thus not research on the GR as such, neither is it research or development on the roots if they are propagated without further breeding or other modification.[6]

2.7 *Technological Application*

2.7.1 Studying Mechanisms of a Lawfully Acquired GR

A pharmaceutical company has marketed a medication based on GR for a long time starting before 2014. After that date it wants to study the mechanism of action of the compound. Is research on its own medication within scope if the only intention is to characterize its pharmacological action?

In principle research on the mechanisms of the compound is part of R&D because R&D includes "any technological application that uses biological systems [. . .] or derivatives thereof, to make or modify products" (Art. 2 (d) NP). Concerning the fact that the utilization of the GR started before 2014, it can be argued that the research is only aimed at better understanding a process, which one is entitled to use anyway so that no new R&D is undertaken. The situation would be different if the research was aimed at developing new uses (which one would expect to normally be the case).[7]

2.7.2 Purchase of Derivatives

Company C in Italy acquires a batch of orange essential oil from an intermediary based outside the EU. The batch of oils is transferred without PIC and MAT. The company analyses the composition of the oils to identify known and new chemical structures and to determine their properties with a view to create a new food flavor.

---

[6] Same view Commission Guidance, para. 2.3.3.2., Annex 2 paras. 8.1 and 8.5.

[7] Contrastingly Commission Guidance para. 2.2. opines that even if there is a new utilization on material that is out of temporal scope (i.e. obtained before 12.10.2014) utilization will still be out of scope.

The Commission has proposed a criterion of continuum of R&D from first steps of research to the technological application of R&D results implying that the continuum is cut if a market transaction (such as the purchase from an intermediary) intervenes.[8] There can be doubts about this criterion. The requirement of continuum could too easily be circumvented by simple introduction of non-R&D steps. It is agreeable that there must be a link between the GR and the final derivative, but the link should be substantial, not procedural. Even if a market transaction intervenes the derivative is still a derivative from a GR. Therefore, C performs R&D on a substance derived from a GR and is subject to PIC and MAT. In some cases C may by viral clause be included into the PIC and MAT conditions that were obtained by the primary accessor to the GR. This will however not work if the primary accessor was an intermediary without any research ambition. The case nevertheless illustrates that also a substantial continuum needs some kind of cut-off points in order to free R&D from timeless bounds. Perhaps, the number and kinds of utilization steps may serve as a criterion.

2.8 Regularization

2.8.1 Ex Post PIC and MAT

Researcher A took samples of sea water in the territorial sea of a South American country when accompanying a geological marine expedition. At home in an EU country he researches microorganisms contained in the sample.

Taking the sample is access and thus subject to PIC and MAT of the State of origin if it has established an access regime. Researcher A should inform the State and inquire what to do. Some States, and in particular in South America, e.g. Brazil, have mechanisms in force that allow rectification of unlawful situations.

2.8.2 Transfer to Third Parties of Unlawfully Obtained GR

The German pharmaceutical company B intends to collaborate with the German university U on a plant that may have interesting medicinal properties. U's researchers had acquired the sample in 2016 in a provider State not being aware of PIC and MAT requirements set out in the ABS legislation of the provider country.

If the sample is transferred to B, B would not be bound by conditions because U had not concluded MAT which by viral clause could have been extended to B. B should nevertheless urge U to rectify the situation, including a viral clause that binds any third person to which the GR is passed. If U refuses to do this B is required to obtain PIC and MAT from the provider State for itself as B is utilizing a GR which is within temporal and geographical scope.

[8] Commission Guidance, para. 2.3.4.

2.9 *Due Diligence in Ascertaining Provider State Requirements*

2.9.1 Unresponsive Provider State

Researcher R of a Belgian university plans a project involving access to GR. The pertinent provider State does not respond to R's mails and calls inquiring about access requirements and procedures. How many times does R need to repeat this and how much time does he/she need to spend on his/her efforts?

In the EU, the yardstick to apply is due diligence. This is a duty of conduct, not of effect. The required effort cannot be quantified but it appears to be reasonable that the inquiry should be repeated at least two times. We suggest that with the third attempt the researcher should submit a description of how the GR shall be accessed, what R&D shall be applied, if for R&D results IPRs shall be obtained, and whether or not resulting products shall be brought to the market. If the provider State still does not respond the researcher when acquiring the GR should be considered to have acted in compliance with provider State requirements as required by EU standards. Of course, R should keep the pertinent documents and inquiry/request protocol for possible subsequent provider or user State checking. There is a remaining risk for R, though. If at a later stage new information occurs that PIC/MAT are necessary, the user has to go back and get the documents retroactively or stop utilization (Art. 4.5 Regulation (EU) 511/2014).

Modification: The biomaterial includes fruits such as *piper longum* and *cassia fistula* from which natural compounds shall be isolated and genetic information generated for further uses in food industry. The fruits can be purchased e.g. on Belgian markets.

The purchase of biomaterial that is offered on consumer markets is nevertheless access in the ABS sense if access is conducted with the intention of R&D. Therefore, R needs to abide by the access requirements of the provider State. If that State does not respond to mails and calls the same steps as suggested above should be taken. It has been proposed that the researcher should repeat their mailing every three months. It is after all the risk of the provider State not to have responded.

2.9.2 ABS Clearance Declaration

The German company C isolates pure substances from different biomaterials collected in Gemany in order to find out what substances can be used for further R&D. The biomaterials stem from Germany. Although Germany has not established regulations on access to its resources C would like to have a declaration of the competent authority that the materials can be utilized without PIC and MAT. C wishes to use the declaration for documentary purposes *vis-à-vis* its international customers.

According to general administrative law any public authority is bound to advise on current legislation and administrative practices. The competent administrative agency would inform in writing that that German law does not require PIC or MAT for access to its GR, also referring to the fact that there may be other regulation such as of nature protection to be observed. The authority will hardly certify that the specific material utilized does not need PIC and MAT because it is not obliged to check the origin of the material.

2.10 Benefit-Sharing

2.10.1 ABS and Patenting

Company C applies for a patent on a trait of provider country E's plant which codes for a property that can be used for a vaccine. Country E's authority competent for ABS opposes the patent arguing that the plant was acquired without PIC and MAT.

Whether the patent office can provide the patent depends on whether the applicable patent law is in any way related to ABS issues. Some patent laws oblige applicants to provide information about the origin of the utilized GR so that the patent office may and must inform the competent national authority of the user or the provider State accordingly. But this does not hinder the granting of the patent if the normal preconditions for granting patents are fulfilled. This is different in those States which have introduced an additional precondition requiring proof of compliance with provider State requirements.

2.10.2 ABS and Plant Variety Protection

Company C develops and trades in seeds. One variety is used for the manufacture of a pharmaceutical product. The seed was bred out of a wild form sample which was accessed in provider country D. The access was based on PIC and MAT which provided that any benefits should be shared upon *bona fide* mutual negotiation. C applies for and receives a breeder's right on the seed. D asks C to enter into negotiations on benefit-sharing. C now argues that the breeders' exemptions—free cultivation of cultivated seeds and free further breeding—exclude any duty to share benefits.

C errs because its own revenues from sales are not touched by the breeders' right. Only the farmers could argue that their breeders' exemptions overrule any ABS based claim to benefit-sharing. While the mere cultivation of the seed would not qualify as utilization the further breeding does qualify as such. Concerning the conflict with the breeders' exemption different opinions are possible. It is submitted that the two regimes are based on different reasons for exclusive rights – the breeders' right remunerating breeding efforts, the ABS right rewarding the conservation of varieties – which justifies conceiving them as parallel and independent from each other.

Modification: C sells the cultivated seed to plant breeding company B which develops a variety deemed new under the criteria of applicable plant variety protection legislation. Does B need to obtain PIC and MAT from the original provider State, D, and/or share benefits with D?

The Commission regards the variety that is protected by the plant variety right as being new as compared with the accessed original variety specimen. For that reason B's R&D on C's variety shall not be R&D on the accessed GR.[9] We have doubts about this if traits of the original variety are still present in the bred variety and are planned to remain present in the variety under new development. B's R&D is however not to be regarded as access. It is rather a step in the overall valorization chain based on the originally accessed GR. B is, nevertheless, under obligation to share benefits if the relevant PIC and MAT conditions flanked by viral clause also demand the sharing of benefits obtained by third parties. Once more, the case shows that cut-off points should be created after a defined number of steps of further breeding.

2.10.3 Sharing of Benefits Accruing to the Buyer of a Patent on Accessed GR Information

Researcher A discovers in a plant that was lawfully accessed in provider State P a gene coding for antibiotic properties and develops the formulation of a drug on that basis. She obtains a patent on this invention and sells it to company C. C produces the drug and brings it on the market. A shares the revenue from the patent sale with P in accordance with the MAT concluded between A and P. Is C obliged to share benefits from the sales of the drug?

C is not a user because it does not perform R&D on the GR. It is therefore not obliged to make a separate contract with P. In such situation it is advisable for P to include a viral clause into MAT requiring that the obligation to share benefits is transferred to anyone who acquires the right to bring a product of R&D on the market. See however below on cut-off points.

2.11 *Cut-off Points*

Processing of Raw Materials for Subsequent Incorporation into a Product
Company A produces an enzyme based on a gene that was transplanted from a microorganism accessed in provider State D under PIC and MAT with D. A brings the enzyme on the market and pays D the agreed shares of the revenue. Company B buys the enzyme from the market and after some formulation experiments, mixes it into a washing powder and brings it on the market. Does D have a right to a share in the revenue from the sales of the washing powder?

It could be argued that the formulation of substances the properties of which are known is no genuine R&D, in other words no utilization, and in consequence out of scope of the ABS regime. Alternatively, the formulation experiments and final mixture could be regarded as utilization in the form of biotechnology, subtype technological application that uses derivatives of GR to make or modify products.[10] Further alternatively, the final mixture could be classified as application subsequent to utilization. This would mean that if the MAT stipulate the sharing of benefits arising (directly or indirectly) from utilization B must negotiate with

[9] Commission Guidance, Annex II para. 8.4.
[10] Cf Art. 2 (d) NP.

D a fair and equitable share in the revenue. In a more general perspective one could argue that the valorization chain cannot be endless so that at some point the causal contribution of the original GR should be cut off. This could be done after a defined number of steps of utilization. As said above (section 2.7) one major cut-off point the EU Commission proposes is if the developed product is brought to the market. Any further work on the product is then not R&D on the original GR.

Acknowledgements We gratefully acknowledge many helpful comments and suggestions by Thomas Greiber. Of course any errors and disputable opinions are ours.

References

Legal Documents

Commission Guidance document on the scope of application and core obligations of Regulation (EU) No 511/2014 of the European Parliament and of the Council on the compliance measures for users from the Nagoya Protocol on Access to Genetic Resources and the Fair and Equitable Sharing of Benefits Arising from their Utilisation in the Union, OJ 2021/C 13/01

Lacey Act 16 USC (of 1900 as consequently amended)

The Convention on Biological Diversity (1992)

The Nagoya Protocol on Access to Genetic Resources and the Fair and Equitable Sharing of Benefits Arising from their Utilization to the Convention on Biological Diversity (2010)

Gerd Winter was born in Diepholz, Germany, in 1943. He is Research Professor of Public Law at Research Center for European Environmental Law, University of Bremen. He has taught and published extensively on German, comparative, European and international environmental law, including on property, distributional and risk related problems of biotechnology (see https://www.gerd-winter.jura.uni-bremen.de/veroeffchronol.html). He also served as consultant on public and environmental law reform in transition countries as well as counsel in numerous administrative and court proceedings concerning environmental issues. gwinter@uni-bremen.de.

Evanson Chege Kamau Dr. iur., is Senior Research Fellow at the Research Centre for European Environmental Law (FEU), University of Bremen. He is also an Adjunct Lecturer at the CvO University of Oldenburg. His area of expertise is public international law, European law and intellectual property law. His research focuses on genetic resources, environment, development and traditional rights on biological resources and traditional knowledge. Kamau has over 17 years of working experience in the area of ABS in which he has published widely. For many years he has also worked as a freelance legal consultant for governments, African Union (hired by IUCN and Capacity Building Initiative resp.), private industries and research organizations. echege@uni-bremen.de.

CPSIA information can be obtained
at www.ICGtesting.com
Printed in the USA
LVHW081657180222
711484LV00004B/194

9 783030 88710(